Springer Texts in Business and Economics

Springer Texts in Business and Economics (STBE) delivers high-quality instructional content for undergraduates and graduates in all areas of Business/Management Science and Economics. The series is comprised of self-contained books with a broad and comprehensive coverage that are suitable for class as well as for individual self-study. All texts are authored by established experts in their fields and offer a solid methodological background, often accompanied by problems and exercises.

More information about this series at http://www.springer.com/series/10099

Gündüz Ulusoy • Öncü Hazır

An Introduction to Project Modeling and Planning

Springer

Gündüz Ulusoy
Faculty of Engineering and Natural Sciences
Sabancı University
Orhanlı, Tuzla, Istanbul, Turkey

Öncü Hazır
Rennes School of Business
Rennes, France

ISSN 2192-4333 ISSN 2192-4341 (electronic)
Springer Texts in Business and Economics
ISBN 978-3-030-61422-5 ISBN 978-3-030-61423-2 (eBook)
https://doi.org/10.1007/978-3-030-61423-2

This Springer imprint is published by the registered company Springer Nature Switzerland AG
The registered company address is: Gewerbestrasse 11, 6330 Cham, Switzerland

Preface

Although a large number of projects have been conceived, planned, and completed since the times of ancient civilizations, it was only in the 1950s that project management emerged as a discipline with the introduction of the Critical Path Method (CPM) and the Project Evaluation and Review Technique (PERT). Drawing upon theoretical developments contributed by numerous researchers and practice over a wide spectrum of application areas distributed globally, the knowledge base of project management has been growing ever since at an increasing rate. The advances in Operation Research methods and the capabilities of the optimization solvers have contributed considerably to the theoretical development of project management. Interest and activity in areas like resource constrained project scheduling, uncertainty, and robustness have been growing rapidly as attested by the large number of journal papers published globally each year. The practice of project management in organizations has also been growing steadily, supported by the emergence of matrix type and project-based organizations. The organizations dealing with new technologies and business models emerging in the last decades organize many of their activities such as new product development, innovation management, and app development as projects.

Project management offers a promising field for theoretical and practical applications and developing related software and Decision Support Systems. It attracts not only academicians and software providers, but also professionals who take part in project management activities. Considering this diverse audience, this book aims at explaining the basic concepts and methods of project management, especially project planning and scheduling with an emphasis on mathematical modeling.

The motivation for this textbook is based on two observations: First, courses dealing with project management in some context are proliferating in engineering schools and becoming attractive for engineering students as it has been the case for some time now for business students. Second, particularly the developments in quantitative methods achieved in the last decades, especially the very significant increase in the power of optimization and simulation software, need to find their way into undergraduate and master's classrooms to become in due time part of routine professional practice.

This textbook is intended to provide theoretical and practical aspects of project planning, modeling, and scheduling to undergraduates and first year graduate students in engineering programs, as well as to practitioners seeking to solidify their knowledge of project management. Most of the material covered in this textbook emerged from the class notes shared with the students at both undergraduate and graduate levels and has been improved over the years by their remarks and suggestions, for which we are grateful.

The models and algorithms presented in the following chapters establish a good starting point to study the relevant literature and pursue academic work in related fields. Theory is discussed at the introductory and moderate levels, and related references are given for those who want to pursue more detailed analysis. A central feature of the book is the support provided for the theoretical concepts through models with detailed explanations, application examples, and case studies. Discussions and analyses are based on real-life problems.

All chapters include exercises that require critical thinking, interpretation, analytics, and making choices. A solution manual for these exercises will be provided online. The textbook is organized to enable interdisciplinary research and education, interactive discussions, and learning. Learning outcomes are defined, and the content of the book is structured in accordance with these goals.

Chapter 1 introduces the basic concepts, methods, and processes in managing projects. This chapter constitutes the base for defining and modeling project management problems.

Chapter 2 discusses project management from an organizational perspective, dealing with the fundamentals of organizing and managing projects. Project-based organizations have gained increasing importance in the last decades. Many firms are organized around projects and operate in a multi-project environment. Issues related to project team formation, the role of project managers, and matrix organization are discussed. Recent developments in the organization of projects, such as the adoption of agile techniques and establishing project management offices, are discussed.

Chapter 3 is devoted to project planning and network modeling of projects, covering fundamental concepts such as project scope, Work Breakdown Structure (WBS), Organizational Breakdown Structure (OBS), Cost Breakdown Structure (CBS), project network modeling, activity duration and cost estimating, activity-based costing (ABC), and data and knowledge management.

Schedules determine when projects can be delivered, intervals when each activity needs to be performed, and how resources are assigned to activities and present multiple decision problems. In Chap. 4, we deal with a decision environment where we assume that the decision-makers know the problem parameters precisely and no constraints are imposed on the resource usage. Models employing time-based and finance-based objectives are introduced. We use Operations Research techniques to formulate and solve these deterministic scheduling models for constructing the time schedules as mathematical programming problems. The CPM is covered, resulting in a baseline project schedule, the project duration, the critical path, and the critical activities. The unconstrained version of maximizing Net Present Value (NPV) is also treated here together with the case of time-dependent cash flows. Continuous tracking of the project plan and associated data and relationships allow it to be revised through replanning whenever deemed necessary. A case study on installing a plant biotechnology lab is presented.

In project management, it is often possible to reduce the duration of some activities and expedite the completion of the project by incurring additional costs. The resulting time/cost trade-off problem has been widely studied in the literature, and Chap. 5 addresses this important topic for both the continuous and discrete cases.

Chapter 4 introduced project scheduling in a deterministic decision environment. However, in reality, project managers have at best limited information on the parameters of the decision environment in many cases. For this reason, Chap. 6 discusses models and methods of scheduling under uncertain activity durations. PERT is introduced for minimizing the expected project duration and extended to the PERT-Costing method for minimizing the expected project cost. Simulation is presented as another approach for dealing with the uncertainty in activity durations and costs. To demonstrate the use of the PERT, a case study on constructing an earthquake resistant residential house is presented.

Another important topic for a more realistic representation of the decision environment is the integration of the resource constraints into the mathematical models since in almost all projects, activities compete for scarce resources. Classifications of resource and schedule types are given in Chap. 7, and exact and heuristic solution procedures for the single- and multi-mode resource constrained project scheduling problem (RCPSP) are presented. The objective of maximizing NPV under resource constraints is addressed, and the capital constrained project scheduling model is introduced.

In Chap. 8, resource leveling and further resource management problems are introduced. Two problems associated with resource leveling are discussed, namely total adjustment cost and resource availability cost problems. Various exact models are investigated. A heuristic solution procedure for the resource leveling problem is presented in detail. In addition, resource portfolio management policies and the resource portfolio management problem are discussed. A case study on resource leveling dealing with the annual audit project of a major corporation is presented.

Project contract types and payment schedules constitute the topics of Chap. 9. Contracts are legal documents reflecting the results of some form of client–contractor negotiations and sometimes of a bidding process, which deserve closer attention. Identification and allocation of risk in contracts, project control issues, disputes, and resolution management are further topics covered in this chapter. A bidding model is presented to investigate client–contractor negotiations and the bidding process from different aspects.

Along with planning and scheduling, project monitoring and control are important responsibilities of project managers. Whereas monitoring the project includes actions to collect, record, and present project realization data, control covers the analysis of the data with the purpose of preparation of action plans to eliminate the deviations from the current plan. Chapter 10 focuses on processes and methods for monitoring and control. Earned Value Management is studied to measure the project performance throughout the life of a project and to estimate the expected project time and cost based on the current status of the project. How to incorporate inflation into the analysis is presented.

Risk management is essential for effective project management. In Chap. 11, qualitative and quantitative techniques including decision trees, simulation, and software applications are introduced. Risk phases are defined and building a risk register is addressed. An example risk breakdown structure is presented. The design of risk management processes is introduced, and risk response planning strategies are discussed. At the end of the chapter, the quantitative risk analysis is demonstrated at the hand of a team discussion case study.

Uncertainty in project schedules has been a topic of great interest for both researchers and practitioners, accumulating a rich literature over the last decades. Chapter 12 covers several models and approaches dealing with various stochastic aspects of the decision environment. Stochastic models, generation of robust schedules, and use of reactive and fuzzy approaches are presented. Sensitivity and scenario analysis are introduced. In addition, simulation analysis, which is widely used to analyze the impacts of uncertainty on project goals, is presented.

Chapter 13 addresses repetitive projects that involve the production or construction of similar units in batches such as railway cars or residential houses. Particularly in the construction industry, repetitive projects represent a large portion of the work accomplished in this sector of the economy. A case study on the 50 km Gebze–Orhangazi Section of the Gebze–Izmir Motorway Project is used for demonstrating the handling of repetitive project management.

How best to select one or more of a set of candidate projects to maintain a project portfolio is an important problem for project-based organizations with limited resources. The project selection problem is inherently a multi-objective problem and is treated as such in Chap. 14. Several models and solution techniques are introduced. A multi-objective, multi-period project selection and scheduling model is presented. A case study that addresses a project portfolio selection and scheduling problem for the construction of a set of dams in a region is presented.

Chapter 15 discusses three promising research areas in project management in detail: (*i*) Sustainability and Project Management, (*ii*) Project Management in the Era of Big Data, and (*iii*) the Fourth Industrial Revolution and the New Age Project Management. We elaborate on the importance of sustainability in project management practices, discuss how developments in data analytics might impact project life cycle management, and speculate how the infinite possibilities of the Fourth Industrial Revolution and the new technologies will transform project management practices.

In the several years of teaching and research on project management, we are indebted to many researchers and practitioners whose work motivated and guided us in our efforts to write this book. We are particularly grateful to Professor Reha Uzsoy, who has been extremely generous with his time going over several versions of the manuscript and making a large number of suggestions for improvement. We have also benefited greatly from the critical remarks made by Professor Kuban

Altınel and thank him for that. We also would like to thank two PhD students, Nozir Shokirov and Ahmed Faran, for their contributions to the case studies and the Solution Manual. The responsibility for any remaining errors obviously rests solely with us.

Istanbul, Turkey
Rennes, France
September 2020

Gündüz Ulusoy
Öncü Hazır

Acknowledgement

With love and gratitude to my wife Nefise Ulusoy and my parents Zeynep and Halit Ulusoy

With love to my wife Çilem Selin Hazır

Contents

1 Introduction to Project Modeling and Planning ... 1
- 1.1 What Is a Project? ... 1
- 1.2 Project Management in Today's World ... 4
- 1.3 Challenges in Keeping to Scope, Cost, and Schedule Targets ... 5
- 1.4 A Life Cycle Concept for Projects ... 6
  - 1.4.1 Conceptual Design and Project Definition ... 7
  - 1.4.2 Planning ... 8
  - 1.4.3 Implementation, Monitoring, and Control ... 9
  - 1.4.4 Evaluation and Termination ... 10
- 1.5 Post-Project Analysis ... 10
- 1.6 Uncertainty and Risks in Projects ... 12
- 1.7 Project Modeling and Planning Tools ... 13
- 1.8 Resource Constrained Modeling ... 16
- 1.9 Agile Project Management ... 17
- Exercises ... 18
- References ... 19

2 Organizing and Managing Projects ... 23
- 2.1 Projects and Organizational Structure ... 23
  - 2.1.1 Functional Organizations ... 24
  - 2.1.2 Project Organizations ... 25
  - 2.1.3 Matrix Organizations ... 25
- 2.2 How to Choose the Project Management Structure? ... 27
- 2.3 Stakeholder Management ... 28
- 2.4 Project Management Culture ... 29
- 2.5 Organizing Projects ... 31
  - 2.5.1 Steering Committee ... 32
  - 2.5.2 Project Sponsor ... 32
  - 2.5.3 Project Manager ... 33
  - 2.5.4 Project Team and Team Members ... 35
  - 2.5.5 Project Meetings ... 38
- 2.6 Project Management Office ... 38
- 2.7 Program Management ... 40

2.8 Project Portfolio Management ... 41
2.9 Agile Project Management ... 43
2.10 Conclusions, Recent Developments, and Some Future Research Directions ... 45
Exercises ... 45
References ... 46

3 Planning and Network Modeling of Projects ... 49
3.1 Project Scope ... 49
3.1.1 Project Charter ... 50
3.2 Work Breakdown Structure ... 51
3.2.1 Decomposition of WBS into Work Packages ... 52
3.2.2 Types of WBS Representation ... 54
3.2.3 Gantt Chart Representation of Work Packages ... 55
3.2.4 Organizational Breakdown Structure ... 58
3.2.5 Cost Breakdown Structure ... 59
3.3 Network Modeling of Projects ... 61
3.3.1 Types of Models ... 61
3.3.2 Network Modeling of Projects ... 62
3.3.3 Activity on Arc Type Representation ... 64
3.3.4 Activity on Node Type Representation ... 68
3.3.5 Generalized Precedence Relationships ... 69
3.3.6 Level of Aggregation in Project Networks ... 71
3.3.7 Network Modeling of Multiple Projects ... 74
3.4 Estimating Activity Durations ... 76
3.4.1 Tools for Estimating Activity Durations ... 77
3.5 Cost Management ... 80
3.5.1 Tools for Estimating Activity Cost ... 80
3.5.2 Types of Project Costs ... 83
3.5.3 Activity-Based Costing ... 84
3.6 Data, Information, and Knowledge Management ... 86
3.6.1 Which Data to Collect? ... 87
3.6.2 With Which Frequency to Collect Data? ... 88
3.6.3 With Which Tools to Collect and Integrate Data? ... 89
3.6.4 With Which Tools and How to Analyze Data? ... 90
3.7 Conclusions, Recent Developments, and Some Future Research Directions ... 91
Exercises ... 91
References ... 99

4 Deterministic Project Scheduling with No Resource Constraints ... 101
4.1 Introduction ... 101
4.2 Critical Path and Its Determination ... 102
4.2.1 Critical Path Method ... 103
4.3 Linear Programming Formulation ... 111

4.4 Maximizing the Net Present Value of a Project Under No Resource Constraints 115
4.4.1 A Zero-One Mathematical Programming Formulation 116
4.4.2 Time Dependent Cash Flows 120
4.5 Keeping the Project Plan Current 122
4.6 Conclusions, Recent Developments, and Some Future Research Directions 124
4.7 Case Study: Installation of a Plant Biotechnology Lab 124
Exercises 130
References 140

5 The Time/Cost Trade-off Problems 141
5.1 Problem Definition 141
5.2 Time/Cost Trade-Off in Projects 145
5.2.1 Project Crashing 145
5.2.2 An Algorithm for Project Crashing 146
5.3 Continuous Time/Cost Trade-off Problems 154
5.3.1 Activity-on-Arc Representation 154
5.3.2 Activity-on-Node Representation 155
5.4 Discrete Time-Cost Trade-off Problems 156
5.4.1 The Deadline Problem 157
5.4.2 The Budget Problem 157
5.5 Conclusions, Recent Developments, and Some Future Research Directions 158
Exercises 159
References 164

6 Stochastic Project Scheduling with No Resource Constraints 167
6.1 Introduction 167
6.2 Representation of the Randomness of Activity Durations 168
6.3 Determining the Critical Path 170
6.4 Shortcomings of the PERT Method 174
6.5 A Model for Managing Uncertainty in Activity Costs 177
6.6 Monte Carlo Simulation Approach to Handle Uncertainty 178
6.7 Conclusions, Recent Developments, and Some Future Research Directions 183
6.8 Case Study: Constructing an Earthquake Resistant Residential House 184
Exercises 187
Appendix 6A: Area Under the Normal Curve 196
References 197

7 Resource Constrained Project Scheduling 199
7.1 Introduction 199
7.1.1 Categories of Resources 200
7.1.2 Resource Parameters 201
7.2 Objective Functions 203
7.2.1 Regular and Non-regular Objective Functions 203
7.2.2 Time-Based Objectives 204
7.2.3 Financial Objectives 206
7.2.4 Quality Based Objectives 206
7.2.5 Resource Based Objectives 207
7.3 A Mathematical Programming Formulation of the Resource Constrained Project Scheduling Problem 207
7.4 Exact Solution Procedures Using Branch and Bound 209
7.5 Heuristic Solution Procedures 214
7.5.1 Types of Schedules 214
7.5.2 Priority Rules 217
7.5.3 Serial Schedule Generation 218
7.5.4 Parallel Schedule Generation Scheme 220
7.5.5 A Brief Assessment of Serial and Parallel Schedule Generation Schemes 221
7.6 A Mathematical Programming Formulation with Multiple Modes 222
7.7 A Mathematical Programming Formulation for Maximizing the Net Present Value of a Project Under No Budgetary Constraint 224
7.8 Conclusions, Recent Developments, and Some Future Research Directions 227
Exercises 228
Appendixes 240
Appendix 7A: Computational Complexity 240
Appendix 7B: Branch and Bound 245
References 248

8 Resource Leveling and Other Resource Management Problems . . . 251
8.1 Introduction 251
8.2 Resource Leveling 252
8.2.1 The Burgess and Killebrew Algorithm 254
8.2.2 Other Solution Approaches 261
8.2.3 The Total Adjustment Cost Problem 263
8.3 Resource Availability Cost Problem 265
8.4 Resource Management Policies 267
8.5 The Resource Portfolio Management Problem 268
8.6 Conclusions, Recent Developments, and Some Future Research Directions 269

8.7 Case Study: Conducting Annual Audit for Celik Company 270
Exercises 274
References 278

9 Project Contract Types and Payment Schedules 281
9.1 Introduction 281
9.2 Contract Types 283
9.2.1 Fixed Price Contracts 284
9.2.2 Variable Price Contracts 284
9.2.3 Cost of Managing Contracts 285
9.3 Payment Scheduling Models 286
9.4 Change and Control Issues in Contracts 287
9.4.1 Managing Change Issues in Contracts 287
9.4.2 Control Issues in Contracts 288
9.5 Identification and Allocation of Risk in Contracts 288
9.6 Bidding 290
9.6.1 Estimating the Total Cost of the Project 292
9.6.2 Front-loaded Bidding Strategy 293
9.6.3 A Bidding Model for Determining the Bidding Price 293
9.7 Mathematical Models for Negotiation in Project Management 297
9.7.1 Preparation for a Bid 297
9.7.2 Client-Contractor Bargaining Problem 297
9.7.3 Equitable Payment Schedule Problem 298
9.8 Conclusions, Recent Developments, and Some Future Research Directions 298
Exercises 300
References 308

10 Progress and Cost Control 311
10.1 Introduction 311
10.2 Cost/Schedule Control System Measures 312
10.3 Estimating the Percent Complete 313
10.4 Cost/Schedule Control Performance Indicators 314
10.4.1 The Impact of Inflation 320
10.5 Time and Cost Predictions Using EVM 322
10.6 Limitations of Earned Value Management 325
10.7 Conclusions, Recent Developments, and Some Future Research Directions 325
Exercises 326
References 333

11 Project Risk Management 335
11.1 Introduction 335
11.2 Processes in Project Risk Management 336
11.2.1 Risk Identification 336
11.2.2 Risk Analysis 339
11.2.3 Risk Response Planning 348
11.2.4 Risk Monitoring and Control 350
11.3 Project Risk Analysis: A Team Discussion Case 350
11.4 Conclusions, Recent Developments, and Some Future Research Directions 352
Exercises 352
References 355

12 Project Scheduling Under Uncertainty 357
12.1 Introduction 357
12.2 Robust Scheduling 359
12.2.1 Critical Chain Project Management 359
12.2.2 A Robust Optimization Model for Project Scheduling 361
12.2.3 Robustness Measures and Project Scheduling 365
12.3 Stochastic Scheduling 366
12.4 Sensitivity, Scenario, and Simulation Analysis 369
12.4.1 Sensitivity Analysis 369
12.4.2 Scenario Analysis 370
12.4.3 Monte Carlo Simulation 370
12.5 Conclusions, Recent Developments, and Some Future Research Directions 374
Exercises 375
References 378

13 Planning and Scheduling of Repetitive Projects 381
13.1 Introduction 381
13.1.1 An Analogy to Serial Production Lines 384
13.1.2 Duration – Resource Relation for WPs 384
13.2 A Mathematical Programming Model for the Repetitive Projects 385
13.2.1 The Objective Function 386
13.2.2 The Constraints 387
13.2.3 The Mathematical Programming Model 388
13.3 A Model with an Alternative Total Cost Function 389
13.4 Conclusions, Recent Developments, and Some Future Research Directions 394
13.5 Case Study: 50 Km Gebze-Orhangazi Section of Gebze-Izmir Motorway Project 394
Exercises 400
References 405

14 Project and Portfolio Selection 407
14.1 Introduction 407
14.2 A Multi-criteria Scoring Model 408
14.3 The Analytic Hierarchy Process 410
14.3.1 Sensitivity Analysis 417
14.4 Mathematical Programming Models 418
14.4.1 A Single Objective Mathematical Programming Model 419
14.4.2 Multi-objective Mathematical Programming Models . . . 421
14.4.3 Multi-objective Linear Programming Approach 423
14.4.4 Goal Programming Model 429
14.4.5 Multi-objective, Multi-period Project Portfolio Selection and Scheduling Model 433
14.5 Conclusions, Recent Developments, and Some Future Research Directions 436
14.6 Case Study: Project Portfolio Selection for the Construction of a Series of Dams 437
Exercises 444
References 455

15 Recent Developments and Some Promising Research Areas 457
15.1 Sustainability and Project Management 457
15.2 Project Management in the Era of Big Data 460
15.2.1 Meaning and Techniques of Big Data Analytics 460
15.2.2 Application Areas and Opportunities in Project Management 462
15.3 Fourth Industrial Revolution and the New Age Project Management 463
15.3.1 Evolution of the Project Environments 463
15.3.2 Transformation of the Project Management Practices . . 464
References 467
References for Section Sustainability and Project Management 467
References for Section Project Management in the Era of Big Data 468
References for Section Fourth Industrial Revolution and the New Age Project Management 469

Abbreviations 471

Mathematical Notations 473

Index 477

Introduction to Project Modeling and Planning

1

Learning Outcomes
Upon successful completion of this Chapter, the reader will be able to:

1. Describe the characteristics of projects.
2. Assess the need for organizing tasks as projects for competitive advantage.
3. Describe project management, its principles and importance.
4. Define project life cycle management.
5. Explain the basic project planning methods and tools.

1.1 What Is a Project?

"Project" is a common word in our everyday vocabulary. In its daily use, the word often refers to an imprecise idea, a plan, a desire, an invention, a mission, or an assignment. However, in business "project" has a different, well-defined meaning.

In business, a "project" refers to the entire set of interrelated tasks that must be completed within a defined time frame - with a clear beginning and end - to achieve a set of goals with limited resources. Thus, projects are temporary in nature, have distinctive features that make them unique and are undertaken to achieve predefined goals. For a more formal characterization, we adopt the definition of the Project Management Institute (PMI), one of the recognized international organizations of project managers. Emphasizing the above-mentioned distinguishing characteristics; PMI defines projects as "a temporary endeavor undertaken to create a unique product, service or result" (PMI 2017).

Based on this definition, we can give many examples of projects from different fields of activity. For instance, we can definitely consider the construction of the Gotthard Base Tunnel, the longest railway tunnel in the World, as a project. Going back many years in time, we could cite the building of the Eiffel Tower or even the Great Pyramid. The construction of these spectacular structures has all the characteristics given in the definition: they are distinctive, unique accomplishments undertaken over a specified time frame with time, cost and quality-based goals. In

G. Ulusoy, Ö. Hazır, *An Introduction to Project Modeling and Planning*, Springer Texts in Business and Economics, https://doi.org/10.1007/978-3-030-61423-2_1

the same vein, the design of the Airbus A380 civilian airliner, which can carry more than 500 passengers or the reconstruction of the giant Camp Nou Stadium with a capacity of more than 90,000 seats can all be viewed as projects.

While these examples illustrate extremely large projects that require massive resources and take years to complete, other projects may be much less complex, requiring a few simple resources and little time to complete. Examples of such smaller scale projects might include organizing a concert in the university spring festival, developing a simple computer game or a mobile application, procuring and installing solar panels on the roof of your house or editing a new book. Regardless of how many activities are involved or the business sectors in which they are undertaken, these examples all represent temporary, unique enterprises that aim to deliver a product or a service.

In business, designing and launching new products or delivering new services are typical examples of projects. However, mass production of a certain product such as an automobile or a washing machine requires performing the same operations for each product in a repetitive manner and hence does not meet the definition of a project. Designing an electric car, for example, would be considered as a project, but the continuous production of the resulting vehicle on the assembly line would not.

Both the above definition and the examples indicate that a project has some specific goals. Organizations or people initiate and carry them out with specific purposes or intentions. For instance, in a project to design a new electric car, delivering the prototype of the new car model within a time frame could be the principal goal. Operational definitions of the goals constitute objectives. The degree to which the objectives are achieved determines the performance of the project. Therefore, the objectives must be concrete and measurable. For example, in the electric car design project, a time limit, such as 2 years, by which time the prototype should be delivered, could be imposed.

Projects have been called the stepping-stones of corporate strategy. An organization's overall strategic vision is the driving force behind its project development. Although projects are well-defined entities, they are thus part of a larger system. An organization's projects need to be strategically aligned with its vision, objectives, strategies, and goals and should support the strategic alignment between these. To maintain a competitive strategy of cost leadership, for example, the company should focus on cost reduction projects. On the other hand, a company following a competitive strategy of differentiation might allocate more of its resources, for example, to in-house technology development projects or new product development projects. The principles and values on which these projects are based should not be in conflict with those of the organization. All the above statements indicate a need for effective coordination and communication between the organization and the ongoing projects as well as among the ongoing projects (Maylor 2003).

In almost all projects, several objectives must be taken into account simultaneously. Most projects seek to optimize some combination of time, cost, quality, and recently also environment-based targets. For example, an electric car design project might aim at delivering the prototype on time, without exceeding the budget limits while meeting quality and environmental standards. Both the latter, in turn, would require specific, measurable definitions such as miles traveled without breakdown or impacts on emissions.

The management literature has grouped project objectives along the three major (direct) performance dimensions: time, cost, and scope – usually referred to as the triple constraints. The third dimension, scope, specifies the boundaries of the work content and the outcomes to be delivered. PMI's Book of Knowledge (PMBOK) defines the scope as "the sum of the products, services, and results to be provided as a project" (PMI 2017). The quality and environmental standards are associated with the scope of the project. Meeting the time, cost, and scope objectives is the main challenge facing the project manager (PM) making project planning and management a multi-dimensional decision problem whose objectives are expressed in terms of target values. The nature of these target values, whether deterministic, stochastic, or fuzzy, determines the nature of the multi-dimensional decision problem the PM will face. Unless otherwise stated, all three of these performance dimensions are considered of equal importance.

These three project objectives are closely interacting with each other and affecting each other. For example, trying to reduce the project duration through the use of additional resources would increase the cost. Increasing the scope would in general be expected to increase the project duration and cost. Reducing the scope of the project, on the other hand, would in general be expected to reduce them. Whether increased or decreased, the scope changes should be acceptable to all parties involved in the project.

The firm's ability to deliver the required product/service or meet all of a project's multiple success criteria is usually threatened by various sources of uncertainty that projects face. Availability of resources might not be certain; time and cost estimates might not be correct. The requirements, scope, project team, and even the project objectives might also change during the course of the project's execution. For instance, in an electric car design project, some of the engineers on the design team might leave the company. Due to new legislation or technological improvements, standards for the car engine or battery system could change. Any or all of these unexpected events or change requirements might delay or prevent the desired level of success. Hence exposure to uncertainty and change is an essential feature of projects, highlighting the importance of risk and change management. We refer to Hazır and Ulusoy (2020) for a classification of major sources of uncertainty in projects and for a review of methods to manage uncertainty.

Having defined the characteristic features of a project and listing some important success criteria, we now investigate the relevance of project organizations in today's world and study the processes involved in effective project management. We introduce a life cycle concept for the projects and follow a process-based approach.

1.2 Project Management in Today's World

Today's organizations have become more and more project oriented. Regardless of the industrial or economic sector in which they operate, it has become common for organizations of all types and sizes to organize tasks as projects. One of the main reasons for this is that modern management orientation has shifted organizations from hierarchical to horizontal structures, and projects are perceived as a means to organize shared activities and teamwork between different functional departments of the firm, and even among various entities of the larger supply chain (Shtub et al. 2014). Another factor is the harsh competitive pressure and the resulting quest for excellence in realizing organizational goals. In the quest for competitive advantage, faster product life cycles and increasingly global supply chains require firms to organize many steps of once routine operations, such as manufacturing and production processes, as projects (Gunasekaran and Ngai 2012).

The increasing scope and complexity of the endeavors undertaken, combined with the frequent need to complete them in a short time to effectively support corporate strategies, has led to an increasing need to define, plan, and execute them in more formal ways. The increasing interconnectedness of the project stakeholders results in increasing complexity of managing projects. The increasing complexity, on the other hand, is also reflected as a higher chance of failure of the project.

In recent decades we observe increasing importance of knowledge as a factor of production, accompanied by an accelerating pace of technological change and activity. Most of these activities take place within relatively small research and development (R&D), engineering, and technology firms including so-called start-ups or spin-offs located mostly within knowledge clusters such as universities, science parks, techno-parks, and industrial parks. The majority of these firms operate through planning and executing projects and hence, are classified as project-based firms.

Another important sector of the economy that makes extensive use of project management is infrastructure construction and management. Due to the increasing volume of global trade, projects involving the construction and management of infrastructure such as telecommunication systems, roads, airports, bridges, and port facilities for facilitating the movement of goods, people, and information have gained in importance.

All these developments have increased the importance of coordination and control in management practice. Project based organizational structures help to focus responsibility and authority on achieving the organizational goals, facilitate control and coordination of the activities, and stimulate communication and better customer relationships (Meredith and Mantel 2011). Because of these advantages, many organizations have adopted project centered business models, rendering project management skills and the associated techniques crucial to successful business practice.

Each project involves performing many tasks that require time and various types of resources, such as labor, equipment, and money. These smaller elements of the projects are called activities. Performing these activities in a coordinated manner is critical for the successful delivery of the project as a whole and requires managerial support and leadership. Project management refers to the efforts required to ensure that these activities are carried out in an organized way and the overall project moves in the direction of the targets. It covers planning, organizing, controlling, and coordinating the activities taking place throughout the life cycle of the project from the inception of the idea motivating the project until its completion or termination. Obviously, not every project reaches the completion stage. For example, some projects might fail to reach their objectives, or the need for the project might no longer exist. An important task in project management is to identify deviations from the initially set targets through project control. When no recovery seems possible as a result of these deviations, the project is either terminated by winding up the project in the best manner possible or reformulated to be continued further following a new project plan. The termination of a project would have serious financial and organizational consequences for the organization and would not look good on its record but, on the other hand, it would free resources possibly for better and more effective use in other endeavors and further create an opportunity to learn from its failure. A termination might result in a blow on the career path of the team members as well.

The increasing demand for project management has transformed the practice into a developing profession with its own standards, procedures, methodologies, and regulations. To sustain this interest in project management without compromising the quality of service it provides, more people trained in the necessary tools and techniques are needed. Project management professionals in different countries are organized in professional societies, the largest being the Project Management Institute (PMI) located in Pennsylvania, USA (https://www.pmi.org) with over 540,000 global members and 295 chapters internationally (PMI Today 2018).

The above arguments indicate that project management as an effective methodology and as a profession will continue to thrive in the future. For that, those who develop methodology and those who implement them have to try to comprehend the increasing pace and depth of change occurring globally and have to be in close contact and cooperate. Platforms like PMI need proliferate in this context. Another practice supporting the proliferation of project management as a profession and improving its quality would be certification through professional organizations. The impact of this practice would increase with the increasing use of these certificates by the organizations as a reference for employment.

1.3 Challenges in Keeping to Scope, Cost, and Schedule Targets

Several empirical studies examine the root causes of the challenges project management faces in meeting scope, cost, and schedule targets (Parvan et al. 2015). Data has been gathered through surveys of clients, contractors, and design personnel to assess the magnitude of cost and schedule overruns and their root causes. The major findings are:

- The quality and the extent of early design and planning are critical for the assurance of project performance.
- The factors that influence the quality of task implementation are critical contributors to overall performance.
- Changes requested by clients during the implementation of the project cause changes in the scope of the project and often lead to many ripple effects that impact the project well in excess of the direct cost of those changes.
- The leadership and team structure and incentives moderate the impact of project-specific factors on performance.

The latent impact of design errors on the implementation phase has been analyzed by several studies, particularly for construction projects. Burati et al. (1992) reported that design deviations make up on average 78% of the total number of deviations. The direct costs associated with rework (including redesign), repair, and replacement were considered. The design defects were responsible for 79% of total change costs, and 9.5% of total project cost. Analyses of the data indicated that design deviations accounted for an average of 12.4% of the total project costs. Lopez and Love (2012) showed that the mean direct and indirect costs due to the design errors were 6.85 and 7.36% of the contract value, respectively. Conducted in the same sector -construction- these studies seem to have reached comparable results.

Hanna et al. (1999) concluded that change orders have a negative impact on labor efficiency. A further conclusion was that the later a change order occurs in the life of a project the greater its negative impact on labor efficiency. Hanna et al. (2002) found that design errors accounted for 38–50% of change orders in the projects they studied.

To summarize, changes are inevitable during the execution of projects, making robustness of the project plans and agile management techniques important aspects of project management. These two topics will be studied in detail in the next sections.

1.4 A Life Cycle Concept for Projects

Each project has a life cycle, passing through certain identifiable phases over the course of its life between initiation and completion or termination. These phases are usually sequential but can also be overlapping and iterative. The main phases constituting the project life cycle are:

(i) Conceptual design and project definition,
(ii) Planning,
(iii) Implementation, monitoring, and control,
(iv) Evaluation and termination.

We should note that the above list of phase names is not unique. Depending on the type and scope of the project different numbers of phases and names can be employed. For example, in a defense acquisition project the life cycle phases could be (i) determination and mission need, (ii) concept exploration and definition, (iii) demonstration and validation, (iv) engineering and manufacturing development, (v) production and deployment.

The concept of the life cycle we will investigate in this section is mainly for project environments with low or moderate uncertainties around project requirements. Most of the planning is accomplished early in the project. Although iterations occur within and across project phases, the phases are executed mainly through a sequential process. The project life cycles for cases with high levels of uncertainty around their project requirements are formulated through agile approaches, which will be treated later in this chapter.

Next, we investigate each phase (i) through (iv) above in detail.

1.4.1 Conceptual Design and Project Definition

Conceptual design identifies the need for the project and establishes the basic principles that will be referred to in the later phases. At this point, the project requirements usually are not yet fully clear. Initial feasibility and risk analyses must justify the investment decision and reveal the inherent risks. In case they don't, abandonment is always an option and may often be more rational not to initiate the project and hence, reject the project proposal.

Once a project is formulated conceptually, economic and risk studies are completed and the decision to invest is taken, the project definition process is initiated. At this point; strategies, objectives, and scope should be defined clearly. While defining the scope and specifying the outcomes to deliver, it is important to agree on the boundaries of the work content, and hence, what the project will exclude. This allows for focusing on the required tasks and effective resource allocation. If the scope is not clear or too broad, resources are more likely to be used ineffectively, increasing the likelihood of failure.

An important part of the project initiation is the identification of constraints, assumptions, and the dependencies between them as well as with other projects within and outside the organization. In addition, main risk items are identified, and initial risk analysis is performed. *Stakeholders*, i.e., persons and organizations (see Section 2.3), who are affected by the decisions taken during the design and execution of the projects, are identified and their priorities and expectations clarified and analyzed. Finally, in the definition phase, the project organization is established, a PM is assigned, and the project team is selected.

1.4.2 Planning

Once the objectives, scope, and constraints are specified, the planning process can take place. A fundamental rule of planning is to integrate the project team into the planning process as soon as possible so that by participating in the planning effort they develop ownership of the plan, and hence, a stake in its success or failure of the project.

Project planning requires the allocation of limited resources, especially the project budget, among different tasks in the project so as to realize the specified objective(s). Hence, resource allocation serves as a major input to managerial decision-making. The resources should support the execution of the project strategy that defines the future evolution of the project. On the other hand, tactics detail what must be done to achieve the objectives. Within the specified scope, plans are made for reaching the objectives. These initiation steps require a strategic project management approach, consisting of the steps taken by the project to progress towards its objectives, the competitive advantages of the firm, and decision-making processes.

Concrete plans are prepared to accomplish the predefined project objectives. A crucial aspect of the planning phase is the development of key performance indicators (KPIs) to be able to answer questions like "To what degree have we accomplished these objectives?", "How successful was the project management?", "How successful were the PM and the project team members?" These KPIs should be defined and agreed upon before the implementation phase of the project starts. The selected KPIs should be easily understandable, relevant to the organization, measurable, and their established target values achievable. The stakeholders, PM, and the project team should be fully informed about the KPIs before project implementation starts. It is good practice for the team members to be involved in the development of the KPIs.

During the planning phase, the work content of the project is divided into work packages (WPs), which are combinations of related project activities. An *activity* is a particular unit of work that consumes time and requires resources during project execution. For example, in the construction of a single story residential house electrical works would be organized as a WP with activities such as laying the cables to the outdoor lighting and placing the light switches in the building (see Chap. 3 for a detailed analysis of WPs and activities).

Defining and grouping the activities requires knowledge of the characteristics and requirements of the work content involved. The level of detail at which the WPs and the activities are specified is an important decision that requires considerable expertise and team discussions.

The appropriate level of detail depends on the specific project, its objectives and expected outcomes. To give an example, the activities in a manufacturing plant capacity expansion project could be divided into three WPs: procurement of the new equipment, installation and testing. However, for a software development project, a more detailed breakdown, with WPs specific to the development processes such as conceptual design, requirements analysis, software design, software development, integration and testing might be necessary.

All these WPs and the activities that they contain are not independent of each other. Technological or economic constraints, legal requirements or resource conflicts may cause some activities to wait for the completion of some others. In the software development project, the specifications must be defined before starting the design and development activities, i.e., the definition of specifications activity must precede the design and development activities. These precedence relationships restrict the sequence in which the activities can be executed. Resource relationships among activities that must share the same resources, such as the same designers or developers, impose additional dependencies among project activities. All these relations must be considered in project planning, which is an important phase of the project life cycle.

Having defined the activities and their relationships, time and resource requirements and the resulting costs are estimated, and scheduling is performed. This involves the preparation of time and resource allocation plans that determine when the project will be delivered, when activities will be started and completed, and how resources will be assigned to these activities over time. They establish a baseline for measuring and assessing project progress. Various project scheduling problems together with suitable modeling and solution methods, make up an important portion of this book.

Once time schedules and financial plans are prepared, and the baselines for performance evaluations are established, projects are ready for execution to begin. However meticulously plans have been prepared, unexpected events such as variations in resource usages or availabilities, irregularities in cash flows, or deviations from the planned performance are widely encountered in practice jeopardizing successful achievement of the project objectives. To prevent significant deviations from the targets, projects must be continuously monitored and controlled.

1.4.3 Implementation, Monitoring, and Control

During project implementation deviations from the plan are prevalent rather than the exception, and quick corrective action is essential to minimize their adverse impacts or take advantage of unforeseen opportunities. The essential components of any such feedback mechanism are monitoring and control. *Monitoring* is the collection and preparation of information necessary to assess project performance, permitting timely identification of deviations from plan and initiation of corrective actions such as rescheduling, or additional resource allocation by PMs. Formally, project controlling involves comparing observed performance with the targets and performing corrective or preventive activities. In many business cases, monitoring and control have been shown to be critical to successful project completion. Therefore, PMs will benefit greatly from establishing a structured monitoring and control system. An effective control system should facilitate data analysis by integrating various data presentation tools to support decision-makers in determining when, under what conditions, and how to intervene in projects (Hazır 2015).

1.4.4 Evaluation and Termination

Finally, in the last phase of the project life cycle, the project performance is assessed, final reports and further documents are prepared, the deliverables are transferred to the customer and the project team is dissolved.

The project performance can be assessed in both the process and the product (the deliverables) dimensions. The process assessment focuses on the performance of the PM and the project team members throughout their stay in the project and the phases of the project. The performance metrics employed might vary from organization to organization but the metrics reflecting the deviations from the initially estimated project duration and the project budget are essential performance metrics. The performance metrics associated with the deliverables reflect the requirements imposed on the deliverables such as quality standards.

Each phase has its requirements and is managed through a set of processes: Initiation processes, planning processes, execution processes, monitoring and controlling processes, closing processes (PMI 2017). These processes are applied at differing levels and scope in different phases of the project life cycle. It is enlightening to observe the similarity between the processes and the phases of the project life cycle. For each phase, different performance metrics may be defined, priorities may change, and different managerial techniques may be applied. However, all these stages involve complex decision problems such as the planning of interdependent activities and allocation of resources to these activities over time.

1.5 Post-Project Analysis

Effective management requires that results be recorded, analyzed and lessons learned from the project documented. It is important to distinguish between the assessment of project performance and the post-project analysis. *Post-project analysis* (PPA), in contrast, seeks to document the good and bad practices observed during various project phases and come up with opportunities for future improvement. Although PPA is often neglected, it is essential for developing and sustaining a learning organization by providing valuable information to future PMs, while PPA reports would serve as guidelines in the new projects. Before planning a new project, PMs can benefit from reviewing the lessons learned that are documented in these reports.

An effective PPA requires a good data collection system. What data to collect and the process by which to collect it depend on which aspects of the project we will analyze for this purpose, which may include (Lientz and Rea 2011):

- Project planning, management, and control,
- Project management tools and methods used,
- Organization structure employed,
- Effective training level of the PM and team members,
- Methods employed in the project,
- Tools used by the project team and their relationship to the methods,
- Communication and coordination with the rest of the organization,
- Communication and coordination with the stakeholders.

Both good and bad practices relative to these aspects are archived and shared within the organization. Good practices should be implemented in later projects, and bad practices avoided as far as possible in the future. Action plans need to be devised and implemented for any areas of improvement that are identified.

If the results of a PPA are not expected to enhance the organization's project management process, the organization might decide not to perform a PPA. The budget allocated for this purpose might also limit the number of PPAs executed. But once the organization has decided to perform one, a PPA team must be formed. As projects have unique characteristics, it is usually a better practice not to maintain a standing committee for this purpose but instead to constitute a new group appropriate to the size and nature of the project. The organization might prefer to employ professional facilitators for this purpose. For its conclusions to be credible, the PPA team should be impartial and have no conflict of interest with the PM and the project team and be impartial. The PM and the team must interact with the PPA team to provide their views and opinions but should not be part of the PPA team.

The PPA reports need to be organized in an archive that is maintained as part of the organization's information system and serves as a decision support system to assist during the planning process of new projects. The design of the archive among others should allow for easy and quick reference to good and bad practices in past projects to be taken into account when planning the current project proposal. Depending on the size of the organization, there should be a person or a unit responsible for the management of the archive. As suggested in Chap. 2, the firm's Project Management Office would be a good candidate for ownership, if one exists in the organization.

Next, we examine uncertainty and risks in projects.

1.6 Uncertainty and Risks in Projects

Uncertainty, as defined by Meredith and Mantel (2011), refers to having "only partial knowledge of the outcome and the situation". Uncertainty is unavoidable in decision-making, and lack of information makes taking decisions more difficult. Events such as machine failures, incorrect time estimates, quality problems, delays in cash inflows, or the arrival of last minute change requests create a work environment in which we do not have complete information, and which can prevent the achievement of the project and organizational goals. Therefore, strategies need to be developed to avoid unforeseen problems in projects and to limit at least their negative consequences.

Uncertainty in projects can be investigated from different perspectives. De Meyer et al. (2002) studied project uncertainty under four different headings, each of which requires different management techniques:

(i) *Variation* refers to random deviations, which are present in every production or service system. Short duration machine breakdowns, minor quality problems and absenteeism due to a seasonal flu epidemic are some examples.
(ii) *Foreseen uncertainty* addresses the cases where some probable events could be identified and anticipated. Previous experiences and data might suggest that some possible side effects are likely to be encountered in a drug development project before initiating it. However, the magnitude and timing of events are often hard to predict.
(iii) In contrast to foreseen uncertainty, events and their possibilities are unknown in *unforeseen uncertainty*. As they could not be anticipated, contingency plans cannot be integrated into planning. For instance, a drug developed to treat a specific health problem could be observed to aid in solving other illnesses, which have not been targeted.
(iv) *Chaos,* a state of utter confusion, is the hardest case to manage because even the project structure might radically change during the period of chaos. Crisis management is required in these situations. Leadership, experience, and rapid decision-making become more important than planning. Natural disasters, such as earthquakes and hurricanes, are typical examples of chaotic situations.

Lack of information inevitably imposes some risks on projects. Although uncertainty and risk are related to each other, they are not the same. Uncertainty characterizes the unknowns. Project risk, on the other hand, can be defined as "an uncertain event or condition that, if it occurs, has a positive or negative effect on one or more project objectives" (PMI 2017). In this definition, positive effects represent the opportunities for enriching the results of the project, whereas the negative ones the threats hindering the success of the project. To give an example for a negative effect, in many cases, payments received from the client firms might be uncertain. The time of the payment, the amount, or both might not be known with certainty since they are largely dependent on the financial position of the client. Any delays in payments might significantly affect the financial viability of the projects and create financial risks for the contractors; perhaps driving them to bankruptcy. An example for a positive effect would be the financial impact of a favorable change in the foreign exchange rate for the imported machinery in an investment project.

Another important difference between risk and uncertainty is that the risks are usually expressed in terms of the impacts of the realization of events. In contrast, uncertainty might or might not have a known effect. Moreover, the word "risk" usually evokes negative consequences in the business world.

Another distinction in the literature is based on probabilities of event realizations. Risk refers to cases where the probability of events can be predicted accurately, and uncertainty refers to those where event probabilities cannot be reliably estimated (March and Simon 1958). As a consequence; in decision-making under risk, an alternative that generates the highest expected utility is chosen; whereas in decision-making under uncertainty, only criteria like maximin, minimax, maximax are employed (Clemen and Reilly 2001).

Analysis of risk items and their management are shown to be a critical dimension of effective project management. Project success can be enhanced by establishing a risk management culture that facilitates the integration of risk analysis tools into project management practice (Voetsch et al. 2004).

Risk management will be discussed in detail in Chap. 10. Proactive approaches to uncertainties and risks will be emphasized throughout this book. These include planning and re-planning techniques, robust modeling approaches and project monitoring and control techniques. We first focus on time-based objectives and introduce the widely used time modeling and planning tools.

1.7 Project Modeling and Planning Tools

Mathematical modeling of projects requires the partitioning of the project scope into non-overlapping WPs, each made up of specified groups of activities that collectively constitute the entire work content of the project.

Schedules are essential tools of project managers for time and budget planning and control. A schedule is specified by the starting and ending times of the WPs or their constituent activities together with the project milestones, if any exists. Project milestones are tasks of zero duration representing a significant achievement or a point in time in the project such as the delivery of the prototype in a new product development project. The project schedule prepared before project execution begins is called the *baseline schedule*. A good baseline schedule is an important aid to planning and coordination of the activities, such as the procurement of materials, planning equipment, and staff usage. Moreover, in practice, milestones and due dates on events including the completion of the project are usually set based on these schedules. Mathematical models have been developed for preparing these schedules. The critical inputs of the scheduling problems are usually activity durations, costs and resource requirements, precedence relationships among activities, and resource availabilities. Decision variables are usually the start or end times of activities.

The duration and the sequence of activities as well as their current status can be illustrated by using a time scaled bar graph called a *Gantt chart*, first developed by Henry L. Gantt in 1917 for a World War I military requirement (Moder and Phillips 1970). Gantt charts continue to be the most frequently used visualization of time plans and the progress of a project. They are useful in simplifying complex projects and presenting a quick overview of the project with all its milestones and due dates. A major deficiency of a Gantt chart, however, is that it might not be able to totally display the precedence relations among project activities, which might establish critical constraints on scheduling. When an activity B immediately follows another activity A on the Gantt chart, this does not necessarily mean that there is a precedence relationship between these activities. Hence, the Gantt chart alone does not provide information on activities that can be delayed. This information is crucial for the PM, particularly when rescheduling the project. It is important that any graphical representation of projects displays these precedence relations in total, particularly in complex projects. Such a graphical tool is provided by the project network diagram.

Project network diagrams consist of arcs and nodes, and they are drawn from left to right to reflect the time sequence. Two alternative network representations are used in the literature and practice. In an *activity-on-node* (AON) *representation*, nodes represent the activities and arcs define the precedence relationships among the nodes. On the other hand, in an *activity-on-arc* (AOA*) representation*, the nodes represent events such as the completion of the activities, whereas the arcs correspond to the activities. An activity that must be completed before another activity can start is called its *predecessor*. An activity that can start after the completion of another activity is referred to as its *successor*.

In addition to graphical aids for scheduling, the *Critical Path Method* (CPM) and The *Program Evaluation and Review Technique* (PERT) are widely used network analysis techniques in project management and are integral part of available project scheduling software.

CPM has been used to define the time schedule and identify the critical activities and the critical path. It was introduced in the late '50s in the United States and first employed by the Du Pont Chemical Company for planning and maintenance of chemical plants (see Kelley and Walker (1959) for historical development). The *critical path* refers to the longest path in the project networks. Since all activities need to be completed, it also establishes the earliest completion time for a project. Project networks contain multiple paths, each representing a set of activities to be processed in the given order, between the start and end nodes of the project. Some of these paths might have equal length; hence there may be more than one critical path in projects. The activities on the critical path(s) are called *critical activities*. Any delay in the completion of a critical activity results in a delay of the same amount in the project completion. Therefore, monitoring and controlling of the critical activities is essential to avoid late project completion. Thus, besides establishing the activity and project completion times, CPM serves to determine the critical activities.

PERT estimates the project completion time and the starting time of each activity under uncertainty in activity durations. PERT was first developed for the Polaris Missile Program of U.S. Navy in 1958 by representatives of Lockheed Aircraft Corporation, the Navy Special Projects Office, and the consulting firm Booz Allen Hamilton (see Moder and Phillips 1970; Sapolsky 1972; Kerzner 2013). It is a well-known method incorporating uncertainty into activity durations and is seen as a stochastic alternative to CPM that can be used to calculate the expected completion times and to estimate the probability of on time project completion. It estimates the mean and variance of the activity durations based on the most likely, optimistic and pessimistic estimates of the activity durations.

Like any other modeling tool, CPM and PERT have their limitations. The *Graphical Evaluation Review Technique* (GERT) provides several additional capabilities besides probabilistic activity durations for a more realistic representation of projects. (i) Activities are allowed to have several outcomes whose probabilities of occurrence add up to 1.00 rather than only a single deterministic outcome. (ii) An activity might need to be repeated because, for example, its output is not acceptable. (iii) The output of an activity might be a signal to another activity requesting the execution of all consecutive activities connecting these two activities resulting in a loop (Moore and Clayton 1976). By means of these additional capabilities GERT modeling facilitates the treatment of a whole new set of problems through project network analysis. Such a set of problems would be, for example, management of R&D projects, where the outcomes of activities, in general, are of a probabilistic nature involving loops as well as possible repetition of certain activities (Whitehouse 1973). GERT modeling coupled with GERT simulation provides considerable insight to project managers in both the planning as well as implementation phases. Fixed and variable costs of activities can also be taken into account in GERT modeling and simulation allowing trade-off analysis between project time and cost. An extension of GERT modeling is called Q-GERT modeling. In addition to most of the capabilities and features of GERT, Q-GERT contains special queue nodes allowing for modeling of situations in which queues form in front of service nodes (Pritsker 1977; Taylor III and Moore 1980).

Finally, the *Critical Chain Project Management* (CCPM), introduced by Goldratt (1997), applies the Theory of Constraints (Goldratt and Cox 1984) to project management. Unlike CPM, the resource relationships are also considered. The critical chain, which is the focus of CCPM, is defined as the sequence of both precedence and resource dependent activities that determines the project completion time. The methodology also underlines the importance of buffers, usually extra time inserted into the schedule before starting a new activity. They provide protection against various sources of uncertainties. The buffers are generally added at the end of the project or end of the critical chains. During the project execution, they need to be controlled carefully to achieve project targets.

1.8 Resource Constrained Modeling

CPM and PERT consider only the project network structure and activity processing times. The resource requirements of the activities and how these requirements are met over time are not addressed, essentially assuming infinite availability of resources. In practice, resources differ from each other in significant ways and are usually available only in limited quantities at any point in time. It is thus necessary to explicitly incorporate resource constraints into the mathematical analysis. The execution of project activities requires a variety of resources ranging from workers, machinery, and equipment to financial instruments. In the scheduling literature, resources are classified in several different ways (Demeulemeester and Herroelen 2002). We will limit ourselves with the following three: renewable, non-renewable and doubly constrained. Further detail is provided in Chap. 7.

A *renewable resource* is available in a constant amount in all periods throughout the planning horizon. Machines, equipment and staff are examples of renewable resources. Some renewable resources may only be available in certain periods; these are called *partially renewable*. In contrast, the available quantities of non-renewable resources, such as construction materials, decrease with consumption. *Doubly constrained resources* have limited availability in some periods of the planning horizon and also a limited total consumption over the entire horizon. For instance, project budgets might limit the consumption of monetary resources in both a specific time interval and also over the complete project period. In those cases, these resources impose double constraints.

In contrast to CPM and PERT, the *Resource Constrained Project Scheduling Problem* (RCPSP) explicitly incorporates resource constraints. In this problem, project completion time is minimized assuming resource availability over the planning horizon is known with certainty. Both precedence relationships among the activities and constant resource availability constraints are integrated into the analysis. The RCPSP has been extensively studied in the literature in many different versions (Özdamar and Ulusoy 1995; Herroelen et al. 1998; Kolisch and Padman 2001; Hartmann and Briskorn 2010). Several combinatorial optimization problems have been shown to be special cases of RCPSP (Hartmann 1999). An example is the job shop scheduling problem (Sprecher 1994).

RCPSP assumes a single resource mix for the execution of an activity and a fixed resource allocation, but in practice, there are sometimes alternative resource mixes of processing the activities. Each of these alternative resource mixes, which are characterized by known duration and resource requirements, is called a *mode* in the scheduling literature. These alternatives result through the use of different technologies, different types and quantities of resources. For example, 3 workers might accomplish a certain amount of work in 8 h. On the other hand, 2 workers with the use of equipment can finish the same amount of work in 6 h. Hence, in this example, we have 2 modes: Mode 1 (3 workers, 8 h); mode 2 (2 workers + equipment, 6 h). The extension of RCPSP to multi-mode settings results in the *Multi-Mode Resource Constrained Project Scheduling Problem* (MRCPSP). This problem addresses assigning exactly one mode to each activity so that the project completion time is minimized while satisfying all precedence and resource constraints. MRCPSP explicitly models the use of *renewable, nonrenewable and doubly constrained* resources. A special case of MRCPSP that models only a single nonrenewable resource (usually money) is called the *Discrete Time/Cost Trade-off Problem* (DTCTP). DTCTP seeks the minimum cost for reducing the project duration by compressing some of the activities at the expense of increased cost.

Another problem of practical relevance in project scheduling is the *Resource Leveling Problem.* In contrast to the RCPSP, in the Resource Leveling Problem the project duration is fixed and cannot be exceeded, while resource requirements vary over time. Since no availability limit on the resources is taken into account, the resource fluctuations might become large and be costly. For example, varying demand for workers might require frequent signing of new short-term contracts for certain time periods, which would usually be relatively costly. It might also be risky as well because there might be a shortage of certain skills/talents at certain time periods. Therefore, baseline schedules must take measures to reduce the variability of total resource needs over time. The solution procedures for the Resource Leveling Problem seek to minimize some measure of the variation of resource utilization over time in order to level the resource requirements over the planning horizon.

An extension of resource leveling involves the resource utilization cost, which is taken to be the sum of the costs of increasing and decreasing resource levels throughout the project implementation. The problem is called the *Total Adjustment Cost Problem.* This problem is of practical relevance. For example, the hiring and firing of human resources would be a major application area.

The *Resource Availability Cost Problem* (RACP), like the Resource Leveling Problem, also addresses resource requirements throughout the project execution. However, unlike the Resource Leveling Problem, which aims to minimize the variation in resource utilization over time; the RACP seeks to minimize the resource cost given the maximum resource demand. RACP tries to find a feasible project schedule that allows the completion of the project by its deadline and minimizes the total cost, which is a function of the peak resource demand.

Finally, we present a conceptual map in Fig. 1.1 to highlight the important concepts and relations in project modeling.

1.9 Agile Project Management

We will deal with agile project management in some depth in Sect. 2.9 and hence, we will briefly emphasize some of the main points here.

Replanning, rescheduling, or redesign as a result of scope changes have been commonly encountered by many managers in projects. The increasing frequency of change requirements has led managers to redefine the project life cycle, reconsider the planning needs and adopt more adaptive approaches such as agile project management.

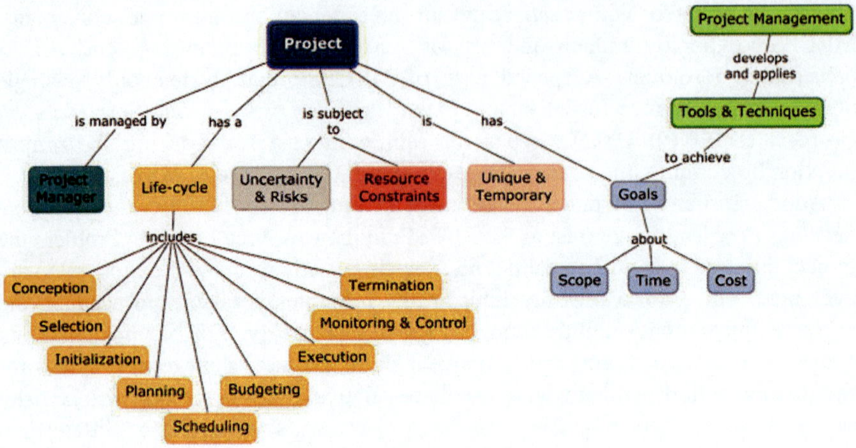

Fig. 1.1 A conceptual map of project planning and modeling

Project life cycles can vary from totally planned driven or predictive approaches to adaptive or to agile ones (PMI 2017). In a predictive life cycle, project scope and deliverables are clearly defined when the project is started, and scope changes are tried to be managed during project execution. On the other hand, agile life cycles are adaptive and iterative; the deliverables are finalized through multiple iterations. In other words, the scope is detailed at each iteration. This iterative approach allows managers to incorporate changes more easily and facilitate stakeholder engagement in product/service design.

We note that it would be almost impossible to define the project scope clearly at the beginning of the project in business environments that incorporate a high level of uncertainty. Project scopes evolve during executions in these environments. As the agile approach does not require agreeing on the scope at the early stages but rather focuses on refining the scope through iterations, it has been increasingly adopted in several industries; especially where projects are delivered in highly uncertain, rapidly changing environments.

Exercises

1.1 For each item below state whether it can be formulated a project or not and justify your answer briefly.
 (a) Organizing a conference
 (b) Preventive maintenance
 (c) Construction of a concert hall
 (d) Billing clients
 (e) Producing a movie
 (f) Relocating a factory

1.2 Explain uncertainty, risk, and the relation between risk management and project management?

1.3 Describe the project life cycle and explain the stages briefly. Give some examples.

1.4 Consider an investment project to produce electricity in a university campus using renewable energy resources.
   (a) Design this project conceptually. Write down the possible goals, risks and define the scope of the project.
   (b) Consider the goals that you defined and formulate the project strategy.

1.5 Discuss what a baseline schedule is and why it is important to prepare effective baseline schedules.

1.6 Explain the importance of work breakdown structure (WBS) in project planning.

1.7 Prepare a survey of major project management software used in practice.

1.8 What would you consider the major evaluation criteria to be when selecting project management software?

1.9 Explain the reasoning behind using resource levelling in projects.

1.10 What is the post-project analysis and how does its objective differ from the performance evaluation of the project manager and project team members?

1.11 Explain how an organization would make use of post-project analysis and its expected benefit from this practice.

1.12 Explain in your own words the strategic alignment between a firm's projects and its basic vision, objectives, strategies, and goals and provide an example.

1.13 What would you consider a disadvantage of Gantt charts? Give an example.

1.14 Establishing targets properly is important for the effectiveness of planning, monitoring, control, and performance evaluation of projects. What methods can you suggest for establishing targets? Discuss.

References

Burati, J. L., Farrington, J. J., & Ledbetter, W. B. (1992). Causes of quality deviations in design and construction. *Journal of Construction Engineering Management (ASCE), 118*(1), 34–49.

Clemen, R. T., & Reilly, R. (2001). *Making hard decisions*. Duxbury.

De Meyer, A., Loch, C. H., & Pich, M. T. (2002). From variation to chaos. *MIT Sloan Management Review, 43*, 60–67.

Demeulemeester, E., & Herroelen, W. (2002). *Project scheduling-a research handbook* (International Series in Operations Research and Management Science) (Vol. 49). Kluwer Academic Publishers.

Goldratt, E. M. (1997). *Critical chain*. Great Barrington: The North River Press Publishing Corporation.

Goldratt, E. M., & Cox, J. (1984). *The goal*. Croton-on-Hudson: North River Press.

Gunasekaran, A., & Ngai, E. W. T. (2012). The future of operations management: An outlook and analysis. *International Journal of Production Economics, 135*(2), 687–701.

Hanna, A. S., Russell, J. S., Gotzion, T. W., & Nordheim, E. V. (1999). Impact of change orders on labor efficiency for mechanical construction. *Journal of Construction Engineering Management (ASCE), 125*(3), 176–184.

Hanna, A. S., Camlic, R., Peterson, P. A., & Nordheim, E. V. (2002). Quantitative definition of projects impacted by change orders. *Journal of Construction Engineering Management (ASCE), 128*(1), 57–64.

Hartmann, S. (1999). Project scheduling under limited resources. models, methods, and applications. In *Lecture notes in economics and mathematical systems* (Vol. 478). Berlin: Springer.

Hartmann, S., & Briskorn, D. (2010). A survey of variants and extensions of the resource-constrained project scheduling problem. *European Journal of Operational Research, 207*(1), 1–14.

Hazır, Ö. (2015). A review of analytical models, approaches and decision support tools in project monitoring and control. *International Journal of Project Management, 33*(4), 808–815.

Hazır, Ö., & Ulusoy, G. (2020). A classification and review of approaches and methods for modeling uncertainty in projects. *International Journal of Production Economics. 223*, https://doi.org/10.1016/j.ijpe.2019.107522.

Herroelen, W., De Reyck, B., & Demeulemeester, E. (1998). Resource-constrained project scheduling: A survey of recent developments. *Computers and Operations Research, 25*(4), 279–302.

Kelley, E. J. Jr, Walker, M. R. (1959). *Critical-path planning and scheduling*. In Proceedings Eastern Joint Computer Conference, Boston, 1–3 December, 160–173.

Kerzner, H. (2013). *Project Management: A systems approach to planning, scheduling, and controlling*. New York: Wiley.

Kolisch, R., & Padman, R. (2001). An integrated survey of deterministic project scheduling. *Omega, 29*(3), 249–272.

Lientz, B. P., & Rea, K. P. (2011). *Project management for the 21st century* (3rd ed.). New York: Routledge.

Lopez, R., & Love, P. E. D. (2012). Design error costs in construction projects. *Journal of Construction Engineering Management (ASCE), 138*(5), 585–593.

March, J. G., & Simon, H. A. (1958). Cognitive limits on rationality. *Organizations*, 136–171.

Maylor, H. (2003). *Project management* (3rd ed.). Essex: Pearson Education Ltd.

Meredith, J. R., & Mantel, S. J., Jr. (2011). *Project management: A managerial approach*. New York: Wiley.

Moder, J. J., & Phillips, C. R. (1970). *Project management with CPM and PERT* (2nd ed.). New York: Van Nostrand Reinhold Company.

Moore, L. J., & Clayton, E. R. (1976). *GERT modeling and simulation: Fundamentals and applications*. New York: Petrocelli/Charter Publishing Company.

Özdamar, L., & Ulusoy, G. (1995). A survey on the resource-constrained project scheduling problem. *IIE Transactions, 27*, 574–586.

Parvan, K., Rahmandad, H., & Haghhani, A. (2015). Inter-phase feedbacks in construction projects. *Journal of Operations Management, 39–40*, 48–62.

PMI. (2017). *A guide to the project management body of knowledge (PMBOK® Guide)* (6th ed.). Newtown Square: Project Management Institute.

PMI Today. (2018). *Statistics through 30 Sept 2018*. Newtown Square: Project Management Institute.

Pritsker, A. A. B. (1977). *Modeling and analysis using Q-GERT networks*. New York: Wiley.

Sapolsky, H. M. (1972). *The polaris system development*. Boston: Harvard University Press.

Shtub, A., Bard, J., & Globerson, S. (2014). *Project management: Processes, methodologies, and economics* (New International Edition). New York: Prentice Hall.

Sprecher, A. (1994). Resource-constrained project scheduling: Exact methods for the multi-mode case. In *Lecture Notes in Economics and Mathematical Systems* (Vol. 409). Berlin: Springer.

Taylor, B. W., III, & Moore, L. J. (1980). R&D project planning with Q-GERT network modeling and simulation. *Management Science, 26*(1), 44–59.

Voetsch, R. J., Cioffi, D. F., & Anbari, F. T. (2004). Project risk management practices and their association with reported project success. In *Proceedings of sixth IRNOP project research conference, Turku, Finland* (pp. 680–697).

Whitehouse, G. E. (1973). *Systems analysis and design using network techniques*. Englewood Cliffs: Prentice Hall.

2 Organizing and Managing Projects

Learning Outcomes
Upon successful completion of this Chapter, the reader will be able to:

1. Distinguish between functional and project organizational structures.
2. Explain the characteristics of project organizations and discuss their advantages.
3. Describe and compare the different forms of matrix organizations.
4. Explain the criteria for choosing a particular project management structure.
5. Recognize the importance of stakeholder management.
6. Recognize the roles of a project sponsor, project manager, and project team.
7. Assess the needs for creating project management offices.
8. Distinguish between project, program, and portfolio management.
9. Compare process-based project management with agile project management.

2.1 Projects and Organizational Structure

A key question that looms large following the decision to undertake a project is how to link the project to the organizational structure of the parent organization. The answer to this question shapes the span of control, communication, and coordination in a project. The ability of the project manager to acquire the necessary resources, to mobilize the best qualified individuals, to motivate the project personnel, to resolve conflicts, in other words, the eventual success or failure of the project, depend on how the project is configured within the parent organization. The best manner to accomplish this depends on the structure of the parent organization and the specifics of the project in question. An effective structuring balances the requirements of the project and the parent organization.

To this end, we will introduce the three main organizational structures of parent organizations (Larson and Gray 2018): *functional, project-based*, and *matrix organizations*, and discuss different ways to integrate a project as well as advantages and disadvantages of each one. Composite structures that combine different aspects

G. Ulusoy, Ö. Hazır, *An Introduction to Project Modeling and Planning*, Springer Texts in Business and Economics, https://doi.org/10.1007/978-3-030-61423-2_2

of these principal structures have been also adopted by many organizations to facilitate the management of complex organizational entities.

2.1.1 Functional Organizations

Functional organization has been the traditional way of organizing work. In this form, project work is implemented within functional departments, where people performing similar activities work together, allowing them to develop in-depth expertise in that particular area. To give an example, the finance department employs accounting and finance experts who consult with several projects as well preparing financial reports for general management. In functional organizations, project tasks are performed under the responsibility of a functional unit. Project work is divided into work packages (WPs) that require resources from several departments. Coordination of the activities, performance control, and resolution of conflicts among departments become critical and are usually undertaken by top management. This function or department-based structure built on the division of labor and specialization offers several advantages to the parent organization as well as some disadvantages regarding coordination and control.

Firstly, it usually requires minimal organizational change, since this organizational structure allows projects to be configured within the existing functional hierarchy. Functional managers have the authority to assign their employees to different projects and the flexibility to move them to other projects or tasks. When projects are terminated, employees can start working on other projects or operations. This transferability makes the future of employees independent of the life or budget concerns of the project and allows the employees to continue their careers within the departments (Meredith and Mantel 2012).

Despite the advantages of departmental focus, expertise development and employee transferability, there are several disadvantages, mainly the need for careful and continuous allocation and coordination of the resources and the project activities. As departments specialize in particular functions, their members focus only on specific tasks. Given the complexity and numerous requirements of today's projects, coordination among various departments and integration of the work performed becomes difficult to manage. As many departments need to communicate and collaborate, decision-making can be slow compared to fully project-based organizations where project managers have stronger authority and responsibility. Delays in decision-making might also slow responses to changing client needs.

Another major problem might be differences among the departments regarding project priorities or objectives. Departments might have different short/medium term goals leading them to prioritize the projects differently. Some departments might not pay enough attention to some projects while focusing on other operations. Focusing on project objectives and following a holistic approach to managing the project work becomes difficult (Meredith and Mantel 2012; Shtub et al. 2014). Departments inevitably focus on their own areas of interest. Even if all of them offer high quality service, project success might not be guaranteed. The overall requirements of the project should be viewed as an interconnected system that contains many resource dependencies and requires careful coordination of interdisciplinary work. Managing this system effectively requires a high level of dedication and communication within the project teams, which will be easier to achieve in project-based structuring.

2.1.2 Project Organizations

If functional organizations lie at one end of the structural spectrum, project organizations, in which dedicated project teams are established, lie at the other extreme (Larson and Gray 2018). In practice, the financial sustainability of the parent organizations often depends on the success of the projects undertaken. In these cases, even though top management tries to exercise some level of control (usually financial), authority and responsibility are mainly delegated to project managers. This structure allows PMs to take decisions more quickly and facilitates the establishment of a coherent team whose members are motivated to reach the project targets. Project teams can work independently, communicate easily and report directly to their PMs.

As there are usually several projects that must be managed simultaneously within the organization, this structuring can result in duplication and underutilization of resources. PMs usually demand that some resources be fully allocated to their projects and avoid sharing them with other project teams, potentially leaving them underutilized. Decisions regarding the allocation of these resources to the projects can create conflict among different project teams. In practice, top management prioritizes some projects over others in the allocation of scarce resources, which might demotivate other employees, and even result in conflicts between teams.

Unlike functional organizations, project-based organization can hinder development of expertise in specific areas or implementation of certain technologies that require high level of focus on some areas. However, it facilitates a holistic approach to optimizing the outputs of the project. PMs can create project environments where the team members are attached to each other and internalize the project objectives. However, this structuring makes the career of the team members highly dependent on the projects, since to a large extend they have only their performance in the projects to show. Project-based organizations usually keep the number of full-time employees to a minimum. The project work is mostly accomplished by a temporary work force that is disbanded or is assigned to other projects when the project is completed.

2.1.3 Matrix Organizations

The matrix organization structure seeks to combine the project-based and functional structures in such a way that their desirable features are preserved, and their disadvantages avoided (Meredith and Mantel 2012). Duplication of resources can be avoided by assigning specific staff (resources) from each functional unit to projects and report directly to the PMs. This hybrid structuring requires project and functional managers to collaborate, negotiate and work as partners. As a result, organizations can benefit from the advantages of both functional and project-based structures; however, management becomes more challenging.

A matrix management structure would be appropriate in the following three situations (Mullins 1999):

(i) There is more than one orientation to the activities of the operation, e.g., multiple customers or geographical differences in markets served;
(ii) There is need for simultaneously processing large amounts of information;
(iii) There is need to share resources, i.e., one function or project cannot justify the expenditure for a dedicated resource.

As the unity of command principle is violated, employees report both to PMs and functional managers, who are both involved in decision-making. Hence, there is a *double-boss* problem for the PM to manage. Functional managers choose the employees to be assigned to the project teams and based on area specific requirements and expertise, they can be involved in choosing the processes, methodologies, and technology to be implemented. On the other hand, PMs define the work requirements and perform the scheduling of the employees. In terms of the relative powers of PMs and functional managers, a matrix structure can take on a variety of specific forms:

(i) Weak (lightweight) matrix,
(ii) Balanced matrix,
(iii) Strong (heavyweight) matrix.

At one end of the spectrum, the weak matrix resembles functional structuring, with the distinction of a formally assigned PM (Larson and Gray 2018). Functional managers remain powerful, while the PM acts as a coordinator of the work. On the other extreme, PMs are much more powerful in strong matrix structures while functional managers serve as expert consultants in their domain of expertise. Balanced approaches represent the classical matrix structuring, where both functional managers and PMs are actively involved in decision-making. However, difficulties in managing the resources emerge since the power balance is usually delicate and sharing of authority and responsibility is complex (Meredith and Mantel 2012).

Considering the various possibilities, choosing the organizational structure is a difficult and critical managerial decision. Before taking this decision, the current organizational structure should be carefully analyzed, and strategic objectives, advantages and disadvantages of each structure should be taken into account.

2.2 How to Choose the Project Management Structure?

There is no single, universally applicable best way to implement a project. Before deciding on a particular organizational structure, organizational and project requirements need to be carefully studied. For complex, high value products, and systems, for example, the nature, composition, and scale of the product influence considerably the organizational structure (Hobday 2000). As these requirements change over time, the organizational structure might also change. Shtub et al. (2014) and Larson and Gray (2018) list some of the critical factors that affect the structuring decision.

- The projects that the organization undertakes; their number, budgets, sizes, and their relative importance for the organization,
- Resource availabilities of the parent organization, budget and time constraints for the projects,
- Organizational culture, managerial practices, need for change,
- Level of uncertainty in the projects, monitoring and control needs,
- Type of technology used, level of technical expertise needed,
- Novelty and innovation brought by the projects,
- Complexity of projects.

If the relative importance of the projects for the success of the firm is much higher than the other work, monitoring and controlling these projects become critical, and giving more authority to PMs would be beneficial, suggesting fully project-based or strong matrix structures. Organizations that invest mostly in projects and perform project work do better to adopt these structures rather than the functional structure. This choice also supports the management of projects that require quick decision making, such as new product development projects, or that require constant contact with clients such as software development projects.

On the other hand, if there are tight constraints on resource availabilities, and many resources need to be shared, then matrix organizations may be a better fit compared to establishing dedicated project teams. If technical expertise in some functional area is critical to the success of a project, functional organizations or weak matrices might be preferred. However, if the project involved is an innovation project, since innovation projects in general require autonomy, a project organization would be recommended (Maylor 2003).

The learning capability of individuals and organizations clearly affect the organizational structure of the projects. Managers adapt past experiences, failures and successes to the design of the organization for the new project at hand. Similarly, experience gained or changes such as changes in project goals, or in project scope occurring in earlier phases of a project can also be used for reorganization purpose.

2.3 Stakeholder Management

Stakeholders are persons or organizations, such as shareholders, employees, customers, and suppliers, who are actively involved in the project or whose interests may be positively or negatively affected by its performance or outcome (Pinto 2016). Stakeholders have diverse views and interests (Caron 2015). Integrating their opinions in forecasting and planning processes through continuous interaction can lead to beneficial results. With the additional knowledge acquired from them, the accuracy of the forecasts and the robustness of the project plans could be improved.

There are several different definitions for stakeholders available in the literature (Ward and Chapman 2008). Stakeholders may exert influence over the project, its deliverables, and PM and the project team members sometimes with a considerable positive or negative impact on the fate of the project. This can indeed be the case, for example, in large infrastructure investments such as power stations and airports, where non-profit organizations and surrounding communities can be actively involved in all phases of the project starting with the design phase. Stakeholders may become a major source of uncertainty in projects such as, for example, causing unexpected delays in the project timeline.

The project management team must identify both internal and external stakeholders in order to determine the project requirements and expectations of all parties involved. We can distinguish between internal stakeholders, such as top management, functional managers, project sponsor, team members, PMs of other projects, and internal clients; and external stakeholders like suppliers, competitors, shareholders, regulatory bodies, end users, and external clients. We would expect internal stakeholders to be supportive of the project, but this need not always be the case. Some units within the organization might be adversely affected by the expected results of the project, and hence, might not cooperate willingly.

The PMs need to maintain good relations with the functional managers. The quality of team members and their time allocated to the project is usually negotiated between the functional manager and the PM. Securing the smooth flow of data/information from different functional groups is another important responsibility of the PM.

The owner(s) of the project deliverables are referred to as external clients. Although the final approval of the deliverables of the project rests with the external client, the external client can also exert influence in different ways during project execution. The management of this relationship is an extremely important, and often delicate, issue.

A key policy issue for the PM is to build trust between the project organization and the stakeholders. Establishing and maintaining open communication channels with both internal and external stakeholders not only supports the building of trust but is also essential for effective and efficient project management. Throughout the project process, the PM must manage the interests and expectations of the stakeholders, which requires a good judgment of their relative power, intentions, and their attitude towards the project and each other. The proper management of the project stakeholder expectations is critical for success.

2.4 Project Management Culture

Project management culture, which refers to the values, standards, formal or informal rules, is an essential pillar of project management that can take a long time and persistent policies to build in an organization. To mold a sound project management culture top management must have a good understanding of project management, its tools, methodologies, and processes. It is this foundation upon which top management builds a project management culture over time and by example. Considering that projects are executed within organizations a strong relationship between organizational culture and project management culture would be expected.

The project management culture, style, and structure influence how projects are performed. Project management cultures and styles – typically known as cultural norms – may have a strong influence on a project's ability to meet its objectives. The cultural norms include common knowledge regarding how to approach getting the work done, what means are considered acceptable for getting the work done, and who is influential in facilitating the work getting done. How the organization deals with issues like risk, stakeholder involvement, communication, and project prioritization are all part of project management culture. In order to get the best out of project management's tools and methodologies, organizations may need to undergo a cultural shift.

According to Larson and Gray (2018) organizational culture *(i)* creates a sense of identity, which can lead to commitment and loyalty to the organization; *(ii)* helps to define the authority relationships, explains the reasons why some people have managerial authority; *(iii)* helps to clarify what are appropriate and inappropriate behaviors, shapes the social norms and relations within the organization and project teams.

All these functions of the culture affect the performances of the projects of the organizations.

Culture shapes the formation of the project teams and affects many other factors such as the authority and responsibilities of the project managers, communication within the team and with stakeholders.

Most organizations have developed their own unique project management cultures with more emphasis on different characteristics, but the following need to be part of any project management culture:

Effective communication: Open channels of communication within the project team as well as with the rest of the organization and with the stakeholders enhances information flow and the quality of decision-making.

Trust: A project management culture based on openness and honesty builds mutual trust between the organization, the PM, the project team, and the stakeholders.

Teamwork: The project team needs to function as a unit, with team members committed to project goals and objectives and operating in an environment that provides for good communications, information flow, and transparency.

Cooperation: Working together towards a common goal creates synergy and allows the organization to attain higher levels of productivity and unity. Lack of cooperation would support the well-protected silos in the organization.

Shared vision and values: Shared vision helps the members of the organization to interpret their position and function within the organization. That the vision and values are understood, accepted, and shared by all the members of the organization is essential for creating unity within the organization directing it to a common goal.

Policies, methods, and procedures: The policies, methods, and procedures need to be documented and shared in order to establish the standards for managing projects in the organization and updated to reflect the best practices.

View of authority relationships: The way authority is exercised, and things do get done in an organization is related to the prevailing culture in that organization. Organizations with fluid boundaries and multiple supervisors for employees require more than a direct exercise of authority to bring out the best of the team members. Exercising formal authority might suppress the possible benefits to be obtained from the diversity of a team – a characteristic much sought for creative teamwork.

Work ethic: A strong work ethic among the employees leads to a more productive environment where hard work is not considered as a burden but rather a way to improve one's character, morale and strengthen one's abilities.

Going over the above characteristics we observe that indeed they are closely related to each other and interacting. For example, effective communication plays a major role in building trust and in increasing cooperation among the actors in the project environment.

There are different types of project management cultures based on the nature of the business, the amount of trust and cooperation, and the competitive business environments in which the firm must operate (Kerzner 2013a, b):

In *Cooperative Cultures*, effective internal and external communication, cooperation, and trust are the underlying fundamentals. Decisions are taken to balance the benefits for all stakeholders. The management style is more informal than formal.

Non-cooperative Cultures: In this type culture, mistrust is dominant over the relations between various entities of the project environment. This might lead to less engagement of the parties and an eventual failure due to the lack of engagements.

Competitive Cultures: Here, different project teams compete for the organization's available resources. In cases where employees can be assigned to multiple projects simultaneously, some PMs might force employees to focus more on the project they manage leading to conflicts and mistrust.

Isolated Cultures: This type of culture might occur in large companies where the management allows the functional units or business units to develop their own project management cultures.

Fragmented Cultures: Geographical separation of the project team can lead to fragmented project management culture, especially in multinational projects where there is a difference in the maturity of project management culture between the home office and the foreign team(s). The different national or regional cultures can also contribute to this situation.

When a project involves external entities as part of a joint venture or partnering or branches in different countries, the project might be influenced by more than one project management culture. Such projects are managed by *virtual teams*. Two main challenges encountered in managing a virtual project team are developing trust and establishing effective communication (Lipsinger and DeRosa 2010). Online meetings could be less effective to develop trust compared to face to face meetings. To reflect the emotions, videoconferencing would be more effective than communicating with emails or voice calls. However, it might not be as effective as face to face communication to develop trust. In many projects, time zone differences between the locations of the team members can create limitations in organizing the team meetings effectively.

It would be expected that cultural differences between different countries can impact on the project management style organizations adopt. Zwikael et al. (2005) evaluated the nine classical project management areas as defined by PMBOK and the organizational support required for a proper project management structure by investigating two different cultures – the Israeli and the Japanese. Israeli project managers were found to be more focused on performing scope and time management processes, assisted by project management software, whereas Japanese project managers were observed to concentrate more on formal communications and cost management. They found that Japanese organizations used clear and measurable success measures for each project, while project objectives were often less precise in the Israeli case. Also, the project owners' selection of criteria to measure project success in both countries were relatively different.

The influence of cultural differences is also detected when trying to manage cultural diversity in global projects (Huang 2016). It is a big challenge for any project manager to manage team members and stakeholders from diverse cultures with different ways of thinking and working.

2.5 Organizing Projects

As mentioned above, the organizational structure for a project determines how the project is positioned within the parent organization. For managing the project, on the other hand, whether small or large, its effective management requires an organization comprised of well-defined units with clearly delineated roles and responsibilities. These units must be linked by effective communication channels among themselves and, to an appropriate degree, with the parent organization and the stakeholders.

2.5.1 Steering Committee

In complex projects involving several organizations, the upper body of the project organization is the *Steering Committee* (also called *the Executive Committee*). The Steering Committee provides strategic direction, advice, and guidance, oversees the progress of the project and seeks to ensure the successful delivery of the project outcomes. Its key members are representatives of the participating organizations, who do not usually work on the project themselves. Individuals with valuable experience or expertise might also be assigned to the Steering Committee as members or consultants. The PM is a member of the Steering Committee providing communication between the Steering Committee and the project team and also managing the Steering Committee's secretariat.

The tasks the Steering Committee should include the following:

- Providing input during the development of the project, including the evaluation strategy;
- Providing advice on the budget;
- Defining and helping to achieve the project outcomes;
- Identifying the priorities in the project – where the most energy should be directed;
- Identifying and monitoring potential risks;
- Monitoring the project schedule and milestones;
- Monitoring the quality of the project as it develops;
- Providing advice/making decisions about changes to the project as it develops.

It is a good practice to establish a Steering Committee, even if only a single organization is involved when the project has a long time span and/or is complex in nature. In such a case, its members can also include individuals with valuable experience or expertise to serve as members or consultants, and members from some of the stakeholder organizations as well as from potential clients of the intended project outputs.

2.5.2 Project Sponsor

The *project sponsor* is an individual (often a manager or executive) with overall accountability for the project. The project sponsor plays a vital role in the successful conclusion of the project with the business results achieved as intended and is involved in the initiation, planning, implementation and control, and closing of the project.

Initiation phase: The project sponsor assigns the PM and conveys clearly his/her role, responsibilities, and level of authority; makes sure that the PM understands the relevance, importance, and priority of the project for the parent organization and how it is related to the strategy of the parent organization; contributes to the formulation of the project charter and to the formation of the project organization; participates in the project kick-off meeting; and has the final say on the decision to begin or cancel execution of the project.

Planning phase: The project sponsor makes sure that the plans are realistic and feasible with no unrealistic commitments; all parties involved have given enough time and effort; all issues and concerns are raised and openly discussed; prioritization of the project within the parent organization and within the project are assessed and decided upon in line with the strategies and expectations of the parent organization.

Implementation and control phase: The project sponsor works with the PM to ensure effective implementation of the project consistent with the objectives, processes, standards, and rules set for the project; provides feedback; empowers and motivates PM and team members to solve the project problems by identifying the underlying factors and root causes.

Closing phase: The project sponsor evaluates the project performance; participates in the post-project analysis and ensures that the process is a constructive one and the results are made available for use in future projects; makes sure that the project is properly terminated with the deliverables handed over to the client.

The role of the project sponsor covers the following areas (Schibi and Lee 2015):

- Provides business context, expertise, and guidance to the PM and the team;
- Champions the project, including "selling" and marketing it throughout the organization to ensure capacity, funding, and priority for the project;
- Acts as a channel to higher management for decisions and issues that are beyond the authority of the PM;
- Acts as an additional line of communication and observation with team members, customers, and other stakeholders; and
- Acts as the liaison between the project, the business community, and strategic level decision-making groups.

The Steering Committee, the project sponsor, and the PM have to be strategically aligned and in good cooperation and communication for successful project management.

2.5.3 Project Manager

In managing projects, even though the responsibility and authority are shared and communicated among the project team, PMs have the principal responsibility for the effective management of the project. They play an active and fundamental role in establishing and leading the project team; formulate objectives and strategy for the project and implement tools and methods to ensure that projects achieve their goals. In addition to formulating objectives and strategy, the PM needs to link these objectives and strategy to the objectives and strategy of the parent organization (Thomas et al. 2001).

Projects cover a wide range of areas and hence, we see PMs with distinctly different backgrounds and roles. Effective PMs usually have work experience and training in relevant fields. A set of desirable skills for a good PM would include (Lockyer and Gordon 2005):

- Technological understanding.
- An understanding of project economics.
- Knowledge of human resources management.
- Competence in system design and maintenance
- Competence in planning and control.
- Financial competence.
- Competence in procurement.
- Good personal communication abilities.
- Negotiation and conflict resolution skills.

More recently, the teams of high technology projects have become more multi-disciplinary. To manage such cases, the PMs need to develop multi-disciplinary management competencies.

Facilitating communication, organizational learning, use of technology, and teamwork are the key success factors in managing the projects.

Even though the role of PMs in the success and failure of projects is significant, team members are the performers of the activities. Therefore, establishing a competent and coherent team becomes a crucial role for PMs. Even though functional area or top managers might have the final say on allocating the employees to the projects, PMs usually negotiate with them to acquire experienced and motivated members.

Having established the team, the PM must then lead the team to perform the project tasks as required and keep them motivated until project termination. In this regard, the technical skills alone are not sufficient, and human relations become critical. Political and negotiation skills play an important role in project success (Meredith and Mantel 2012). The team members and the PM must trust each other, which requires participation in the decision-making process, information sharing and transparency.

In addition to negotiating with other managers and communicating with team members regularly, the PM informs the project sponsor on a regular basis about the progress and ensures their consistent, adequate support for the project. The PM also works with clients and communicates with other stakeholders to inform them regularly of scope changes, project progress and outcomes.

PMs have started to control more projects simultaneously as organizations are undertaking a large number of projects simultaneously. To enhance competitiveness, many companies have been launching more new products in shorter time periods and have been operating in international markets (Pajares et al. 2017). As a result, they require experienced and flexible PMs. All these new product development or international infrastructure projects need expertise in many different areas such as engineering, production and quality management, finance, human resource management, etc. Therefore, cross-functional teamwork is essential for the success of these projects and is critical for the PM's performance.

2.5.4 Project Team and Team Members

Project teams play a major role in the success of the projects in reaching their objectives. Projects differ extensively in their size, scope, complexity, and the environment in which they are realized. The local project management culture also heavily influences how projects are managed in an organization. Hence, it is difficult to state general guidelines for the formation and running of project teams to meet stakeholder expectations, but at least some common ones can be stated as follows.

- The team members need to be informed on how their project fits with the company strategy.
- There should be clearly stated objectives understood and accepted by the team members.
- Each team member should be clear about their work objectives and responsibilities together with the authority vested and resources made available to achieve these work objectives.
- The rules, norms, and procedures to be adhered to by the team members should be stated to them clearly in writing. In the event of changes during the project lifecycle, the change process should be transparent and open, informing all team members in a timely manner of any changes taking place.
- A crucial set of rules, norms, and procedures are those associated with the performance evaluation of the team members. The individual performance measure(s) and the associated target values should be clearly conveyed to each team member. The penalty and reward mechanism should be known beforehand by the team members and be applied in a transparent way.
- Rules, norms, and procedures have to be in place for disbanding the project team.
- An environment of openness, mutual trust, cooperation, cohesion, inclusiveness, constructive criticism, continuous improvement, and innovativeness must be developed.
- Teams need to be provided with the necessary resources and skills to achieve the task.
- The team's relationship with other teams and entities in the parent organization, as well as to stakeholders, needs to be clearly defined.
- The size and composition of the team might change throughout the project lifecycle. Care should be exercised to inform all the team members in a timely manner of changes taking place.
- The team size should be adjusted to the workload and needs to encompass all the skills and skill levels required for the tasks to be achieved. When deciding on the size of the team, the need for efficient and effective communication should be a

prominent consideration; a large team will place a heavy communications burden on the project.
- In case smaller teams not chaired by the PM need to be formed within the project team, they should be given necessary support and status, with their deliverables well defined and incorporated into the project process.
- The accomplishments of both individual team members and the team should be adequately recognized and shared to maintain motivation.

Initially, when the team is formed, team members might not know each other, or even have no experience of working together. This situation can even become more accentuated for the case of geographically distributed project teams. The leadership and management capability of the PM must mold this group of individuals into a team based on mutual trust, cooperation, and harmony. The job of the PM will be easier if the individuals in the group have a certain set of skills and characteristics, on which we briefly elaborate in the next section.

2.5.4.1 Desirable Skills and Characteristics of Team Members

Usually, the PM is responsible for bringing together the project team in consultation with the project sponsor if there is one. It is a well-established good practice to build a cross-functional team (Scott-Young and Samson 2008). The skills and characteristics expected of the team members include:

- Capable of good communication with the PM, within the team as well as with the individuals/units outside the team.
- Remains loyal to the team, maintaining the unity of the team even in the face of failure.
- Adopts an innovative and problem-solving approach.
- Seeks cooperation from outside the team.
- Open to risks so as to share the risks the team is facing.
- Supportive of other team members, particularly during times of intense work.
- Able and be willing to share the leadership.

In a large, complex project the team might require the support of functional specialists who will suffer from the double-boss syndrome mentioned earlier. They join and leave the team as the project progresses and demand on different specialties change. It is the PM's responsibility to secure proper integration of these specialists into the team and to create a harmonious relationship with their functional boss.

Some important but unpleasant decisions concerning team members may have to be taken by the PM. One such is to fire a team member if it becomes apparent that this person cannot work as part of the team (Lewis 1993). Such a decision must be based on concrete data and information before presenting it for management approval to minimize the risk of rejection of the PM's proposal. If the person stays on the team, a relatively safe approach for the PM would be to assign that person to a less important tasks. Once the decision is approved, the PM has to share the decision with the rest of the team. This whole process will be a test of the leadership of the PM.

In today's dynamic business environment, the importance of information exchange and networking is evident for project management professionals. In this regard, project management professionals from organizations all over the world, have been gathering under the umbrella of the Project Management Institute (PMI) since 1969. They have established an international network of communication and organization. PMI also organizes or accredits various project management training programs so that project management professionals can collaborate internationally; work together using the same terms, concepts and methods. In this context, knowledge and experience have been tested by several examinations. PMP (Project Management Professional) certificate has become popular along with collaborations among different sectors increasing day by day in the business world and with the internationalization of the projects. Having taken the certification, project management professionals further need to participate in professional development activities such as seminars. Certification relates to professionalism. Although results linking certification with performance are yet tentative at best, to increase the efficiency of the selection process, recruiters increasingly use voluntary professional certification as a signal of applicant competencies and likely future performance (Farashah et al. 2019). Other well-known organizations are Association for Project Management (https://www.apm.org.uk) and International Project Management Association (https://www.ipma.world).

In addition to the framework and process definitions of PMI, a process-based methodology PRINCE2 (PRojects IN Controlled Environments), has been taking the attention of project management professionals (see e.g., http://prince2.wiki/PRINCE2). It is United Kingdom based but has been internationally recognized and applied. Like in the case for PMI framework, many project management professionals attend the trainings and work to get PRINCE2 certification.

2.5.4.2 Geographically Distributed Project Teams

With the increase in telecommunications, digital infrastructure, and global supply chains, distributed project teams whose members are from different companies and their subsidiaries located at different locations as a consequence of global cooperation and expansion of global supply chains have become common. The same is also valid for companies with different branches in different locations. In addition to the difficulties of managing such a distributed project team towards a common goal and securing proper communication, a further source of difficulty may arise from the different company and project management cultures of the team members from different companies. An additional dimension of complexity for multi-national teams is the different national cultures represented in the project team. Furthermore, the ability to create compliance through the exercise of formal authority diminishes with distance requiring a different approach of managing the authority relationships beyond formal authority between the managers and employees.

2.5.5 Project Meetings

The PMs will often have to chair meetings. Although it might appear an easy task, it most certainly is not. It requires strategy and planning. What is it that one wants to achieve through this meeting and how should the meeting be conducted to achieve the intended results? In spite of extensive preparations things can still go wrong due to many factors beyond control but this does not diminish the value of planning the meeting. Some guidelines are (Maylor 2003):

- Confirm the purpose of the meeting. The main agenda item should be made clear so that attention would not be distracted from this item.
- Identify the list of participants to be included in the meeting.
- The pre-meeting preparation: The location, time, agenda preferably with time frames for each item, and any background information on the topic under discussion. Make sure participants are prepared to address agenda items.
- When running the meeting it is important to provide a forum for constructive and innovative debate but at the same time to limit the discussions so that the meeting does not adjourn without the intended results being achieved.
- Post-meeting follow-up with the distribution of the minutes. The minutes should cover action items and their assignment to individuals, which are expected to be realized till the next meeting and be reported as part of the next meeting's agenda.

Meetings can be effective means of communication within the team and with parties outside it. They serve multiple purposes such as assessment of the current situation in the project, action planning for the next phase of the project, work assignment to individuals based on action points decided upon in the meeting, and conflict resolution. It is good practice to reserve a meeting for the major issue on the table instead of having a crowded agenda in order to avoid distraction. In line with this approach, only those associated with this major issue should be invited to the meeting.

2.6 Project Management Office

As mentioned earlier, projects have become a major way of doing work in today's organizations. With the proliferation of projects in organizations, matrix and project-based structures are encountered more frequently. All these developments have created the need for an organizational unit to assist PMs and teams throughout the organization in implementing project management principles, practices, methodologies, tools, and techniques (Ward 2000). To this end, *Project Management Offices* (PMO)s have been created and integrated into the organizational structure, particularly in project organizations. PMI (2017a) defines PMO as "a management structure that standardizes the project-related governance processes and facilitates the sharing of resources, methodologies, tools, and techniques".

PMOs can play an important role in coping with the managerial and administrative complexities and improving the project management capabilities of an organization. They do no not manage projects and typically do not have profit and loss responsibility for projects. They are service units supporting PMs and other units of the organization as requested.

PMO's have a variety of responsibilities including (see, e.g., Meredith and Mantel 2012; Mulcahy 2011).

- Development and maintenance of the organization's project management methodology and standards for reporting, bidding, and risk analysis.
- Definition of requirements for the project management tools to be employed in the organization.
- Definition and development of training material and processes for the PMs and other related personnel.
- Management of intellectual property related to projects run by the organization.
- Providing support in choosing projects that build up or reinforce the strategic positioning of the organization; recommending termination of projects that are no longer viable.
- Serving as a center of communication and information exchange to manage the relationships among projects, programs, portfolios.
- Development and maintenance of the organization's Knowledge Management System, especially the archive containing the results of the post-project analyses.
- Administration of post-project analyses for the identification and diffusion of best practices in project management within the organization.
- Providing guidance to PMs and supporting project termination, reporting and performance assessment.

In order to perform these tasks successfully, PMOs need strong support from top management. It is also essential that the job descriptions of PMOs are clearly defined and well communicated within the organizations.

In some organizations, certain PMO staff members are assigned as PMs to new projects. The basic rationale behind such a policy is to benefit from the project management expertise this personnel has accumulated. They are expected to follow rules, regulations, standards, and good practices in the project management domain as well as within the organization.

According to the structure and needs of the organizations, PMOs might be formed in various ways. Based on the authorities delegated and responsibilities given, they can be basically classified as supporting, controlling, or directive (managing) (Mulcahy 2011). From supporting to managing, the powers and responsibilities of PMOs increase.

Supporting PMOs assist the PMs and the team in choosing and implementing project management methodologies and in report writing. They inform them of organizational policies, provide them with templates and organize training. Controlling PMOs, on the other hand, are more involved in decision-making and seek to ensure that organizational policies are followed, and best practices adopted by the organization are effectively implemented. Managing PMOs represent the case where PMOs are fully responsible for project success, and hence, exercise considerable control over PMs. In all three cases, the PMO's role in ensuring compatibility of projects with company strategies, communication with PMs, coordination between projects, and efficiency in resource utilization are critical for their success.

Singh et al. (2009) reported the results of a study examining the challenges of implementing a PMO for managing information technology (IT) projects. The top-three challenges were (1) rigid corporate culture and failure to manage organizational resistance to change, (2) lack of experienced PMs and PMO leadership and (3) lack of appropriate change management strategy. They suggested several actions that can be taken to overcome these challenges: (1) Having a strong PMO champion, (2) starting small and demonstrating the value of the PMO, (3) obtaining support from opinion leaders, (4) bringing the most talented PMs into the PMO implementation team, (5) adopting a flexible change management strategy, and (6) standardizing processes prior to PMO implementation. Although the study was limited to the IT sector, the results are likely to be representative of the situation of the PMOs in other sectors as well.

2.7 Program Management

A *program* is defined as a group of related projects managed in a coordinated way to obtain benefits and control not available from managing them individually. Programs may include related work elements outside of the scope of the individual projects (PMI 2017b). This group of projects and the related work forming a program contribute to a common, higher order objective and have common outputs. A program serves as a framework providing strategic direction to a group of projects seeking to achieve higher order strategic or developmental change for the organization that could not be delivered by any of the individual projects involved. (Turner and Müller 2003).

Programs extend over longer time horizons. The Polaris Missile Program mentioned in Sect. 1.7 was a highly complex endeavor and the US Navy created the Special Projects Office (SPO) to manage it. Another example is the Apollo Program, which was executed by the National Aeronautics and Space Administration (NASA) through the years 1961–1975 with the objective of landing the first humans on the moon. This was first realized in July 20, 1969. A more recent example of a major program is the International Space Station (ISS) Program, which is a multi-national collaborative program initiated and managed by five participating space agencies: NASA(USA), Roscosmos (Russia), JAXA (Japan), ESA (Europe), and CSA (Canada). Starting in 1998 with its first component, additional modules have been added over time with the last one added in 2011 (Wikipedia, June 10, 2020).

Since programs generally involve multiple complex projects directed towards a common goal with a time horizon of 5 years or more, the coordination and control of these projects become the critical element in program management. A typical characteristic of programs is that individual projects are accomplished through contracts whose cost, quality, and timing must be closely controlled. Exceeding the targeted costs will lead to problems in budget management. Rework associated with poor quality will cause delays and if not detected in time, might even lead to costly consequences affecting the progress of the program. Any delay in one project might delay the program since the projects are interrelated. A major task for program management is the proper scheduling of projects which make use as input the output of some other project.

Considering the number of individual projects and thus contractors, subcontractors, and project teams involved, a staggering amount of work is involved in managing these complex systems. A program structure must be established to perform these tasks to achieve the intended deliverables of the program. A typical program structure would be a matrix structure with different functional units and the program management office, headed by the program manager, managing the different projects. Personnel with different specialties would then be borrowed from their functional units by the program management office for temporary assignment to different projects.

Many concepts from project management would be expected to apply to program management as well, though some need to be modified to deal with the larger magnitude of work involved (Nicholas 2001). Obviously, considering the longer time horizon of a program, turnover of the key personnel including the program manager becomes a serious concern. Hence, human resources management – particularly training and replacement of personnel – often takes on extreme importance.

2.8 Project Portfolio Management

According to PMI (2017c), "a *portfolio* is a collection of projects, programs, subsidiary portfolios, and operations managed as a group to achieve strategic objectives". To achieve these strategic objectives project portfolio management (PPM) includes identifying, prioritizing, authorizing, managing, and controlling projects, programs, and/or other related work. The emphasis is on successfully executing the most beneficial projects for the organization. Thus, the alignment of the project portfolio to the organization's strategy and the identification and prioritization of the projects are of fundamental importance. PPM is actually more than managing multiple projects. A portfolio of projects needs to be assessed in terms of its business value and adherence to strategy. Hence, it involves the selection and support of projects or program investments guided by the organization's strategic plan and available resources (PMI 2017a). The project portfolio should be designed to achieve a defined business objective or benefit.

Another definition is given by Turner and Müller (2003). They define a project portfolio as an organization (temporary or permanent), in which a group of projects is managed together to coordinate interfaces and prioritize resources between them and reduce uncertainty. Blichfeldt and Eskerod (2008), on the other hand, focus on PPM and describe it as the managerial activities that relate to (1) the initial screening, selection, and prioritization of project proposals, (2) the concurrent reprioritization of projects in the portfolio, and (3) the allocation and reallocation of resources to projects according to priority.

The projects or programs of a portfolio do not have to be directly related. However, by virtue of sharing the same resource pool, the projects in the project portfolio can be managed jointly for the efficient utilization of the resource pool. They can address functional areas of the organization such as R&D, supply chain, and information systems so as to encompass the whole organization and its environment.

Four main characteristics of a project portfolio are given as follows (PMI 2006):

1. All components such as the projects or programs included in a project portfolio represent investments made or planned by the organization.
2. These components should be aligned with the organization's strategic goals and objectives.
3. All components typically have some distinguishable characteristics allowing the organization to cluster them for more effective management; usually, this is done in programs.
4. All components of a project portfolio are quantifiable, i.e., can be measured, ranked, and prioritized. This is critical because an important mission of PPM is making decisions about the allocation of resources or choosing between different alternatives. These decisions should be based on objective (measurable) data.

In the initial phases of its development, the emphasis in project portfolio management was on project selection and particularly on choosing the right number of projects. Mathematical models were developed for choosing the optimal project portfolios (see, e.g. Henriksen and Traynor 1999; Archer and Ghasemzadeh 1999). According to Petit (2012), the most important literature on PPM was developed to address new product development portfolios. Over time it has expanded into a complete managerial approach as it can be followed from the evolving set of PPM Standards issued by PMI over the years (PMI 2006, 2008, 2013, 2017c). Indeed, the dynamic nature of a project portfolio with new projects arriving and some existing ones finishing presents considerable challenges for planning and management. In this regard, Araúzo et al. (2010) focus on resource planning in project portfolios and address the dynamic resource scheduling problem using simulation. A project portfolio being a multi-project system contains many interconnected elements, such as the projects and resource pools, and dynamic changes, which make this system complex to model and produce analytical solutions. Claiming that mathematical programming approaches might not handle the complexity of the real cases, they adopted a multi-agent decision-making system and defined project managers and resource managers as agents. The decision-making system they propose focuses on the allocation of resources dynamically to the projects and decides on accepting or rejecting the new projects, based on their impacts on the profitability, schedule, and operational planning of the existing portfolio.

Although the emphasis has traditionally been on determining the right number and mix of projects in the project portfolio, it has been observed that often PPM covers only a subset of on-going projects in an organization. Additional projects although not included in the project portfolio may tie up resources initially dedicated to the projects in the project portfolio creating points of possible conflict (Blichfeldt and Eskerod 2008). Resource allocation has always been a sensitive and critical issue in organizations but such a misjudgment by management further exasperates the issue resulting in conflicts in the organization, contention for limited resources, and delayed projects.

PPM takes place in a dynamic environment, which is another source of complication (Petit 2012). To overcome the potential disturbances arising from operating in such an environment, PMI suggests two changes (PMI 2008): The first of these is periodic review of the portfolio performance to make sure that the projects in the portfolio still support the achievement of the strategic goals for which they were originally conceived. In order to achieve this, components must be added, reprioritized, or excluded based on their performance and ongoing alignment with the defined strategy in order to ensure effective management of the portfolio. Secondly, if the dynamics of the internal and external environment of the organization is such that significant changes occur over time leading the organization to new strategic directions, then it is suggested that the criteria for determining the composition and direction of the portfolio may also change (PMI 2008). When the need for new criteria is observed, then it is recommended that the current criteria in the strategic plan are examined and appropriate changes are made, usually focusing first on categorization. In case of a stable environment with no change in the strategic directions of the organization, then the efforts should focus on portfolio balancing.

The Association for Project Management (APM) suggests that a complete review cycle might not always be needed. For such cases, APM suggests a change in terms of adjustments of the portfolio with regard to the constraints, risks, and returns anticipated, and in the light of developing circumstances around the portfolio (APM 2006).

2.9 Agile Project Management

Even though following a process-based management approach to projects and standardization of processes is crucial for many projects to reach the time and budget targets, some projects, especially those where technology use, uncertainty and need for creativity are high, may require different managerial approaches. These require flexibility, dynamic decision-making, and incremental development due to the requirements of the dynamic environment, so the constraints of the process-oriented management must be reconsidered.

Traditional process-based approaches, also called as *waterfall* in practice, emphasize the preplanning and effective control and might be ineffective in projects undertaken under high uncertainty due to the need for frequent scope changes. These requirements have led to the adoption of a new more flexible approach, *agile project management*, by many organizations.

The most important distinguishing feature of the agile managerial systems is their approach to planning. Traditional systems are plan-focused (waterfall) and it is important to plan the tasks, prepare a WBS at the beginning. Agile approaches emphasize progression to targets step-by-step, with the help of iterative development cycles. Instead of developing a product/process upfront, design and development are continuously adopted. Features of the outputs and customer requirements are emphasized. The project starts with a high-level project scope, that is much less detailed compared to the waterfall approach. Project requirements are defined with small functional units, namely features, which are to be satisfied by the work of small teams within a short period development cycle. These cycles, which take less than 4 weeks, are called *sprints*. In other words, project outputs are delivered as a consequence of evolving work and connected sprints. Feedback between the iterations; verification, validation, and integration of the customer voice to development continuously are the main requirements of the agile approach (Larson and Gray 2018). Communication between the project team and the client, teamwork and feedback between the different parties are critical for success in agile project management as they are in traditional project management. An important deliverable of traditional project management is the Earned Value Management (EVM) (see Chap. 10). It should be noted though that the traditional EVM is not suitable for use in agile project management. Instead different, relevant metrics need to be developed.

Due to the continuous evolution of the project work, agile management is ideal for projects where requirements might change within time, or there is a lot of uncertainty, or the use of new techniques is vital. Software development, technological product design, game developments are some suitable areas. It has been observed that the agile approach is more effective in small projects where small teams are involved (Larson and Gray 2018).

Today especially in the software sector, *hybrid methods* combining waterfall and agile approaches become more common. Iterative sub-product/feature development processes are combined with structural planning. This hybrid approach tries to balance the control requirements, especially for the budget, and the flexibility of the continuous development. One possible hybrid approach would be implementing a phase-based approach: using the plan-driven waterfall or the agile approach depending on the project phase. Imani et al. (2017) give the example of using the planned driven approach in requirement analysis, testing, and release phases, and the agile for design and development phases.

We predict that in the future more attention will be given to Agile Project Management and the related hybrid approaches. The increased need for rapid new product development and the use of new technologies, flexibility requirements in management are the main drivers in adaptation of these approaches. This prediction is also supported by the demand of PMs for certification. PMI has been offering the Agile Certified Practitioner (PMI-ACP) program since 2012 and it is among the most popular certification programs, after the PMP and CAPM (Certified Associate in Project Management).

2.10 Conclusions, Recent Developments, and Some Future Research Directions

How to define the success of their project is a critical practical question that PMs should ask themselves. Most of them would refer to the goals of finishing the projects in a timely manner, within the budgeted cost and scope. However, we emphasize that projects are tools for executing strategic decisions in organizations. Organizational goals and strategies should not be overlooked while managing the projects. It is essential to take into account the overall objectives of the organization.

In order not to put long-term strategic goals at risk, it might be necessary that PMs change the project scope during project execution. In some cases, it might be crucial that PMs ask that a project be abandoned. These decisions can be taken effectively only if the organizational goals and strategies are clearly defined. Ensuring the alignment of project objectives with these strategies, and using appropriate managerial approaches and tools are a critical role of PMs. PMs and project teams need to be assessed based on their performance relative to both operational and strategic goals (Shenhar et al. 2001). In this regard, developing strategic project management models that will guide PMs has become an important research and practice area (Shenhar 1999).

Strategic project management is also directly linked to project portfolio management (Meskendahl 2010), which has also been increasingly taking the attention of the researchers and practitioners. Portfolio building is crucial for scarce resources to be invested in the right projects. For this reason, many organizations have been establishing PMOs to ensure effective project monitoring and control. Integrated portfolio management software also looms large as a tool to manage multiple projects effectively towards attaining strategic targets. Currently, there is a need for integrated multi-project planning and control models, and user-friendly decision support and software tools.

Strategic management also requires organizations to quickly adapt to dynamic business environments. In this regard, organizations have been increasingly working to implement agile management systems today. Especially in software development, agile project management techniques have been commonly implemented. To benefit from the advantages of the both waterfall and agile approaches, many organizations have started to adopt hybrid methods (Rahmanian 2014). How to combine these two approaches and implement a hybrid methodology is a promising research and practice question at stake.

Exercises

2.1 What could be the reasons for the increase in the number of organizations that are adopting project-oriented structures?

2.2 Company *Flux* initiated a new product development project and appointed Ali as the project manager. Ali negotiated with the head of the Manufacturing

Department, Deniz, to assign a production engineer for the project. After their interview, Ali called Emre, who has been working at the department as an engineer for 4 years. He informed Emre about the project and told him to allocate 65% of his time to the project and 35% of his time to the duties of the manufacturing department. He also told Emre that after the completion of the project he would continue working 100% for the department as before.

(a) Which structure is used in Company *Flux* to link the project to the parent organization?

(b) Write down 2 advantages and 2 disadvantages of organizing projects in such a way.

2.3 What could be the reasons for the increase in the number of organizations that are establishing PMOs?

2.4 Which matrix structure could be advantageous for managing projects in which integration and tight control are significantly important for achieving the project and organizational targets? Why?

2.5 Discuss the role of stakeholder management in project planning.

2.6 Discuss how organizational culture impacts the performance of the projects. Explain with some examples.

2.7 Give some real-life examples of functional, project-based, and matrix organizational structures. Elaborate on the industries of the companies and their organizational and project characteristics.

2.8 Consider a software development project. Discuss how traditional plan driven, waterfall and iterative agile methods could be combined.

2.9 Discuss the relationship between strategic management and portfolio management. Why is strategic project management important for organizations?

2.10 Discuss some of the challenges in managing virtual project teams.

2.11 Discuss why the preparation of a project charter is essential for a PM. Find out a project charter and list the information given in the charter.

2.12 Discuss some of the traits and skills that project managers need to develop to be effective in managing the projects.

References

Araúzo, J. A., Pajares, J., & Lopez-Paredes, A. (2010). Simulating the dynamic scheduling of project portfolios. *Simulation Modelling Practice and Theory, 18*(10), 1428–1441.

Archer, N. P., & Ghasemzadeh, F. (1999). An integrated framework for project portfolio selection. *International Journal of Project Management, 17*(4), 207–216.

Association for Project Management. (2006). *APM body of knowledge* (5th ed.). Cambridge: APM.

Blichfeldt, B. S., & Eskerod, P. (2008). Project portfolio management – There's more to it than what management enacts. *International Journal of Project Management, 26*, 357–365.

Caron, F. (2015). Data management in project planning and control. *International Journal of Data Science, 1*(1), 42–57.

Farashah, A. D., Thomas, J., & Blomquist, T. (2019). Exploring the value of project management certification in selection and recruiting. *International Journal of Project Management, 37*(1), 14–26.

Henriksen, A. D., & Traynor, A. J. (1999). A practical R&D project-selection scoring tool. *IEEE Transactions on Engineering Management, 46*(2), 158–170.

Hobday, M. (2000). The project-based organisation: An ideal form for managing complex products and systems? *Research Policy, 29*(7–8), 871–893.

Huang, J. (2016). The challenge of multicultural management in global projects. *Procedia – Social and Behavioral Sciences, 226*, 75–81.

Imani, T., Nakano, M., & Anantatmula, V. (2017). Does a hybrid approach of agile and plan-driven methods work better for IT system development projects? *International Journal of Engineering Research and Application, 7*(3), 39–46.

Kerzner, H. (2013a). *Project management best practices*. Hoboken: Wiley.

Kerzner, H. (2013b). *Project management: A systems approach to planning, scheduling, and controlling* (2nd ed.). Hoboken: Wiley.

Larson, E. W., & Gray, C. F. (2018). *Project management: The managerial approach* (7th ed.). Irwin: McGraw-Hill.

Lewis, J. P. (1993). *How to build and manage a winning project team*. New York: American Management Association.

Lipsinger, R., & DeRosa, D. (2010). *Virtual team success: A practical guide for working and leading from a distance*. San Francisco: Jossey-Bass.

Lockyer, K., & Gordon, J. (2005). *Project management and project network techniques* (7th ed.). Essex: Pearson Education Limited.

Maylor, H. (2003). *Project management* (3rd ed.). Essex: Pearson Education Limited.

Meredith, J. R., & Mantel, S. J., Jr. (2012). *Project management: A managerial approach* (4th ed.). New York: Wiley.

Meskendahl, S. (2010). The influence of business strategy on project portfolio management and its success – A conceptual framework. *International Journal of Project Management, 28*(8), 807–817.

Mulcahy, R. (2011). *PMP exam prep: Rita's course in a book for passing the PMP exam*. Burnsville: RMC Publications.

Mullins, A. J. (1999). *Management and organizational behaviour* (5th ed.). London: Financial Times Pitman Publishing.

Nicholas, J. M. (2001). *Project management for business and technology: Principles and practice* (2nd ed.). Upper Saddle River: Prentice Hall.

Pajares, J., Poza, D., Villafañez, F., & López-Paredes, A. (2017). Project management methodologies in the fourth technological revolution. In *Advances in management engineering* (pp. 121–144). Cham: Springer.

Petit, Y. (2012). Project portfolios in dynamic environments: Organizing for uncertainty. *International Journal of Project Management, 30*, 539–553.

Pinto, J. K. (2016). *Project management: Achieving competitive advantage* (4th ed.). Upper Saddle River: Pearson/Prentice Hall.

PMI. (2006). *The standard for portfolio management* (1st ed.). Newtown Square: Project Management Institute.

PMI. (2008). *The standard for portfolio management* (2nd ed.). Newtown Square: Project Management Institute.

PMI. (2013). *The standard for portfolio management* (3rd ed.). Newtown Square: Project Management Institute.

PMI. (2017a). *A guide to project management body of knowledge. PMBOK guide* (6th ed.). Newtown Square: Project Management Institute.

PMI. (2017b). *The standard for program management* (4th ed.). Newtown Square: Project Management Institute.

PMI. (2017c). *The standard for portfolio management* (4th ed.). Newtown Square: Project Management Institute.

Rahmanian, M. (2014). A comparative study on hybrid IT project management. *International Journal of Computer and Information Technology, 3*(5), 1096–1099.

Schibi, O., & Lee, C. (2015). Project sponsorship: Senior management's role in the successful outcome of projects. In *Paper presented at PMI® Global Congress 2015-EMEA (London).* Newtown Square: Project Management Institute.

Scott-Young, C., & Samson, D. (2008). Project success and project team management: Evidence from capital projects in the process industries. *Journal of Operations Management, 26*(6), 749–766.

Shenhar, A. J. (1999). Strategic project management: The new framework. In *Management of engineering and technology, 1999. Technology and Innovation Management, PICMET'99, Portland international conference* (pp. 382–386).

Shenhar, A. J., Dvir, D., Levy, O., & Maltz, A. C. (2001). Project success: A multidimensional strategic concept. *Long Range Planning, 34*(6), 699–725.

Shtub, A., Bard, J., & Globerson, S. (2014). *Project management: Processes, methodologies, and economics*. Prentice Hall: New International Edition.

Singh, R., Keil, M., & Kasi, V. (2009). Identifying and overcoming the challenges of implementing a project management office. *European Journal of Information Systems, 18*(5), 409–427.

Thomas, J., Delisle, C. L., Jugdev, K., & Buckle, P. (2001). Mission possible: Selling project management to senior executives. *PM Network, 15*(1), 59–62.

Turner, J. R., & Müller, R. (2003). On the nature of the project as a temporary organization. *International Journal of Project Management, 21*, 1–8.

Ward, J. L. (2000). *Project management terms: A working glossary*. Arlington: ESI International.

Ward, S., & Chapman, C. (2008). Stakeholders and uncertainty management in projects. *Construction Management and Economics, 26*, 563–577.

Wikipedia. International Space Station. Accessed on 10 June 2020.

Zwikael, O., Shimizu, K., & Globerson, S. (2005). Cultural differences in project management capabilities: A field study. *International Journal of Project Management, 23*(6), 454–462.

Planning and Network Modeling of Projects 3

Learning Outcomes
Upon successful completion of this Chapter, the reader will be able to:

1. Build a sound knowledge base for project planning.
2. Develop the project scope and the project charter for a project
3. Obtain the Work Breakdown Structure of a project.
4. Represent activity relationships using network models.
5. Decide on the level of aggregation at which to model projects.
6. Represent the precedence relations through different types of time lags among activities.
7. Obtain consistent estimates of the durations and costs of the work packages and activities.

We establish the basis of project modeling and planning in this Chapter by introducing a number of fundamental concepts: Project scope, Work Breakdown Structure (WBS), Organizational Breakdown Structure (OBS), Cost Breakdown Structure (CBS), project network modeling, activity duration and cost estimating, data and knowledge management.

3.1 Project Scope

PMI (2017) defines *the project scope* as "the entire body of work performed to deliver a product, service, or result with the specified features and functions". Project scope may also specify the starting date, target delivery date, major milestones, and target budget.

Performing the work defined in the project scope is not sufficient for its successful completion. Besides the content of the work to be performed, the scope statement should also include the standards and performance measures in order to assess the quality of work and define acceptable completion.

G. Ulusoy, Ö. Hazır, *An Introduction to Project Modeling and Planning*, Springer Texts in Business and Economics, https://doi.org/10.1007/978-3-030-61423-2_3

The project scope is an initial statement that can be subject to changes during the implementation of the project. Whether they enlarge the scope (usually called scope creep) or reduce it, changes can lead to deviation from the initially defined outputs and performance levels, causing conflicts among stakeholders. Frequent, major changes in the scope will put pressure on the budget and the time schedule and will make successful termination of the project difficult for the PM and the project team. Accumulated scope changes may eventually lead to unsuccessful termination or a major redefinition of the project.

The project scope needs to be managed to ensure that the work initially proposed is accomplished successfully. But this should not mean that change is to be avoided under any circumstances. Throughout the implementation of the project, changes may occur in the stakeholders' expectations and points of view, the available technologies and resources, new and more accurate data may become available, etc. Hence, the project plan might be changed, and the scope redefined. A proper project scope management based on a well-defined change management process should see to it that the resulting changes are performed in a systematic manner without getting out of hand.

Project scope generally defines what is to be included in the project, but if there is potential for conflict among the parties involved, it may also be good practice to state explicitly what will not be included in the project as a precautionary measure.

3.1.1 Project Charter

The *project charter* stemming from the project scope is the primary document used to define a project and to establish the general framework for its implementation (Haugan 2002). It should be the result of mutual agreement between the PM, the organization, and the other stakeholders. Its approval by senior management implies that all units within the organization concur with its contents and providing the PM with a clear framework for his interactions with these entities throughout the project lifecycle. It is the responsibility of the PM to ensure that the project charter is adhered to by all parties involved throughout the project lifecycle and updated appropriately whenever necessary.

Although in some organizations the PM is assigned after the charter is approved and announced, it is good practice for the PM to be on the planning team preparing the project charter. In other cases, the PM is already assigned and prepares the project charter in close cooperation with the project sponsor, and it becomes effective after approval by top management.

The contents and format of the project charter can vary depending on the organization, the nature of the project, and the environment in which it will be executed. The project charter and its elements are covered widely in the literature (see, e.g., PMI (2017), Kerzner (2009)). Possible items that can be included in a project charter may include:

(i) *Project purpose*. Business justification; the business need satisfied through this project.

(ii) *Project objectives*. Objectives stated in measurable terms such as percent reduction in cost, together with success criteria.

(iii) *General project information*. Project name; project sponsor; project key stakeholders; expected impact of the project.

(iv) *Project description*. Project scope; deliverables (description, quantity, quality, performance); project starting and ending dates; milestone schedule; budget; resources to be made available to the project.

(v) *Project manager and the team*. Project manager (authority, responsibility, communication and coordination plan, reporting requirements); team members.

(vi) *Preliminary risk analysis*. Major risks involved and their possible impact, together with preliminary plans for impact reduction or elimination.

(vii) *Constraints*. Any limitations imposed on the project team with respect to resources, personnel, project schedule, and others.

(viii) *Project closure and delivery of end product to the client*. Person(s) or unit with authority to decide on successful or unsuccessful termination of the project; evaluation criteria for both cases; rules and requirements for the delivery of end product to the client; the disbanding of the project team.

(ix) *Ratification*. The signing of the project charter by parties involved as a sign of approval and commitment to cooperation.

The project charter is generally a relatively short document of a few pages based on the project management culture and formal procedures already prevalent in the organization. One such procedure, for example, is the change management procedure. Change is an essential characteristic of the project process and the project on hand should be no exception. Hence, there is no need to reiterate such issues in the project charter. It should be concise, clear, and complete.

Small projects may not need a project charter if a verbal agreement suffices. However, depending on the circumstances, it might still be helpful for the PM to have one ratified by all parties involved.

3.2 Work Breakdown Structure

The other important input to project planning is the *Work Breakdown Structure* (WBS). This serves to clearly identify the required tasks, divide the work content into *Work Packages* (WPs) that consist of activities, and establish the main framework for planning and control. WBS is central to project planning, scheduling, and costing, and is an input for many other processes such as risk planning, bid preparation, and proposal development.

Preparing the WBS with care is critical for effective project planning and control. The WBS also sets the level of detail, which is a difficult decision to take in project planning. The project work should be decomposed into activities in such a way that managers can estimate the time and resource requirements of these activities accurately (PMI 2006).

Using information generated by WBS, an *Organizational Breakdown Structure* (OBS) and a *Cost Breakdown Structure* (CBS) are prepared for a project. These are complementary components of project planning that will be discussed in detail in the following sections.

3.2.1 Decomposition of WBS into Work Packages

The WBS is a deliverable oriented, hierarchical decomposition of a project into its components. The first level of the hierarchy contains the deliverables, which are then broken down into work packages (WPs). Depending on the scope of the project, a deliverable might be broken down into its sub-deliverables, creating an additional level. A typical structure, for example, with 3 deliverables is given in Fig. 3.1. In general, each branch in the hierarchy may have a different number of levels.

Each WP represents a selected group of well-defined, related activities or tasks whose execution is the responsibility of a particular person, organizational unit, or group of organizational units. The WBS defines and groups a project's discrete work elements in a way that helps to organize and define the total work scope of the project. As work at a higher level of the hierarchy (*parent*) is decomposed into its lower level components (*children*) giving increasingly detailed definitions of the work involved, all the work at the higher level should be accounted for in the lower level components, with no work from the higher level left out. Similarly, as one moves from a lower level to a higher level in the WBS structure, one moves from a detailed work definition to an aggregate one through the integration or assembly of

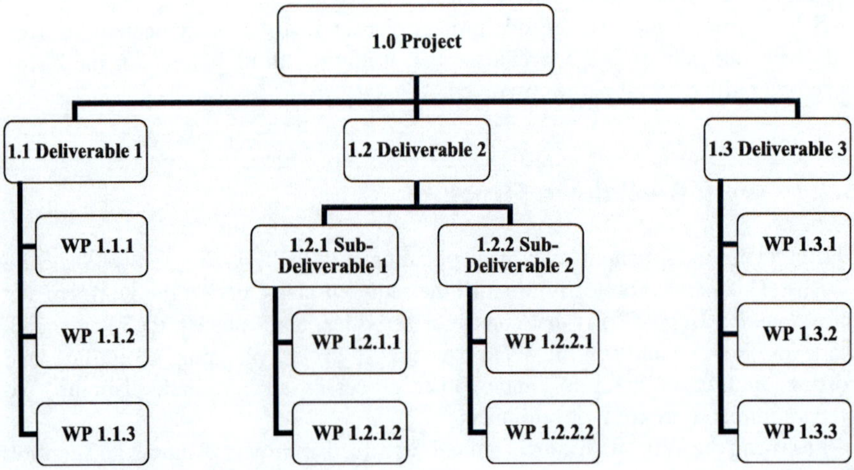

Fig. 3.1 A typical WBS hierarchy

the deliverable components or outputs, i.e., WPs. Each level down the hierarchy represents a more detailed description of the work involved. As one moves from a lower level to a higher one, all in the lower level should be accounted for at the higher level, with no additional work added. This two-way total decomposition rule is called the *100% rule*. The scope of work assigned to a WP should be well defined and should terminate in a deliverable product or service. As we move from lower to higher levels of the WBS hierarchy, the scope, schedule, and cost status reporting of the project will be organized for progressively higher levels of management.

The WBS hierarchy can be obtained using one of the following methods:

The top-down method: We start with the description of the total work required to produce the project's deliverables. The total work is broken down into relatively large subtasks creating the first level down. Then these parts are further broken down to obtain the lower levels of the hierarchy until a specified level of detail is reached.

The bottom-up method: Here, one starts at the lowest level of the hierarchy with all the activities and tasks for the completion of the project listed and then groups them into meaningful entities to obtain the next higher level in the hierarchy, proceeding in this manner at each level until the level associated with the total work is reached.

The different WPs at a given level of the WBS hierarchy may vary in size. We do not have an established level of detail for a WP. It depends on the scope of work assigned to that WP. Hence, size is not a restriction on WPs, but it is important from the point of project monitoring and control. The larger the scope represented by a WP, the more aggregate its contents become and the more prone it becomes to inaccuracies due to % Complete and further estimates needed for monitoring and control.

WPs constitute key elements of the WBS and act as the control points for the scope, budget, and schedule. They contain information on scope, costs, schedule, and resources. Each WP then needs to be monitored and controlled. In order to monitor we need to be able to measure what we want to monitor with an accuracy that is required by the process/task we monitor. Consider, for example, a WP that is comprised of five activities and we want to monitor the raw material cost. If the WP is not further broken down into its activities, then we have one cost item. But, if we monitor the cost at the activity level, then we need five cost items – one for each. Furthermore, establishing these individual cost items will require more effort and most probably will require more attention to detail relative to the cost item at the WP level. Once we measure the cost, we would like to control the cost; in other words, we would like to be able to manage the cost by taking actions so as to direct it towards its planned pre-specified value. Hence, instead of one cost item we will monitor and control five cost items requiring more effort and resources.

A project can often be broken down into its WPs and activities in several different ways. The essential consideration is the level of detail at which we want to plan and control the project. As shown above, a more detailed analysis requires more managerial effort and resources. Thus, there is a trade-off between the level of detail and the effort of planning and management.

Table 3.1 Collection sheet for the work package "Data Analysis of Customer Inquiries"

WP Identification Code	1.3		
Description	Data Analysis of Customer Inquiries		
Date Issued	June 3, 2019		
Responsible Person	Turgay Argun		
Responsible Unit(s)	Customer Relations		
Scheduled Start Date	June 6, 2019		
Scheduled Finish Date	July 26, 2019		
Activities	Budgeted Work (man-days)	Unit Cost ($)	Budgeted Cost ($)
Data Dictionary Preparation	12	110	1320
Data Collection	28	85	2380
Frequency Analysis	12	110	1320
Key Words' Selection	12	110	1320
Total Budgeted Man-Days	64	Total Budgeted Cost	6340

Table 3.1 gives the typical data associated with a WP titled "Data Analysis of Customer Inquiries" whose scope is the collection, classification, analysis, and presentation of data on customer inquiries over a pre-specified period. The data is grouped under a number of keywords to facilitate access to data as well as analysis and presentation. With the principal measure of work being the number of customer inquiries and complaints, we see that it is easily measurable. It is also controllable in the sense that there is a clear relation between work delivered and resources employed.

The WP identification code can be as simple as the level number followed by its order in that level such as (1.3), where the first digit refers to the level number and the second to the order in level 1. Assume WP1.3 is further broken down into 2 WPs. Then their codes will read as WP1.3.1 and WP1.3.2. With this coding, each WP will have a unique code.

Table 3.1 provides information on the estimated (scheduled) start and finish dates that will be employed during the implementation phase for controlling the progress in the schedule associated with this WP by comparing them to the corresponding observed values. The person and the unit(s) responsible for this WP are indicated. Similarly, budgeted work, expressed in man-days, and budgeted cost are also reported. Since these are estimated values the realized values of these quantities may differ. These deviations of the actual values from their initial estimates are used for controlling the cost performance realized during the project implementation.

3.2.2 Types of WBS Representation

We will deal here with *process-based* and *product-based* representations of WBS, giving examples of both. These types are not substitute for each other, but the choice between the two depends on the project.

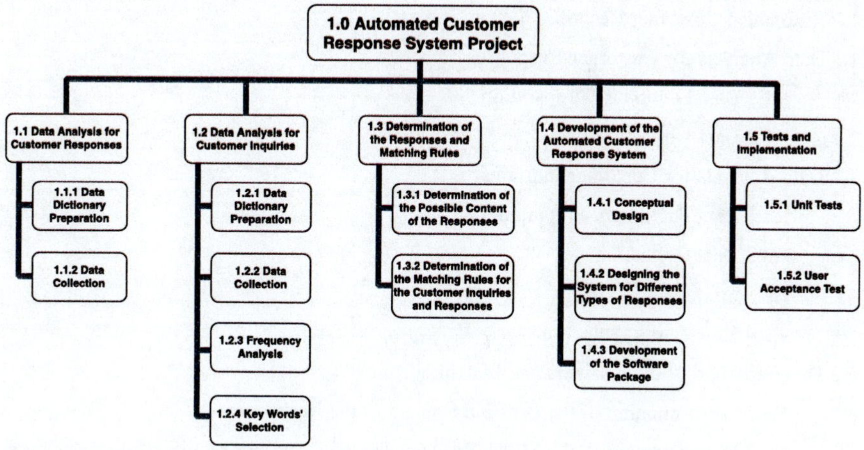

Fig. 3.2 Process-based WBS: The automated customer response system project

An example of a process-based WBS representation is given in Fig. 3.2. It presents the breakdown of the scope for an automated customer response project being implemented by an electricity distribution company. The project deliverable is the automated customer response system. A process-based representation is adopted here since the deliverables at the first level can be represented as the sub-processes of the overall process. The resulting WBS is shown in a *tree format* in Fig. 3.2.

The five phases of the project constitute the first level of the WBS, below which the WPs form the second level. Note that the number of WPs under each phase is not the same. If further detail is required, the WPs can be further broken down into their activities to form a third level, and so on. As stated earlier, each level down the hierarchy represents a more detailed description of the work involved.

We could have also presented the WBS for the same project in a *list format* as shown in Fig. 3.3 that provides the same information as the tree format used in Fig. 3.2.

The second type of WBS representation is the product-based representation. This type of representation is preferred in projects with tangible outcomes whose deliverable is a product. The first level consists of the major components of the product, service, or results. An example for a product-based WBS in tree format is given in Fig. 3.4, displaying a possible work breakdown for a new product development project, where only Deliverable 1.2 Design is further broken down to its WPs.

3.2.3 Gantt Chart Representation of Work Packages

Figure 3.5 shows the Gantt chart displaying the schedule for the WPs derived from the process-based WBS in the automated customer response system project. Recall that when the Gantt chart was introduced in Sect. 1.7, we pointed out that its major deficiency is that it does not explicitly reflect the precedence relationships between

1.0 Automated Customer Response System Project

1.1 Data Analysis for Customer Responses

- 1.1.1 Data Dictionary Preparation
- 1.1.2 Data Collection

1.2 Data Analysis for Customer Inquiries

- 1.2.1 Data Dictionary Preparation
- 1.2.2 Data Collection
- 1.2.3 Frequency Analysis
- 1.2.4 Key Words' Selection

1.3 Determination of the Answers and Matching Rules

- 1.3.1 Determination of the Possible Content of the Responses
- 1.3.2 Determination of the Matching Rules for the Customer Inquiries and Responses

1.4 Development of the Automated Customer Response System

- 1.4.1 Conceptual Design
- 1.4.2 Designing the System for Different Types of Responses
- 1.4.3 Development of the Software Package

1.5 Tests and Implementation

- 1.5.1 Unit Tests
- 1.5.2 User Acceptance Test

Fig. 3.3 WBS for the automated customer response system project: List format

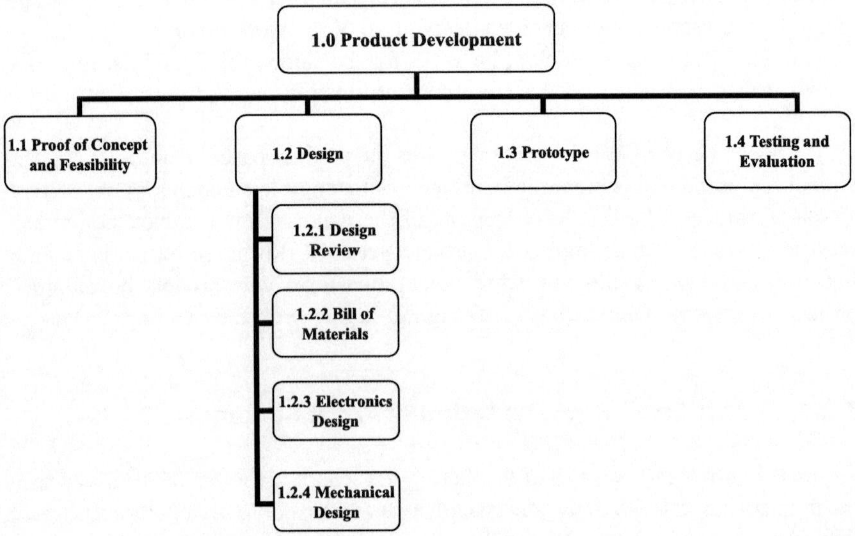

Fig. 3.4 WBS for a new product development project

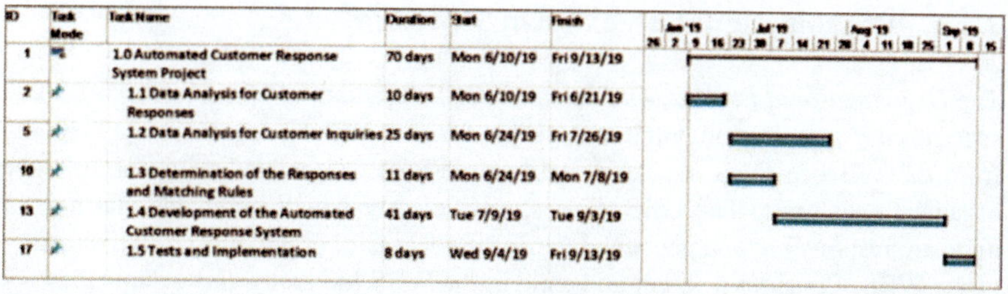

Fig. 3.5 Gantt chart for the WBS elements (WPs) of the automated customer response system project

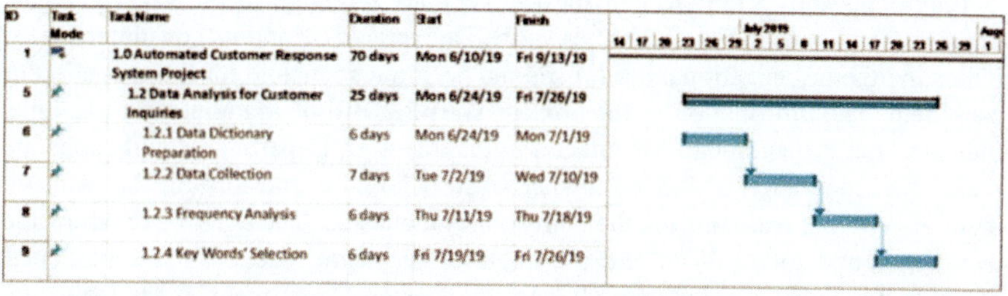

Fig. 3.6 Gantt chart for the WP1.2 data analysis for customer inquiries

the project activities, so the sequence displayed might not present correct information on precedence relationships. An example for this case is provided by in Fig. 3.5. By examining this Gantt chart, we cannot be sure whether WP1.2 is scheduled right after WP1.1 because WP1.1 is a predecessor of WP1.2, or whether they are processed in this order because there are not enough resources available to schedule both WPs simultaneously, or due to some other managerial preference by the PM. This can lead to confusion later when rescheduling is necessary, making it difficult to determine which activities can be delayed without impacting the project schedule and which cannot.

The WP1.2 Data Analysis for Customer Inquiries is further broken down into its activities whose schedule is displayed as a Gantt chart in Fig. 3.6. We observe that the breakdown of WP1.2 is organized by its phases. The activities follow each other and encompass the total scope of work of their parent – WP1.2. The trade-off between the level of detail and the cost of planning and management can easily be observed in Fig. 3.6: Four additional entities whose duration, resource and budget requirements need to be estimated, and which need to be scheduled, monitored, controlled, and, sometimes, rescheduled.

3.2.4 Organizational Breakdown Structure

The *Organizational Breakdown Structure* (OBS) is a natural extension of WBS that specifies the organizational unit responsible for each WP. Hence, there is a mapping from the WBS to OBS, i.e., each entity in WBS is associated with one or more organizational units. The OBS must specify clearly which organizational unit or units are responsible and accountable for the successful completion of which entity of the WBS. Consider a parent and child in the WBS hierarchy and assume they are assigned in the OBS to two different organizational units or to two different individuals from the same organizational unit. The resulting hierarchical structure establishes a line of communication and reporting as well as a line of authority.

Figure 3.7 shows the OBS of the Deliverable1.2 Design in the WBS given in Fig. 3.4. The boxes in Fig. 3.7 represent the Deliverable1.2 and its constituent WPs, indicating the organizational unit(s) and the person responsible for the execution of each item. The information in the box for WP1.2.2 Bill of Materials, for example, indicates that Procurement, Manufacturing, and Design Departments will cooperate under the supervision of the WP owner Aysu, who is responsible for the management of WP1.2.2 and securing the deliverables requested. The other boxes show the teams assigned to the WPs and the name of the team leader. In this particular example, the team members are all from the Design Department. If the teams are mixed with members from different organizational units, then these units have to be indicated as well.

An important supporting document for effective project management is *the Responsibility Matrix* (RM) that shows the Deliverables and WPs, the organizational units or persons responsible, and those with approval authority. The RM facilitates the monitoring and assessment of how well the responsibilities are performed. Various modes of association with the project can be indicated on this matrix such as responsibility, approval, support, and notification. Notification implies formal communication of information, i.e., in the form of a report, memo, etc., which is needed for the progress of the receiving entity. The RM for the OBS depicted in Fig. 3.7 is provided in Table 3.2. Ada, the Head of the Design Department, being responsible for Deliverable1.2, is in the position of approving the work being

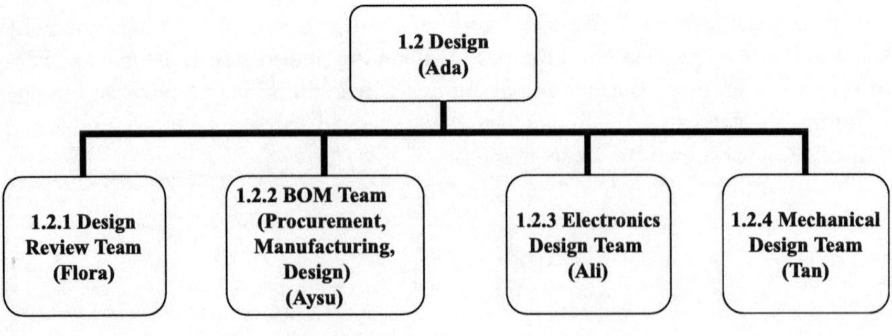

Fig. 3.7 OBS of the Deliverable1.2 Design

Table 3.2 The responsibility matrix

	Ada	Flora	Aysu	Ali	Tan
1.2 Design	RESP[a]				
1.2.1 Design Review	APP	RESP	SUP	SUP	SUP
1.2.2 Bill of Materials	APP	NOT	RESP		
1.2.3 Electronics Design	APP	NOT	SUP	RESP	
1.2.4 Mechanical Design	APP	NOT	SUP		RESP

[a]*RESP* responsible, *APP* Approve, *NOT* Notification, *SUP* Support

performed in all the WPs. All the team leaders are responsible for their own WPs. Flora, being the owner of WP1.2.1 Design Review, is responsible of sharing information resulting from design reviews with the other WPs through notifications. Aysu, the owner of WP1.2.2 Bill of Materials, is asked to give support to the remaining WPs. Ali and Tan, the owners of WP1.2.3 and WP1.2.4, respectively, are given the task of providing support to WP1.2.1 Design Review.

3.2.5 Cost Breakdown Structure

The *Cost Breakdown Structure* (CBS) is associated with both the WBS and OBS. Recall that WPs constitute key elements of the WBS and act as the control points for the scope, budget, and schedule since they contain information on scope, costs, schedule, and resources. The CBS provides the information on costs by assigning costs to the cost account(s) of the organizational unit(s) associated with the WP under consideration.

Each organizational unit associated with a WP of the WBS has a cost account associated with this WP. If there is more than one organizational unit involved, the cost of that WP is distributed among the cost accounts of the units involved in proportions usually agreed on at the start of the project. These proportions may be determined by agreement among the managers concerned, or by the organization's cost accounting procedures. For example, the cost incurred for the WP1.2.2 BOM Team in Fig. 3.7 is shared between the Procurement, Manufacturing, and Design Departments where each is represented by a separate cost account. Hence, the amount charged to Manufacturing Department is the cost accrued by this Department for the realization of WP1.1.2 BOM Team.

3.2.5.1 Bottom-Up Costing - Cost Aggregation

The assignment of costs to cost accounts at the lowest level WPs allows for bottom-up aggregation of cost to determine those for the higher level WPs. For example, let the deliverable D1.2 Design in Fig. 3.4 be a parent. It has four children WP1.2.1, WP1.2.2, WP1.2.3, and WP1.2.4, each with a realization cost. Then the cost of the deliverable D1.2 Design is the sum of the costs of its children. Similarly, by going up one level we obtain the total cost for the whole project, i.e., for "1.0 Product Design". With "1.0 Product Design" being the parent of four children, namely, the deliverables D1.1 Proof of Concept and Feasibility, D1.2 Design, D1.3 Prototype, and D1.4 Testing and Evaluation, its cost is obtained by summing the costs of its children.

3.2.5.2 Cost Accounts

In a WP, the cost assigned to the cost account of an organizational unit is an aggregate value and, if preferred, can be further broken down into cost types for closer monitoring and control by the project management or the organizational unit involved. This requires more detailed cost data, increasing the workload involved in data collection and analysis. Again, we are faced with the trade-off between the higher cost of data collection and processing vs. closer monitoring and control of the project and more accurate performance measurement.

To facilitate cost data collection and aggregation, cost account codes are employed. There are a large number of coding systems mostly proprietary. In the following, we will present an example for direct cost account codes from the construction of a petrochemical complex. The code of accounts for *direct items* consists of two groups of digits:

XX-XXXX

(i) The first two digits of the first group indicate the area of construction.

    XXXX ________Ethylene Unit
    XXXX ________Polyethylene Unit
    XXXX________Vinyl Chlorine Monomer Unit
    XXXX________ Polyvinyl Chloride Unit

    10- XXXX________Offsites

(ii) The first digit of the second group indicates the type expenditure whether for direct labor, direct materials or subcontract as follows.

    XX-1XXX________Direct Material
    XX-2XXX________Direct Subcontract Material and Labor
    XX-3XXX________Direct Subcontract Labor Only
    XX-4XXX________Direct Labor Only

(iii) The second digit of the second group indicates the major category of work or items and are standardized so that these mean the same for each area.

    XX-X1XX________Vessels
    XX-X2XX________Heat Transfer Equipment
    XX-X3XX________Mechanical Equipment
    XX-X4XX________Civil Engineering

    XX-X9XX________Miscellaneous Equipment and Items

(iv) The third digit of the second group is assigned to Various Types of Items within the major category of work.

(v) The fourth digit of the second group applies to specific different types of the major category of work.

For example, the cost account code 01-4461 can be decoded as follows:

01: Ethylene Unit; 01-4 Direct labor in Ethylene Unit; 01-44 Direct labor in Ethylene Unit for a Civil Engineering item; 01-446 Direct labor in Ethylene Unit for a Civil Engineering item using Structural Steel; 01-4461 Direct labor in Ethylene Unit for a Civil Engineering item using Structural Steel in Steel Structure.

Hence, the cost account code 01-4461 records the cost of direct labor in the Ethylene Unit in the Civil Engineering category of work of constructing the steel structure using structural steel.

The cost account codes facilitate the cost aggregation approach employed in Sect. 3.2.5.1 above. Given a WP is defined in the WBS for the construction of the steel structure in the Ethylene Unit, its total cost will be calculated by aggregating over the digits X in 01-XXX1.

At a more aggregated level of the same WBS consider a WP defined for all the work in the Ethylene Unit. Its total cost will be obtained by aggregating over the digits X in 01-XXXX.

3.3 Network Modeling of Projects

3.3.1 Types of Models

Models are abstractions of reality we employ in our efforts to understand and conceptualize the problematic situation at hand, explain the relationships involved, predict how they behave and interact and then act on it (Ackoff 1978). Being an abstraction, the model is never a direct, completely accurate representation of reality. Models are classified in different ways depending on the purpose of use. Three types of models employed frequently will be briefly introduced next: iconic, analog, and symbolic models.

Iconic models, although differing in scale, usually look like what they represent. Examples would be maps, photographs, model airplanes, ships, and automobiles.

In *analog* models, one set of properties are used to represent another set of properties. An example would be the contour lines on maps, which are analogs of elevation of the terrain.

Symbolic models are mostly expressed in the form of mathematical terms and relationships, such as the well-known law in physics $F = ma$.

When dealing with a project, depending on the size and complexity of the project, we might make use of more than one of these types of models. For example, in a residential building construction project, a 3-D model can be built to display the final appearance of the building to potential customers. The structural analysis of the building is based on mathematical models. For the planning and control of the construction, more specifically for time and resource planning of the project activities, other, different mathematical models are employed.

As in the construction example above, models can be used as a means of communication about the project and the way we conceive it. Using the 3-D model of the building, we communicate to the project owners and stakeholders how we understand their expectations and how we plan to meet these expectations. New ideas about the project can be generated through the discussions around the 3-D model. An earthquake simulation model applied to the building, on the other hand, can predict how the building will behave under different circumstances allowing these to be considered in the design of the building.

The network representation of projects is essentially the result of a modeling process. It can be defined as a model that represents the project through its events, activities, precedence relations among activities and serves as a project planning tool considering the duration, the cost, the resources, and other project related factors. Over time project networks have become by far the most widely preferred and employed planning and management tool for the following reasons:

(i) Allows easy visualization of the relationships between activities in a clear and easily understandable way, facilitating the presentation of the project to different parties.
(ii) Easy to represent on the computer, supporting the formulation of complex mathematical programming models.
(iii) Supports the identification of critical paths and critical activities and effective planning and monitoring around them.
(iv) Easy to observe bottlenecks resulting from the delay or slower pace of certain activities.
(v) Allows for the planning and control of more than one project at any given time.
(vi) Allows for easy updating of the project.

In the following sections, we will give the details of the network modeling of projects.

3.3.2 Network Modeling of Projects

Project network models are based on the assumption that a project can be represented using well-defined events, activities and relationships among the activities. Once we have gone through the first phase of breaking the project scope down into events and activities, establishing the WBS and the precedence relationships among these activities, we are ready to represent this information in the form of a network.

The events represent the start and end of one or more activities. An event is an instantaneous point in time and is represented by a node. The parameters associated with each activity are its *duration*, its *resource requirements*, and its *precedence relationships* with other activities, which may be technological or managerial in nature. Technological precedence reflects a sequence among activities due to technological requirements. For example, the pressure testing of a pipe can only be performed after the pipe is properly placed and secured. A managerial precedence constraint, on the other hand, may arise when a PM prefers, for example, a team with a certain skill set to work on two different activities with no technological precedence requirement between them. S/he might impose a precedence relationship between these activities so that they cannot be executed in parallel, allowing her/him to assign the same team to both activities so that they can execute them in sequence.

More formally, a project network consists of three entities: $\{N, A, \Phi\}$, where N stands for the set of nodes, A the set of arcs, and Φ the set of precedence relationships.

In order to map the precedence relationships into the network diagram, we ask the following questions while considering the activities one by one:

(i) Which activities must be completed immediately before this activity can start?
(ii) Which activities must follow this activity?
(iii) Which activities can be processed concurrently with this activity?

The average number of precedence relations per activity in a project reflects the density of relations between activities and is accepted as a measure of *network complexity* (*NC*). It is given by

$$NC = (\text{Number of precedence relations})/n$$

where n is the number of activities. In general, the higher the *NC* the more complex it is to manage the project since a larger number of constraints must be considered to determine a feasible schedule.

There are certain conventions upon which the network representation is based:

(i) The project network has only one start and one end node. If that is not the case, then the missing start or end node are created using dummy activities.
(ii) The length and shape of an arc on the project network are governed only by the convenience of illustration and have no additional implication. A rare exception is the time-scaled project network where the length of an arc is proportional to its length (see, e.g., Moder and Phillips 1970).
(iii) The project network has to be acyclic, i.e., a network with no loops. A cycle in the precedence relations would imply that an activity can be processed simultaneously with its own successor and predecessor, which is clearly inconsistent. An algorithm for detecting projects containing cycles is provided in Sect. 3.3.3.1 together with an example.

There are two types of project network representation: (*i*) the activity on arc (or arrow) (AOA) project network and (*ii*) the activity on node (AON) project network. Both result in a directed acyclic graph. Most commercial computer packages employ the AOA representation, for reasons more historical than practical. The first project management tools developed used the AOA representation, which over time has become almost the standard. Only after the practicality of AON was recognized have some commercial computer packages adopted that representation. AON is also preferred in mathematical programming formulations of project scheduling problems due to its simplicity.

3.3.2.1 Hammock Activities

In addition to the type of activity defined in the previous section, a further type of activity called a *hammock activity* is used to represent a project as a project network. We can also refer to it as a task or link to differentiate it from the description of activity in the context of project networks. A hammock activity links two event points but has no pre-specified duration. It is flexible in the sense that its beginning as well as its ending time might change depending on the occurrences of its start and end events. An example of a hammock activity is the project control activity throughout the project, which will extend between the start and end events of the project. If this task is given to a consultant who charges a fee per period, then its cost will be determined by its realized duration and the consultant's per period fee. Thus, hammock activities have the useful property of carrying cost and/or usage throughout their duration. For example, the hiring and use of rental equipment in a certain portion of a project can be represented through a hammock activity.

3.3.3 Activity on Arc Type Representation

3.3.3.1 Network Construction Rules

In AOA networks, the activities are represented by arcs and events by nodes. Each activity should be represented by one and only one arc. The *start node* (the tail) indicates the start of an activity and the *finish node* (the head) its completion. These nodes correspond to the events of the activity starting and ending, respectively.

Let us consider two activities that must be accomplished in series, i.e., the first activity precedes the second, such as preparing a package (activity A) and mailing it (activity B). Using directed arcs to indicate precedence relations, we can represent these two activities on an AOA type project network as shown in Fig. 3.8.

If two activities can be carried out in parallel, they are said to be *concurrent*. However, if these activities share some resources and there are not enough resources to execute them simultaneously, they will have to be scheduled in some order, one before the other, without violating the limits on resource availability. Scheduling activities under limited resource availability will be the topic of Chap. 7.

Consider the three activities of "making the package" (A), "writing the label" (B), and "placing the label on the package" (C). Activities A and B can be processed concurrently but must both precede C. Although activities A and B, can finish at different times, they must both be completed before the event "ready for label fixing" occurs and activity C can start. The AOA network representation of this case is depicted in Fig. 3.9, where event 3 is called the *merge event*.

Sometimes an event may permit two or more activities to start. For example, let activity A be "prepare material list". Once the material list is prepared, then activity

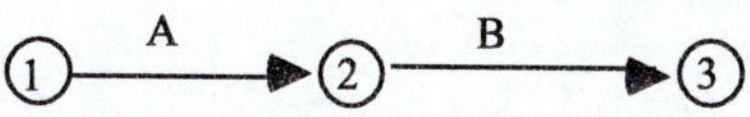

Fig. 3.8 Representation of activities in series

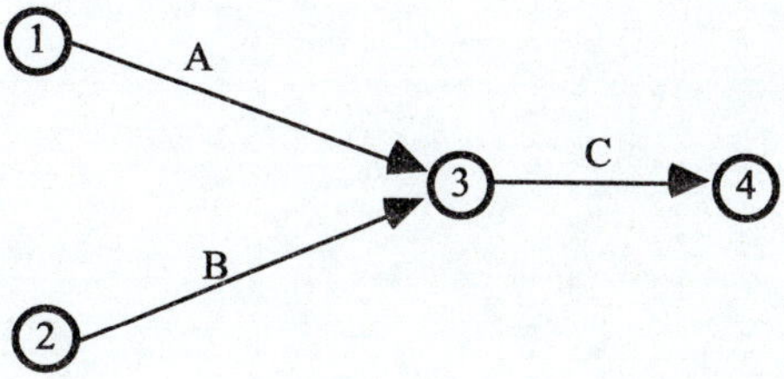

Fig. 3.9 Representation of concurrent predecessors – *the merge event*

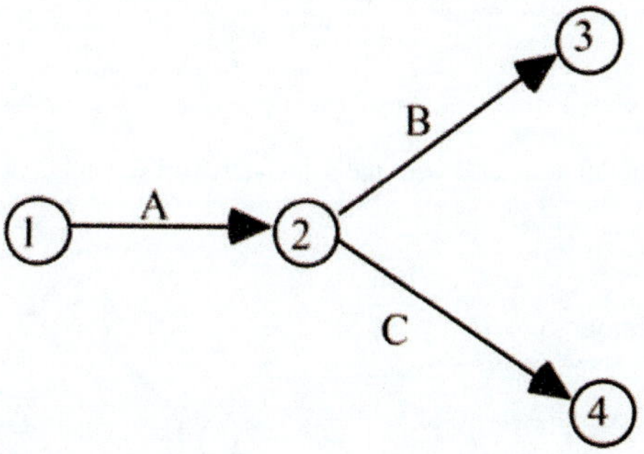

Fig. 3.10 Representation of concurrent successors – *the burst event*

B "procure pipes" and activity C "procure valves" can start. Activity A precedes both activities B and C, and activities B and C are concurrent. For this case, the network will be as in Fig. 3.10, where event 2 is called the *burst event*.

While representing concurrent activities starting from the same event and leading to the same event, we must be careful not to use parallel arcs. In other words, no two activities should be identified by the same two events, as this would create confusion with regard to the occurrence time of the finish event. In order to avoid such a situation, dummy activities, whose duration and resource requirements are zero, are added. An example with two activities A and B having the same start and finish event nodes is given in Fig. 3.11a. To avoid this situation, a dummy activity is added to either activity A or B as a predecessor. The cases of appending a dummy activity to activity A as a successor or predecessor are depicted respectively in Fig. 3.11b, c.

Dummy activities are also employed when several activities preceding an activity X are also predecessors of another activity Y, where the predecessors of X constitute a proper subset of the predecessors of Y. In such a case, the termination information of the activities in the proper subset (at the start node of X) is captured by adding a dummy activity to the start node of Y as illustrated in Example 3.1.

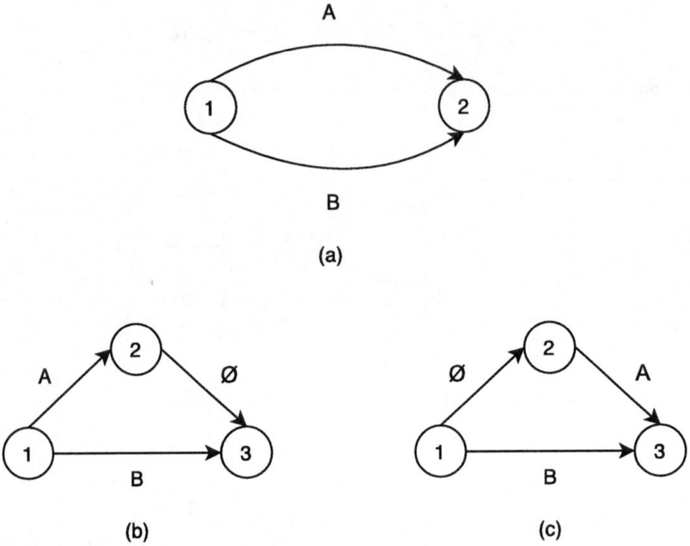

Fig. 3.11 Avoiding parallel activities with the same start and finish events

Fig. 3.12 Precedence feasible AOA representation

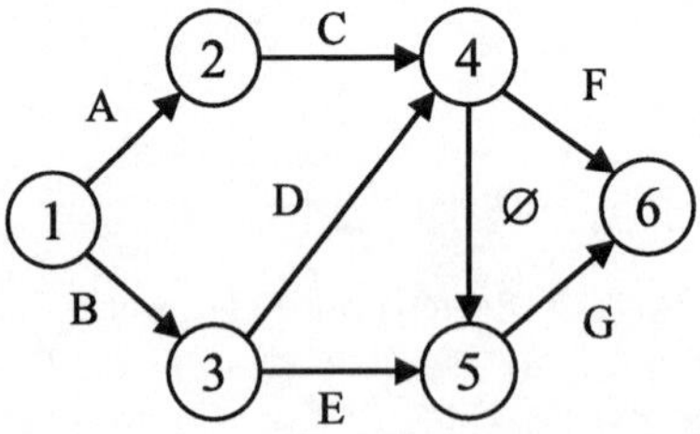

Example 3.1 Consider a project with 7 activities and the following precedence relations:

A precedes C,
B precedes D and E,
C and D precede F,
C, D, and E precede G.

Draw the AOA project network representing these activities and their precedence relations.

Solution
These precedence relationships result in the project network shown in Fig. 3.12. The dummy arc Ø carries the information of termination of activities D and E to the start node of activity G, allowing activities D and E to become predecessors of activity

G. Note that activities C and D preceding activity F constitute a proper subset of another set of activities (C, D, E) preceding a different activity, G.

3.3.3.2 Checking for Cyclic Project Networks

Recall that the network representation assumes that the project network has to be acyclic. Therefore, it becomes essential that we discover whether a project network is cyclic or not, especially when using computers to process very large project networks that are hard to verify manually. The following algorithm also serves the dual purpose of *numbering the nodes (events)* such that activities are always directed from lower to higher numbered events. This property of a project network, referred to as topological ordering, is preferred for computational reasons.

An algorithm for *checking for cyclic project networks* is given below:

Step 1. *Initialization.* Search for nodes with no incoming arc. If there is only one such node; call it the start node. If there is more than one such node; add an artificial start node to the project network and connect it to all nodes with no incoming arc. Label the start node as node 1 and let $k = 1$.

Step 2. *Iterations.* Delete all arcs outgoing from node k. If there is no node with an incoming arc, go to Step 3. Else, choose a node with no incoming arc, label it as node $k + 1$, and set $k = k + 1$ and repeat this step.

Step 3. *Cyclic project network detection.* If any nodes remain unlabeled, terminate the algorithm. Declare the project network on hand as a cyclic project network. Else, continue.

Step 4. *Termination.* Search for nodes with no outgoing arc. If there is only one such node, call it the terminal (end) node. If there is more than one such node, add an artificial terminal node and connect all nodes with no outgoing arc to the artificial terminal node. Terminate the algorithm and declare the project network on hand an acyclic project network.

Example 3.2 Apply the algorithm for checking for cyclic project networks to the project network in Fig. 3.13a.

Solution

Step 1: The event node from which the activities (arcs) A and B are incident from is designated as the start node and is labelled as event node 1 on the project network.

Step 2: Delete the activities A and B. There is only one event node with no incoming activity. Set $k = 2$ and label it as event node 2.

Step 2: Delete the outgoing activity C from event node 2. There is only one event node with no incoming activity. Set $k = 3$ and label it as event node 3.

Step 2: Delete the outgoing activities D and I. There is no event node with no incoming activity, so we go to Step 3.

Step 3: Not all event nodes are labelled. We terminate the algorithm and declare the project network a cyclic project network demonstrating the infeasibility of the precedence relations.

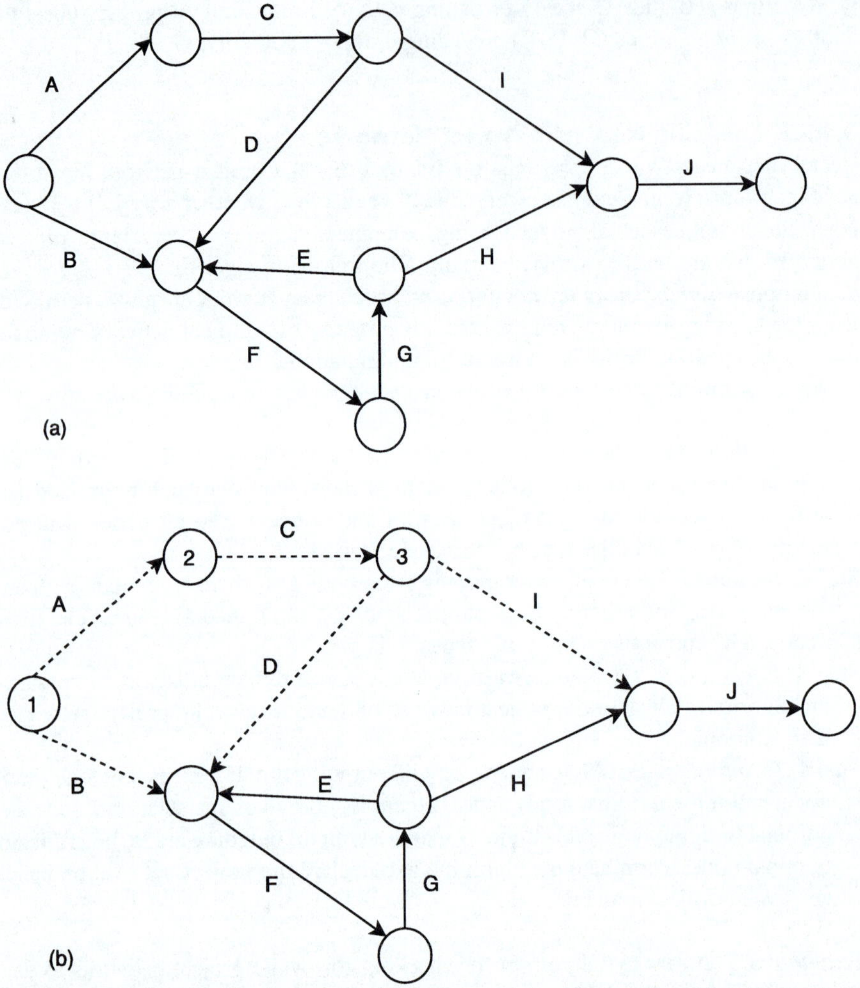

Fig. 3.13 (**a** and **b**) An example for a cyclic project network with infeasible precedence relations

The partial solution reached is presented in Fig. 3.13b. The dotted activities represent the eliminated activities.

3.3.4 Activity on Node Type Representation

In the AON representation, the activities are represented by nodes and the arcs capture precedence relationships among the activities. The drawing of AON type networks is straightforward since there is no need for dummy activities. The AON

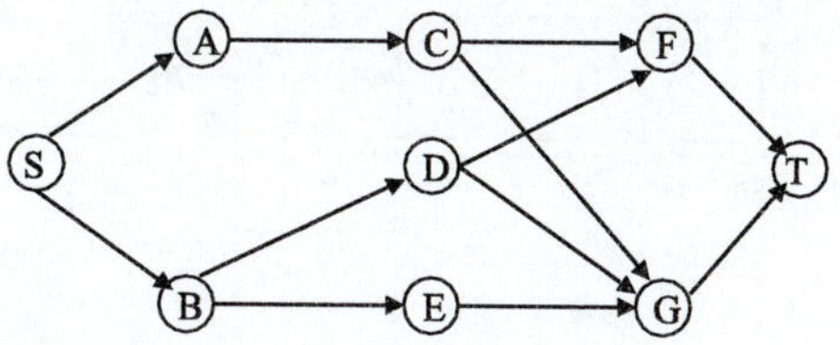

Fig. 3.14 Precedence feasible AON representation of project network given in Fig. 3.12

representation of the project network in Fig. 3.12 is given in Fig. 3.14, where S and T represent the start and terminal nodes, respectively.

Whether the AOA or AON representation is used, once the project network is obtained, we should apply post-processing to the resulting project network. We have two major sources of error to check. One relates to the data employed. Data should be checked to ensure that it is correct and is correctly represented in the project network.

In addition, each precedence relation should be checked for whether it is correct in the first place and whether it is correctly represented in the project network. Erroneous precedence relations might lead to cycles in the project network, which need to be detected and corrected before proceeding further. The node numbering algorithm can be used for this purpose.

3.3.5 Generalized Precedence Relationships

The precedence relationship discussed so far implies that once an activity is finished, its successor(s) can be started immediately if the PM chooses to do so. However, it is also possible that once an activity is finished, its successor(s) can only be started after a minimal time lag of FS periods. We can also have overlaps of activities such as, for example, an activity can start only a minimal time lag of SS periods after the other activity has started. There are four such generalized precedence relationships (GPRs) with *minimal time lags* (Moder and Phillips 1970; Elmaghraby and Kamburowski 1992) which we illustrate in Fig. 3.15a–d.

(i) *Finish-to-Start Precedence*: In this case, if activity i precedes activity j, j can start no less than FS_{ij} periods after i is finished. In other words, the difference between the start time of j and the finish time of i must be at least FS_{ij}. For example, in a building construction project you cannot start the activity of erecting the columns on a floor (activity 2) before FS_{12} periods passes after the pouring of the concrete (activity 1) is completed for the concrete to solidify and cure.
(ii) *Start-to–Start Precedence*: In this case, given activity i precedes activity j, j can start no less than SS_{ij} periods after i has started. For example, assume the steel construction of a single-story factory building has been completed. The contractor starts with the brick wall (activity 1). The activity of plastering the wall (activity 2) can start after SS_{12} periods once a certain portion of the wall is completed. The time between the start of building the wall and the start of the

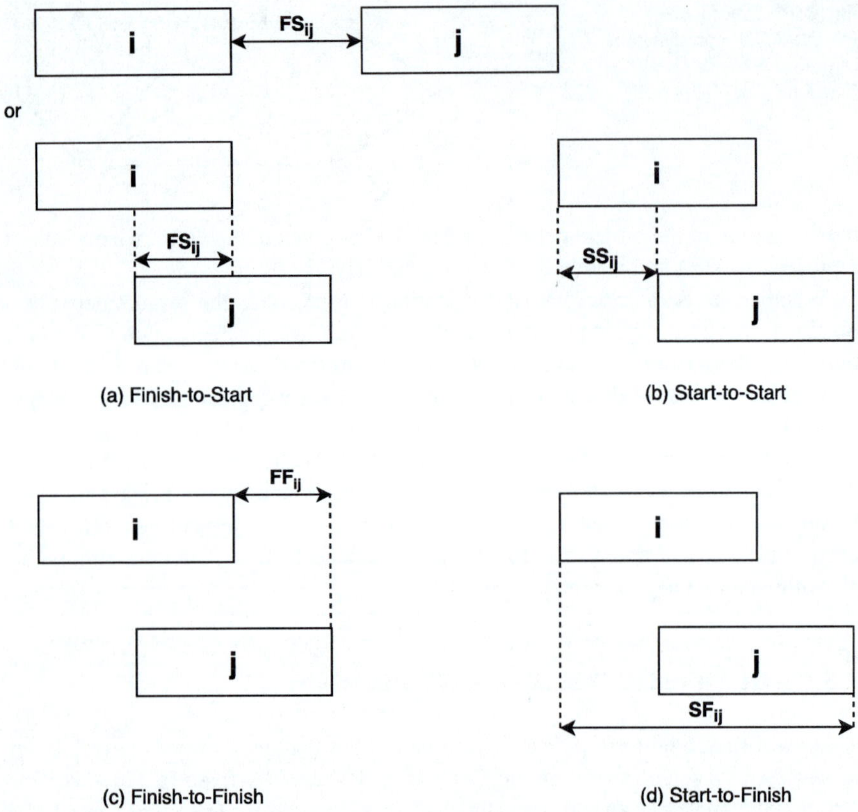

Fig. 3.15 (a–d) Gantt chart representation of the GPR. (Source: Dodin and Elimam 1997)

plastering depends on the rate of work of these two activities and it must be adjusted such that plastering is always kept behind building the wall so as to provide enough space between these two activities for proper execution of both.

(iii) *Finish-To-Finish Precedence*: In this case, given i precedes j; j can finish no less than FF_{ij} periods after i is finished. For example, consider a project for the make-to-order production of a die. Two pieces of the die are to be assembled where piece 1 is to be processed at high temperature (activity 1). Before the production of piece 2 (activity 2) is completed and the two pieces can be assembled, piece 1 has to be cooled down slowly to room temperature for FF_{12} periods. In line with the just-in-time approach, the earliest time piece 2 is ready for assembly is when piece 1 is cooled down to room temperature.

(iv) *Start-to-Finish Precedence*: In this case, given i precedes j, j can finish no less than SF_{ij} periods after i has started. Consider the die assembly example in (*iii*) above. Given the production of piece 1 (activity 1) has started, the production of piece 2 (activity 2) can finish no earlier than SF_{12} periods after the production of piece 1 has started, counting the duration needed for the cooling of piece 1.

All of the above GPRs can be transformed into each other (Bartusch et al. 1988). Consider, for example, the case of finish-to-start and start-to-start. Let d_i represent the duration of activity i. Then, $FS_{ij} = d_i - SS_{ij}$, if activities i and j overlap and $FS_{ij} = SS_{ij} - d_i$, if they do not. Note that this transformation property applies only for problems with single mode of execution; multiple modes of execution will be discussed in Chap. 7.

The same four GPRs can be defined analogously for *maximal time lags* (see, e.g., Neumann et al. 2003). Consider, for example, the case for finish-to-start precedence. Given i precedes j; j should start at most FS'_{ij} periods after i is finished. In other words, the difference between the start time of j and the finish time of i cannot exceed FS'_{ij}. The start-to-start, finish-to-finish, and start-to-finish maximal time lags are given by SS'_{ij}, FF'_{ij}, and SF'_{ij}, respectively. (See Exercises 3.15–3.17 for the inequalities associated with these four GPRs.)

The use of GPRs extends our ability to model the project environment more realistically, allowing us to represent concurrency in the project environment to obtain reduced project duration. For example, when there is a finish-to-start relationship between two activities i and j with i being the predecessor, the duration from the start of activity i to the completion of activity j would be shorter than the sum of the durations of both activities even if we assume zero time lag between the activities. Consider, for example, a new product introduction project from the concept development phase till the initiation of production involving the contribution and cooperation of several functions of the organization such as design, manufacturing engineering, human resources management, and production. The work of these functions is mostly formulated as sub-projects. Concurrency among these sub-projects is an essential tool for the reduction of the duration of this process (Wognum and Trienekens 2015).

3.3.6 Level of Aggregation in Project Networks

We can envision a hierarchy of project networks differing in terms of size and complexity, where higher-level networks represent more aggregate networks. Aggregation of project networks is particularly useful for the management of large complex projects. We can model both the aggregate and the detailed project plans with the project network representation. The level of detail is determined by the number of activities of the project network. We might be interested in a project master plan. In that case, each activity of the master plan might correspond to a sub-network consisting of several activities. The network we are currently dealing with might be the detailed representation of an activity of that master plan.

Different levels of network aggregation serve different levels of management. For monitoring the progress of the project, the top management would like to see a clear, concise presentation of the project in its aggregate components together with associated overall KPIs reflecting the current status of the project. For lower levels of management, on the other hand, the portion of the project under their responsibility needs to be presented in more detail than other parts of the project. The level of aggregation should be adjusted such that it provides a sound basis for the targeted decision-maker.

The level of aggregation depends on many factors such as project complexity, size, available resources, organizational culture, and structure (Buchtik 2010). As the size or complexity of the projects increase, managers need a more detailed decomposition. The less experience or technical knowledge managers have about the project, the more detailed planning will be needed to facilitate effective control. According to Buchtik, tasks that are outsourced would contain fewer details, hence can be more aggregated, as the managers cannot directly control this work. Finally, redefining some activities by dividing them into smaller activities (sub-activities) creates flexibility in scheduling some of these sub-activities concurrently with others. It also allows managers to monitor and control project progress more accurately through the activity-based time and cost deviations. However, this requires collecting and processing more activity data and more detailed analysis of the data collected, which can make planning and scheduling more complicated and costly for managers.

In the following, we will examine two means of project network aggregation: milestones and summary activities.

3.3.6.1 Milestones

Recall that a *milestone* is defined as the scheduled start and completion of major deliverables and key external interfaces (PMI 2004). "1.1 Data Analysis for Customer Responses" phase of the automated customer response system project (Fig. 3.2) we dealt with, for example, constitutes a major deliverable for the project. The start and completion dates can then be specified among the milestones of the project. Assume, for example, another major deliverable "1.4 Development of the Automated Customer Response System" of the same project is subcontracted to a software company. Based upon the contract signed there is a pre specified date for the payment event for the services of the software company. This payment is a transaction between the organization and the software company and refers to an external interface between the parties. Hence, the pre-specified date for the payment event can be indicated as another milestone of the project.

Milestones can be employed for a higher level of aggregation of a project. A project schedule expressed in terms of its milestones is usually called a *milestone schedule*. This is the terminology we employed in Sect. 3.1.1 when summarizing the possible contents of a project charter.

3.3.6.2 Summary Activities

A *summary activity* can be defined as a single aggregate activity representing a sub-network of an AOA project network consisting of one or more paths between two event nodes of the network. The early start of the summary activity is

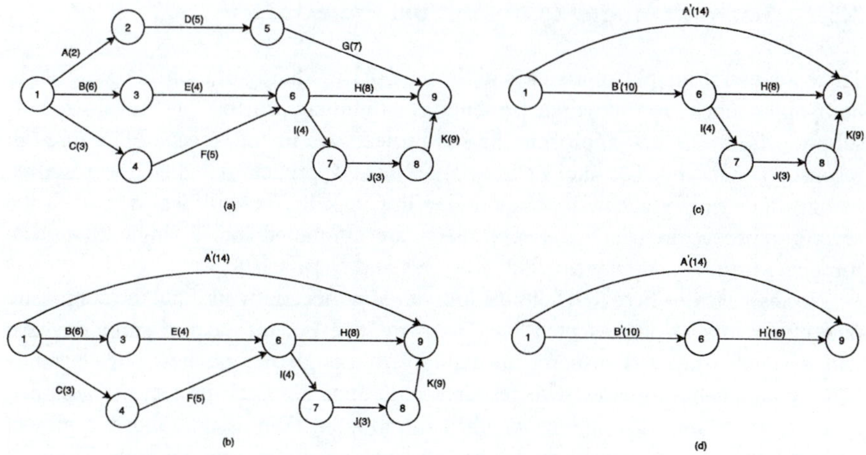

Fig. 3.16 (a–d) Different levels of aggregation for a project network

determined as the earliest of the early starts (ESs) of the activities it represents. Similarly, its early finish (EF) is given by the latest of the EFs and its late finish (LF) by the latest of the LFs of the activities represented by the summary activity. The duration of the summary activity corresponds to the maximum length of all the paths in the sub-network it represents.

Example 3.3 Define a sequence of summary activities to reduce the size of the project network depicted in Fig. 3.16a.

Solution

A summary activity A' can be defined between event nodes 1 and 9, representing the activities A, D, and G (Fig. 3.16b). Since the summary activity A' consists of a single path, its duration is 14, the length of path {A, D, G}. A summary activity B' replaces the sub-network consisting of activities (B, C, E, F) (Fig. 3.16c). Its duration is equal to 10, determined by the longest path between event nodes 1 and 6. Similarly, activities H, I, J, K can be reduced to the summary activity H' defined between nodes 6 and 9 (Fig. 3.16d). Its duration is 16, determined by the longest path between nodes 6 and 9. Hence, we can represent a set of activities through a summary activity thus reducing the size and complexity of a project network.

Obviously, the examples above are relatively small networks employed for the purpose of demonstration. Based on the relative scale between the project networks above, if we view the project network in Fig. 3.16a as a detailed project network, the one in Fig. 3.16d could be viewed as a project network of summary activities.

3.3.7 Network Modeling of Multiple Projects

There are two principal approaches to the network modeling of multiple projects. In both approaches the networks presenting individual projects are considered as subprojects. In the first approach, the sub-projects are linked to each other through resource constraints (Besikci et al. 2013). This approach is called the resource management policy and will be addressed in Chap. 8. We will deal here with the second approach in which the subprojects are combined into a single *composite project* (Kurtulus and Davies 1982; Kurtulus and Narula 1985).

We shall assume here that both the individual project networks and the composite project network are represented as AOA networks. For the sake of simplicity, we will assume unlimited resource availability, although the method introduced is equally applicable to cases with resource requirements. Each project j is assigned an *early start time EST* before which it cannot start. We assume that the project networks involved all have a single start node and a single terminal node. The way we bring together these project networks into a composite project network depends on the precedence relationships among the project networks and also whether their early start times are zero or positive. The resulting composite project network has a single start node and a single terminal node.

We now introduce different cases of combining a given set of project networks with possible structural properties and early start times. Obviously, the following cases are not exhaustive and other ways of forming composite project networks exist.

Case 1. The project networks have no precedence relationship among each other, and their early start times are all zero. In this case, the start node of each project is connected to the dummy start node of the composite network by dummy arcs. In order to reduce the network size, one of the start nodes can be employed to replace the dummy start node. The terminal node of each project is connected to the dummy terminal node with a dummy arc.

Case 2. If there is a finish-to-start precedence relationship with zero time lag between two projects A and B with project A preceding project B, then the arc from the dummy start node to the start node of project B is replaced by a dummy arc from the terminal node of project A to the start node of project B. In the case of positive time lags, then the duration of the dummy arc is equal to the time lag value.

Case 3. The project networks have no precedence relationship among each other, but some have positive *EST*. Then the start node of one of the projects with an *EST* of zero is selected as the start node of the composite project. An arc goes from the start node of the composite project to the start node of each of the remaining projects. These arcs have durations equal to the corresponding project's *EST*.

Case 4. One of the projects, say project A, has to be finished before the others can start. In such a case, the start node of the composite project becomes the start node of project A and the terminal node of project A is connected to the start node of every other project by a dummy arc.

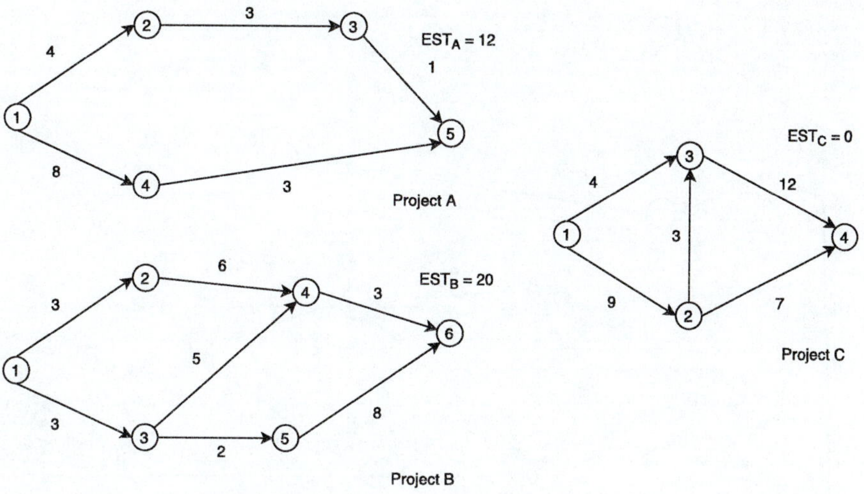

Fig. 3.17 AOA project networks for 3 projects A, B, and C

Case 5. One of the projects, say project Z, has to be finished last. In such a case, a dummy arc from the terminal node of all the projects to the terminal node of project Z is added. The terminal node of project Z also serves as the terminal node of the composite project network.

Example 3.4 Let us consider the 3 projects, whose AOA networks are displayed in Fig. 3.17. The duration of each activity is indicated on the corresponding arc. For each project *j*, the earliest starting time denoted by EST_j is given. Note that this example corresponds to Case 3 above.

Solution

Figure 3.18 shows the composite project network. Node 1 becomes the dummy start node with arcs (1,5) and (1,10) of zero duration and zero resource requirement linking the dummy start node to the start nodes of projects A and B, respectively. We don't have an arc for project C since its start time is zero. The end nodes of each project are connected with a dummy arc with zero duration and resource requirement to the dummy terminal node, node 16, of the composite project network. After the composite network structure is obtained using the original node numbers, the algorithm for checking for cyclic project networks given in Sect. 3.3.3.2 is applied to number the nodes in the composite network.

To extend the example problem further, assume that the deliverables of project C form an input (predecessor) to activity (5,6) of project B. This situation is represented with a dummy arc leading from node 4 to node 14 of the composite network and eliminating the dummy arc (4,16).

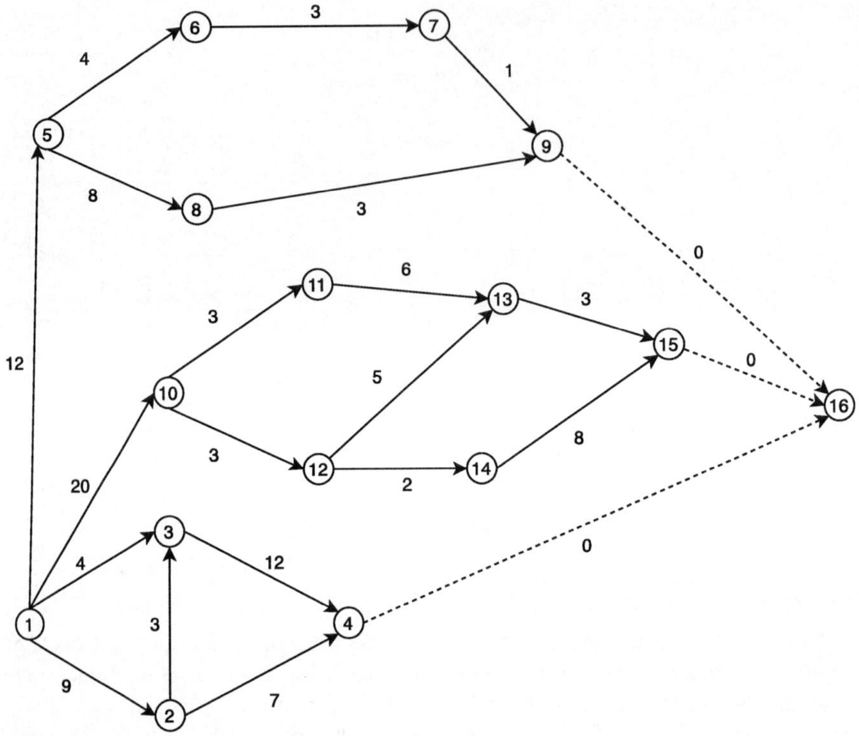

Fig. 3.18 Composite network

3.4 Estimating Activity Durations

The starting point for estimating activity durations is the estimation of the work content of that activity expressed in terms of man-hours or man-days, whichever is appropriate. This requires detailed knowledge of the content need to know the contents of the activity in detail. In general, activity duration is an outcome of the resources employed to realize it. Different combinations of resources, referred to as modes of execution or just modes, lead to different durations, and are discussed in detail in Sect. 7.6. For example, one mode for realizing an excavation activity might be performed by 1 worker and 1 excavator in 5 days. Modes represent the input (resource combination) and the output (duration and/or cost).

The duration of each activity should be estimated independently of its preceding and succeeding activities and should reflect the duration expected under normal conditions using standard productivity rates. Any risks that might be encountered in the execution of the activity should not be accounted for in the estimate of its duration under normal conditions.

3.4.1 Tools for Estimating Activity Durations

There are several tools employed in practice to estimate the activity durations in a project. Some major ones will be presented below based on PMI (2017):

Expert judgment: This is probably the most common approach to estimate durations. Experts are expected to be knowledgeable individuals in the domain area (s) covered by the project, who preferably have training and/or previous experience in similar projects. It is good practice to acquire estimates from more than a single expert. The usual practice is to arrive at a final value, or range of values, through an iterative decision-making process where the experts produce estimates and then update them based on the estimates obtained from other experts. This process can be conducted through organized meetings such as focus groups where the experts come together or through some form of correspondence where the experts go through the iterations without meeting. The latter format is used in the Delphi method which is widely employed for obtaining consensus from panels of experts (Wikipedia 2020).

Analogous estimating: Previous projects conducted by the organization can be a good source of information for activity duration estimation. Analogous estimating generates a rough project duration estimate early on, which can be employed in the early phases of planning. We will mention here the so-called Square Root Rule (Rad 2002):

$$PD_p = PD_b\left(C_p/C_b\right)^{1/2}$$

where PD_p is the project duration of the proposed project to be estimated, PD_b is the known project duration of the benchmark project, C_p is the estimated cost of the proposed project, and C_b is the cost of the benchmark project. Note that this approach assumes that a cost estimate for the proposed project has already been obtained. The benchmark project used should be a relatively recent, representative project with similarities to the proposed project. More than one benchmark project might be employed to estimate the project duration.

If there are, say, two cost values estimated for the proposed project, then the square root rule can be applied twice using each of the cost estimates. Then the two duration estimates obtained can be finalized by simple averaging, weighted averaging, or by some other means of assessment. Example 3.5 below demonstrates this approach.

Example 3.5 Consider a proposed project to build a house with 180 m^2 of floor space and 5 rooms. We will employ two cost estimates for the proposed house project obtained under analogous cost estimating in *Example 3.6*: \$130,119 and \$140,818. The benchmark project has 120 m^2 of floor space with 3 rooms, a cost of \$96,000, and a duration of 28 weeks.

Solution

Using the first cost estimate as the index, we get:

$$\mathrm{PD_p} = 28(130{,}119/96{,}000)^{0.75} = 35.17 \text{ weeks.}$$

Using the second cost estimate as the index, we get:

$$\mathrm{PD_p} = 28(140{,}818/96{,}000)^{0.75} = 37.32 \text{ weeks.}$$

Taking the simple average, 36.25 weeks can be used as a rough initial estimate for the proposed project's duration.

Analogous estimating can be refined further. Projects implemented under similar conditions and environments might allow durations to be estimated based on comparative assessment considering factors such as technological developments, learning accomplished in that area in the organization, and changes in the human, technological and information resources available to the organization. This approach to cost and activity duration estimation underlines the importance of systematic accumulation of historical data and information by the organization for future use and for organizational learning purposes.

Parametric estimating: This approach involves statistical analysis based on historical data available in the technical literature as well as data accumulated in the organization. In a parametric model, project (activity) duration is treated as a dependent variable which is related to a number of independent variables. Parametric estimation can be applied to the whole project as well as to sub-projects or activities. Estimates obtained using historical project data can be scaled with one or more multiplicative factors to adjust for changes in technology and methods and in labor and material requirements of the activity under consideration. Furthermore, as more data becomes available during the project implementation, more accurate estimates might be calculated using the same model.

Labor productivity-based estimation of activity durations: Another quantitative method involves determining activity duration by multiplying the amount of work performed by the number of labor hours per unit of work. The number of labor hours per unit of work is the labor productivity and is closely related to the technology applied and resources employed. An example of labor productivity would be 30 m^2 of tile set per man-day. Hence, if the tile setting activity requires 120 m^2 of tile set, the duration of this activity would be estimated as 4 man-days for 1 worker. The actual duration specified would require a decision on the number of workers to be assigned to this activity. If automated machinery executes an activity, the duration would be determined by the processing rate of the automated machinery. Whether manual or automated processing, a parallel can be drawn with the standard times employed in manufacturing.

Three-point estimating: Since all projects are subject to various uncertainties. Uncertainty around the activity durations has been an active area of research where activity durations are treated as random variables (Hazır and Ulusoy 2020). One of the major reasons for this analysis is its impact on project duration, which is the most widely employed performance measure. Three-point estimating approach is based on the premise of an underlying probability distribution governing the activity durations, and is used to improve the accuracy of single-point estimates, providing estimates of both the expected value for the activity duration and its variance. The uncertainty around the activity durations will be covered later in detail in Chaps. 6 and 12.

These estimates are:

(i) The most likely estimate representing the duration estimate expected to occur under normal conditions (tM);
(ii) Duration estimate representing the best-case (optimistic) scenario for the activity (tO);
(iii) Duration estimate representing the worst-case (pessimistic) scenario for the activity (tP).

Depending on whether the distribution of the actual duration is assumed to be *triangular* or *Beta, the expected activity duration* (tE) is calculated as follows:

$$\text{tE (Triangular distribution)} = (\text{tO} + \text{tM} + \text{tP})/3,$$

$$\text{tE (Beta distribution)} = (\text{tO} + 4\text{tM} + \text{tP})/6.$$

Using a three-point estimation with an assumed distribution requires more effort than the single point estimates above, but an expected value and a range around it are obtained in return. Three-point estimation assuming a Beta distribution is also employed in PERT for handling the uncertainty in activity duration in Chap. 6.

Bottom-Up Estimating: The basic idea of bottom-up estimating is to obtain the estimates for the lowest level of the WBS and then progressively aggregating these values to obtain the durations for WPs/activities each time at a higher level of the WBS.

Documentation: PMI (2017) cites the following requirement for the supporting documentation of duration estimates. "Regardless of the level of detail, the supporting document should provide a clear and complete understanding of how the duration estimate was derived." This can be considered as a minimum requirement.

The supporting documentation should include the following: (*i*) How the estimate was developed; (*ii*) assumptions made; (*iii*) constraints considered; (*iv*) identified risks taken into account; and, (*v*) if known, the range around the estimate and the confidence level of the estimate, and how these were estimated.

The duration estimation for an activity obtained through the use of one or more of the tools briefly explained above results in a single point estimate. It is meant to provide an estimate for the duration expected to be realized under normal conditions. But the uncertainties and risks prevailing during the implementation of the project might force deviations from this estimate. In cases, where it is concluded, to make up for this contingency (for example, if the three-point estimate approach is employed, the range resulting might be assessed to be too large), an allowance or reserve might be put aside as part of the budget. The reserve can be for a particular activity or for the whole project. In the latter case, as additional data and information become available during the progress of the project, the reserve can be monitored to reflect the current assessment of uncertainty and risks.

3.5 Cost Management

In this section, we will deal with the components of cost management, specifically the principal categories of cost estimation, documentation, types of project costs and determination of activity cost for activities realized. A further integral part of cost management is cost control, which will be treated in detail in Chap. 10.

In the following section, we will try to define various types of costs, which will be useful in estimating the activity costs and building mathematical models involving costs.

3.5.1 Tools for Estimating Activity Cost

There are several tools employed in practice to estimate the costs in a project. Some major ones will be presented below based on PMI (2017). As in the case of estimating activity durations, here also detailed knowledge of the contents of the activity and the manner in which it will be executed will contribute to the accuracy of the estimates.

In the following, we will cover the basic tools already been explained in Sect. 3.4.1 for the case of estimating costs.

Expert judgment: This approach is probably the most common approach to estimate not only durations but costs as well. It is good practice to acquire estimates from more than a single expert. The usual practice is to arrive at a final value -might be defined within a range- through iterations of estimations provided by the experts at each iteration. How to conduct this group decision-making process depends on the decision environment such as time available and relative importance and accuracy or robustness of the data to be generated and also, on the relative status of the experts among themselves and their attitudes towards negotiation and compromise.

Analogous estimating: As suggested earlier, analogous estimating is the simplest form of estimating based on the use of previous projects conducted by the organization. It generates early on cost estimate for the overall project at the 1.0 level of the CBS, which is a rather rough estimate to be employed in the early phases of planning. The equivalent of the Square Root rule for costs is the Three-quarters Rule (Rad 2002):

$$C_p = C_b\left(S_p/S_b\right)^{3/4}$$

where C_p is the cost of the proposed project to be estimated, C_b the known cost of the benchmark project with known size or capacity S_b, and S_p the size or capacity of the proposed project.

The benchmark project should be a relatively recent and representative project that is similar to the proposed project. More than one benchmark project might be employed, and the results be assessed to estimate the project duration.

It is good practice to use, if possible, more than one index for estimating the cost of the proposed project such as the number of rooms or the floor space (m^2) in a house project. The cost estimates obtained can be finalized by simple averaging, weighted averaging, or by some other means of assessment.

Example 3.6 Consider a proposed house project with 180 m^2 floor space with 5 rooms. The benchmark project has 120 m^2 floor space with 3 rooms and has a cost of $96,000.

Solution

Using first the floor space as the index, we get:

$$C_p = 96{,}000(180/120)^{0.75} = \$130{,}119.$$

Using the number of rooms as the index, we get:

$$C_p = 96{,}000(5/3)^{0.75} = \$140{,}818.$$

If the decision-maker decides to use simple averaging, then she would accept $135,469 as a first rough estimate for the cost of the proposed house project.

As *Example 3.5* and *Example 3.6* show us, analogous estimating is indeed a rough estimate useful in the early phases of the planning process.

Projects implemented under similar conditions and environments might allow for cost estimating based on comparative assessment accounting for changes in relative unit labor and material costs, inflation and foreign exchange rates, and changes in labor productivity through learning and technological developments in the organization such as automation and digitalization. Once again the importance of systematic accumulation of historical data and information for future use and for organizational learning purposes becomes apparent.

Parametric estimating: Parametric models to estimate costs are the same in principle as those used for estimating durations. Statistical analysis based on historical data is applied with cost as the dependent variable determined by one or more independent variables. Project organizations should develop their own proprietary parametric estimate models. In such a case, the models developed need to be regularly evaluated, validated, calibrated, and customized for accuracy and appropriateness (Rad 2002).

A parametric model calculates the output variables based on the input variables using sets of input and output data. The input variables for a construction project would be, for example, project type, frame material, exterior material, roof type, ground conditions, desired floor space, equipment types and the output variables would be design cost, structure cost, size and cost of equipment, crew size, labor cost, duration of structure construction, duration of equipment installation, and project duration (Rad 2002).

Labor productivity-based estimating: The cost of a WP or activity would be determined as the sum of the direct cost and the overhead cost. The direct cost is the sum of direct labor cost and direct material cost. Let us continue with the example given for the duration case: Consider a tile setting activity requiring 120 m^2 of tile laid and let the labor productivity be 30 m^2 of tile set per man-day. Hence, the duration of this activity would be 4 man-days for 1 tile setter. Let the unit direct cost of labor (tile setter) be \$160 per man-day and the unit direct material cost (tile + auxiliary material) be \$75 per m^2. Then the direct cost estimate is $4 \times 160 + 120 \times 75 = \$9640$.

In case automated machinery executes an activity, then the cost would be determined by the operating cost such as energy cost and the overhead cost including maintenance cost and the depreciation cost of the automated machinery.

Three-point estimating: Similar to the approach based on the randomness of the activity durations, three-point estimating for the activity cost is based on the randomness around it. Three-point estimates are requested from the experts to arrive at the expected value for the activity cost basing it on the assumed underlying probability distribution around the activity cost. These estimates are:

(i) The most likely estimate representing the cost estimate expected to occur under normal conditions (cM);
(ii) cost estimate representing the best-case scenario for the activity (cO);
(iii) cost estimate representing the worst-case scenario for the activity (cP).

Depending on whether the assumed distribution covering the range of the three estimates is triangular distribution or Beta distribution, *the expected activity cost* (cE) is calculated as follows:

$$\text{cE (Triangular distribution)} = (\text{cO} + \text{cM} + \text{cP})/3,$$

$$cE\ (\text{Beta distribution}) = (cO + 4cM + cP)/6.$$

As suggested earlier, we put more effort to apply this tool but in return we obtain besides an estimate also for its variance. Three-point estimation with Beta distribution is employed in PERT Costing when dealing with uncertainty in activity costs (see Sect. 6.5).

Bottom-Up Estimating: The basic idea of bottom-up estimating is to obtain the estimates for the lowest level of the WBS and then aggregating these values to obtain the costs for WPs/activities each time at a higher level of the WBS.

Documentation: Proper documentation is an essential component of cost management. PMI (2017) cites the following requirement for supporting documentation of cost estimates. "Regardless of the level of detail, the supporting document should provide a clear and complete understanding of how the cost estimate was derived." This should be considered as a minimum requirement.

As suggested in the case of estimating the activity durations, the supporting documentation should include the following: (*i*) How the estimate was developed; (*ii*) assumptions made; (*iii*) constraints considered; (*iv*) identified risks taken into account; and, (*v*) if known, the range around the estimate and the confidence level of the estimate.

The cost estimation for an activity obtained through the use of one or more of the methods briefly explained above results in a single point estimate. It is meant to provide an estimate for the cost expected to be realized under normal conditions. But the uncertainties and risks prevailing during the implementation of the project might force deviations from this estimate. In cases, where it is concluded, to make up for this contingency (for example, if the three-point estimate approach is employed, the range resulting might be assessed to be too large), an allowance or reserve might be put aside as part of the budget. The reserve can be for a particular activity or for the whole project. In the latter case, depending on the data and information becoming available during the progress of the project, the reserve can be monitored to reflect the current assessment of uncertainty and risks.

3.5.2 Types of Project Costs

Direct labor cost: The labor that directly contributes to the realization of an activity and can be directly and easily traced to that activity is the direct labor and and its cost is referred to as the direct labor cost. For example, let the activity be "painting the interior walls of the first floor of a house construction". The cost of painters executing this activity is a direct labor cost. It can be calculated by multiplying the direct labor rate paid to the painters by the total number of hours the painters spent executing this activity.

Direct material cost: Materials that are employed for realizing an activity are direct materials and the cost associated with these materials is the direct material cost. Consider, for example, the painting activity above. All the paint and thinner and other related material put on the wall for the realization of this activity would be the direct material cost.

Project overhead cost: This cost category includes all the costs incurred in the management of that project, except the direct labor cost and direct material cost, that can be traced to that project alone. Indirect material cost, indirect labor cost, overtime payments, repair and maintenance costs and depreciation of facilities and equipment used for the realization of the project, quality control, rents for facilities used in the project, taxes, security, travel, heating, ventilating, and lighting are major components of the project overhead cost.

General overhead cost: General overhead costs should be distinguished from the project overhead cost. Cost items involved in general overhead costs are not directly attributable to the project. They cover the costs beyond the direct cost accrued in the management of the overall organization. General overhead costs include costs like general administrative costs, research and development (R&D), selling and marketing, and financing costs. General overhead costs are distributed over all the projects of the organization in proportion subject to the organization's cost accounting procedures.

3.5.3 Activity-Based Costing

When determining activity costs, traceability becomes a central issue. Consider, for example, a WP in the WBS defined as the pouring of the concrete of the first floor in a construction project. Water, aggregate (rock, sand, or gravel) and cement are the basic ingredients of concrete and are all direct materials. Their costs are classified under direct material cost. Labor involved in pouring the concrete is direct labor and its cost is classified as direct labor cost. In addition, we have overhead costs such as indirect labor, indirect material, depreciation, insurance, and inspection. In order to calculate the cost of the WP realized, namely the cost of pouring of the concrete of the first floor, we need to know the amount of each direct material and their unit costs, the duration of work by each type of direct labor and their unit costs and the overhead cost to be allocated to this activity.

The traceability of the different cost components incurred to the WPs/activities can be challenging. For this purpose, data collection systems need to be designed and implemented to capture data particularly on the activity, duration, and costs involved. For example, a direct material cost item can be charged to a WP if the information on the code of that WP, the code of the direct material, and the amount of direct material used during the realization of that WP can be captured. Obviously, the same process applies to direct labor as well.

The project overhead cost items, which are not traceable, need to be charged to associated WPs/activities. Raz and Elnathan (1999) claim that typical overhead ratios in projects are in the range of 20 to 40% of direct costs and include the costs of functions such as marketing, purchasing, personnel management, general management, etc. Hence, the proper allocation of overhead costs to WPs/activities is crucial for accurate pricing of projects, which has a direct impact on the competitiveness of the organization. A more recent methodology for the allocation of overhead cost items is the *activity-based costing* (ABC) (see, e.g., Cooper et al. 1992). In ABC, so-called *cost drivers* are employed for allocating project cost overhead items to individual WPs/activities. Cost drivers can be conceived as factors that affect the consumption of resources. In traditional cost models, volume or volume related proxy cost drivers are employed such as the number of direct labor hours or number of units. In ABC, on the other hand, the set of potential cost drivers are directly related to the activities performed (Raz and Elnathan 1999). For example, for allocating procurement cost a typical cost driver would be the number of requisitions; or for inspection cost, it would be the amount of work accomplished; or for schedule planning or control it would be the project duration. An example of the application of ABC to component-based software development is provided by Fichman and Kemerer (2002).

The process of allocating project overhead costs can be stated as follows:

(i) Identify project overhead cost items with relatively high cost.
(ii) Assign respective costs to cost items.
(iii) Identify a cost driver for each cost item.
(iv) Allocate each cost item's cost to the associated WPs/activities using the cost driver.

Example 3.7 Determine the cost of a certain WP2.3 in a project. The direct labor cost and the direct material cost for WP2.3 are reported as $22,280 and $86,470, respectively. Data for overhead cost items is provided in Table 3.3.

Solution

The relatively larger cost items among the project overhead cost items, the associated cost drivers, the cost driver amounts for the project and for WP2.3, the cost driver cost for the project are given in Table 3.3.

Based on these data the cost driver cost for WP2.3 is calculated and reported in the last column of Table 3.3. As an example, let us go through the calculation of the energy cost for machinery:

$$\text{Energy cost for machinery} = (925/27{,}750) \times 425{,}250 = \$14{,}175.$$

The sum of the last column of Table 3.3 is $27,194. Now we can calculate the realized cost of WP2.3 as follows:

$$\text{The realized cost of WP2.3} = 22{,}280 + 86{,}470 + 27{,}194 = \$135{,}944.$$

Table 3.3 Project overhead cost data for WP2.3

Overhead cost item	Cost Driver (CD)	CD amount for the project	CD amount for WP2.3	CD cost for the project ($)	CD cost for WP2.3 ($)
Energy cost for machinery	Machine hours	27,750 h	925 h	425,250	14,175
Inspection cost	Inspection hours	120 h	3 h	126,000	3150
Meals and related costs	Labor hours[a]	164,000 h	5400 h	245,800	8093
Purchasing	Purchase requisitions	144 purchase requisitions	12 purchase requisitions	21,312	1776

[a]The labor hours for the meals and related costs is taken here as the sum of direct and indirect labor

Whether general overhead cost should be included in the WP/activity cost together with the project cost is a decision to be made by the organization. The organization might prefer to ignore the allocation of general overhead cost items to projects if the level of project activities is much smaller than the level of process type activities within the organization.

This section on estimating the costs of WPs/activities and determining their realized costs underlines clearly the importance of preparing the WBS properly, which has been emphasized earlier in Sect. 3.2 as well. The level of detail is set by the WBS. The relationship between this level of detail and the burden it puts on the managers to estimate the time and resource requirements of these WPs/activities accurately has to be taken into account when planning the level of detail of the WBS.

In the following section, we will deal with data, information, and knowledge management in project management. In the age of rapidly developing data acquisition technologies and data analytics, we would expect improved cost estimation and more precise determination of cost realizations. The energy cost for machinery in Example 3.7, for example, can be captured at each machine, which would result in more accurate energy consumption values.

3.6 Data, Information, and Knowledge Management

Reliable, accurate, and timely data is a great asset for successful project management. Having a data collection system in place to obtain such data is essential for successful project management. Data is collected for four purposes: (*i*) Planning, (*ii*) monitoring and controlling, (*iii*) evaluation and termination, and (*iv*) archiving and post-project analysis. The data collected constitutes the input into the decision-making and reporting processes throughout all phases of the project. The major decisions to be taken for this purpose are:

(i) Which data to collect?
(ii) With which frequency to collect data?
(iii) With which tools to collect and integrate data?
(iv) With which tools and how to analyze data?

Before focusing on their applications to project management, it is essential to distinguish the three related terms: data, information, and knowledge. Davenport and Prusak (1998) define data as "a set of discrete, objective facts about events", and information as "a message, usually in the form of a document or an audible or visible communication". Hence, information is obtained by processing data and extracting a message from it. On the other hand, knowledge goes beyond processing to assimilate the message implicit in the data, interpreting, synthesizing or re-constructing it; hence, it involves significant elements of learning and experiencing. Davenport and Prusak (1998) define knowledge as follows: "Knowledge is a fluid mix of framed experience, values, contextual information and expert insight that provides a framework for evaluating and incorporating new experiences and information. It originates and is applied in the minds of knowers. In organizations, it often becomes embedded not only in documents or repositories but also in organizational routines, processes, practices and norms."

The above given definition implies that knowledge is not always explicit or codified. It has also a tacit component (Polanyi 1966) that is transmitted by observing or interacting rather than by codification in a written form. Both the codified and the tacit knowledge generated in a project are of central importance, not only for the project's own success but also for that of subsequent projects. Even though projects are generally unique, many individual activities are repetitive or subject to similar technical requirements or standards. Therefore, knowledge management is an indispensable part of project management.

Managing knowledge in a project means recording data, storing information, and implementing effective mechanisms to share, use and synthesize them. Some of these tasks, especially post-project reports or "lessons learnt" memos are frequently perceived as non-productive tasks by some in the organization (Leseure and Brookes 2004). Nevertheless, they are vital for making more effective decisions, performing appropriate qualitative and quantitative analyses, learning from previous experiences and mistakes, and constructing a construct, a common knowledge base that helps PMs in their future projects.

3.6.1 Which Data to Collect?

All the mathematical models formulated in this book require some data as input. Project planning models require data on activities, resources, and system constraints such as activity durations, activity costs, precedence relationships, resource requirements, project deadline or budget. The commonly used guideline (PMI 2017) defines a framework for data collection and information requirements that focuses more on documents and standards from a project information systems and knowledge management perspective.

According to PMI (2017), organizational knowledge repositories for storing and retrieving information include but are not limited to:

- *Configuration management knowledge*: Versions of software and hardware components and baselines of all performing organization standards, policies, procedures, and any project documents;
- *Financial data*: Information such as labor hours, incurred costs, budgets, and any project cost overruns;
- *Historical information and lessons learned*: For example, project records and documents, closure information and documentation, the results of previous project selection decisions and previous project performance information, and information from risk management activities;
- *Issue and defect management data*: Issue and defect status, control information, issue and defect resolution, and action item results;
- *Data for metrics*: Used to collect and make available measurement data on processes and products; and
- *Project files from previous projects*: For example, scope, cost, schedule, and performance measurement baselines, project calendars, project schedule network diagrams, risk registers, risk reports, and stakeholder registers.

The last item in the above list constitutes an important component of organizational knowledge and serves as an archive to support post-project analysis (see Sect. 1.5) for improved planning and management of projects.

We emphasize that data collection and information storage is a continuous process over the entire project life cycle. Some data types are usually static, such as the activity precedence relationships defined by technical requirements; however, in many cases, data types are dynamic. For example, activity costs change with price changes in the market. According to Vanhoucke et al. (2016), dynamic project data covers all types of data needed to model and analyze the project progress. These data need to be collected during project execution to track and assess the performance and to support the PM in his decision making, especially in predicting the final project duration and cost, and taking preventive/corrective actions.

3.6.2 With Which Frequency to Collect Data?

As projects are undertaken in dynamic environments where events evolve continuously and the processes and environmental factors change over time, do the PMs need to gather data continuously? More specifically, should data such as the money spent for the activities or their completion percentages be recorded continuously or at periodic intervals? What should be the frequency of data collection and recording?

There is no single answer to these questions, whose answers will vary across industries and projects. Data collection in some research projects, such as spacecraft design, can be extremely expensive, whereas for automotive designs it is relatively less costly. Even within a single project the ideal frequency of data collection can be different for each data type. For instance, *financial data* needs to be collected more frequently than the *data for metrics* defined in Sect. 3.6.1. On the other hand, a *cost/benefit analysis* is required for the frequency of data collection. The frequency can be increased until the incremental cost of collecting data more frequently outweighs the additional gains from improved analysis and decision making. Moreover, in a given project, decisions on data collection are contingent on project performance, monitoring and control needs. If the PM considers that project is behind schedule, s/he can decide to increase the monitoring and control efforts. At the same time, s/he might need to update and analyze data more frequently. Finally, available data collection technology has an impact on the frequency of data collection. Examples of such technology tools are cited in the next section.

3.6.3 With Which Tools to Collect and Integrate Data?

Data is collected through observations and recording derived from technical standards, historical records and expert opinions. Emerging technologies and managerial approaches offer new ways to gather and integrate data. Some examples could be given as follows:

- Global Positioning System (GPS) receivers provide real-time information about the location of expensive pieces of equipment for their availability check, transportation and scheduling; or real-time information about logistics, such as the materials needed at activity sites could be tracked and delivery times precisely estimated.
- Building Information Modeling (BIM) systems in construction (see, e.g., Eastman et al. 2008) allow construction engineers, among others, to access and work on different virtual models, integrate information coming from subcontractors, and perform simulations and impact analyses. An important notion in this regard is the interoperability of different systems and to support sharing information between project team members rapidly and accurately.
- Sensors and automatic monitoring tools enable gathering of precise data on the progress of the activities.
- Barcodes, QR codes, radio frequency identification (RFID) tags and readers, voice recognition and other scan and data storage technologies allow practitioners to capture, share and access data easily.
- Electronic libraries and digital sources such as smart phones are becoming more common ways to access and share the required data.
- Work and quality standards have been continuously developing in many areas. New focus groups among academicians and practitioners have been emerging. New communication technologies are making meetings and information sharing

easier and less costly among these focus groups, which support them in revising the existing standards or developing new ones more effectively.

3.6.4 With Which Tools and How to Analyze Data?

Once the data is gathered, it needs to be consolidated, integrated and analyzed. Data collection, storage and processing technologies have been rapidly developing. To analyze data, statistical methods have been used for decades. Today, in the era of big data analytics, data mining and machine learning techniques are used more and more (see Chap. 15, for application areas of big data analytics in project management). Data-intensive analysis tools of operations research, computer science, and artificial intelligence enable practitioners to perform complex analytical computations much faster (Vanhoucke et al. 2016). Consequently, PMs have many more opportunities to support their decision-making using data analytics and decision support systems (DSS).

DSS help analyzing data; provide analytical models and tools that contribute to the selection of alternatives (Shim et al. 2002). Project management problems, from selection to planning and control, are suitable application areas for DSS, as they are unstructured or semi-structured decision-making problems that usually require several alternative solutions to be considered (Hazır 2015). These DSS usually include a model base, which contains data analysis tools, simulation and optimization models, and data presentation and graphical interfaces.

While analyzing the data and making inferences, there are several limitations that PMs need to keep in mind. Rees (2015) lists these limitations in the context of distribution fitting and simulation analysis. Their warnings, which are given below, are important to consider when using any tool for project data analysis.

- Business systems are dynamic; processes can change over time, so that data might not be comparable, even though they measure the same process. To give an example, over time companies might have improved the efficiency or quality of the processes, or restructured the contractual conditions, the market environment might have changed.
- Data might need to be transformed or processed: To give an example, if data on commodity prices is studied over a long time window, inflation patterns could also be considered and adjustments or transformations might be performed accordingly.
- If data measured concerns a compound process or multiple processes, each process needs to be isolated separately.

3.7 Conclusions, Recent Developments, and Some Future Research Directions

In this Chapter, we have established a basis for project scheduling through the WBS and project network models which serve as our vehicles for the remainder of the topics in this book. The WBS represents the project deliverables and how the work to achieve them is organized, whereas the network models show the activities to be performed to create these deliverables and the temporal relations among them. While introducing this basis for project scheduling, we discussed the data requirements and the importance of data analysis and management in projects.

The claim put forward by Globerson (1994) that although the WBSs are the backbones of project planning, the literature has not given enough importance to the proper preparation of WBS appears still to be the case today. Project teams need to consider many factors such as the organizational structure, managerial experience, industrial facts, and project characteristics. Preparation of the WBS is a critical and challenging task. More research on WBS design will contribute to project management literature. To give examples, the trade-off between work package size and managerial costs could be studied. Testing the impact of WBS preparation on project success, exploring the industrial differences are also promising research directions (Hall 2016).

Availability of data is one of the more important factors that affect the success of managerial practices. This concerns the planning, control, and risk analysis practices of project managers. Even though big data and data analytics are closely related to project life cycle management, research on this topic is scarce. Researchers have lately started to investigate big data applications and their impacts in project, program, and portfolio management.

Managers can benefit from the accumulation of knowledge through the analysis of big data sets. At this point, there is a need for studies that give insights to organizations and managers on how to apply big data tools and technologies to create value. Analysis of level of coordination and information sharing among stakeholders and its impact on project success is a related and interesting area, as well.

Regarding the analysis of project data, availability of DSS is important to support managers in decision making. Software support is also indispensable. Developing DSS and user-friendly software on many aspects of project life-cycle management, project bidding to post-project analysis, remains an interesting research and practice area.

Exercises

3.1 The following list of activities is relevant to almost any project. Identify the phase (planning, organizing, implementing, and controlling) in which each is typically performed, and order them in the correct sequence.

- Developing the network structure
- Selecting participating organizations

- Developing a calendar
- Developing corrective plans
- Executive activities
- Developing a budget
- Designing a project
- Recommending improvement steps
- Monitoring actual performance
- Allocating resources to activities
- Managing the configuration
- Developing the WBS (Work Break-down Structure)
- Estimating the LCC (Life-Cycle Cost)
- Getting the customer's approval for the design
- Establishing milestones
- Estimating the activity duration

3.2 Consider the following auditing project in an Accounting Company with its work packages and their precedence relations given in the table below. Develop a WBS for this project and present it in list format.

Work package no	Work packages	Predecessors
1	Planning 2016 audit	–
2	Preliminary discussions with the company	1
3	Review internal control	2
4	Compliance tests	3
5	Yearend procedures	4
6	General audit procedures	9
7	Audit cash	9
8	Audit receivables	7
9	Observation of inventory	5
10	Inventory pricing	9
11	Audit other current assets	6
12	Audit fix assets	11
13	Obtain schedule of liabilities	10
14	Audit capital stock and R/E	11
15	Audit sales	8
16	Audit cost of goods sold	10
17	Audit other revenues and expenses	15, 16
18	Lawyer's letter	27, 32
19	Management's letter	12, 14
20	Subsequent review	12, 14
21	Prepare financial statements	18, 19, 20
22	Prepare tax returns	18, 19, 20
23	Partner/manager review	21, 22
24	Mail confirmations	13

(continued)

25	Test pension plan	13
26	Vouch selected liabilities	13
27	Test accruals and amortization	26
28	Process confirmations	24
29	Reconcile interest expense to debt	25, 26
30	Verify debt restriction compliance	29
31	Investigate debit balances	28
32	Review subsequent payments	30, 31

3.3 A chemical installation is to be constructed within a factory. The work packages and the precedence relations between them have been determined and reported in the table below. Develop a WBS for this project and present it in tree format.

Work package no	Work packages	Predecessors
1	Clear site	–
2	Excavation and foundations	1
3	Install support steelwork	2
4	Install stairways	3
5	Install platforms	3
6	Install handrails and platform angles	3
7	Install equipment	4, 5, 6
8	Calibrate equipment	7
9	Install prefabricated ductwork	7
10	Install prefabricated process pipework	4, 5, 6
11	Make process pipework connections to equipment	8, 10
25	Install pressure gauge on process pipework	11
12	Install prefabricated service pipework	4, 5, 6
13	Make service pipework connections to equipment	8, 12
26	Install pressure gauge on service pipework	13
14	Install instrumentation	4, 5, 6
15	Make instrument connections to equipment	8, 14
16	Install electrical apparatus	4, 5, 6
17	Make electrical connections to equipment	6, 8
18	Lay equipment	8
19	Lay pipework	25, 26
20	Lay ductwork	9
21	Testing of the complete equipment	18, 19, 20
22	Painting	21
23	Clean up	22
24	Submit post-project report	23

3.4 A residential house is to be constructed in a suburb of Istanbul. The following set of work packages and the precedence relations between them have been determined and reported in the table below. Develop a WBS for this project and present it in tree format.

Work package no	Work packages	Predecessors
1	Placement	–
2	Temporary plumbing	–
3	Temporary electricity	–
4	Foundation 1	1, 2, 3
5	Hardening 1	4
6	Foundation 2	5
9	Rough plumbing	10
10	Hardening 2	6
11	Assembling scaffolding	9, 10
12	Framing	11
13	Exterior woodwork	12
14	Roofing	12
15	Plumbing (gas)	12
17	Wiring	12
18	Interior woodwork	14, 15, 17
19	Sheet metal work	13
21	Wire mesh and exterior plasterer	19
23	Exterior painting	21
24	Finish plasterer	18, 23
25	Doors	18
26	Interior painting	18
27	Gutter	24
28	Tiles	26
30	Disassembling scaffolding	27
31	Balcony	30
32	Fusuma partition	25, 28
33	Carpet	28
34	Electric appliance	28
36	Boiler	33
37	Kitchen	33
39	Cleaning	32, 34, 36

3.5 Devise an algorithm specifying the order for aggregating the cost using the cost account system given in Sect. 3.2.5.2.

3.6 The method to estimate the duration of a summary activity is given in Sect. 3.3.6.2. Explain the reasoning behind this method suggested.

3.7 Draw the activity-on-arc network for the following set of activities with the precedence relations as indicated.
- A, B, C are concurrent and begin at the start of the project.
- D and E are preceded by A.
- F and G are preceded by B and D.
- C precedes H, K, I, J.
- G precedes K and J.
- E and F precede H
- M is preceded by I, J, H.
- I and J precede L.
- M, K, L are terminal activities.

3.8 Consider the following set of precedence relationships. Draw both the AON and AOA project networks representing these relations.
- A, B, C can start simultaneously
- A and B precede D
- B precedes E, F, and H
- F and C precede G
- E and H precede I and J
- C, D, F, and J precede K
- K precedes L

3.9 Draw the AOA and AON project network for the following project description and apply the node numbering algorithm:
- O is the start of the job
- O precedes A and B
- A precedes G and H
- B precedes E, F, and C
- C, E, and G precede M
- F precedes G
- H precedes I
- M and I precede J
- J is the end of the job

3.10 Consider the following set of precedence relations.

Activity	Immediate predecessor
A	–
B	–
C	A
D	A
E	C
F	C, D
G	C, D
H	B, D, E
I	B, D, E, G
J	F
K	I, J
L	G, H, K

(a) Determine, if any, false (unnecessary) immediate predecessor relationship (s) exist.
(b) Draw the AOA project network.
(c) Draw the AON project network.

3.11 Consider the following precedence relationships defining a project network.

- Activities C and D are preceded by activity A.
- Activities E and F are preceded by activity B.
- Activities G and H are preceded by activity D.
- Activity G is preceded by E.
- Activity I is preceded by activities C and G.
- Activity K is preceded by activities E, F, and G.
  (a) Draw the AOA project network. Apply correct node numbering for easier CPM calculations.
  (b) Redraw the arrow diagram using the AON convention.

3.12 Draw the AOA and AON project networks employing the following precedence relations.

- A, B, C, the first activities can start simultaneously.
- D, E, F start after A is completed.
- I and G start after B and D are completed.
- H starts after both C and G are completed.
- K and L succeed I
- J succeeds both E and H
- M and N succeed F but cannot start until E and H are completed
- O succeeds M and I.
- P succeeds J, L, and O
- K, N, and P are the terminal activities of the project.

3.13 Consider the project network data below.

Activity	Immediate predecessor	Duration (days)	Additional precedence relations
A	–	3	None
B	–	3	None
C	A	2	C can start earliest 3 days after A has been finished. C and E should start simultaneously
D	B	2	None
E	A	3	C and E should start simultaneously
F	C, D	3	F requires 2 days after H has been completed
G	B	7	The last 1 day work of G and I must be performed simultaneously
H	E	5	None
I	F, H	5	The last 1 day work of G and I must be performed simultaneously.

(a) Draw the activity-on-arc project network reflecting the additional precedence relations. Determine the project duration and the critical path. When determining early and late starts for the activities, assume that once an activity has been started, it has to be completed without interruption, i.e., we have non-preemptive scheduling. (You can indicate the early and late occurrences for the events on the nodes on your project network drawing.)
(b) Specify the generalized precedence relations employed in the above project network data in words.

3.14 Consider the following project network

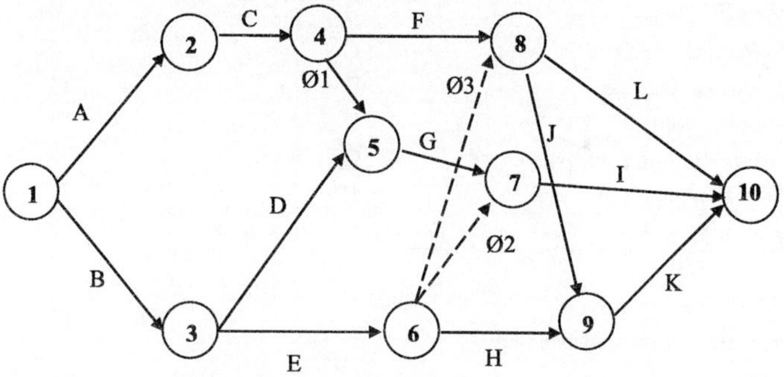

(a) Write the precedence relations among the activities
(b) Draw the AON project network.
(c) For AOA representation, represent on the project network the case where activity K cannot start earlier than 3-time units after activity J has finished.
(d) For AOA representation, represent on the project network the case where there is a minimum of 17-time units between the start of activity C and finish of activity D. Let the durations of C and D be 9- and 12-time units, respectively.

3.15 In Sect. 3.3.5, it is stated that all the four cases of GPRs with minimal time lags can be transformed into each other. Show that this is indeed so.

3.16 Write the associated inequalities for all four cases of GPRs with maximal time lags.

3.17 Show that $FF_{ij} \geq 0$ for all precedence pairs (i, j) implies that only FS_{ij} or SS_{ij} applies.

3.18 Consider the following pipeline renewal problem with the relevant data given in the table below.

(a) Draw the AOA representation of this project. Note that activity D can start only after activity C has been in progress for at least 15 days; activity J can start only after at least 2 days activity I has been completed; activity H can start at least 2 days after activity G has started; activity M can start only 1 day after activity L is finished.

Specify the node numbers such that an arc always goes from a smaller node number to a larger node number. Keep the number of dummy activities at a minimum.

	Activity description	Immediate predecessor	Duration (days)
Q	Lead time for preparations/planning	–	10
R	Make line available	Q	30
A	Measure and sketch	R	2
B	Develop material list	A	1
C	Procure pipes	B	30
D	Procure valves	B	45
E	Prefabricate sections	C	5
F	Deactivate the line	E	1
G	Erect scaffold	F	4
H	Remove old pipe and valves	G	6
I	Place new pipe	H	6
J	Weld pipe	I	2
K	Place valves	D, I	1
L	Fit up pipe and valves	K, J	1
M	Pressure test for valves and pipe	L	1
N	Insulate	M	4
O	Remove scaffold	N	1
P	Clean-up	O	1

(b) Suggest a WBS for this project.

3.19 Consider the following data for an AOA project network.

Activity	Activity description	Imm. Pred.	Duration (days)
A	Equipment installation	–	13
B	System development	–	10
C	Position recruiting	–	4
D	Equipment testing and modification	A	6
E	Manual testing	C	1
F	Job training	B, E	6
G	Orientation	D, F	7
H	System training	C	3
I	System testing	H	3

Draw the project network considering the following generalized precedence relations:
Activity B can be started only after 2 days activity A has been started.
Activity E can be finished only after 1 day activity B has been finished.
Activity D can be finished only after 8 days activity F has been started.

3.20 When managing a project, a Project Manager is involved, among others, in purchasing, planning and control, leadership and communication, and customer relations all resulting in overhead cost. Suggest a cost driver for each overhead cost item.

References

Ackoff, R. L. (1978). *The art of problem solving*. New York: Wiley.

Bartusch, M., Möhring, R. H., & Radermacher, F. J. (1988). Scheduling project networks with resource constraints and time windows. *Annals of Operations Research, 16*, 201–240.

Besikci, U., Bilge, U., & Ulusoy, G. (2013). Resource dedication problem in a multi-project environment. *Flexible Services and Manufacturing Journal, 25*(1–2, SI), 206–229.

Buchtik, L. (2010). Secrets to mastering the WBS in real-world projects. *Project Management Journal, 41*(4), 132.

Cooper, R., Kaplan, R. S., Maisel, L. S., Morrissey, E., & Oehm, R. M. (1992). *Implementing activity-based cost management: Moving from analysis to action*. Montvale: Institute of Management Accountants.

Davenport, T. H., & Prusak, L. (1998). *Working knowledge: How organizations manage what they know*. Boston: Harvard Business Press.

Dodin, B., & Elimam, A. A. (1997). Audit scheduling with overlapping activities and sequence dependent set-up costs. *European Journal of Operational Research, 97*(1), 22–33.

Eastman, C., Teicholz, P., Sacks, R., & Liston, K. (2008). *BIM handbook. A guide to building information modeling*. Hoboken: Wiley.

Elmaghraby, S. E. E., & Kamburowski, J. (1992). The analysis of activity networks under generalized precedence relations (GPRs). *Management Science, 38*(9), 1245–1263.

Fichman, R. G., & Kemerer, C. F. (2002). Activity-based costing for component-based software development. *Information Technology and Management, 3*, 137–160.

Globerson, S. (1994). Impact of various work-breakdown structures on project conceptualization. *International Journal of Project Management, 12*(3), 165–171.

Hall, N. G. (2016). Research and teaching opportunities in project management. In *Optimization challenges in complex, networked and risky systems* (pp. 329–388). Hanover: INFORMS.

Haugan, G. T. (2002). *Project planning and scheduling*. Vienna: Management Concepts.

Hazır, Ö. (2015). A review of analytical models, approaches and decision support tools in project monitoring and control. *International Journal of Project Management, 33*(4), 808–815.

Hazır, Ö., & Ulusoy, G. (2020). A classification and review of approaches and methods for modeling uncertainty in projects. *International Journal of Production Economics, 223*. https://doi.org/10.1016/j.ijpe.2019.107522.

Kerzner, H. (2009). *Project management: A systems approach to planning, scheduling, and controlling* (10th ed.). Hoboken: Wiley.

Kurtulus, I. S., & Davis, E. (1982). Multi-project scheduling: Categorization of heuristic rules' performance. *Management Science, 28*(2), 161–172.

Kurtulus, I. S., & Narula, S. C. (1985). Multi-project scheduling: Analysis of project performance. *IIE Transactions, 17*(1), 58–66.

Leseure, M. J., & Brookes, N. J. (2004). Knowledge management benchmarks for project management. *Journal of Knowledge Management, 8*(1), 103–116.

Moder, J. J., & Phillips, C. R. (1970). *Project management with CPM and PERT* (2nd ed.). New York: Van Nostrand Reinhold Co..

Neumann, K., Schwindt, C., & Zimmermann, J. (2003). *Project scheduling with time windows and scarce resources* (2nd ed.). Berlin: Springer.

PMI. (2004). *A guide to project management body of knowledge. PMBOK guide* (3rd ed.). Newtown Square: Project Management Institute.

PMI. (2006). *Practice standard for work breakdown structures* (2nd ed.). Newtown Square: Project Management Institute.

PMI. (2017). *A guide to project management body of knowledge. PMBOK guide* (6th ed.). Newtown Square: Project Management Institute.

Polanyi, M. (1966). The logic of tacit inference. *Philosophy, 41*(155), 1–18.

Rad, P. F. (2002). *Project estimating and cost management.* Vienna: Management Concepts Inc..

Raz, T., & Elnathan, D. (1999). Activity based costing for projects. *International Journal Project Management, 17*(1), 61–67.

Rees, M. (2015). *Business risk and simulation modelling in practice: Using excel, VBA and @ RISK.* New York: Wiley.

Shim, J. P., Warkentin, M., Courtney, J. F., Power, D. J., Sharda, R., & Carlsson, C. (2002). Past, present, and future of decision support technology. *Decision Support Systems, 33*(2), 111–126.

Vanhoucke, M., Coelho, J., & Batselier, J. (2016). An overview of project data for integrated project management and control. *Journal of Modern Project Management, 3*(2), 6–21.

Wikipedia. *Delphi method.* Accessed 29 June 2020.

Wognum, N., & Trienekens, J. (2015). The system of concurrent engineering. In J. Stjepandić et al. (Eds.), *Concurrent engineering in the 21st century* (pp. 21–49). Cham: Springer.

4 Deterministic Project Scheduling with No Resource Constraints

Learning Outcomes
Upon successful completion of this Chapter, the reader will be able to:

1. Determine the critical path and critical activities for both Activity-On-Arc (AOA) and Activity-On-Node (AON) type project networks.
2. Determine the various types of slacks and employ them in further decision-making.
3. Formulate the project scheduling problem as a Linear Programming (LP) problem.
4. Determine a schedule that maximizes the Net Present Value (NPV) of cash flows arising from a project.
5. Keep the project plan current.

4.1 Introduction

The deterministic project scheduling problem under no resource constraints is the most basic project scheduling problem. The environment in which the project is planned and executed is assumed to be deterministic, and resource availability to be unlimited. Hence, the resource requirements of the activities can be ignored when scheduling the activities.

We will approach this project scheduling problem as an optimization problem with one or more objective functions and constraints.

When managing the project, a project manager (PM) might consider a single objective or several objectives addressing aspects of duration, cost, and quality simultaneously. These objectives might have different importance rankings or weights, reflecting the priorities of the PM, and might change over the course of the project's implementation. In this chapter, we will focus on two main groups of objectives: time-based (temporal) and financial.

G. Ulusoy, Ö. Hazır, *An Introduction to Project Modeling and Planning*, Springer Texts in Business and Economics, https://doi.org/10.1007/978-3-030-61423-2_4

Time-Based Objectives A frequently adopted temporal objective is the minimization of the project duration, i.e., the makespan. The makespan is defined as the time elapsing from the start of the first activity to the completion of the terminal one, i.e., the termination of the entire project. If a target date is set for the completion of the project, then another popular temporal objective is the minimization of tardiness, i.e., the amount of time by which the project exceeds the target date. A bonus might be awarded for finishing early and a penalty imposed for exceeding the target date.

Financial Objectives The minimization of the project cost or the maximization of the profit accrued from the project are common financial objectives. In projects with a substantial duration, the Net Present Values (NPVs) of these quantities are considered in order to take into account the time value of money as well as the timing of the cash flows over the project's duration (see, e.g., Park 2007).

A further example of a financial objective is minimizing the maximum cash outlay in any period throughout the project. This objective does not take into account the time value of money but can be of relevance in case of limited access to operating capital. Another example would be minimizing the payback period, defined as the amount of time it takes until the receipts (cash inflows) accrued over time exactly match the payments. This particular objective also does not take the time value of money into consideration.

Constraints The major set of constraints to be considered here is the *precedence relations* among the activities.

In this chapter, the minimization of the project duration and the maximization of the project profit will be addressed in two separate mathematical models. We first present a Linear Programming (LP) formulation of the deterministic unconstrained project scheduling problem with the objective of minimizing the project duration. This is followed by a binary (0–1) integer programming formulation for maximizing the NPV of cash flows throughout the project.

4.2 Critical Path and Its Determination

The project duration is determined by the longest path from the start node to the terminal node of the project network, called the *critical path*. Note that there can be more than one critical path present on a project network. The concept of critical path was first introduced by Kelley Jr and Walker (1959) and later developed by Kelley Jr (1961 and 1963). The activities on the critical path are called the *critical activities* since a delay in any critical activity results in a delay of the same amount in the project completion. Hence, the critical activities become *bottlenecks* of the project and must be implemented without any delay. Activities that do not lie on a critical path are said to be *non-critical*. By implication, PMs must monitor the progress of the critical activities closely and devote a large portion of their efforts to the planning, execution, and control of such activities. In the terminology of the Theory of Constraints (see, e.g., Goldratt 1990), a single constraint – the critical path – is the bottleneck to completing the project on time. Hence focusing management resources on the critical activities would support this objective through relaxing this constraint. On the other hand, directing the resources to improve the execution of non-critical activities would not contribute as much to completing the project on time. Another analogy could be drawn with the ABC analysis of inventory management (see e.g., Nahmias 2009, 283–284).

If there is only one critical path, then all critical activities are equally critical. When there is more than one critical path, some activities might be more critical than others. If, for example, there are 4 critical paths in a project network and a particular critical activity is located on 3 of these, then its criticality is higher than another critical activity which is located on fewer of the critical paths (Elmaghraby 2000).

An activity that is not on the critical path can be delayed for some time without delaying the project completion. It is important to know the amount of this additional permissible delay, called the *slack*, and how the delay of a non-critical activity affects the succeeding non-critical activities. The amount of the available slack we consume by allowing delays in these non-critical activities can impose different limitations on the start and finish times of the succeeding activities. Different types of slack are distinguished based on the type of these limitations. In particular, delaying a non-critical activity might create a new critical path. Further planning issues like the time-cost trade-off and resource allocation problems and associated sensitivity analyses depend on the flexibility that we have with regard to scheduling the non-critical activities.

4.2.1 Critical Path Method

In this section, the Critical Path Method (CPM) will be introduced for both AOA and AON project networks. In the following, for the sake of simplicity, the start time of the project is set equal to zero without loss of generality. The purpose of the algorithm is to identify the critical paths(s) and compute the slacks associated with each activity in the project.

4.2.1.1 Critical Path Method on Activity-on-Arc Type Project Networks

Assume that there are n activities (arcs) and m events (nodes) in the project network under consideration. In order to apply the algorithm, the event nodes need to be numbered such that all arcs are drawn from a lower-numbered node to a higher-numbered one. This can be accomplished using the node numbering algorithm given in Sect. 3.3.3.2. The critical path, i.e., the longest path, in a project network can be calculated using the recursive relationships of dynamic programming. At each step of the recursion, an additional node of the network (i.e., an event) is included in the partial solutions generated so far. The node included may or may not be on the critical path. The method of dynamic programming has its basis in the *Principle of Optimality* proposed by Richard Bellman (1952), which can be stated as follows (Denardo 1982):

"***The Principle of Optimality*** An optimal policy has the property that whatever the initial node (state) and the initial arc (decision) are; the remaining arcs (decisions) must constitute an optimal policy with regard to the node (state) resulting from the first transition."

Application of the *Principle of Optimality* yields a set of recursive relationships leading to the longest path on the project network. The execution of these recursive relationships in increasing node numbers is called the *forward pass*, and computes the earliest possible occurrence time of each event in the network as well as the project duration. The latest occurrence times are obtained by employing the recursive relationships in decreasing node numbers, which is called the *backward pass*. These give the latest time at which an activity can complete without causing a delay in the project completion.

The recursive relationships for the forward pass are given as:

$$E_1 = 0 \tag{4.1}$$

$$E_j = max_i\{E_i + d_{ij}\} \quad j = 2, \ldots, m \tag{4.2}$$

where E_j denotes the earliest occurrence time of event node j, d_{ij} the duration of activity (i, j), and the maximization is over all event nodes i from which an activity $(i, j) \in A$ is incident into event node j, where A is the set of arcs of the project network.

Note that E_i corresponds to the longest path from the start node 1 to event node i. Hence, the forward pass builds the solution by extending the partial path at each recursion by adding another event node not yet covered until the terminal node m is reached and the earliest occurrence time for the terminal node, E_m, obtained. The earliest occurrence time E_i corresponds to the earliest start time, ES_{ij} of all activities (i, j) incident from node i. The earliest finish times EF_{ij} for these activities are calculated as $EF_{ij} = ES_{ij} + d_{ij}$.

In addition to the longest path, we also would like to compute a time window for each event, such that the event can occur at any time within this interval without increasing the project duration. This time window is defined by the earliest and latest occurrence times of that event, and its duration is called the *event slack*. Events with zero slack are called critical events. Starting with the project duration as the latest occurrence time for the terminal node, we proceed backward through the network to the start node using a recursive relationship resulting in the latest occurrence time for each event. The recursive relationships for this backward pass are given as:

$$L_m = E_m \tag{4.3}$$

$$L_i = min_j\{L_j - d_{ij}\} \quad i = 1, \ldots, (m - 1). \tag{4.4}$$

where L_i is the latest occurrence time for event node i, the minimization is over all event nodes j into which an activity (i, j) is incident from event node i.

Note that the latest occurrence time L_j corresponds to the latest finish time of all activities (i, j), incident into event node j, namely LF_{ij}. The latest start times LS_{ij} for these activities are calculated as $LS_{ij} = LF_{ij} - d_{ij}$.

Recall that each and every activity on the critical path is called a critical activity. We can now give a formal definition of critical activities as follows.

Definition 4.1: An activity (i, j) is a critical activity if and only if all of the following hold:

(i) $E_i = L_i$
(ii) $E_j = L_j$
(iii) $E_j - E_i = L_j - L_i = d_{ij}$

Relations (i) and (ii) imply that the start and end events of critical activities are critical events. Relation (iii), together with the recursive relationships, implies that any delay in activity (i, j) leads to an equal increase in the project duration (see Exercise 4.2 for a proof).

The impact of activity delays on the possible delay of succeeding activities and on the project duration is important for project planning and control. As mentioned earlier, the amount of delay that can be incurred without compromising the timely completion of the project is called the slack. We will elaborate on two types of slack for an activity (i, j): total slack TS_{ij} and free slack FS_{ij}. Two additional types of slack, safety slack SS_{ij} and independent slack IS_{ij}, are also defined although rarely employed: (see Exercises 4.3 and 4.4).

The total slack TS_{ij} for an activity (i, j) corresponds to the maximum amount by which it can be delayed without delaying the project, and is given by

$$TS_{ij} = L_j - E_i - d_{ij}. \tag{4.5}$$

Hence, by definition, the total slack for a critical activity is zero. As shown in Fig. 4.1, the total slack for a non-critical activity (i, j) is the duration of the time window $(L_j - E_i)$ minus the activity duration d_{ij}. Activity (i, j) can be started anywhere in the interval $[E_i, L_j - d_{ij}]$ without delaying the project. The total slack is shared by a sequence of consecutive non-critical activities. If, for example, a non-critical activity is delayed absorbing all the slack available, then there is no slack

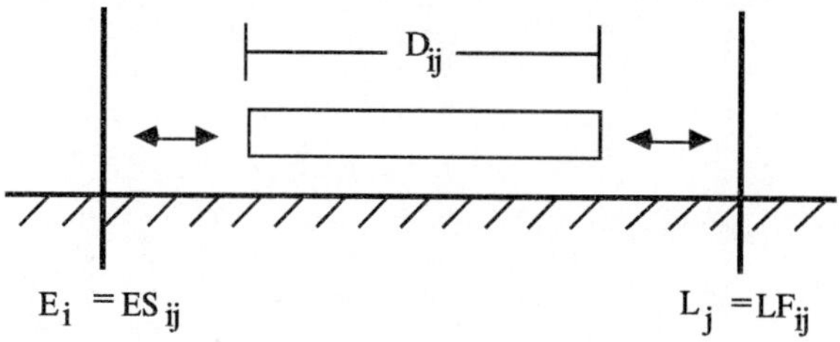

Fig. 4.1 Time window associated with total slack TS_{ij}

left for succeeding non-critical activities. Note that we can write the total slack expression in several different ways using the relationships we have seen earlier.

$$TS_{ij} = LF_{ij} - E_i - d_{ij} = LS_{ij} - E_i = LS_{ij} - ES_{ij} = LF_{ij} - ES_{ij} - d_{ij} \\ = LF_{ij} - EF_{ij} \tag{4.6}$$

Free slack (or free float) FS_{ij} is defined as the maximum amount by which an activity (i, j) can be delayed such that its finish time does not exceed the earliest occurrence time of its end event j. It is expressed as follows:

$$FS_{ij} = E_j - E_i - d_{ij} \tag{4.7}$$

Note that by changing the endpoint of the time window in Fig. 4.1 from L_j to E_j, we obtain the time window associated with free slack FS_{ij}. FS_{ij} can be written as:

$$FS_{ij} = E_j - \left(E_{ij} + d_{ij}\right) = E_j - EF_{ij} \tag{4.8}$$

By definition, critical activities have zero free slack, but the reverse is not true, i.e., an activity with zero free slack is not necessarily critical. Hence, a non-critical activity can have zero free slack. In contrast to total slack, a non-critical activity's free slack is associated only with that non-critical activity. In other words, if a non-critical activity absorbs some or all its free float and is delayed accordingly, this does not affect the free float of other activities.

The following two conditions determine the relation of FS values to TS values.

Condition 1: If for a non-critical activity (i, j) the event j is on the critical path, then $FS_{ij} = TS_{ij} > 0$.

Condition 2: If for a non-critical activity (i, j) the event j is not on the critical path, then $0 \leq FS_{ij} < TS_{ij}$. FS_{ij} is zero if E_j is determined by EF_{ij}. It is in the range $0 < FS_{ij} < TS_{ij}$, if E_j is not determined by EF_{ij}.

Example 4.1 Let us apply forward and backward pass to the network in Fig. 4.2 to obtain the earliest and latest occurrence times for each event. The label and the mean duration time in weeks for each activity are indicated on the corresponding activity.

Solution:

Forward pass:

$$E_1 = 0$$

$$E_2 = E_1 + d_{12} = 3$$

$$E_3 = E_1 + d_{13} = 2$$

$$E_4 = max_{\,i=2,3}\{E_i + d_{i4}\} = \max\{3 + 5, 2 + 4\} = 8$$

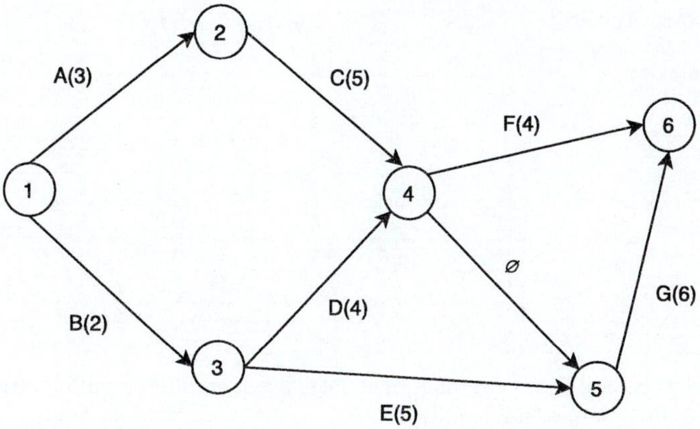

Fig. 4.2 The project network for Example 4.1

$$E_5 = max_{i=3,4}\{E_i + d_{i5}\} = \max\{2+5, 8+0\} = 8$$

$$E_6 = max_{i=4,5}\{E_i + d_{i6}\} = \max\{8+4, 8+6\} = 14$$

Hence, the project duration is found to be 14 weeks. We now proceed with the backward pass to determine the critical activities and the critical path.

Backward pass:

$$L_6 = E_6 = 14$$

$$L_5 = L_6 - d_{56} = 14 - 6 = 8$$

$$L_4 = min_{i=5,6}\{L_i - d_{4i}\} = min\{14-4, 8-0\} = 8$$

$$L_3 = min_{i=4,5}\{L_i - d_{3i}\} = min\{8-4, 8-5\} = 3$$

$$L_3 = \{8-4, 8-5\} = 3$$

$$L_2 = L_4 - d_{24} = 8 - 5 = 3$$

$$L_1 = min_{i=2,3}\{L_i - d_{1i}\} = min\{4-2, 3-3\} = 0$$

Events 1, 2, 4, 5, 6 have equal earliest and latest occurrence times, and hence are critical events with zero slack. The critical path is $\{1, 2, 4, 5, 6\}$ and activities A, C, and G are the critical activities.

Having obtained E_i and L_i values, we can now compute the slacks ES_{ij}, EF_{ij}, LS_{ij}, LF_{ij}, TS_{ij}, and FS_{ij} as shown in Table 4.1. Note that the non-critical activities D, E, and F satisfy *Condition 1*, with their free slacks equal to their total slacks.

Table 4.1 Activity early and late start and finish times and slack values

Activity	D	ES	EF	LS	LF	TS	FS
A	3	0	3	0	3	0	0
B	2	0	2	1	3	1	0
C	5	3	8	3	8	0	0
D	4	2	6	4	8	2	2
E	5	2	7	3	8	1	1
φ	0	8	8	8	8	0	0
F	4	8	12	10	14	2	2
G	6	8	14	8	14	0	0

Non-critical activity B, on the other hand, satisfies *Condition* 2, and its free slack is zero although its total slack is positive.

In the network of Fig. 4.2, suppose we delay activity B by its total slack value, i.e., by one unit. The delay by one unit of activity B reduces the total slack of activities D and E by one unit. The total slack of F remains unchanged since E_4 and L_6 are not affected by this delay.

4.2.1.2 Critical Path Method on Activity-on-Node Type Project Networks

In activity-on-node (AON) project networks, the recursive relationships for forward and backward formulations are similar to those for AOA project networks. Recall that for a project with n activities, the AON project network contains $(n + 2)$ nodes with the addition of two dummy nodes, the start node 0 and the terminal node $(n + 1)$.

The recursive relationships for the forward pass are as follows:

$$E_0 = 0 \tag{4.9}$$

$$E_j = max_i\{E_i + d_i\} \quad j = 1, \ldots, n+1; \quad i \in P(j) \tag{4.10}$$

where $P(j)$ is the set of predecessors of activity j.

The recursive relationships for the backward pass are:

$$L_{n+1} = E_{n+1} \tag{4.11}$$

$$L_i = min_j\{L_j + d_j\} \quad i = 0, 1, \ldots, n; \quad j \in S(i) \tag{4.12}$$

where $S(i)$ is the set of successors of activity i.

The total and free slack for activity i are given by:

$$TS_i = LS_i - ES_i = LF_i - EF_i \tag{4.13}$$

$$FS_i = \min_j\{ES_j\} - EF_i, \; j \in S_i \tag{4.14}$$

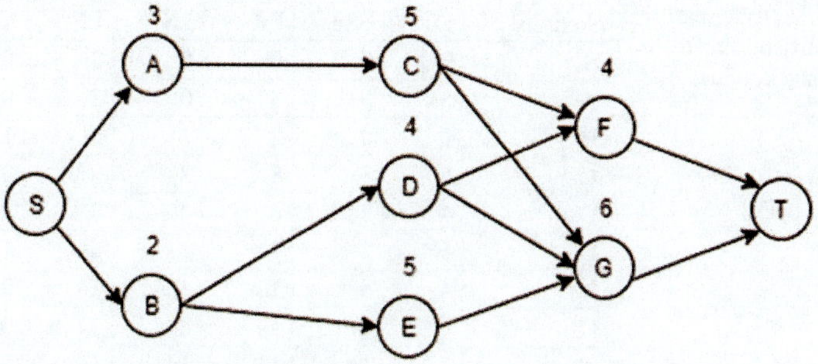

Fig. 4.3 AON project network for Example 4.2

Example 4.2 Let us apply the forward and backward pass computations to the AON project network for the project data in Example 4.1, with S and T denoting the start and terminal nodes of the project network (Fig. 4.3).

Solution:
Forward Pass:

$$ES_S = 0$$

$$ES_A = 0$$

$$ES_B = 0$$

$$ES_C = ES_A + d_A = 0 + 3 = 3$$

$$ES_D = ES_B + d_B = 0 + 2 = 2$$

$$ES_E = ES_B + d_B = 0 + 2 = 2$$

$$ES_F = max\{ES_C + d_C; ES_D + d_D\} = max\{3 + 5; 2 + 4\} = 8$$

$$ES_G = max\{ES_C + d_C; ES_D + d_D; ES_E + d_E\} = max\{3 + 5; 2 + 4; 2 + 5\} = 8$$

$$ES_T = max\{ES_F + d_F; ES_G + d_G\} = max\{8 + 4; 8 + 6\} = 14$$

Backward Pass:

$$LF_T = LF_G = LF_F = 14$$

$$LF_E = LF_G - d_G = 8$$

$$LF_D = min\{LF_F - d_F; LF_G - d_G\} = min\{14 - 4; 14 - 6\} = 8$$

$$LF_C = min\{LF_F - d_F; LF_G - d_G\} = min\{14 - 4; 14 - 6\} = 8$$

Table 4.2 Activity early and late start and finish times and slack values

Activity	D	ES	EF	LS	LF	TS	FS
S	0	0	0	0	0	0	0
A	3	0	3	0	3	0	0
B	2	0	2	1	3	1	0
C	5	3	8	3	8	0	0
D	4	2	6	4	8	2	2
E	5	2	7	3	8	1	1
F	4	8	12	10	14	2	2
G	6	8	14	8	14	0	0
T	0	14	14	14	14	0	0

$$LF_B = min\{LF_D - d_D; LF_E - d_E\} = min\{8-4; 8-5\} = 3$$

$$LF_A = LF_C - d_C = 8 - 5 = 3$$

$$LF_S = min\{LF_A - d_A; LF_B - d_B\} = min\{3-3; 3-2\} = 0$$

To demonstrate how FS values are obtained let us determine FS_D (Table 4.2).

$$FS_D = \min_{F,G}\{ES_F, ES_G\} = \min_{F,G}\{8, 8\} - 6 = 2$$

The critical activities have equal *ES* and *LS* values (or similarly, *EF* and *LF* values) and hence, their total slack values are zero. Using this relationship, we identify activities S, A, C, G, T as the critical activities and the critical path as {S, A, C, G, T}.

4.2.1.3 Criticality and Slacks for the Case of Generalized Precedence Relationships

Under generalized precedence relationships (GPRs) the critical path definition remains the same as in standard CPM; namely, the longest path in the project network. The concept of criticality in project networks with GPRs is slightly different: An activity is said to be critical if either its start or end node lies on a critical path (Elmaghraby and Kamburowski 1992).

Recall our definitions of total slack and free slack for the case of standard CPM: The total slack indicates the maximum amount by which an activity can be delayed without delaying the project. The free slack, on the other hand, is the maximum amount an activity can be delayed without its finish time exceeding the earliest occurrence time of its end event or the minimum of the early start times of the succeeding activities for AOA and AON type of project networks, respectively. These definitions rely on the well defined precedence relationship between two succeeding activities. In GPRs, different from the standard CPM, this relationship is "no longer straightforward because of the inherent difficulty in defining which activity precedes which" (Elmaghraby and Kamburowski 1992), and provide a detailed analysis of the topic which we shall not pursue further here.

4.2.1.4 Sensitivity Analysis

Although uncertainty in activity durations, small or large, exists in practice, CPM ignores this and assumes deterministic durations. Obviously, the results of CPM would reflect reality better closely when the variability of activity durations are smaller. Particularly, if there are several paths in the project network with lengths close to that of the critical path, the length of the deterministic critical path might not turn out to be the true longest path during the implementation of the project; some other path(s) might become the longest in the implementation and hence be the realized critical path. To investigate such possibilities, we can apply sensitivity analysis to the activity durations. We refer the readers to Chap. 12 for a detailed description and example of sensitivity analysis in project planning.

4.3 Linear Programming Formulation

We can formulate the critical path problem as an LP problem. The formulation is based on the occurrence time T_i of event i, where $i = 1, 2, \ldots, m$. There are n activities, and the project network is of AOA-type. The formulation can be stated as follows:

$$\min Z = T_m - T_1 \tag{4.15}$$

subject to

$$T_j - T_i \geq d_{ij} \quad \textit{for all } (i,j) \in A \tag{4.16}$$

$$T_i \textit{ unrestricted} \quad \textit{for all } i, i = 1, 2, \ldots, m \tag{4.17}$$

where A is an n-dimensional column vector of activities. The objective is to minimize the project duration or makespan, the difference between the occurrence times of the terminal event and the start event, $(T_m - T_1)$. The first set of constraints implies that the time difference between the occurrence times of two events connected by an activity cannot be less than the duration of that activity. Note that we have one constraint for each activity. In the above formulation, for the sake of computational simplicity, the occurrence time of event 1 can be set to zero without affecting the final solution, implying $T_i \geq 0$ for all $i = 2, 3, \ldots, m$.

Written in matrix form the above formulation becomes:

$$min\ Z = c^T T \tag{4.18}$$

subject to

$$CT \geq D \tag{4.19}$$

$$T \; unrestricted \tag{4.20}$$

where c is an m-dimensional column vector corresponding to the cost vector of the primal objective function, T an m-dimensional column vector of occurrence times, D an n-dimensional column vector of activity durations, and C the $(n \times m)$ incidence matrix of the project network. The incidence matrix C is defined as follows: An entry (g, h) is equal to 1, if the activity corresponding to the gth entry of A is incident into node h; is equal to -1, if it is incident from h; and 0 otherwise. Hence, in each row of C we have one entry of value (1) and one entry of value (-1).

The dual of this formulation is given as follows.

$$Max\, Z' = w^T D \tag{4.21}$$

subject to

$$C^T w = c \tag{4.22}$$

$$w \geq 0 \tag{4.23}$$

where w is a column vector of the dual variables w_{ij}, one for each activity $(i,j) \in A$.

The dual formulation is a network flow problem with a source with one unit capacity and a sink with one unit of demand. The objective is to maximize the gain as the one unit flows through the acyclic directed network, and yields the longest path of the project network. The dual variables w are expressed as nonnegative continuous decision variables but due to the totally unimodular nature of the incidence matrix the dual variables will be integer taking on value 0 or 1 (see, e.g., Garfinkel and Nemhauser 1972)

Note that the LP method facilitates a systematic sensitivity analysis around the variability of the activity durations. The project manager can use this analysis and come up with some managerial insights.

Example 4.3 Consider the project network in Fig. 4.4. The duration for each activity is indicated on the corresponding arc.

Example 4.3a Formulate the problem of determining the critical path as an LP problem and obtain the optimal solution.

Solution:
Setting $T_1 = 0$, we obtain the following LP formulation.

$$\min Z = T_5$$

subject to

Fig. 4.4 The project network for Example 4.3

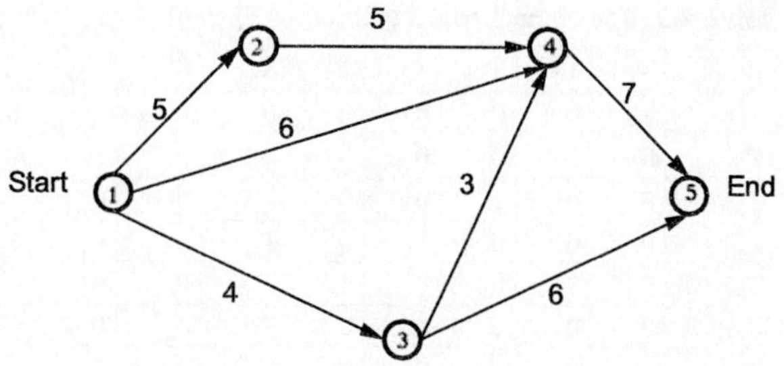

1. $T_2 \geq 5$
2. $T_3 \geq 4$
3. $T_4 \geq 6$
4. $T_4 - T_2 \geq 5$
5. $T_4 - T_3 \geq 3$
6. $T_5 - T_3 \geq 6$
7. $T_5 - T_4 \geq 7$
8. $T_1 = 0$
9. $T_i \geq 0 \quad i = 2, 3, 4, 5$

The slack variables for constraints (1) – (7) are denoted by S_{ij} corresponding to the constraint associated with activity (i, j), and represent the slack value for activity (i, j). It is important to note that S_{ij} does not necessarily correspond to total slack value TS_{ij}.

Solving this problem using an LP solver results in the optimal tableau given in Table 4.3.

The optimal solution is read from Table 4.2 as: $T_1* = 0$, $T_2* = 5$, $T_3* = 7$, $T_4* = 10$, $T_5* = 17$, $S_{13}* = 3$, $S_{14}* = 4$, and $S_{35}* = 4$. The optimal objective function value, i.e., the project duration, is 17.

The reduced costs in the optimal primal tableau are 1, 1, 1 for the constraints 1, 4, and 7, respectively. Thus, the critical path consists of activities (1,2), (2,4), (4,5).

An *alternative optimal solution* is obtained by letting S_{34} become basic and forcing S_{13} out of the basis. Note that the dual price for S_{34} is zero. Then, $T_3* = 4$, $S_{14}* = 4$, $S_{34}* = 3$, $S_{35}* = 7$. The slack of 7 units associated with activity (3,5) is divided in the previous solution between the activities (1,3) and (3,5) as 3 and 4 units, respectively.

Example 4.3b Write the dual formulation for this problem to verify the optimal solution from the optimal primal tableau of part (a).

Solution:

The coefficient vectors and matrices of the dual formulation are as follows:

$$c^T = [-1, 0, \ldots, 0, 1]$$

Table 4.3 The optimal primal tableau for Example 4.3a

Basis	T_1	T_2	T_3	T_4	T_5	S_{12}	S_{13}	S_{14}	S_{24}	S_{34}	S_{35}	S_{45}	RHS
T_1	1	0	0	0	0	0	0	0	0	0	0	0	0
T_2	0	1	0	0	0	−1	0	0	0	0	0	0	5
T_3	0	0	1	0	0	−1	0	0	−1	1	0	0	7
T_4	0	0	0	1	0	1	0	0	1	0	0	0	10
T_5	0	0	0	0	1	−1	0	0	−1	0	0	−1	17
S_{13}	0	0	0	0	0	−1	1	0	−1	1	0	0	3
S_{14}	0	0	0	0	0	−1	0	1	−1	0	0	0	4
S_{35}	0	0	0	0	0	0	0	0	0	−1	1	−1	4
Obj.	0	0	0	0	0	1	0	0	1	0	0	1	17

$$A^T = [(1,2),(1,3),(1,4),(2,4),(3,4)(3,5),(4,5)]$$

$$C = \begin{pmatrix} -1 & 1 & 0 & 0 & 0 \\ -1 & 0 & 1 & 0 & 0 \\ -1 & 0 & 0 & 1 & 0 \\ 0 & -1 & 0 & 1 & 0 \\ 0 & 0 & -1 & 1 & 0 \\ 0 & 0 & -1 & 0 & 1 \\ 0 & 0 & 0 & -1 & 1 \end{pmatrix}$$

$$C^T = \begin{pmatrix} -1 & -1 & -1 & 0 & 0 & 0 & 0 \\ 1 & 0 & 0 & -1 & 0 & 0 & 0 \\ 0 & 1 & 0 & 0 & -1 & -1 & 0 \\ 0 & 0 & 1 & 1 & 1 & 0 & -1 \\ 0 & 0 & 0 & 0 & 0 & 1 & 1 \end{pmatrix}$$

$$D^T = [5,4,6,5,3,6,7].$$

The dual formulation then becomes:

$$\max Z = 5w_{12} + 4w_{13} + 6w_{14} + 5w_{24} + 3w_{34} + 6w_{35} + 7w_{45}$$

subject to

1. $-w_{12} - w_{13} - w_{14} = -1$
2. $w_{12} - w_{24} = 0$
3. $w_{13} - w_{34} - w_{35} = 0$
4. $w_{14} + w_{24} + w_{34} - w_{45} = 0$

Table 4.4 The optimal dual tableau for Example 4.3b

Basis	w_{12}	w_{13}	w_{14}	w_{24}	w_{34}	w_{35}	w_{45}	s_8	RHS
w_{45}	0	0	0	0	0	1	1	0	1
w_{12}	1	0	1	0	1	1	0	0	1
w_{13}	0	1	0	0	−1	−1	0	0	0
w_{24}	0	0	1	1	1	1	0	0	1
s_8	0	0	0	0	0	0	0	1	0
Objective	0	0	−4	0	−3	−7	0	0	17

5. $w_{45} + w_{35} = 1$
6. $w_{12}, w_{13}, w_{14}, w_{24}, w_{34}, w_{35}, w_{45} \geq 0$

The first constraint represents the one-unit flow from the source (start node). The following constraints with zero right hand side correspond to transshipment nodes with no flow accumulation taking place. Constraint 6 represents the one-unit flow into the sink (terminal node).

Reading from Table 4.4, $w_{12}* = 1$, $w_{24}* = 1$, $w_{45}* = 1$ indicating a unit flow from node 1 to node 5 via nodes 2 and 4. This implies a critical path consisting of activities (1, 2), (2, 4), (4, 5).

Note that the reduced costs in the optimal primal tableau (Table 4.4) are 1, 1, 1 for the constraints 1, 4, and 7 respectively. This implies that $w_{12}* = 1$, $w_{24}* = 1$, $w_{45}* = 1$ leading to the above result.

4.4 Maximizing the Net Present Value of a Project Under No Resource Constraints

As discussed earlier, the most popular objective in project management is the minimization of the project duration; in other words, completing the project as early as possible. Another popular objective is the maximization of the NPV of the cash flows occurring during the project. This objective is quite realistic since every project has a budget and aims at creating monetary value to the sponsoring organizations. It thus reflects the financial success of a project. The present value, i.e., the value at $t = 0$, of a cash flow occurring at some specific time point t in the future is calculated by multiplying the cash flow by a discount factor $[(1/(1 + \alpha)^t]$, where α is the discount rate (see, e.g., Park 2007).

Since the calculation of NPV is based on the concept of time value of money, not only the absolute values of receipts and expenditures are of interest, but also their realization times. We adopt the convention that cash outflows are negative (expenditures) and inflows (receipts) positive in sign. The net cash flow amount is then obtained by discounting the individual cash flows. For short duration projects with relatively small discount rates, the discounting might be neglected, allowing the net cash flow to be estimated by subtracting the total expenditures from the total receipts. However, for long duration projects, the discounting process is important to analyze the financial feasibility correctly.

The scheduling question now turns out to be that of determining the start times of the activities so that the NPV of all net cash flows are maximized. The simplest case arises when the only positive net cash flow occurs at the completion of the project with all the other net cash flows being negative. It can be shown that the optimal schedule for this situation is to start all non-critical activities at their late start times. This optimal scheduling procedure is only valid if the project has a positive NPV (Smith-Daniels 1986). However, when there are both positive and negative net cash flows throughout the project duration it is not so obvious how to obtain the optimal solution. A mathematical programming formulation for this more general case is presented in the next section.

4.4.1 A Zero-One Mathematical Programming Formulation

A zero-one mathematical programming formulation for the maximization of NPV under no resource constraints is given below for an AON representation with zero time lag. Continuous compounding is applied, in which case the discount factor f_t^α becomes $e^{-\alpha t}$, where α is the discount rate or the cost of capital. The zero-one decision variables can be defined according to whether the net cash flow amounts associated with an activity are discounted to the start or finish period of the activities; for example, if all the cash flows throughout each activity are discounted to the finish period for all activities, then the zero-one decision variables are defined based on the activity finish period. For the following formulation, this convention is employed and the zero-one decision variables are defined accordingly.

In order to reduce the number of zero-one decision variables, we define a time window for each activity bounded by its early finish and late finish periods such that an optimal solution is never excluded. The early finish period for an activity is provided by the forward pass of the CPM calculations. The backward pass starting from the deadline, on the other hand, determines the corresponding late finish period. The formulation can be stated as follows:

$$\max Z = \sum_{j=1}^{J} \sum_{t=EFT_j}^{LFT_j} f_t^\alpha C_j x_{jt} \tag{4.24}$$

subject to

$$\sum_{t=EFT_j}^{LFT_j} x_{jt} = 1 \quad for\, j = 1, 2, \ldots, J \tag{4.25}$$

$$-\sum_{t=EFT_j}^{LFT_j} t x_{it} + \sum_{t=EFT_j}^{LFT_j} (t - d_j) x_{jt} \geq 0 \quad j = 2, 3, \ldots, J; \; i \in P_j \tag{4.26}$$

$$x_{jt} = 1, \quad \textit{if activity } j \textit{ finishes in period } t; \ 0, \textit{otherwise} \quad \textit{for all } j, t \tag{4.27}$$

where

J = number of activities including dummy activities, if any.
j = activity index.
t = time index in periods.
d_j = duration of activity j.
P_j = set of immediate predecessors to activity j.
α = the discount rate.
f_t^α = the discount factor for the cash flow in period t.
C_j = net cash flow amount for activity j.
EFT_j = early finish period of activity j.
LFT_j = late finish period of activity j.

The objective function (4.24) is to maximize the NPV of all cash flows throughout the project. The constraint set (4.25) ensures that each activity j completes in the range $[EFT_j, LFT_j]$. Constraint set (4.26) models the precedence relations among the activities. Constraint set (4.27) defines the zero-one decision variables.

When a due date D for project completion is imposed, then the following constraint has to be added to ensure that the project completes before the predefined deadline.

$$\sum_{t=1}^{D} t x_{Jt} \leq D \tag{4.28}$$

Obviously, the deadline cannot be less than the earliest project duration time calculated by the CPM. Specifying a deadline greater than the length of the critical path introduces additional slacks and increases the scheduling flexibility.

Example 4.4 Consider a project with 12 activities whose data is provided in Table 4.5. The project network is given in Fig. 4.5. The numbers above the node denote the activity duration and the number below the node the cash flow associated with the activity, which can be positive (receipt) or negative (expenditure). The project deadline is given to be 25 weeks and the discount rate is 1% per week.

Solution: Substituting the data provided into the mathematical programming formulation (4.24), (4.25), (4.26) and (4.27) and employing an integer programming solver we obtain the optimal solution shown in the Gantt chart in Fig. 4.6 with an objective function value of 1073.53. The optimal project duration of 25 weeks is equal to the deadline. Delaying the activities with negative cash flows has led to a break in activity processing in the interval [12–13]. All the activities in the first group have positive cash flows and hence are scheduled as early as possible without violating the precedence constraints. Those in the second group have negative cash flows; those with positive cash flow are their successors.

Table 4.5 Duration and cost information for activities

Activity	2	3	4	5	6	7	8	9	10	11	12	13
Duration (weeks)	6	5	3	1	6	2	1	4	3	2	3	5
Cost	−140	318	312	−329	153	193	361	24	33	387	−386	171

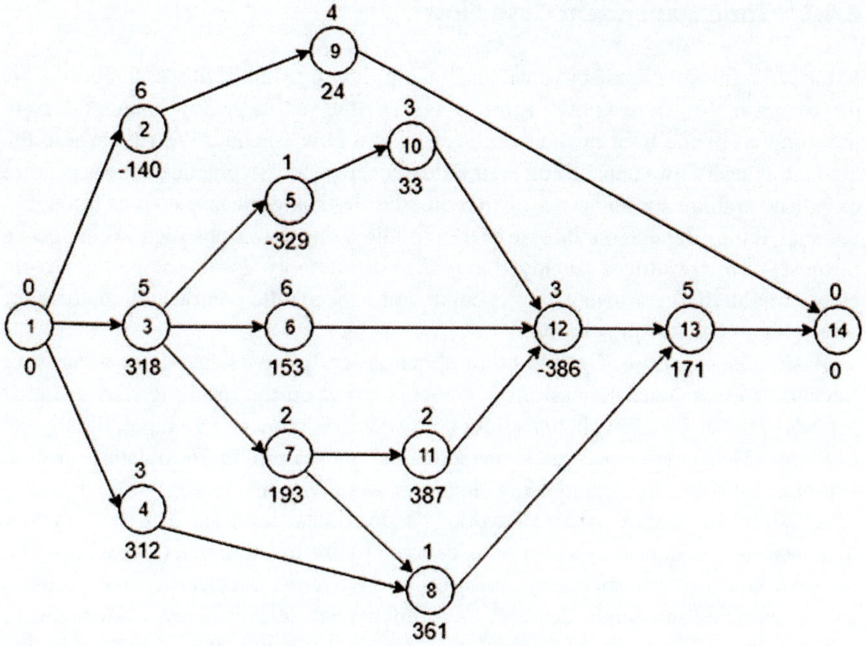

Fig. 4.5 Project network with cash flows (Vanhoucke 2000)

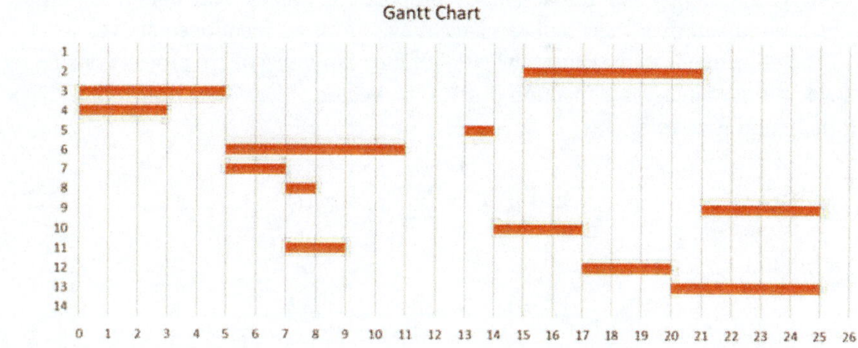

	14	13	12	11	10	9	8	7	6	5	4	3	2	1
Start Date	25	20	17	7	14	21	7	5	5	13	0	0	15	0
Duration	0	5	3	2	3	4	1	2	6	1	3	5	6	0

Fig. 4.6 The Gantt chart for the optimal solution of Example 4.4

Vanhoucke (2000) solved the same problem assuming a 44-week deadline. The optimal value of the completion time is equal to the deadline and the optimal NPV is found to be 1165.90; a larger value than the case above, with a standstill of activities in the interval [12–32]. Note that the start of the standstill period is the same in both cases, and extension of the deadline simply extends the standstill interval.

4.4.2 Time Dependent Cash Flows

In the NPV formulations above we made the following simplifying assumptions: (*i*) the net cash flows are known a priori and (*ii*) the net cash flow magnitudes are independent of the time of their occurrence. We now consider the case where the amount of cash flows depends on their time of occurrence. In practice, unless a future cash flow amount for an activity is fixed earlier by some means, such as through a contract, it may depend on the date of the cash flow due to reasons such as changes in the costs of the resources employed to realize that activity. Furthermore, the project owner might charge a penalty if the contractor exceeds the contractual due date, or give a bonus for finishing early.

A simplistic approach would be to obtain an early start schedule or a late start schedule and evaluate the cash flow amounts based on the resulting start or finish periods of the activities. A better solution procedure is proposed by Etgar, Shtub, and LeBlanc (1996), who present a mathematical programming formulation and an approximate solution method using simulated annealing for the case of time dependent cash flows in AOA project networks. No deadline is imposed. The net cash flow of an activity j occurs at its finish time, denoted by the event time t_j (milestone). The net cash flow is a non-increasing function of the event's occurrence time. In other words, cash inflows might decrease as a function of delay whereas cash outflows might increase. The activity j's finish time t_j must lie in the interval defined by the activity's early (EFT_j) and late (LFT_j) finish times and is denoted by [EFT_j, LFT_j]. EFT_j and LFT_j are determined by applying CPM as described in Sect. 4.4.1. Later, Shtub and Edgar (1997) proposed a branch and bound algorithm, which improved over the simulated annealing algorithm mentioned above.

The objective is to maximize the NPV of the cash flows of the project denoted by *Z(t)* and represented as a function of the time vector of activities' finish times. The formulation is as follows:

$$\max{}_t Z(t) = \sum\nolimits_{j=1}^{J} f_t^{\alpha} a_j(t) \tag{4.29}$$

subject to

$$\sum\nolimits_{t=EFT_j}^{LFT_j} x_{jt} = 1 \quad for\, j = 1, 2, \ldots, J \tag{4.30}$$

$$-\sum\nolimits_{t=EFT_j}^{LFT_j} t x_{it} + \sum\nolimits_{t=EFT_j}^{LFT_j} (t - d_j) x_{jt} \geq 0 \quad j = 2, 3, \ldots, J; \quad i \in P_j \tag{4.31}$$

$$x_{jt} = 1, \quad if\ activity\ j\ finishes\ in\ period\ t; 0, otherwise \quad for\ all\ j, t \tag{4.32}$$

where

J = number of activities including dummy activities, if any.
j = activity index.
t = time index.

d_j = duration of activity j.
f_t^α = the discount factor for the cash flow in period t.
α = the discount rate (cost of capital).
$a_j(t)$ = net cash flow function associated with activity j as a function of its activity finish time
EFT_j = early finish period of activity j.
LFT_j = late finish period of activity j.

Note the the constraint set is the same as the one in the zero-one mathematical programming formulation (4.25), (4.26) and (4.27).

Example 4.5 Let us consider the project data of Example 4.4 with the project network displayed as in Fig. 4.5 but this time let the activity costs increase by 5% per period of delay and activity receipts decrease by 5% per period of delay. The project network is of AOA type. The due date is taken to be 19 periods, which is the length of the critical path.

Solution: As in Example 4.4, substituting the data provided into the mathematical programming formulation (4.25), (4.26), (4.27) and (4.28) and employing an integer programming solver we obtain an optimal solution. Recall that for a critical activity j, $EFT_j = LFT_j$, and hence, their finish times are fixed, reducing the problem size.

A letter designation together with a duration for each non-dummy activity are also provided. The cash flow functions for activities A and B are given below, where activity A has a cash outflow and B has a net cash inflow:

$$\mathrm{a_A}(t) = -140 - (140)(0.05)(t - EFT_A)x_{At}$$

$$\mathrm{a_B}(t) = 318 - (318)(0.05)(t - EFT_B)x_{Bt}$$

The Gantt chart for the solution is displayed in Fig. 4.7. The maximum NPV for the project is found to be 1189.59.

The critical path is {B, E, K, L}. The non-critical activities C, F, G, and J have positive cash inflows and are on sub-paths on which no activity has a negative cash flow. They are all scheduled at their earliest start period to maximize their contribution to NPV. In the non-critical activity pairs {A, H} and {D, I}, an activity with a relatively large negative cash flow precedes a second with a relatively small positive cash flow; with the negative cash flow dominating. Both these activity pairs are delayed as much as possible to minimize their negative contribution to NPV.

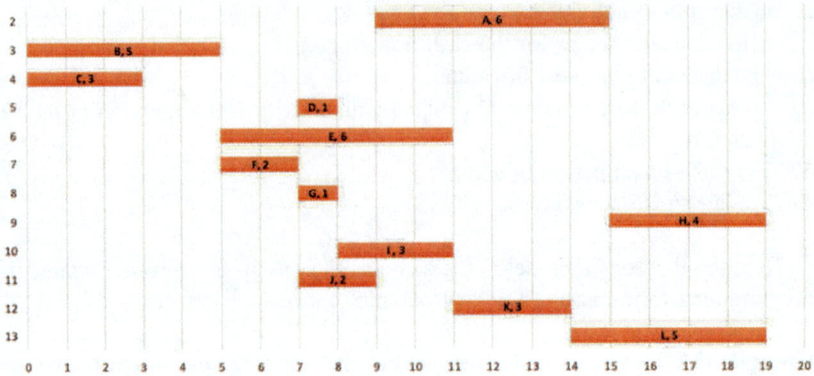

	13	12	11	10	9	8	7	6	5	4	3	2
Start Date	14	11	7	8	15	7	5	5	7	0	0	9
Duration	5	3	2	3	4	1	2	6	1	3	5	6

Fig. 4.7 The Gantt chart for the optimal solution of Example 4.5

4.5 Keeping the Project Plan Current

During the implementation of a project, it is expected that there will be deviations from the plan. It is crucial for the success of the project that these changes are detected in a timely and correct manner and are reflected and documented properly in an updated project plan. If such a control mechanism is not created, then these deviations from the plan will accumulate, leading to greater complications, which would be more difficult to deal with in later stages of the project. Hence, it is imperative that a control mechanism be built into the project management process early on.

We can talk about *minor* and *major updates of the project plan*. Minor updates are those which can be handled locally, meaning with minor adjustments. An example would be those deviations occurring within the slacks provided by the project plan and thus absorbed by the slacks without a need for replanning. Major updates, on the other hand, imply replanning of the project. Note that replanning might also lead to rescheduling.

A major source of change is associated with the estimates made during the planning phase.

Changes in Activity Duration Estimates With additional information gathered and with learning during the implementation of the project better estimates might be obtained for the remaining activities as they are getting closer. Changes in resource availabilities can also lead to changes in activity durations.

Changes in Resource Estimates Resource availabilities might change during the implementation of the project with an impact on activity durations. Resources might get broken without any replacement and the resource budget might change affecting the overall resource level decisions taken during the planning phase. Decreasing resource allocation to activities will increase the corresponding activity durations in general. Similarly, increasing resource allocation will decrease the activity durations. The resource requirements by the activities might also change due to changes in the technology/methods and resource combinations the activities are performed with.

Changes in the Project Network Structure Usually, the project network structure remains essentially the same with the same set of activities and precedence relations when replanning. But as the project progresses there can be changes in those aspects of the project as well. A new activity or a set of new activities might be introduced into the project network, potentially leading to scope change. This new set of activities might be introducing a new way of executing a work package in which case the scope does not change. But, on the other hand, if a new work package is introduced or an existing work package significantly extended, the project scope will change. Similarly, an activity or a set of activities might be deleted changing the network structure and the scope of the project.

Changes in the precedence relations might also occur. Generalized precedence relations might be introduced to expedite a set of activities leading to the realization of an event. The PM might choose to introduce precedence relations to facilitate resource management of the project, or to impose a preferred sequence of the work packages involved. The PM might also choose to remove a precedence relation of managerial nature so as to speed up the execution of the project.

Changes Associated with Risks When planning a project possible risks are identified and analyzed and response plans to reduce their impact are suggested. The risks identified are compiled in a list called *the risk register*. During the implementation of the project, changes would be expected to occur in the risk environment of the project: New risks might emerge; the severity level of the risks already identified might change; new response plan opportunities might arise; current response plans might become void. The risk register and the response plans need to be updated following the occurrence of changes mentioned.

Changes in the Schedule Schedule changes might be needed to take into consideration the major changes mentioned above. For example, rescheduling will be necessary upon a change in the network structure or in resource availabilities. It is also possible that rescheduling itself requires changes in the items discussed. The need to expedite a project would be such an example. It will require, for example, additional resources to decrease the duration of certain activities so that the overall duration will be reduced.

Keeping the project plan current is necessary for the success of the project management. It should be conceived as an integral part of the monitoring and control of the project, which we will deal with in detail in Chap. 10.

4.6 Conclusions, Recent Developments, and Some Future Research Directions

In this chapter, we addressed project planning using network models and mathematical programming formulations. Specifically, we studied the deterministic models with no resource constraints. These studies mostly focus on optimizing the project duration. On the other hand, financial objectives are also important in the literature. Regarding this, we presented mathematical models for maximizing the NPV of time independent and dependent cash flows. Although these models do not incorporate the resource limitations and uncertainty, they serve to establish a baseline to develop more complex models that address the real life instances.

A common assumption in project scheduling studies with NPV criterion is that the amount and timing of cash flows is known in advance. However, in practice, there are often delays in payments. Long payment delays could affect the NPV of the projects significantly. Ignoring the risks of late payments and their effects could put the profitability of projects at risk. In their model, Hazır et al. (2016) integrated the payment delays and their dependence on the financial risk structure of client firms. There is a need for new models that optimize financial objectives while considering different uncertainty cases and risk factors. Regarding the project risks, Hall (2016) underlines that the level of several risk items declines over time as the project is executed and more information is acquired. Therefore, he proposes using risk adjusted discount rates, decreasing them as tasks are completed. He cites developing mathematical models that maximize risk adjusted NPV as a promising research opportunity.

The models presented in this section address time and financial objectives. However, for sustainable business systems, environmental and social criteria need to be considered as well. In this regard, new methodical models could be developed. To give an example, energy consumption and CO_2 emissions could be integrated as objective functions or they could be limited by adding them as constraints in the given models. Project scheduling studies that address environmental objectives are very few in the literature. We refer to Habibi et al. (2019) as an example study that presents a mathematical model integrating these objectives and implementing the model on a real life case study.

4.7 Case Study: Installation of a Plant Biotechnology Lab

The Department of Molecular Biology, Genetics, and Bioengineering at Sabanci University decided to establish a dedicated laboratory for plant biotechnology for the testing and analysis of genetically modified organisms (GMO). After a careful review of the existing space available in the building, it was decided to merge two adjacent rooms, which were previously utilized for administrative purposes. Therefore, in addition to the procurement and installation of necessary equipment, considerable construction work will be needed to convert the selected office space into a plant biotechnology laboratory.

The project starts with the planning phase, which includes activities such as finalizing the layout of the lab, selecting the laboratory equipment, and placing orders for it. After careful analysis of the current needs of the Department, the equipment selected includes analytical balances, microcentrifuges, a light microscope, high-speed centrifuges, an ultracentrifuge, incubators and shakers, spectrophotometers, horizontal and vertical electrophoresis systems, and a climate-controlled chamber. The space selected will be completely redesigned in terms of its layout; therefore a team of graduate students and faculty will finalize its layout before the start of construction works.

The construction work for this project is divided into two phases: the first phase must be completed before equipment installation begins and the second will commence after equipment installation and calibration. Both installation and calibration of equipment will be done by the same team of technicians separate from the construction team. While the first phase of construction is in progress, orders can be placed for the procurement of necessary equipment. However, all installations must take place after completion of the first phase of construction and after the arrival of equipment at the site. After consulting with the contractor, the project manager listed the activities and their durations in working days to estimate the time it will take to complete both phases of construction work, which are tabulated in Table 4.6.

After completion of the second phase of construction and equipment installation, the necessary instrumentation will be installed and connected to the equipment. After all installation is complete, the lab will be prepared for commissioning by a dedicated team. The activities required for the commissioning phase are listed in Table 4.7.

The PM grouped the above activities into six different work packages such that a dedicated team will be responsible for each work package. A meeting of the entire team resulted in a final list of activities to be performed as well as their precedence

Table 4.6 Task description, precedence relation, and duration

Activities	Duration (days)
Construction Work – Phase I	
Clear site and demolish walls to merge rooms	8
Install support steelwork	10
Install stairways	2
Install platforms	8
Install handrails and platform angles	7
Construction Work – Phase II	
Install prefabricated ductwork	22
Install prefabricated process pipework	10
Make pipework connections to equipment	4
Install prefabricated service pipework	7
Make service pipework connections to equipment	3

Table 4.7 Task description and duration for the commissioning phase of the project

Activities	Duration (days)
Install electrical apparatus	9
Make electrical connections to equipment	3
Lay equipment	10
Lay pipework	13
Lay ductwork	6
Testing equipment	4
Painting	7
Clean up	1

relations and durations (Table 4.8). In addition to the precedence relationships cited in Table 4.8, the following Generalized Precedence Relations (GPR) had to be satisfied.

(i) *Cleanup* can only be started one day after *Painting* is finished.
(ii) *Make instrument connections to equipment* can only finish two days after Install electrical apparatus finishes.
(iii) *Install handrails and platform angles* can only start three days after *Install platforms* starts.
(iv) *Install instrumentation* can only start one day after *Install equipment* finishes.

Saturday and Sunday are not regular working days. Daily working hours are 8:00 to 18:00 with a lunch break during 12:00–13:00. The project started on June 01, 2017 at 8:00 hrs. The network diagram was constructed to visualize critical path activities, which are highlighted in red in Fig. 4.8. The PM reported the estimated finish time for this project as **September 25, 2017**.

The project started at its scheduled start time and orders are placed for all equipment and instrumentation. The procurement department had initially informed the PM that all equipment and instrumentation will arrive at the site latest by July 10, 2017. However, on June 20, 2017, the procurement department was informed of a delay in delivery schedules due to an unforeseen strike at the Original Equipment Manufacturer's (OEM) plant. After some negotiations with the OEM a new delivery date was set at August 04, 2017. This delayed the project completion time and resulted in a new critical path, and hence, a new set of critical path activities. The estimated finish time of the project became October 13, 2017. The new critical path is illustrated in Fig. 4.9. Note that under the original plan, installation of equipment was scheduled to start on July 17, 2017, but due to this delay, it could only begin on August 04, 2017.

After the completion of instrumentation installation (WP – E), the PM convened a meeting of all team leaders to review the overall progress on the project, which is graphically illustrated in Fig. 4.10 and Table 4.9.

Table 4.8 Work packages, task description, precedence relation and duration

ID	WBS	Task Name	Predecessors	Duration (days)
1	**WP-A**	**Initialization Stage**		
2	WP-A.1	Finalize the layout of lab		4
3	WP-A.2	Selecting equipment for lab	2	8
4	WP-A.3	Place order for all the equipment and instrumentations	2	7
5	**WP-B**	**Construction - Phase I**		
6	WP-B.5	Clear site and demolish walls to merge rooms	2	8
7	WP-B.1	Install support steelwork	3, 6	10
8	WP-B.2	Install stairways	7	2
9	WP-B.3	Install platforms	7	8
10	WP-B.4	Install handrails and platform angles	7, 9	7
11	**WP-C**	**Equipment Installation**		
12	WP-C.1	Arrival of equipment at site	4	0
13	WP-C.2	Install equipment	8, 9, 10, 12	4
14	WP-C.3	Calibrate equipment	13	7
15	**WP-D**	**Construction - Phase II**		
16	WP-D.1	Install prefabricated ductwork	13	22
17	WP-D.2	Install prefabricated process pipework	8, 9, 10, 14	10
18	WP-D.3	Make pipework connections to equipment	17	4
19	WP-D.4	Install prefabricated service pipework	8, 9, 10, 14	7
20	WP-D.5	Make service pipework connections to equipment	19	3
21	**WP-E**	**Instrumentation Installation**		
22	WP-E.1	Arrival of instruments at site	4	0
23	WP-E.2	Install instrumentation	8, 9, 10, 13, 22	15
24	WP-E.3	Make instrument connections to equipment	14, 23, 26	4
25	**WP-F**	**Commissioning of Lab**		
26	WP-F.1	Install electrical apparatus	8, 9, 10	9
27	WP-F.2	Make electrical connections to equipment	14, 10	3
28	WP-F.3	Lay equipment	14	10
29	WP-F.4	Lay pipework	18, 20	13
30	WP-F.5	Lay ductwork	16	6
31	WP-F.6	Testing equipment	30, 27, 29, 24, 28	4
32	WP-F.7	Painting	31	7
33	WP-F.8	Clean up	32	1

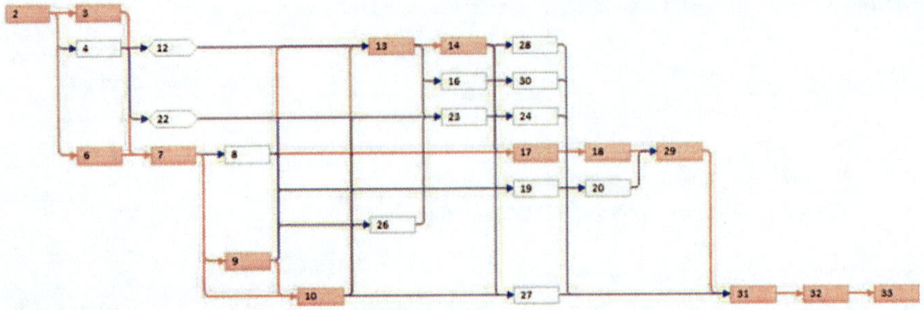

Fig. 4.8 Project network and the initial critical path

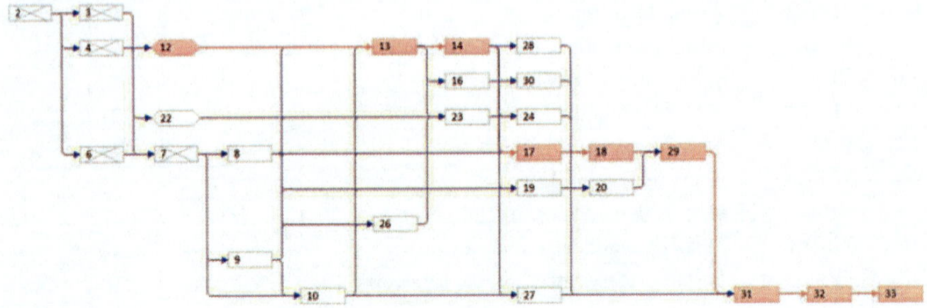

Fig. 4.9 Project network and the critical path after the delay of arrival of equipment and instrumentations

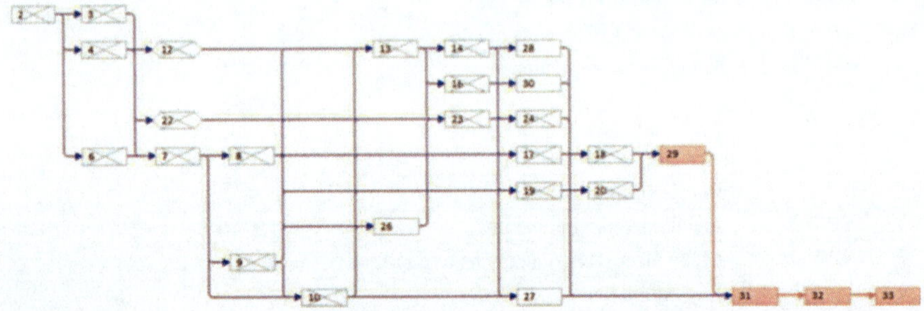

Fig. 4.10 Project network and the critical path after the completion of instrumentation installation

Although some activities were delayed, none of these was on the critical path and hence, the delays in those non-critical activities did not affect the critical path or the project duration. The project manager asked his/her team to take extra precautions on the remaining critical activities to ensure that the project would be completed no later than October 13, 2017.

Table 4.9 Status of work packages and activities

ID	WBS	Task Name	Planned		Actual		Status
			Start	Finish	Start	Finish	
1	**WP-A**	**Initialization Stage**					
2	WP-A.1	Finalize the layout of lab	6/1/2017	6/6/2017	6/1/2017	6/6/2017	F*/S**
3	WP-A.2	Select equipment for lab	6/7/2017	6/16/2017	6/7/2017	6/16/2017	F/S
4	WP-A.3	Place order for all the equipment and instrumentations	6/7/2017	6/15/2017	6/7/2017	6/15/2017	F/S
5	**WP-B**	**Civil Works - Phase I**					
6	WP-B.5	Clear site and demolish walls to merge rooms	6/7/2017	6/16/2017	6/7/2017	6/16/2017	F/S
7	WP-B.1	Install support steelwork	6/19/2017	6/30/2017	6/19/2017	6/30/2017	F/S
8	WP-B.2	Install stairways	7/3/2017	7/4/2017	7/3/2017	7/4/2017	F/S
9	WP-B.3	Install platforms	7/3/2017	7/12/2017	7/3/2017	7/12/2017	F/S
10	WP-B.4	Install handrails and platform angles	7/6/2017	7/14/2017	7/6/2017	7/14/2017	F/S
11	**WP-C**	**Equipment Installation**					
12	WP-C.1	Arrival of equipment at site	7/10/2017	7/10/2017	8/4/2017	8/4/2017	F/D***
13	WP-C.2	Install equipment	7/17/2017	7/20/2017	8/4/2017	8/9/2017	F/D
14	WP-C.3	Calibrate equipment	7/21/2017	7/31/2017	8/10/2017	8/18/2017	F/D
15	**WP-D**	**Civil Works - Phase II**					
16	WP-D.1	Install prefabricated ductwork	7/21/2017	8/21/2017	8/10/2017	9/8/2017	F/D
17	WP-D.2	Install prefabricated process pipework	8/1/2017	8/14/2017	8/21/2017	9/1/2017	F/D
18	WP-D.3	Make pipework connections to equipment	8/15/2017	8/18/2017	9/4/2017	9/7/2017	F/D
19	WP-D.4	Install prefabricated service pipework	8/1/2017	8/9/2017	8/21/2017	8/29/2017	F/D
20	WP-D.5	Make service pipework connections to equipment	8/10/2017	8/14/2017	8/30/2017	9/1/2017	F/D
21	**WP-E**	**Instrumentation Installation**					
22	WP-E.1	Arrival of instruments at site	7/10/2017	7/10/2017	8/4/2017	8/4/2017	F/D
23	WP-E.2	Install instrumentation	7/24/2017	8/11/2017	8/11/2017	8/31/2017	F/D
24	WP-E.3	Make instrument connections to equipment	8/14/2017	8/17/2017	9/1/2017	9/6/2017	F/D

(continued)

Table 4.9 (continued)

			Planned		Actual		Status
ID	WBS	Task Name	Start	Finish	Start	Finish	
25	**WP-F**	**Commissioning of Lab**					
26	WP-F.1	Install electrical apparatus	7/17/2017	7/27/2017	7/17/17	7/27/17	F/D
27	WP-F.2	Make electrical connections to equipment	8/1/2017	8/3/2017	8/21/17	8/23/17	F/D
28	WP-F.3	Lay equipment	8/1/2017	8/14/2017	8/21/17	9/1/17	F/D
29	WP-F.4	Lay pipework	8/21/2017	9/6/2017	9/8/17	9/26/17	F/D
30	WP-F.5	Lay ductwork	8/22/2017	8/29/2017	9/11/17	9/18/17	F/D
31	WP-F.6	Testing equipment	9/7/2017	9/12/2017	9/27/17	10/2/17	F/D
32	WP-F.7	Painting	9/13/2017	9/21/2017	10/3/17	10/11/17	F/D
33	WP-F.8	Clean up	9/25/2017	9/25/2017	10/13/17	10/13/17	F/D

**F* Finished, **Scheduled, ****D* Delayed

The initial plan was to finish the project in time for the start of the Fall semester. Although the project was delayed due to the late arrival of equipment and instrumentation, the lab was still commissioned within one month of start of the Fall semester and some demo sessions are planned by the professors and graduate students this semester, which will make the lab operational as soon as possible.

Exercises

4.1 Use the recursive relationships for E_i and L_i to show that once a critical activity is incident into a node other than the terminal node, there will be one or more critical activities incident from that node.

4.2 Show that Condition (*iii*) of Definition 4.1 together with the recursive relationships imply that any delay in a critical activity (i, j) leads to an increase of the same amount in the project duration.

4.3 The safety slack for activity (i, j) is determined as

- $SS_{ij} = L_j - L_i - d_{ij} = LF_{ij} - L_i - d_{ij} = LS_{ij} - L_i$.
  - (a) Give an interpretation of safety slack. For what purpose would you employ safety slack?
  - (b) For the following project network determine SS_{ij} for all activities (i, j).

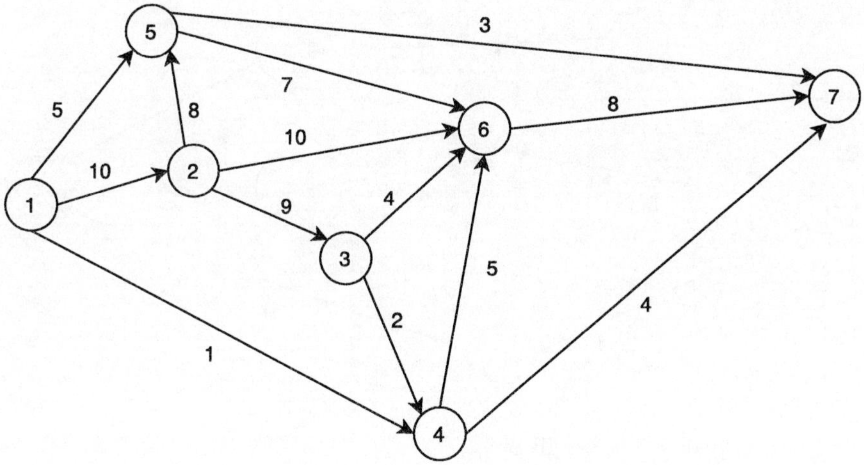

4.4 The independent slack for activity (i, j) is determined as

- $IS_{ij} = \max\{0, E_j - Li - d_{ij}\}$.
  - (a) Give an interpretation of independent slack. For what purpose would you employ independent slack?
  - (b) For the project network in problem 4.3b, determine IS_{ij} for all activities (i, j).

4.5 Show that the following statement is correct: The safety slack for non-critical activities (i, j) is smaller than the corresponding total slack value unless i or both i and j are events on the critical path, in which case they are equal.

4.6 Show that the Condition 1 in Sect. 4.2.1.1 is correct: If for a non-critical activity (i, j) the event j is on the critical path, then $FS_{ij} = TS_{ij} > 0$.

4.7 Show that the following statement is correct (Condition 2 in Sect. 4.2.1.1): If for a non-critical activity (i, j) the event j is not on the critical path, then $0 \leq FS_{ij} < TS_{ij}$ is zero, if E_j is determined by EF_{ij}. It is in the range $0 < FS_{ij} < TS_{ij}$, if E_j is not determined by EF_{ij}.

4.8 Explain why the following statement is true: The critical activities on a project network are equally critical regardless of the number of critical paths on the project network.

4.9 Consider the following project network with deterministic activity durations and an AOA representation:

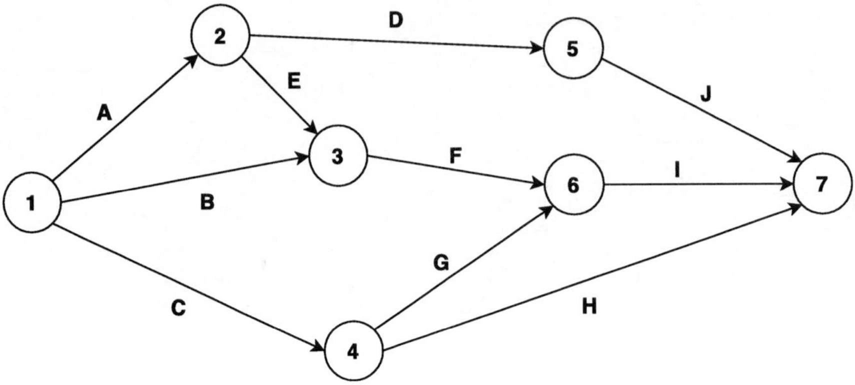

The activity durations are given as:

A	B	C	D	E	F	G	H	I	J
4	3	7	5	3	5	7	4	6	8

(a) Formulate this project as an LP problem with the decision variables being the occurrence times of the events i, T_i, showing both the primal and the dual formulations.

(b) Calculate E_i and L_i for the events i; and, ES_{ij}, EF_{ij}, LS_{ij}, LF_{ij}, and TS_{ij} values for the activities (i, j) using CPM. Show your calculations.

(c) Determine the values for the primal and dual variables and the slack variables using the results of part (b) by letting $T_i = E_i$ for all i.

(d) Repeat part (c) with $T_i = L_i$ for all i. Compare both solutions.

4.10 Consider the following unconstrained max NPV problem. Assume that activity costs are incurred at the completion of the activity.

Activity	Immediate predecessor	Duration	Cost (×1000)
A	–	8	80
B	–	11	40
C	–	3	130
D	A	6	70
E	B	9	60
F	C	12	70
G	D	7	90
H	E	8	30
I	F	5	90
J	F, G, H	4	80

According to the contract between the client and the contractor, the client will make payments at certain predefined events during the project. These are the following events:

- Completion of activity D.
- Completion of activity E.
- Completion of the project.

Payments include all costs incurred by the contractor since the last payment. A mark-up of 25% is added to the cost to determine the amount the client has to pay. The client pays to the contractor the amount remaining after holding 10% retainage. The last payment includes in addition the retainage from the two payments made earlier by the client. The client pays no interest on the amount kept as retainage.

(a) Calculate the NPV for the contractor. The activities with a slack are to be assigned so as to increase the resulting NPV. The discount rate is given as 6% per period.

(b) If the contractor does not use her funds and only employs bank loan to manage the project and receives bank loan at an interest rate of 4% per period, then what would be the NPV resulting from the project?

4.11 Consider the following project network and the corresponding data. Determine the primal and dual formulations for this problem and give an interpretation of the resulting dual problem and the dual variables.

Activity	Activity description	Immediate predecessors	Duration
A	Develop brochure plan	–	3
B	Develop training plan	–	4
C	Draft brochure	A	8
D	Select trainees	B	3
E	Print brochure	C	5
F	Deliver sample products	A	2
G	Conduct training course	D	3
H	Initiate advertising	E, F, G	4

4.12 Consider a project with 8 activities. Information concerning duration and cost of activities are provided together with the project network below. The client and contractor sign a contract agreeing upon a lump-sum payment of $120,000 to be paid at the successful termination of the project on or before the due date of 25 weeks. The cost for an activity is incurred at the termination of that activity. The discount rate is given as 0.25% per week. Using any LP solver, determine the project schedule maximizing the profit of the contractor. Draw the Gantt chart.

Activity	A	B	C	D	E	F	G	H
Duration (weeks)	8	12	3	10	4	11	5	6
Cost ($'000)	10	12	6	12	8	17	9	9

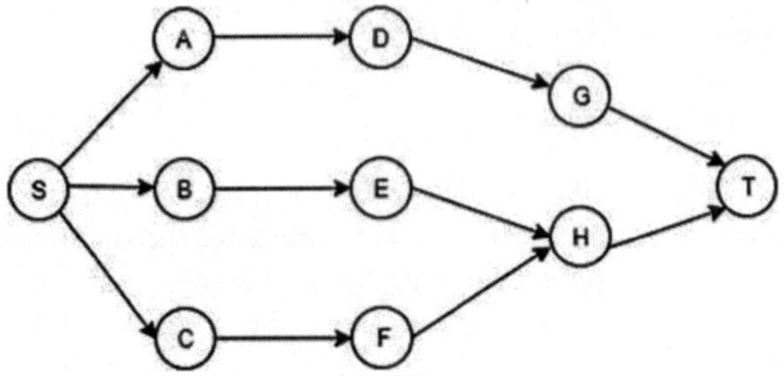

4.13 Consider the project network with the predecessor and duration data given. Determine the project duration and total and free slacks using the AON convention.

Activity	Predecessor	Duration (weeks)
A	–	7
C	A	6
D	A	3
E	B	6
F	B	4
I	C	8
B	–	7
G	D, E	9
H	F	11

4.14 Consider the following project network data.

Activity	Predecessors	Duration
A	**	5
B	**	3
C	**	6
D	A	8
E	A	11
F	A	3
G	B, D	16
H	C, G	9
I	B, D	14
J	E, H	7
K	I	9
L	I	12
M	E, H, F	15
N	E, H, F	10
O	M, I	18
P	J, L, O	2

(a) Draw the activity on arrow and activity on node diagram.
(b) Calculate its critical path and its length.
(c) Calculate *ES*, *LS*, *EF*, and *LF* for each activity in the project.
(d) Calculate *TS*, *FS*, *SS*, and *IS* for each activity in the project.

4.15 Consider the following project network. The activity durations in weeks and the precedence relations are given in the following table.

Activity	Immediate predecessor	Duration
A	–	3
B	–	4
C	A	9
D	B	6
E	B	10
F	C, D	4
G	A	7
H	C, G	8
I	E, F, H	5

(a) Using AOA representation and CPM determine the critical path and critical activities together with their *TS* and *FS* values.
(b) If activity A is started with 2 weeks delay, how is this going to impact the project duration and what will be the new set of *TS* values of the other activities?

4.16 In a project the generalized precedence relation and the activity durations in weeks are given as:

- A > B; B > D; B, C > E; C > F; D, E, F > H; D, E > G
- E can only start 2 weeks after B has finished.
- C can start only 1 week after A has started.
- D can finish only 1 week after E has finished.

Activity	A	B	C	D	E	F	G	H
Duration (weeks)	6	7	8	9	10	12	11	5

(a) Draw the project network (AOA) and apply the node numbering procedure.
(b) Find the project duration, *ES*, *EF*, *LS*, *LF*, *TS*, and *FS* and the critical path.
(c) Solve this project scheduling problem using any LP solver. Compare the result you obtain with the one you obtained using CPM.

4.17 The project network appeared originally as shown below, where the activity durations are indicated on each arc and ϕ designates dummy activity. After 15 days the project has been initiated, the following activities have been completed: (2,6), (2,4), (2,10). Activity (4,16) has 5 more days left; activity (6,8) has 10 days remaining; activity (8,18) has not started but its duration is re-estimated as 4 days; and all remaining activities have not yet started.

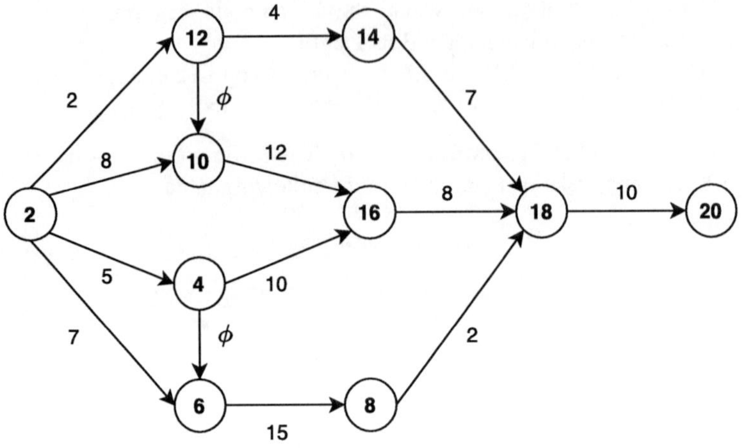

(a) Draw a new project network and evaluate the position of the company with respect to its original scheduled completion.
(b) Discuss whether one has to recalculate the critical path, or could one work with the slacks on the network and make the updating?
(c) Determine the precedence relationships of the original project network.

4.18 Consider the following project planning problem.
(a) Formulate the above problem as a linear programming problem and solve using any LP solver. Give an interpretation of the slack variables.
(b) Formulate the dual problem and solve using any LP solver. Give an interpretation of the dual variables.

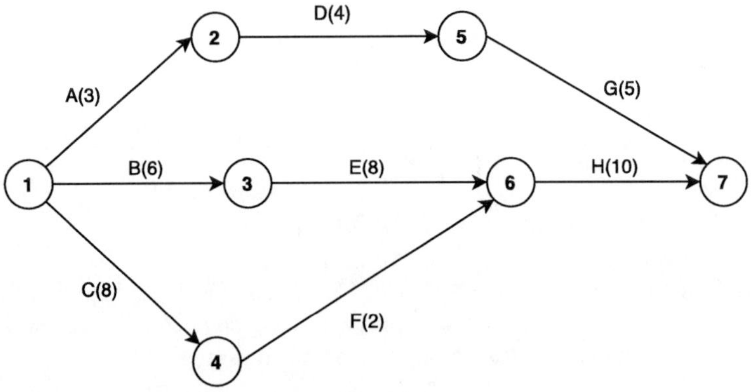

4.19 Consider the following project network.
(a) Determine the critical path and its length.
(b) Calculate *ES*, *LS*, *EF*, and *LF* for each activity in the project.
(c) Calculate *TS*, *FS*, *SS*, and *IS* for each activity in the project.

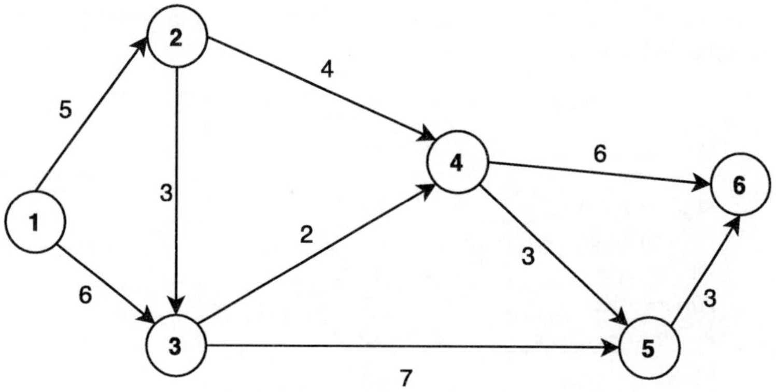

4.20 Let us consider a restaurant arrangement and construction. The activities are as follows:

Activity description	Predecessor	Duration (days)
A. Purchase coaches	–	10
B. Purchase restaurant equip.	–	3
C. Hire personnel	–	1
D. Select and purchase site	–	2
E. Obtain license and permits	D	7
F. Site preparation	E	3
G. Move coaches onto site	A, F	5
H. Install utilities	G	4
I. Install equipment	B, H	3
J. Decorate	B, H	6
K. Stock bar and kitchen	I, J	6
L. Advertisement	G	3
M. Train personnel	C, I	4
N. Undertake trial operation	K, L	7

(a) Draw the AOA representation of this project. Indicate the duration of each activity on the corresponding arc. Specify the node numbers such that an arc always goes from a smaller node number to a larger node number. Keep the number of dummy activities at a minimum. Determine the critical path and critical path length.

(b) Draw the AON representation of this project. Indicate the duration of each activity on the corresponding node. Determine the critical path and critical path length.

4.21 Consider the following pipeline renewal problem with the relevant data given in the table below.

Act	Activity description	Immediate Predecessor	Duration (days)
Q	Lead time	–	10
R	Line available	–	30
A	Measure and sketch	Q	2
B	Develop material list	A	1
C	Procure pipes	B	30
D	Procure valves	B	45
E	Prefabricate sections	C	5
F	Deactivate line	R, B	1
G	Erect scaffold	B	2
H	Remove old pipe and valves	F, G	6
I	Place new pipe	H, E	6
J	Weld pipe	I	2
K	Place valves	D, F, G	1
L	Fit up pipe and valves	K, J	1
M	Pressure test	L	1
N	Insulate	K, J	4
O	Remove scaffold	L, N	1
P	Cleanup	M, O	1

Assume each activity has a cost associated with it, which you can further assume to occur at the completion of the corresponding activity. How would you employ total slack for minimizing the NPV of the total cost? Would the policy you propose cause any disadvantage? Explain.

4.22 For the AOA project network below. The durations are indicated in parenthesis on the arcs:

(a) Construct a table showing the earliest start, latest finish, latest start, earliest finish.
(b) Determine the total slack and free slack values for the activities.
(c) Specify the critical path and the critical activities.

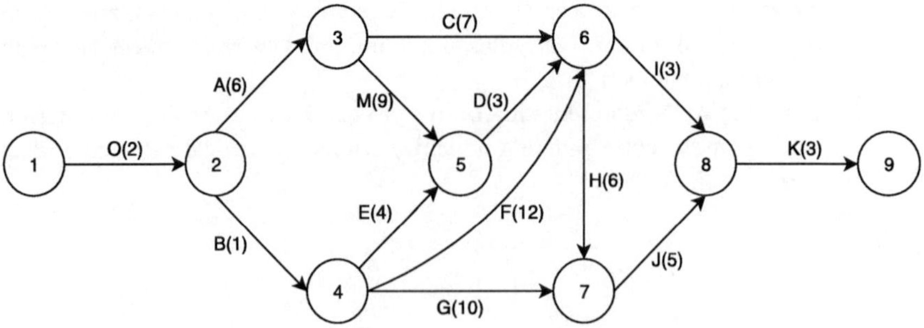

4.23 For the AOA type of project network given below with durations indicated on each activity:

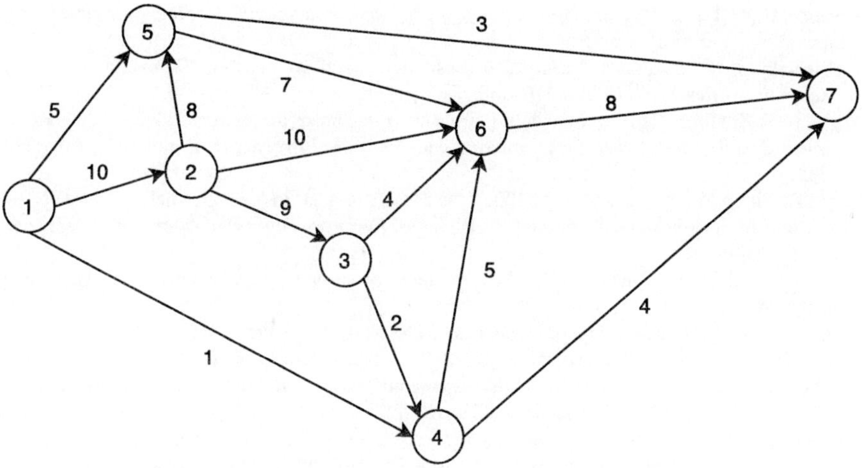

(a) Find the critical path and the critical activities.
(b) Find total float, free float, and independent float for each activity.
(c) How will your result change, if at all, if the duration of activity (5,6) is updated as 8 units of time?
(d) Draw the project network using the AON convention and apply CPM to this network.

4.24 Consider the project network information provided in Table below. Durations are in weeks. There is one initial payment of 60,000TL and one final payment of 140,000TL.

Activity	Duration	Predecessor	Cash flow occurs	Nature	Value (10^3 TL)
A	8	–	Beginning of activity	Cost	42
B	3	–	End of activity	Cost	67
C	4	A	End of activity	Receipt	55
D	2	B	End of activity	Cost	45
E	7	B	Beginning of activity	Cost	39
F	4	C, D	Beginning of activity	Receipt	40
G	9	E	End of activity	Cost	62

You are asked to determine the project schedule so as to maximize the profit for the contractor expressed in terms of its NPV. The interest rate is 1% per week.

(a) Write down the objective function in summation format. Explicitly define the decision variables.
(b) Write down the precedence relations for activity F in summation format.

References

Bellman, R. E. (1952). On the theory of dynamic programming. *Proceedings of the National Academy of Sciences, 38,* 716–719.

Denardo, E. V. (1982). *Dynamic programming models and applications.* New Jersey: Prentice Hall, Inc.

Elmaghraby, S. (2000). On criticality and sensitivity in activity network. *European Journal of Operational Research, 127*(2), 220–238.

Etgar, R. & Shtub, A. (1999). Scheduling projects to maximize net present value — the case of linear time-dependent cash flows. *International Journal of Production Research, 37*(2), 329–339.

Etgar, R., Shtub, A., & LeBlanc, L.J. (1996). Scheduling projects to maximize net present value — the case of time-dependent, contingent cash flows. *European Journal of Operational Research, 96*(1), 90–96.

Garfinkel, R. S. & Nemhauser, G. L. (1972). *Integer programming.* New York: John Wiley and Sons, Inc.

Goldratt, E. M. (1997). *The critical chain,* Mass: The North River Press.

Habibi, F., Barzinpour, F., & Sadjadi, S. J. (2019). A mathematical model for project scheduling and material ordering problem with sustainability considerations: A case study in Iran. *Computers & Industrial Engineering, 128,* 690–710.

Hall, N. G. (2016). *Research and teaching opportunities in project management.* In Optimization Challenges in Complex, Networked, and Risky Systems, INFORMS, 329–388.

Hazır, Ö., Kilic, O. A., Tunc, H. (2016). *A dynamic programming model for project scheduling under financial risks.* 15th International Conference on Project Management and Scheduling (PMS2016), Valencia, Spain, 239–242.

Kelley, J. E., Jr (1961). Critical path planning and scheduling: Mathematical basis. *Operations Research, 9,* 296–321.

Kelley, J. E. Jr. (1963). The critical-path method: Resources planning and scheduling. In: Muth, J. F. & G. L. Thompson (Eds.), *Industrial scheduling* (pp. 347–365). New Jersey: Prentice Hall, Inc.

Kelley, E. J. Jr & Walker, M. R. (1959). *Critical-path planning and scheduling.* In Proceedings Eastern Joint Computer Conference, Boston, 1–3 December, 160–173.

Nahmias, S. (2009). *Production and operations analysis* (6th ed.). New York: McGraw Hill Education.

Park, C. S. (2007). *Contemporary engineering economics* (4th ed.). Pearson Prentice Hall: Pearson Education.

Shtub, A., Etgar, R., & LeBlanc, L. J. (1996). Scheduling projects to maximize net present value-the case of time dependent, contingent cash flows. *European Journal of Operational Research, 96,* 90–96.

Shtub, A. & Etgar, R. (1997). A branch-and-bound algorithm for scheduling projects to maximize net present value: the case of time dependent, contingent cash flows. *International Journal of Production Research, 35*(12), 3367–3378.

Smith-Daniels, D. E. (1986). *Summary measures for predicting the net present value of a project.* Minnesota: Working Paper, College of St. Thomas, St Paul.

Vanhoucke (2000). *Exact algorithms for various types of project scheduling problems: Non-regular objectives and time/cost trade-offs.* PhD Dissertation, Department of Applied Economics, Katholieke Universiteit Leuven, Belgium.

Vanhoucke, M., Demeulemeester, E., & Herroelen, W. (2001). Maximizing the net present value of a project with linear time-dependent cash flows. *International Journal of Production Research, 39*(14), 3159–3181.

The Time/Cost Trade-off Problems 5

Learning Outcomes
Upon successful completion of this Chapter, the reader will be able to:

1. Define time and cost relationship in projects.
2. Describe the basic approaches and models used in modeling the time and cost relationship.
3. Implement project crashing.
4. Define continuous and discrete time/cost trade-off problems.
5. Model multimode project scheduling problems mathematically.

5.1 Problem Definition

Projects are subject to various sources of uncertainties that often negatively impact activity durations. Unexpected events such as machine breakdowns might prevent completing the activities on time. Contracting organizations may also revise the project time targets during execution and ask PMs to deliver the project outcomes earlier than originally planned. In both cases PMs must revise the project plans, which often requires expediting some activities in order to meet the project time goals.

Activity durations can be reduced in several different ways. The most straightforward way is to use more resources to complete the same activity, which will increase the activity and the project cost. A second way is to change the means (e.g., the technology) of accomplishing the activity. This might require a fundamentally different skill set and resource base, leading to new investments. Hence, changing the process and/or technology may be difficult to implement. A third alternative is to reduce the scope of the activity, which might require the consent of the stakeholders. However, the scope changes necessary to reduce the activity duration may not always be acceptable to the client, making this option infeasible. Finally, all else remaining the same, an activity can be completed earlier than its normal time by downgrading its quality requirements. As in the previous case, this option may not

G. Ulusoy, Ö. Hazır, *An Introduction to Project Modeling and Planning*, Springer Texts in Business and Economics, https://doi.org/10.1007/978-3-030-61423-2_5

always be feasible, as an inferior quality may not be acceptable to the client or other stakeholders.

In this chapter, we focus on the first two alternatives in a deterministic context with deterministic activity durations and costs, reducing the activity duration by allowing increase in cost. This trade-off, inherent in the two main classes of project objectives (cost and time), has been widely studied since the introduction of the critical path method (CPM) in the 1950s. We scrutinize this trade-off by investigating both the continuous and discrete descriptions of the time/cost trade-off (Içmeli et al. 1993; De et al. 1995) after examining the project cost components in some detail in order to model their time dependency.

Management accounting classifies costs according to the immediate needs of management (Park 2007). In the context of time/cost trade-off problems, we need to distinguish between *direct costs* and *overhead costs* at the activity level. The sum of these two major cost components, direct and overhead costs, constitutes the total project cost.

Types of project costs are treated in detail in Sect. 3.5.2. We will briefly define them here. *Direct costs* include direct labor and direct material costs. These are costs that are traceable to work performed during the execution of the activity. Direct costs of activities are usually increasing in decreasing activity durations because as more labor, equipment, and materials are assigned to an activity, it would be expected to complete the activities earlier unless restricted by some constraints such as technical restrictions. Based on this observation, we will assume that direct project cost is increasing in decreasing activity duration.

Overhead costs are considered to cover all the costs associated with a project in addition to the direct costs as defined above. It covers both the project overhead cost and the general overhead cost. The project overhead cost is related to the management of the project whereas the general overhead cost is related to the management of the overall organization. Formally, those costs that are incurred as a result of activities that support the project in general rather than contributing directly to the project goal(s), can be classified as overhead costs (Badiru and Pulat 1995). They include administrative costs and various expenses, such as maintenance, security, heating, ventilating, and lighting, inventory holding cost, etc., that increase with project duration. Therefore, the overhead cost is assumed to be an increasing function of the project duration in the remainder of this chapter.

We illustrate the total project cost, its components, and the time/cost relationships in Fig. 5.1. We observe that the total project cost reaches a minimum value at a certain time. This level is called the *optimal project duration* as it minimizes total project cost. However, this level might not be desirable for the clients; they might have different preferences. To give an example, in infrastructure projects clients often wish to start operating the facilities earlier and usually prefer early deliveries of the project outcomes. Consequently, two parties, client and contractor, negotiate the project deadline, finalize it before starting the project, and make it legally binding through a contractual agreement. The contractual issues are discussed in some detail in Chap. 9 Project Contract Types and Payment Schedules.

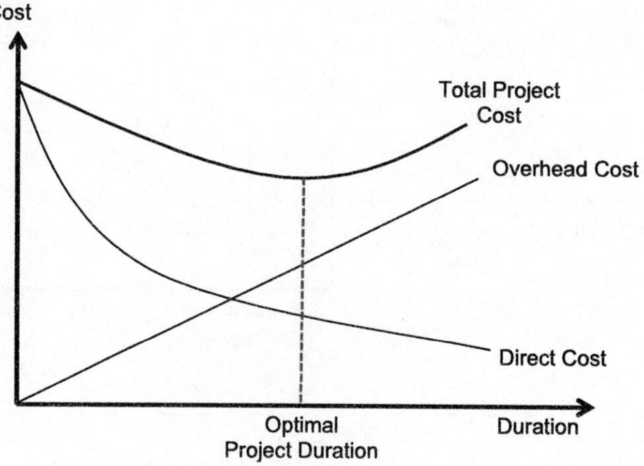

Fig. 5.1 Total project cost and its components

Contracts usually include penalties for delays and, in some cases, bonuses for early completion. In fixing the contractual clauses specifying the deadline and penalties, the trade-off between total cost and project duration plays an important role. The levels of penalty and bonus certainly depend on the time/cost trade-off functions illustrated in Fig. 5.1. Having emphasized its importance in contract management in projects, we next describe how to perform time/cost analysis in a project.

Our analysis first considers the direct costs, which constitute a significant portion of the budget. Overhead costs will be addressed later and integrated into the mathematical analysis. While studying the direct costs, we consider two main processing modes, normal and crash, and formulate the corresponding parameters.

Under the normal setting, we assume the activity is performed using the regular resource assignments operating at standard productivity levels. The normal activity duration, D^N, and the corresponding normal direct cost, C^N, are estimated. Crash mode represents the maximum possible time compression. Crash duration, D^C, thus refers to the minimum possible activity duration and the crash direct cost, C^C, to the corresponding activity cost. Given these parameters, time/cost analysis aims at specifying the activity durations in the range, $[D^C, D^N]$ in a manner that minimizes the total project cost.

Having defined the relevant time and cost parameters, we first assume that direct costs increase at a constant rate as duration decreases; i.e., direct cost is a linear function of time. Figure 5.2 represents a linear time-cost relationship that is continuous both in duration and direct cost.

Materials or electrical energy are examples of these types of continuous resources. However, many other resources, especially equipment and labor, are discrete in nature and will be discussed in Sect. 5.3.

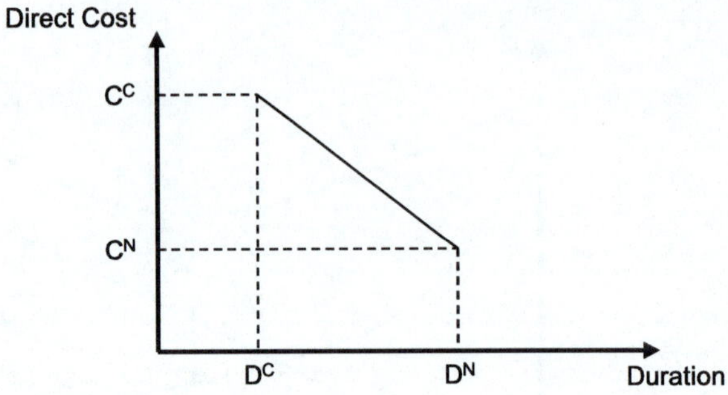

Fig. 5.2 Time-cost relationship: A linear function

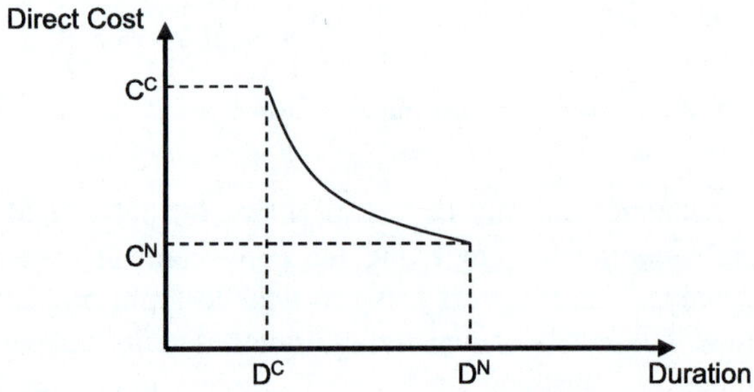

Fig. 5.3 Time-cost relationship: A convex function

Distinct from continuity, the linearity assumption might also not be valid for many activities in practice. The marginal cost of a reduction in duration might vary over the execution of the activity; it may become more costly due to technical reasons or decreases in the productivity of the allocated resources. For example, the productivity of overtime workers decreases as they work more; overtime might also create a need for additional workers or equipment (Klastorin 2004). This case, where the marginal cost of reducing the duration increases at an increasing rate, is illustrated by the convex function in Fig. 5.3.

There are also cases where marginal cost increases at a decreasing rate. An investment in new technology might lead to higher costs for the initial implementation but lower incremental costs once the installation is complete (Klastorin 2004). Learning might also reduce the marginal costs as experience with the task or the technology accumulates. This case is illustrated by the concave function in Fig. 5.4. In all the three cases, linear, convex, and concave, the project direct cost is a non-increasing function of the project duration.

We note that non-linear cost curves can be approximated using linear functions, which may be helpful in solving the complex planning problems. Figure 5.5 demonstrates a piece-wise linear direct cost curve approximating a convex direct cost curve.

Fig. 5.4 Time-cost relationship: A concave function

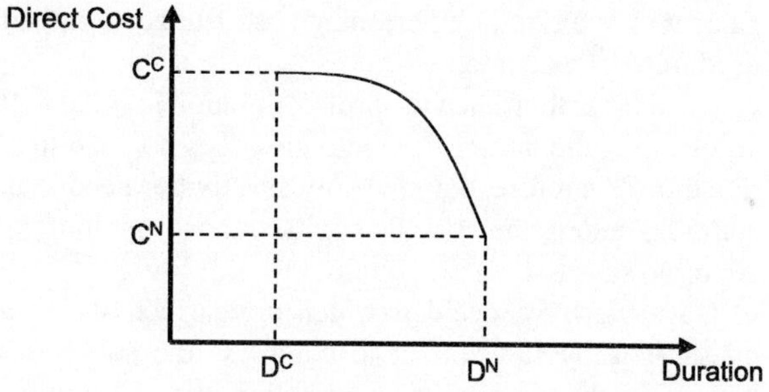

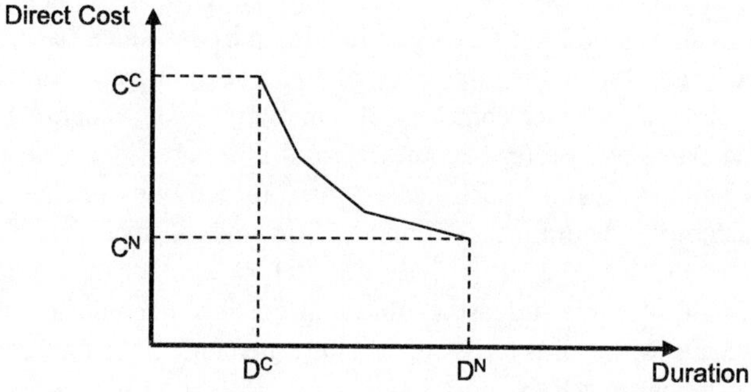

Fig. 5.5 Piecewise linear direct cost function

In the next sections, we will reconsider the time-cost functions and develop optimization models and algorithms to speed up projects by reducing their duration in a cost-effective manner. We first introduce the basic principles to consider and develop an algorithm in the next section.

5.2 Time/Cost Trade-Off in Projects

5.2.1 Project Crashing

In Sect. 5.1, we noted that that expediting the projects by employing additional resources, referred to as project crashing, is a widely encountered practice in projects. However, activities can only be compressed within limits imposed by the technological, economic, and managerial considerations. The minimum activity duration thus specified is called the *crash limit*. In addition, crashing is costly, as it requires the allocation of extra resources. Therefore, PMs must apply crashing selectively, choosing the activities to crash in order to achieve the desired time gains with the minimum crashing cost. Budget constraints and technical limits both need to be considered.

Remember that since the project duration is equal to the length of the critical path, accelerating the non-critical activities does not lead to a decrease in project completion time. Therefore, only the critical activities need to be considered for crashing. In addition, among the critical activities, those resulting in the minimum crashing cost are to be selected.

Once we have decided which critical activity or activities to crash, the next question is by how much to compress the selected activity(ies). There are two bounds on the amount of compression, the *crash buffer* and *free slack buffer*. The *crash buffer* is the difference between the current activity duration and the crash limit, while the *free slack buffer* is the value of the current smallest positive free slack of the non-critical activities. Crashing within these bounds ensures that the current critical path, which we work on, remains critical without some other path(s) becoming critical. We employ free slack for this purpose since basing this buffer on total slack might reduce the total slack of some of the succeeding activities and hence, might prevent a better compression alternative from emerging later in the process. If the free slack buffer becomes zero, it is ignored when determining the compression amount. In the special case when all activities become critical, the minimum crash buffer determines the compression amount since all free slack values are zero.

The observations above suggest a solution procedure for a basic crashing problem, which assumes the direct cost of crashing per unit time for each activity is constant over the entire activity duration.

We distinguish here between two types of time-cost trade-off problems. In the first of these, we consider only the trade-off between the direct cost and the project duration, ignoring the overhead cost, and seek the *shortest project duration* obtained at *minimum direct cost* for the problem. In the second, we search for the *minimum total project cost* resulting in the *optimum project duration*.

5.2.2 An Algorithm for Project Crashing

In the following algorithm, linear time-cost relationship is assumed. The analysis will be extended to continuous activity durations in Sect. 5.3. We present an algorithm for compressing the project to a project duration value targeted by the decision-maker or to its shortest duration at minimum direct cost. The durations will be compressed in integer numbers.

Step 1: *Initialization.* Given the normal and crash duration and cost values for the activities, calculate the cost of crashing per unit time for each activity. Determine the critical path(s) and the critical activities. Calculate the project direct cost. Set iteration counter $k = 1$. Continue from Step 3.

Step 2: *Termination.* If the targeted project duration is reached or all activities on at least one critical path are at their crash limit, then stop. Otherwise, set $k = k + 1$ and continue.

Step 3: *Iteration k.* If the critical path is unique, then determine the critical activity to be compressed as the one with the least cost of crashing per unit time among the critical activities that have not yet reached their crash limit. Determine its crash buffer. If there is more than one critical path, then determine the set of critical activities with the least total cost of crashing per unit time among the critical activities, which have not yet reached their crash limit, and such that their simultaneous compression by a certain amount will result in the compression of the project duration by the same amount. The crash buffer of this set of activities is the minimum of all the crash buffers of the activities in the set. Determine the free slack buffer and let the compression value be the minimum of the crash buffer and the free slack buffer. Compress the critical activity(ies) by the compression value. On the resulting project network determine the critical path(s), the critical activities, and the project duration. Calculate the current project direct cost. Go to Step 2.

The algorithm might terminate in several different ways.

(i) If the algorithm terminates because the targeted project duration is reached, then it does not correspond to the shortest project duration whose further reduction is possible if so desired.
(ii) If the algorithm terminates because the targeted project duration is reached and all activities on at least one critical path are at their crash limit, then the targeted project duration corresponds to the shortest project duration.
(iii) If the algorithm terminates because all activities on at least one critical path are at their crash limit, then the shortest project duration is reached and is larger than the targeted project duration, making the targeted project duration infeasible.

Based on linear time-cost relation, we denote the direct cost of compressing per unit time for each activity *(i,j)* by a_{ij} and formulate it as follows:

$$a_{ij} = \frac{C_{ij}{}^{C} - C_{ij}{}^{N}}{D_{ij}{}^{N} - D_{ij}{}^{C}} \tag{5.1}$$

Note that a_{ij} is also the absolute value of the slope of the linear time-cost relation displayed in Fig. 5.2, which we will refer to as the *slope* from here on.

Example 5.1 Consider the following project compression problem, where the PM wants to know all possible project durations with their associated minimum direct project costs. The activity data is given in Table 5.1. The project network is displayed in AOA format in Fig. 5.6.

Solution
Step 1: *Initialization.*

Using (5.1) we calculate the direct costs of crashing per unit time, i.e., the slopes (Table 5.2).

Table 5.1 The duration and cost data for the Example Problem 5.1

	Normal		Crash	
Activity	Duration	Cost	Duration	Cost
(1,2)	8	100	6	200
(1,3)	4	150	2	350
(2,4)	2	50	1	90
(2,5)	10	100	5	400
(3,4)	5	100	1	200
(4,5)	3	80	1	100

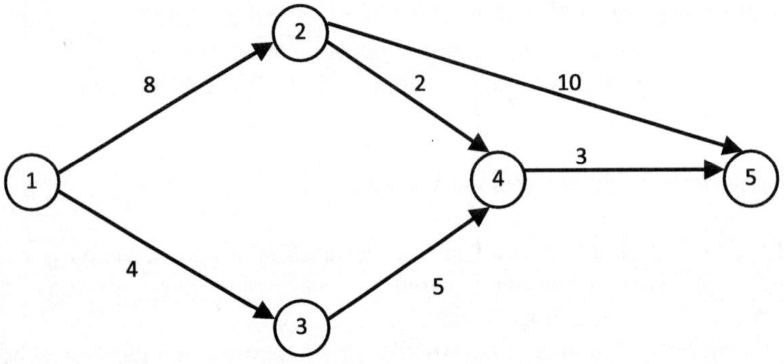

Fig. 5.6 The AOA project network for the Example Problem 5.1

Table 5.2 The slopes for the activities

Activity	(1,2)	(1,3)	(2,4)	(2,5)	(3,4)	(3,5)
Slope	50	100	40	60	25	10

Table 5.3 Initial CPM output table

Activity	ES	EF	LS	LF	TS	FS
(1,2)	0	8	0	8	0	0
(1,3)	0	4	6	10	6	0
(2,4)	8	10	13	15	5	0
(2,5)	8	18	8	18	0	0
(3,4)	4	9	10	15	6	1
(4,5)	10	13	15	18	5	5

We apply the forward pass and backward pass to obtain the earliest and latest occurrence times from which the information in Table 5.3 is generated.

The project duration obtained as a result of the forward pass is 18 time units. The critical path is determined as {1–2-5} with (1,2) and (2,5) as the critical activities. The project direct cost is calculated as 580 using the normal direct cost values from Table 5.1. $k = 1$.

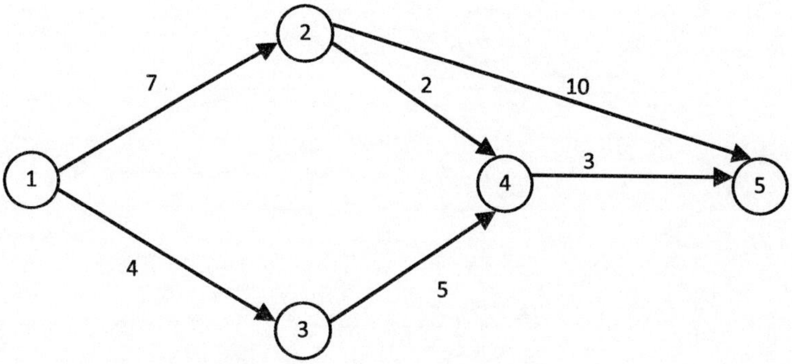

Fig. 5.7 The project network after the first compression

Table 5.4 CPM output table resulting from the first iteration

Activity	ES	EF	LS	LF	TS	FS
(1,2)	0	7	0	7	0	0
(1,3)	0	4	5	9	5	0
(2,4)	7	9	12	14	5	0
(2,5)	7	17	7	17	0	0
(3,4)	4	9	9	14	5	0
(4,5)	9	12	14	17	5	5

Step 3: *First Iteration* ($k = 1$):

We observe from Table 5.2 that activity (1,2) has a smaller slope (marginal cost of compression) than activity (2,5) and hence, is chosen for compression.

Crash buffer (1,2) = 8–6 = 2.

Free slack buffer = min{1,5} = 1.

Compression value = min{2,1} = 1.

Hence, activity (1,2) is compressed by 1 time unit giving the project network depicted in Fig. 5.7. We calculate the new project direct cost value to obtain:

Project direct cost = 580 + (50) = 630.

We apply forward pass and backward pass to the resulting project network to generate the output presented in Table 5.4.

The project duration is crashed to 17 weeks. The critical path remains as {1–2-5} with (1,2) and (2,5) being the critical activities. We go to Step 2 to check for termination.

Step 2: Since not all activities on the critical path are at their crash limit, we set $k = 2$ and continue.

Step 3: *Second Iteration* ($k = 2$):

We know from the previous iteration that activity (1,2) has a smaller slope than activity (2,5) and hence it is chosen for compression.

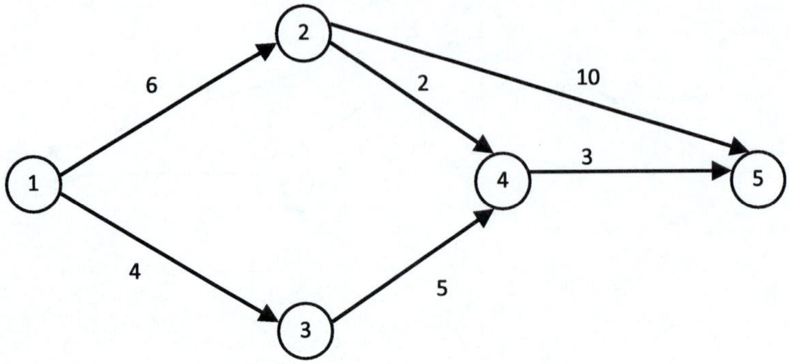

Fig. 5.8 The project network after the second compression

Table 5.5 CPM output table resulting from the second iteration

Activity	ES	EF	LS	LF	TS	FS
[a](1,2)	0	6	0	6	0	0
(1,3)	0	4	4	8	4	0
(2,4)	6	8	11	13	5	1
(2,5)	6	16	6	16	0	0
(3,4)	4	9	8	13	4	0
(4,5)	9	12	13	16	4	4

[a]indicates activities, which have reached their crash limit

Crash buffer = 7–6 = 1.

Free slack buffer = 5.

Compression value = min{1,5} = 1.

Hence, activity (1,2) is compressed by 1 time unit and reaches its crash limit. The resulting project network is displayed in Fig. 5.8. We calculate the current project direct cost.

Project direct cost = 630 + 50 = 680.

We apply forward pass and backward pass to the resulting project network to generate the output presented in Table 5.5.

The project duration is reduced to 16 time units. The critical path remains {1–2-5}. (1,2) with (2,5) as the critical activities.

Step 2: Since not all activities on the critical path are at their crash limit, we set $k = 3$ and continue.

Iterations continue in this manner. Let us skip to the results of iteration $k = 6$. The project network after the sixth compression is shown in Fig. 5.9.

We apply forward pass and backward pass to the resulting project network to generate the output presented in Table 5.6.

We now have two critical paths; {1–2-5} and {1–3–4-5} with lengths of 12 time units.

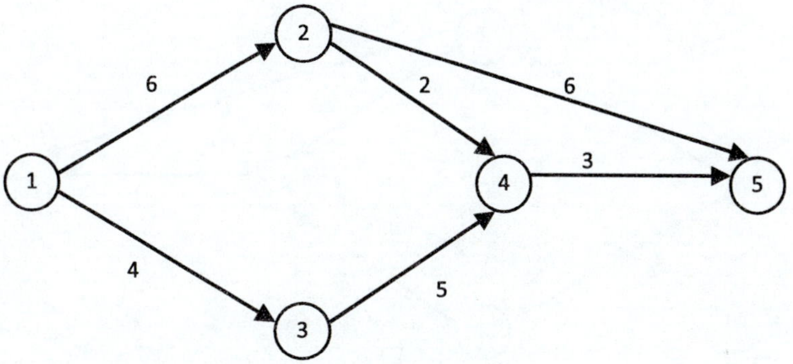

Fig. 5.9 The project network after the sixth compression

Table 5.6 CPM output table for the sixth iteration

Activity	ES	EF	LS	LF	TS	FS
[a](1,2)	0	6	0	6	0	0
(1,3)	0	4	0	4	0	0
(2,4)	6	8	7	9	1	1
(2,5)	6	12	6	12	0	0
(3,4)	4	9	4	9	0	0
(4,5)	9	12	9	12	0	0

[a]indicates activities, which have reached their crash limit

Step 3: *Seventh Iteration* ($k = 7$):

Since we have two critical paths, it is necessary now to reduce the duration of the two critical paths simultaneously. We consider all combinations of critical activities reducing both critical paths' duration and choose the one having the smallest total compression cost per unit time.

For combination (1,3) and (2,5): Compression unit cost = 100 + 60 = 160.

For combination (4,5) and (2,5): Compression unit cost = 10 + 60 = 70.

For combination (3,4) and (2,5): Compression unit cost = 25 + 60 = 85.

Hence, the combination with the least unit cost consists of activities (2,5) and (4,5).

Crash buffer = Min{1,2} = 1.

Free slack buffer = 1.

Compression value = Min{1,1} = 1.

Compressing activities (2,5) and (4,5) by 1 time unit, we obtain the project network in Fig. 5.10. Activity (2,5) has reached its compression limit. The project direct cost is calculated as follows:

Project direct cost = 920 + 70 = 990.

The previous two critical paths are preserved, with the new project duration of 11 time units (Table 5.7).

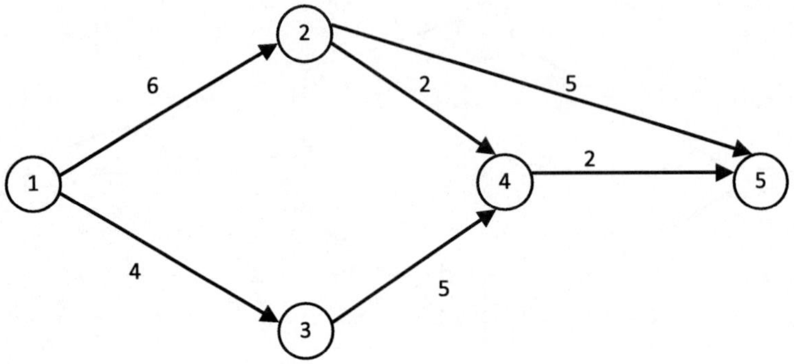

Fig. 5.10 The project network after the seventh compression

Table 5.7 CPM output table for the seventh iteration

Activity	ES	EF	LS	LF	TS	FS
[a](1,2)	0	6	0	6	0	0
(1,3)	0	4	0	4	0	0
(2,4)	6	8	7	9	1	1
[a](2,5)	6	12	6	11	0	0
(3,4)	4	9	4	9	0	0
(4,5)	9	11	9	11	0	0

[a]indicates activities, which have reached their crash limit

Table 5.8 Total project cost and its components for the Example Problem 5.2

Project Duration	Direct Cost	Overhead Cost	Total Project Cost
11	990	605	1595
12	920	660	1580
13	860	715	1575
14	800	770	1570
15	740	825	1565
16	680	880	1560
17	630	935	1565
18	580	990	1570

Step 2: Since all activities on the critical path {1-2-5} have reached their crash limit, we cannot compress the project duration further. The algorithm terminates.

The shortest project duration is 11 time units and the associated project direct cost is 990. The project durations and direct costs generated throughout the compression process are presented in the first two columns of Table 5.8. The decision maker may now choose the most appropriate combination of project duration *vs.* project direct cost among those generated during the compression process.

Let us extend Example 5.1 so as to determine the *minimum total project cost* corresponding to the *optimal project duration*. For that purpose, we will add the overhead cost to the project direct cost to obtain the total project cost.

Example 5.2 Let us assume that in the setting of Example 5.1, the overhead cost is 55 units per unit time; hence, the overhead cost function is linear in the project duration. Note that the overhead cost component can be displayed on the project network in Fig. 5.6 by adding an arc between the start and terminal nodes with duration equal to the completion time of the project and a per unit time crash value a_j equal to the negative of the overhead cost per unit time; (−55 per unit time for this problem). We will now determine the optimal project duration. For each project duration alternative between the initial and the final project durations, the overhead and direct cost components and the total project cost are reported in Table 5.8.

The minimum total project cost is 1560, with the corresponding optimal project duration being 16 time units.

Starting the compression with all the activities at their normal duration, one of the following three outcomes will occur: (*i*) The total project cost increases, and we terminate compression. (*ii*) The total project cost does not change, so we proceed with compression. (*iii*) The total project decreases, and we proceed with compression.

5.2.2.1 Determining the Minimum Total Project Cost

The crashing algorithm in Sect. 5.2.2 can be employed for this purpose as well. Starting with all activity durations at their normal duration, we apply the crashing algorithm and obtain the total project cost resulting from the current crash. The first time we observe the total project cost to increase, we stop and declare the previous project duration as the optimal project duration and the total project cost as the minimum total project cost.

This approach is based on the observation that the total project cost is unimodal, i.e., has a single optimal point – in this case, a minimum point. Unimodality arises here due to the fact that both cost curves making up the total project cost are continuous and monotone increasing for overhead cost and monotone decreasing for direct cost as a function of project duration in both cases. Hence, once we have reached the minimum cost, decreasing the project duration further will lead to an increase in the total project cost.

5.2.2.2 Generalized Precedence Relationships

When the project network contains GPRs, the solution procedure becomes rather involved. Elmaghraby and Kamburowski (1992) study the case where the time-cost functions for the activities are piecewise linear and convex over the interval bounded by the crash and normal durations of the activities. They do not assume that the time-cost functions are non-increasing. They consider three problems: (*i*) The minimum total cost problem where a due date is assigned by which the project needs to be completed. (*ii*) Determining the complete optimal duration – total cost function in an efficient manner. (*iii*) Determining the optimal activity durations for the case where the due date is considered as a target date with rewards and penalties involved for early and late completion of the project, respectively. The interested reader is referred to the paper cited above.

The algorithm presented in this section would not be practical for large project networks. Determining the activity combinations for compression in large networks would become a rather complex process when applying this algorithm. Hence, optimization models or approximation algorithms need to be devised for the generation of activity combinations for compression efficiently for large project networks. The next section will focus on such optimization models.

5.3 Continuous Time/Cost Trade-off Problems

In this section, we address the problem with continuous time and cost variables on both AOA and AON networks.

5.3.1 Activity-on-Arc Representation

We consider a project with n activities and represent the network with an AOA graph $G\ (M, A)$, where M denotes the set of events (nodes) and $A \subset M \times M$ the set of the arcs (activities). Let δ be the project duration specified to realize the project, i.e., the deadline; T_i the occurrence time for event i, $i = 1,\ldots,m$; a_{ij} the direct cost of compressing per unit time for activity (i,j); d_{ij} is the duration of activity (i,j). D_{ij}^{N} is the normal duration of activity (i,j); and D_{ij}^{C} is the crash duration of activity (i,j). Given this notation, an LP model of the *minimum crashing cost problem* for a given project duration δ can be formulated as follows:

$$\text{Min} \sum_{(i,j)\in A} a_{ij}\left(D_{ij}^{N} - d_{ij}\right) \tag{5.2}$$

subject to

$$T_j - T_i - d_{ij} \geq 0 \quad \forall (i,j) \in A \tag{5.3}$$

$$T_m \leq \delta \tag{5.4}$$

$$D_{ij}^{C} \leq d_{ij} \leq D_{ij}^{N} \qquad \forall (i,j) \in A \tag{5.5}$$

$$T_i \geq 0 \quad i = 1, \ldots, m \tag{5.6}$$

The decision variables in this model are compression amounts, i.e., the activity durations. Both T_i and d_{ij} are decision variables. Total crashing cost is minimized (5.2). Precedence constraints (5.3) and the deadline (5.4) must be satisfied without violating crash limits for any activity (5.5).

This model minimizes only the crashing cost; normal costs are the minimum amounts to be paid for all the activities and their sum is constant. Therefore, minimizing the total crashing cost is equivalent to minimizing the total direct costs.

Overhead costs could also be considered by modifying the objective function as follows:

$$Min\left(\sum_{(i,j)\in A} a_{ij}\left(D_{ij}{}^{N}-d_{ij}\right)+OT_{m}\right) \tag{5.7}$$

In the objective function (5.7), O denotes the overhead cost per unit time and is assumed to be constant through the project duration.

5.3.2 Activity-on-Node Representation

We now represent the project network with an AON graph, i.e., G (N, A); where N denotes the set of nodes (activities) and $A \subset N \times N$ the set of arcs representing immediate precedence constraints on the activities. In addition to n project activities, we add two dummy activities corresponding to the project start and end, activity 0 and activity n+1, in the network; i.e. $N = \{0,1,2. \ldots.., n+1\}$.

Defining the decision variables ST_i and d_i, representing the start time and the duration, respectively, of each activity i, an LP model for the *minimum crashing cost problem* for a given project duration δ can be formulated as follows:

$$Min\sum_{i\in N} a_i\left(D_i^N-d_i\right) \tag{5.8}$$

subject to

$$ST_j - ST_i \geq 0 \quad \forall i,j \in N \tag{5.9}$$

$$ST_{n+1} \leq \delta \tag{5.10}$$

$$D_i{}^C \leq d_i \leq D_i{}^N \quad \forall i \in N \tag{5.11}$$

$$ST_i \geq 0 \quad \forall i \in N \tag{5.12}$$

This model for AON networks is similar to that given for AOA networks (5.2–5.6). However, the definitions of the parameters and the variables are different. Nodes represent activities and only one index is used to define the parameters.

As in (5.6), overhead costs can also be considered using the modified objective function:

$$Min\sum_{i\in N} a_i\left(D_i^N-d_i\right)+OST_{n+1} \tag{5.13}$$

In (5.13), total overhead cost is proportional with the project completion time shown as the start/finish time of the dummy activity $n + 1$, ST_{n+1}.

5.3.2.1 Model Solution

Both scheduling models above are LP models and can be solved efficiently for even large-scale networks using commercial LP solvers such as ILOG CPLEX, GAMS etc.

The continuous problem formulated above for both types of network representations, assumes that all time points in the duration range [D^C, D^N] are feasible. However, in practice, due to resource restrictions or technical limitations activities can usually be performed at some discrete points. Any feasible point corresponds to a processing mode and to a (time, cost) pair. The discrete versions of the problem will be investigated in the next section.

5.4 Discrete Time-Cost Trade-off Problems

In the discrete case, we consider a finite number of time/cost pairs for each activity. Each pair refers to a specific resource allocation to perform an activity, referred to as an *execution mode*. Multi-mode scheduling problems, which integrate different processing alternatives and seek to select exactly one execution mode for each activity, are important in practical projects. We refer to the construction project given by Chen and Weng (2009) as an example. This project consists of 37 activities and some of them could be in two alternative modes. For instance, the excavation activity could be processed in two modes. The first mode refers to processing with 2 workers and 2 new excavators, the other one employs 2 workers and 2 old excavators. The first is faster, it takes 10 days but costs $10,000; whereas the second one completes in 12 days with a cost of $9000. The multi-mode scheduling problem involves choosing the mode of this activity and of the other ones. The criticality of the activity, whether it is on the critical path or not, and the tightness of the deadline or the budget are the factors that impact this decision. Considering these practical implications, we investigate a basic multi-mode project scheduling problem in this section; namely, the discrete time/cost trade-off problem (DTCTP).

Three versions of the DTCTP have been studied in the literature: the deadline problem (DTCTP-D), the budget problem (DTCTP-B), and the efficiency problem, (DTCTP-E). In DTCTP-D, given a set of time/cost pairs (mode) and a project deadline, each activity is assigned to one of the possible modes in such a way that the total cost is minimized. The DTCTP-B minimizes the project duration while meeting a given budget, while DTCTP-E is the problem of constructing efficient time/cost solutions (for a definition of efficient solutions, refer to Sect. 7.6) over the set of feasible project durations.

In the following sections, we will focus on the deadline and the budget problems. Let us first formally define the problems.

5.4.1 The Deadline Problem

Given a project deadline of δ time units, exactly one time/cost pair (mode) is to be chosen for each activity so that the total project cost is minimized. An activity i is performed in one of the possible modes, i.e., $m \in M_i = \{\underline{m}_i, \ldots\ldots, \overline{m}_i\}$. If performed in mode m, the activity i is characterized by a processing time d_{im} and cost c_{im}. Without loss of generality, it can be assumed that, for each $j \in N$, $m < m'$ implies $d_{im} > d_{im'}$ and $c_{im} < c_{im'}$. Thus, all modes $m \in M_i$ are non-dominated and are indexed in decreasing order of activity durations and increasing order of activity costs. Given this data, a mixed integer-programming model of DTCTP-D can be formulated as follows:

$$\text{Min} \sum_{i \in N} \sum_{m \in M_i} c_{im} x_{im} \tag{5.14}$$

subject to

$$\sum_{m \in M_i} x_{im} = 1, \forall i \in N \tag{5.15}$$

$$FT_j - FT_i - \sum_{m \in M_i} d_{im} x_{im} \geq 0, \quad \forall i \in N \tag{5.16}$$

$$FT_{n+1} \leq \delta \tag{5.17}$$

$$FT_i \geq 0, \quad \forall i \in N \tag{5.18}$$

$$x_{im} \in \{0, 1\} \quad \forall m \in M_i, \ \forall i \in N \tag{5.19}$$

The non-negative continuous decision variable FT_i denotes the completion time of activity i. The binary decision variable x_{im} assigns modes to the activities, taking a value of 1 if mode m is chosen for activity i, and 0 otherwise (5.19). While minimizing the total cost (5.14), a unique mode should be assigned to each activity (5.15), precedence constraints should not be violated (5.16), and the deadline should be met (5.17).

5.4.2 The Budget Problem

A mixed integer-programming formulation of the DTCTP-B can be given as follows:

$$\text{Min} \quad FT_{n+1} \tag{5.20}$$

subject to

$$\sum_{m \in M_i} x_{im} = 1, \forall i \in N \tag{5.21}$$

$$FT_j - FT_i - \sum_{m \in M_i} d_{im} x_{im} \geq 0, \quad \forall i \in N \tag{5.22}$$

$$\sum_{i \in N} \sum_{m \in M_i} c_{im} x_{im} \leq B \tag{5.23}$$

$$FT_i \geq 0, \qquad \forall i \in N \tag{5.24}$$

$$x_{im} \in \{0, 1\} \qquad \forall m \in M_i, \ \forall i \in N \tag{5.25}$$

In contrast to the deadline problem, the project completion time, FT_{n+1}, is minimized (5.20), without exceeding the budget B (5.23). Note that the start and finish times of the end node (dummy activity $n + 1$) are the same.

Both DTCTP-D and DTCTP-B are *NP*-hard problems (De et al. 1995) (see Appendix 7A for a definition of *NP*-hard problems). Unlike the continuous formulations, it is difficult to solve large instances exactly. Only small and medium size instances with less than 150 activities and 10 modes per activity have been solved to optimality using branch and bound algorithms (Demeulemeester et al. 1996, 1998) or decomposition methods (Hazır et al. 2010a). To solve large real-life instances, several approximate algorithms have been developed (Akkan et al. 2005; Vanhoucke and Debels 2007).

5.5 Conclusions, Recent Developments, and Some Future Research Directions

In this chapter we have studied the continuous and discrete time/cost trade-off problems, formulated mathematical models for them and presented several solution approaches. These problems are practically relevant as it is often possible to reduce the duration of activities by incurring additional expenses, and real life projects often have several activity processing alternatives. However, all these problems assume deterministic activity durations. In practice, while executing a project, the activities might be delayed due to several reasons. The impact of the delays on project completion depends heavily on the structure of the project network (Hazır et al. 2010b).

In this regard, Larson and Gray (2018) focuses on *sensitive* project networks with several critical or near-critical paths. At this point, how to define the (near) criticality becomes a challenging question. To identify the near critical activities, Hazır et al. (2010b) specify a threshold for the slack/duration ratio (SDR). Activities with SDR ratio below the threshold are defined as near critical. It makes sense to assess the activity slacks with respect to the activity's duration, because, we usually expect to observe longer delays with longer duration activities. As a result, the higher the SDR, the higher its capacity for preventing delays.

Since delays in near critical activities probably result in delays in the project completion time, sensitive networks are not preferred regarding robustness. In Chap. 12, we study project uncertainty and robust project scheduling, and extend the time/cost analysis by integrating the near critical activities.

Exercises

5.1 We use free slack rather than total slack in the project crashing algorithm. Give a justification for that.

5.2 Consider the project crashing algorithm given in Sect. 5.2.2. How would you extend this algorithm for the case of finding the optimal project duration?

5.3 A small shipbuilding company building simpler types of small commercial ships is building a towboat. The project consists of 8 major activities. The normal and crash durations of each work package (WP) (in weeks) together with the crash cost per week and the precedence relationships are given in the table below.

WP	Predecessors	Normal duration	Crash duration	Crash cost
A	–	5	3	15
B	–	6	4	20
C	A	6	5	8
D	A	5	4	14
E	B	5	3	9
F	B	5	4	12
G	C, E	4	2	22
H	D, F	3	1	16

Assume that the unit crash cost is the same for all WPs.

(a) Specify all possible crash options available initially.

(b) Order the options in increasing crash cost and crash the first option.

5.4 (a) For the following project network write in explicit form the linear programming formulation of the time/cost trade-off problem for a given project duration, T, to minimize the total cost. Employ AOA representation.

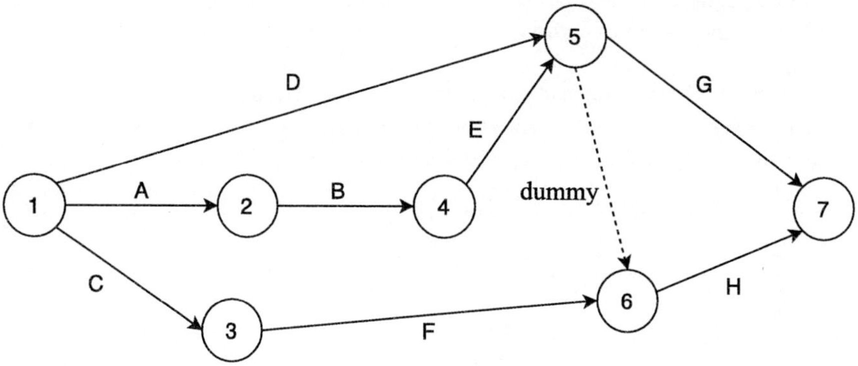

Activity	A	B	C	D	E	F	G	H
Normal duration	3	4	4	5	6	8	7	3
Crash duration	3	3	2	1	6	8	4	1
Unit cost of crashing	–	50	34	45	–	–	39	23

Let T_i be the realization time for event i, $i = 1,\ldots,7$; D_{ij} be the duration of activity (i, j). Let the unit indirect cost be 20 and be a constant overall project duration. Solve the formulation using any LP solver.

(b) Making use of the formulation in part (a), explain how you would obtain the optimum project duration.

5.5 (a) Using the data provided in the table below write down in explicit form the LP formulation for the time/cost trade-off problem with linear continuous compression cost function so as to minimize the project total cost. The project total cost is the sum of the compression costs and indirect cost components. Durations are in weeks. Unit overhead cost is 54 TL/week. Employ activity-on-arc formulation.

(b) Solve the formulation obtained in part (a) using any LP solver.

Activity	Immediate predecessor	Normal duration DN	Crash duration DC	Slope c (TL/week)
A	–	3	1	50
B	–	4	2	43
C	A	9	6	35
D	B	6	3	28
E	C, D	2	2	–

5.6 A project with 6 activities is given and is modeled as an AOA project network. The unit indirect cost is given as 100TL/day. The normal and crash durations as well as their associated costs and the absolute value for the slope of the linear direct compression cost curve are given in the table below. All durations are in days and costs in TL.

For this problem, you can obtain the critical paths by simply obtaining the occurrence times for the event nodes and showing them by the event nodes on the AOA project network you will draw.

Activity	Immediate predecessor	Current duration	DN	DC	CN	CC	c (Slope)
(1,2)	–	6	6	4	40	80	20
(1,3)	–	5	5	1	100	300	50
(2,4)	(1,2)	7	7	6	70	140	70
(3,4)	(1,3)	8	9	7	90	210	60
(3,5)	(1,3)	3	3	2	40	80	40
(4,5)	(2,4), (3,4)	6	6	4	125	385	130

(a) Write down the initial list of compression options for the project above together with their direct costs.
(b) In part (a), which activity or activities will you compress and by what amount? Show your results in detail.
(c) At the end of the second iteration, activities (1,2) and (2,4) are at their crash limit; and d(1,3) = 2; d(3,4) = 8; d(3,5) = 3; d(4,5) = 6. Execute the remaining iterations and report the direct cost and total project cost for the whole compression process.

5.7 The following data were obtained from a study of the activity durations required to conduct a consumer test panel study:

	Crash schedule		Normal schedule	
Activity	Time	Cost	Time	Cost
(1,2)	3	6	5	4
(1,3)	1	5	5	3
(2,4)	5	7	10	4
(3,4)	2	6	7	4
(2,6)	2	5	6	3
(4,6)	5	9	11	6
(4,5)	4	6	6	3
(6,7)	1	4	5	2
(5,7)	1	5	4	2

Note: Costs are given in thousands \$, duration in weeks

Determine the cost of crashing different project durations using the project crashing algorithm. (Stop after the first 2 crash iterations).

5.8 Given the activity diagram and following data. Durations are in days and costs in 100 TL:

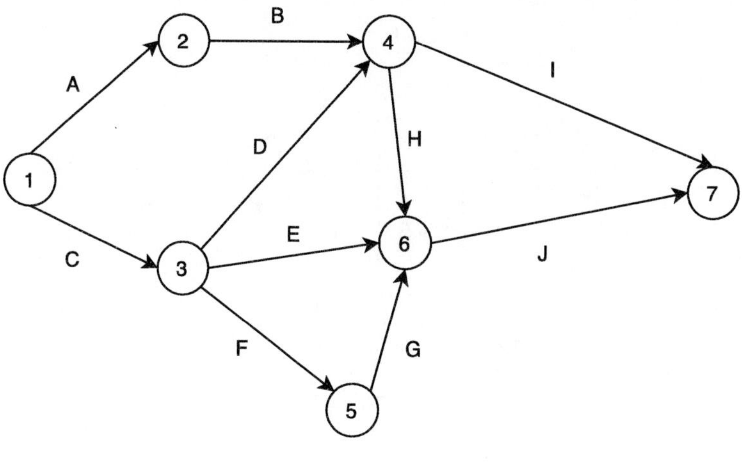

Activity	D^N	D^C	Normal Cost	Crash Cost
A	10	7	700	1000
B	12	9	800	1400
C	7	7	1000	1000
D	13	10	1200	1500
E	3	3	500	500
F	7	6	600	900
G	0	0	0	0
H	2	1	1000	5000
I	7	5	1300	1600
J	4	4	2000	2000

(a) Considering the normal time data, calculate the total and free slacks for activities.

(b) If you wish to crash your schedule to 26 days, what would the minimum direct cost of the project be? Show your work along with the scheduled duration of each activity for the 26-day schedule.

(c) Using linear direct cost curve, what would your estimation for activity B's cost, if it were crashed to 11 days? Assuming a piecewise linear cost curve for activity B with a direct cost of \$900 for a duration of 11 days, how would you refine your diagram and cost figures to show this change? How would this affect your minimum direct cost answer to part b?

(d) If the activity D must be completed in 19 days because of government inspection, write any additional restraints and dummies needed to assume completion on that day.

5.9 (a) Solve the problem using the heuristic approach to find the minimum total project cost. The overhead cost per unit time is given as 10 units. On each arc, (crash duration, normal duration, unit crash cost) are given.

(b) How would the solution change if the feasible time points for the activities (2,3) and (3,4) are given as: (3, 6, 7, 9, 10) and (5, 6, 7, 8, 9, 11).

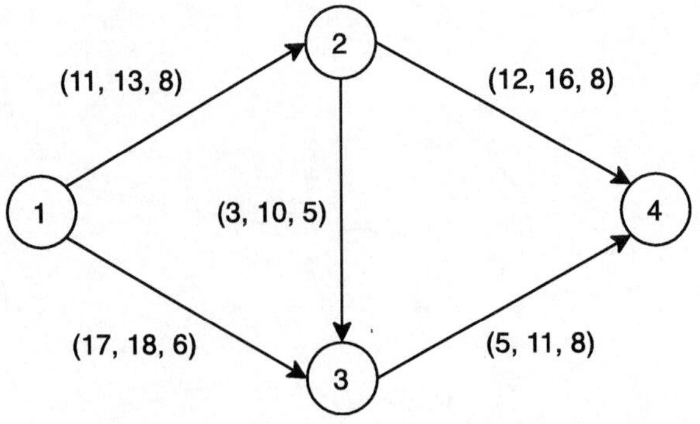

5.10 The duration and the cost data for activities A through H is given in the table. The network below gives the AON representation of the precedence relationship among the activities A through H.

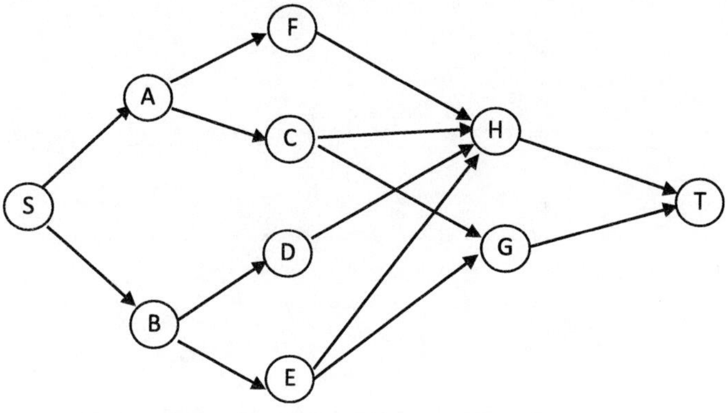

Activity	Normal duration	Crash duration	Normal cost (TL)	Crash cost (TL)
A	7	6	100	170
B	3	2	100	150
C	5	4	25	50
D	5	4	50	75
E	2	1	150	180
F	4	2	100	200
G	6	5	100	136
H	5	4	50	200

(a) Given that the indirect project cost is 50TL per day, obtain the total cost vs. duration curve. What is the optimal project duration?
(b) Write down the mathematical programming formulation for the above problem and solve using any LP solver.

5.11 Draw a cost curve considering the following information regarding excavation work at a construction site.

- Forty man-hours of work are required.
- Operator A is normally assigned.
- Operator B and Operator C can be made available, if necessary.
- First shift operation costs 20 TL/hour; second shift operation costs 20 TL/hour; third shift operation costs 24 TL/hour.
- Operators are assigned for an eight-hour period.
- One compressor is available. Can rent additional compressors at 50 TL per 24-hour period, including delivery and pick-up.
- No operator works overtime (more than 8 hours in any 24-hour period).
- There is no additional expense for second and third shift supervision.

5.12 Given a network for a Human Resources training project with normal times and crash times (in parentheses), find the cost-duration history and report your results in graphic form as well. Assume indirect costs for facilities and equipment are 100 TL per day. The project network and related data are given as follows:

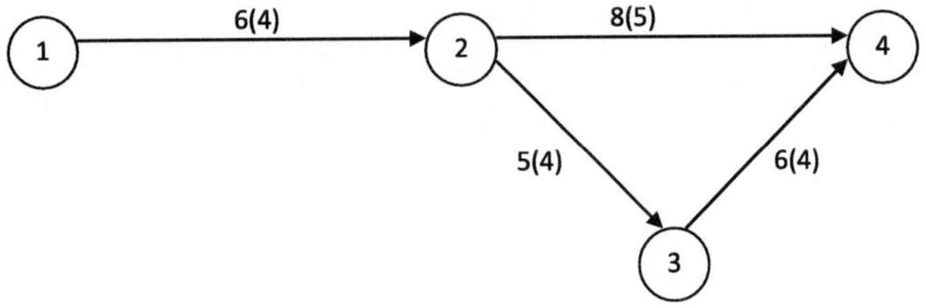

Activity	Time reduction, direct cost per day
(1,2)	30 TL for the first day, 50 TL for the second day
(2,3)	80 TL
(3,4)	25 TL for the first day, 60 TL for the second day
(2,4)	30 TL for the first day, 70 TL for the second day, 90 TL for the third day

References

Akkan, C., Drexl, A., & Kimms, A. (2005). Network decomposition-based benchmark results for the discrete time/cost trade-off problem. *European Journal of Operational Research, 165*(2), 339–358.

Badiru, A. B., & Pulat, S. P. (1995). *Comprehensive project management.* Upper Saddle River: Prentice Hall.

Chen, P. H., & Weng, H. (2009). A two-phase GA model for resource-constrained project scheduling. *Automation in Construction, 18*(4), 485–498.

De, P., Dunne, E. J., Ghosh, J. B., & Wells, C. E. (1995). The discrete time-cost tradeoff problem revisited. *European Journal of Operational Research, 81*(2), 225–238.

Demeulemeester, E. L., Herroelen, W. S., & Elmaghraby, S. E. (1996). Optimal procedures for the discrete time/cost trade-off problem in project networks. *European Journal of Operational Research, 88*(1), 50–68.

Demeulemeester, E., De Reyck, B., Foubert, B., Herroelen, W., & Vanhoucke, A. M. (1998). New computational results on the discrete time/cost trade-off problem in project networks. *Journal of the Operational Research Society, 49*(11), 1153–1163.

Elmaghraby, S. E. E., & Kamburowski, J. (1992). The analysis of activity networks under generalized precedence relations (GPRs). *Management Science, 38*(9), 1245–1263.

Hazır, Ö., Haouari, M., & Erel, E. (2010a). Discrete time/cost trade-off problem: A decomposition-based solution algorithm for the budget version. *Computers and Operations Research, 37*(4), 649–655.

Hazır, Ö., Haouari, M., & Erel, E. (2010b). Robust scheduling and robustness measures for the discrete time/cost trade-off problem. *European Journal of Operational Research, 207*(2), 633–643.

Içmeli, O., Erengüç, S. S., & Zappe, C. J. (1993). Project scheduling problems: A survey. *International Journal of Operations and Production Management, 13*(11), 80–91.

Klastorin, T. (2004). *Project management: Tools and trade-offs*. Hoboken: Wiley.

Larson, E. W., & Gray, C. F. (2018). *Project management: The managerial approach* (7th ed.). Irwin: McGraw-Hill.

Park, C. S. (2007). *Contemporary engineering economics* (4th ed.). Pearson Prentice Hall: Pearson Education.

Vanhoucke, M., & Debels, D. (2007). The discrete time/cost trade-off problem: Extensions and heuristic procedures. *Journal of Scheduling, 10*(4), 311–326.

6 Stochastic Project Scheduling with No Resource Constraints

Learning Outcomes
Upon successful completion of this Chapter, the reader will be able to:

1. Deal with the uncertainty in activity durations.
2. Deal with the uncertainty in activity costs.
3. Obtain estimates for the expected project duration and its variance.
4. Generate probabilities for various project durations.
5. Obtain estimates for the expected project cost and its variance.
6. Generate probabilities for various project cost values.
7. Apply Monte-Carlo simulation to handle uncertainty both in activity duration and activity cost.

6.1 Introduction

The central assumption of CPM is the deterministic nature of activity durations. In some projects, where there is considerable experience with the activities involved, these durations can be estimated relatively accurately, justifying the use of CPM. But when experience regarding the duration of an activity is limited, or activity durations are subject to a high degree of uncertainty by their nature, a single point estimate is no longer a satisfactory representation of the activity duration for planning and control purposes. The *Program Evaluation and Review Technique* (PERT) has been developed for use in such environments (Malcolm et al. 1959). Realizing the importance of effective cost control in addition to time control, extensions to basic CPM/PERT supporting cost control began to be required by large US Government agencies and were designated as *PERT-Cost* (Moder and Phillips 1970). PERT-Cost estimates individual activity costs and applies cost control based on these costs. Hence, the need arises for an accounting system for the purpose of project cost control that is different from the traditional accounting system, which does not group costs from this perspective. As the uncertainty in activity durations is a major problem so is the uncertainty in activity costs. We will introduce a simple model

G. Ulusoy, Ö. Hazır, *An Introduction to Project Modeling and Planning*, Springer Texts in Business and Economics, https://doi.org/10.1007/978-3-030-61423-2_6

for handling the uncertainty in activity costs proposed by Case (1972), which extends the way PERT handles uncertainty in activity duration to activity costs.

6.2 Representation of the Randomness of Activity Durations

The best way to represent uncertainty in the duration of an activity is to determine the underlying probability density function *f(X)*, where *X* is a continuous random variable representing the activity duration. For a given project and a given activity, is the underlying distribution normal, triangular, Beta, uniform, or some other distribution? In order to answer this question, one needs to go through a distribution fitting procedure using a number of observations of the activity duration.

In PERT applications, the lack of sufficient observations for the application of standard statistical techniques to estimate the probability distribution of the activity duration forces us to represent it using a robust approximation based on the *Beta distribution*. The Beta distribution defined over the range [0, 1] is given in Eq. (6.1) (Drake 1967):

$$f(X) = \begin{cases} c(\alpha, \beta)X^{\alpha-1}(1-X)^{\beta-1} & 0 < X < 1 \\ 0 & otherwise \end{cases} \tag{6.1}$$

where $\alpha > 0, \beta > 0,\ \ c(\alpha, \beta) = \frac{(\alpha+\beta-1)!}{(\alpha-1)!(\beta-1)!}$

The expected value, the variance, and the mode of the random variable X are given by:

$$\mathrm{E(X)} = \mu = \alpha/\alpha + \beta \tag{6.2}$$

$$\mathrm{Var(X)} = \sigma^2 = \alpha\beta/(\alpha+\beta)^2(\alpha+\beta+1) \tag{6.3}$$

$$\mathrm{Mode(X)} = (\alpha-1)/(\alpha+\beta-2) \tag{6.4}$$

The Beta distribution is preferred for several reasons. It is a rich distribution in the sense that different values for its two parameters, α and β allow a large number of shapes to be represented as shown in Fig. 6.1a. The cumulative distribution function for the Beta distribution is displayed in Fig. 6.1b.

The Beta distribution can be defined over the unit interval as in Fig. 6.1 but can also be defined over any finite range of its argument; say [a,b]. The Beta distribution with $a \le Y \le b$ can be transformed into one with $0 \le X \le 1$ by letting $X = \left(\frac{Y-a}{b-a}\right)$.

It can be symmetric or skewed to the right or left. When both α and $\beta > 1$, then the function is unimodal, i.e., has a single mode. When $\alpha < \beta$, then the distribution is right-skewed. On the other hand, when $\alpha > \beta$, then the distribution is left-skewed. It becomes symmetric when $\alpha = \beta$.

When modeling activity durations, usually right-skewed distributions are preferred as being more realistic. In right-skewed distributions mean is larger than the mode and hence, most data falls to the right of the mode resulting in a longer tail

Fig. 6.1a The probability density function of the Beta distribution

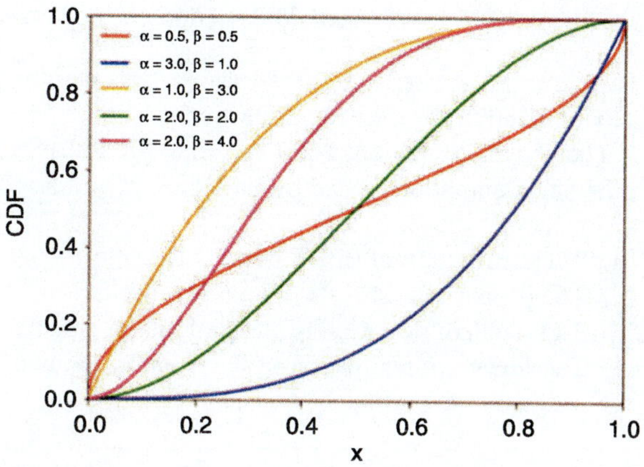

Fig. 6.1b The cumulative distribution function of the Beta distribution

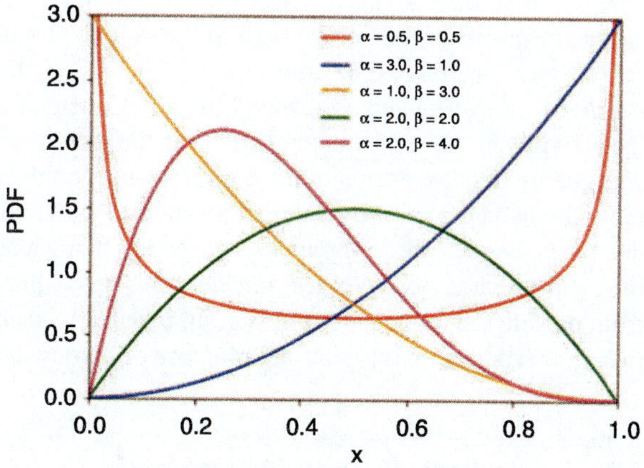

representing larger values for activity duration. A right-skewed distribution allows for longer late completions and shorter early completions, which is usually what would be expected in practice.

To operationalize the use of the Beta distribution we need to estimate the mean and the standard deviation of the activity durations in terms of quantities we know or can estimate. According to Littlefield Jr. and Randolph (1987), "the early workers in PERT reasoned that the standard deviation is about one sixth of the range". Further assuming a linear approximation of the mean by the mode, the following estimates for the mean D and the standard deviation s of the activity duration distribution are obtained:

$$\text{Mean duration} = D = (a + 4m + b)/6 \tag{6.5}$$

$$\text{Standard deviation} = s = (b - a)/6 \tag{6.6}$$

where a and b are defined as the least possible and the greatest possible values the activity duration can take on, respectively.

The above expressions (6.5) and (6.6) are indeed rather easy to use given the following assumptions stated by Littlefield Jr. and Randolph (1987) hold:

(i) The distribution of PERT times is Beta.
(ii) The people can estimate a, m, b very well.
(iii) The standard deviation is one sixth of the range.
(iv) The linear approximation of the mean in terms of the mode is acceptable for estimating the mean.

The three statistics used to estimate the mean and standard deviation of the Beta distribution are the mode and the end points of the estimated range of the activity duration distribution $[a,b]$. The lower end point of the range, a, is called the *optimistic duration* and corresponds to the estimate of the smallest possible duration the activity can take. The upper end point, b, on the other hand, is called the *pessimistic duration* and corresponds to the estimate of the largest possible duration the activity can take. The mode represents the estimate for the *most likely duration*. The quality of these estimates has a direct impact on the quality of the solutions, and they should be provided by the right people and with great care. Hence, the estimates need to be based on *expert opinion* – an expert being a person knowledgeable on the details of the activity involved and can be from within the organization as well as from outside the organization. It is good practice to seek these estimates from more than one expert preferably through the use of a knowledge acquisition method.

6.3 Determining the Critical Path

Using the estimation procedure presented in the previous section, estimates for the mean duration for each activity of a project network can be obtained. Once these estimates are available, calculations for PERT become identical to the CPM calculations.

One has to be cautious when interpreting the results of these CPM calculations in the PERT implementation. Having estimated the project duration, it is easy to forget that the result is an estimate of the expected value of the underlying random variable. Thus, there is a non-zero probability that the actual project duration will be less than or larger than the expected project duration obtained through PERT. Relying on a single point estimate is obviously not correct for decision-making under uncertainty. Suppose that the estimated mean for an activity's duration is 15 days with its optimistic and pessimistic values being 7 and 23 days, respectively. Now consider another activity with the same estimated mean but this time, let its optimistic and pessimistic values be 14 and 16 days, respectively. These two activities have the same estimated mean duration, but their estimated standard deviations are quite different. Instead, we should be interested in answering questions like: "What is the probability that the project will take less than (or more than) a certain number of days?"

To answer such questions, one needs to know about the probability density function of the project duration, i.e., *the critical path length* (CPL). Note that the CPL consists of a sequence of activities, and the sum of the activity durations determines the CPL. Assuming each activity's duration as an independent random variable, the sum will be a random variable as well. *The Central Limit Theorem* states that if a sufficiently large number of random variables are involved in the sum (according to folklore, more than 30 is usually sufficient), the CPL follows a normal distribution with the mean and variance defined as the sum of the means and variances of the individual random variables, respectively.

The Central Limit Theorem Let $X = X_1 + X_2 + \ldots + X_n$ be a random variable with mean $\mu = \mu_1 + \mu_2 + \ldots + \mu_n$ and variance $\sigma^2 = \sigma^2_1 + \sigma^2_2 + \ldots + \sigma^2_n$, where X_1, $X_2, \ldots, X_n$ are independent identically distributed random variables, then the distribution of the random variable $Z = (X-\mu)/\sigma$ approaches the standard normal distribution as n approaches infinity.

The normal approximation for X will be generally good if the number of critical activities $n \geq 30$. On the other hand, for practical applications with $n < 30$, the approximation would be acceptable if the distributions of the activity durations are not too different from a normal distribution (Walpole et al., 1998). Furthermore, the assumption of independent activity durations might not always be valid due to precedence relationships and other interdependencies between the critical activities.

If the normality assumption holds, then CPL would be distributed normally. The *expected project duration* would correspond to the *expected critical path length* (ECPL) and hence, the probability of completing the project no later or earlier than ECPL would be only 0.50. At this point, further analysis is needed to obtain probabilities of not exceeding any specified project duration, which would be useful to PM in decision-making. For this analysis, one would need to estimate the standard deviation of the project duration as well as its mean.

We note that when an activity's duration is a random variable, there is a positive probability that the activity's duration will exceed its mean value. This implies, in turn, a positive probability that the activity duration will exceed the slack, delaying the starting times of the succeeding activities. A common assumption in PERT-type models is that an early-start schedule would be best for any realization of uncertainty. However, this is not necessarily true when objectives other than minimizing makespan are used, such as minimizing NPV of costs (Buss and Rosenblatt 1997) or minimizing the impact of disruptions, for example, on resource availability or on milestone disruptions (Zhu et al. 2005).

When the activity durations are known with certainty, critical activities can be unambiguously distinguished from the noncritical ones. However, when activity durations are random variables, classifying the activities as critical or non-critical becomes ambiguous. Instead of making a strict classification, we can associate activities with a *criticality index* (CI), which will be defined in Sect. 6.4.

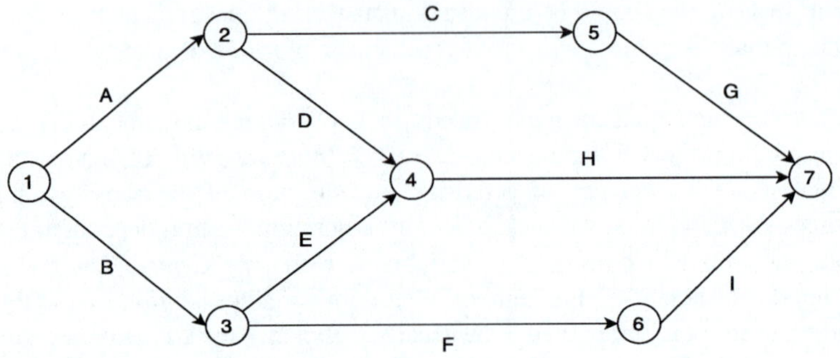

Fig. 6.2 Project network for Example 6.1

Table 6.1 The duration estimates and the mean durations and variances

Activity	a	m	b	D	s^2
A	3	5	7	5	0.4444
B	4	5	12	6	1.7778
C	1	3	5	3	0.4444
D	3	4	5	4	0.1111
E	2	2	2	2	0.0000
F	4	6	14	7	2.7778
G	2	3	4	3	0.1111
H	3	5	13	6	2.7778
I	1	3	5	3	0.4444

Example 6.1 Consider a project consisting of nine activities. The AOA project network is presented in Fig. 6.2. The optimistic (a), most likely (m) and pessimistic estimates (b) for each activity's duration in days are obtained and listed in Table 6.1.

(a) Determine the expected project duration, the corresponding critical path, and the total and free slacks. Estimate the parameters of the Normal distribution representing the project duration.
(b) Find the probability that the project will not take longer than 18 days.

Solution

(a) Using the duration estimates given in the first three columns of Table 6.1, the mean duration and the variance for each activity are calculated. The results are displayed in the last two columns of Table 6.1.

Let us now calculate the earliest and latest occurrence times for the events by employing the forward and backward passes using the mean durations.

$E_1 = 0$

$E_2 = 5$
$E_3 = 6$
$E_4 = max_{2,\,3} = \{5 + 4, 6 + 2\} = 9$
$E_5 = 8$
$E_6 = 13$
$E_7 = max_{4,\,5,\,6} = \{9 + 6, 8 + 3, 13 + 3\} = 16.$

Hence, the expected project duration is found as 16 days. Now we will calculate the latest occurrence times for the events in order to obtain the critical path and the slacks.

$L_7 = E_7 = 16$
$L_6 = 13$
$L_5 = 13$
$L_4 = 10$
$L_3 = min_{4,\,6} = \min\{10 - 2, 13 - 7\} = 6$
$L_2 = min_{4,\,5} = \min\{10 - 4, 13 - 3\} = 6$
$L_1 = min_{2,\,3} = \min\{6 - 5, 6 - 6\} = 0$

The results of the forward and backward passes indicate that events 1, 3, 6, and 7 are on the critical path since their earliest and latest occurrence times are equal.

Early start and finish times, late start and finish times, and total and free slacks for the activities are calculated following the CPM steps (Sect. 4.2) and are reported in Table 6.2. The activities B, F, and I have zero total slack and hence, are critical. The critical path becomes then {B, F, I}. The expected project duration ECPL is determined as the sum of the mean durations of the critical activities on the critical path:

$$\text{ECPL} = D_B + D_F + D_I = 6 + 7 + 3 = 16 \text{ days.}$$

The estimated variance of the project duration is the sum of the variances of the activities on the critical path:

Table 6.2 The early/late start and finish times and the slack values

Activity	D	ES	EF	LS	LF	TS	FS
A	5	0	5	1	6	1	0
B	6	0	6	0	6	0	0
C	3	5	8	10	13	5	0
D	4	5	9	6	10	1	0
E	2	6	8	8	10	2	1
F	7	6	13	6	13	0	0
G	3	8	11	13	16	5	5
H	6	9	15	10	16	1	1
I	3	13	16	13	16	0	0

$$s^2 = s^2_B + s^2_F + s^2_I = 1.7778 + 2.7778 + 0.4444 = 5.0000.$$

Hence, the project duration will be assumed to be normally distributed with a mean of 16 and a variance of 5 (N~(16; 5)).
(b) Find the probability that the project will not take longer than 18 days.

Assuming N~(16; 5) and using the Cumulative Normal Distribution Table (Appendix 6A) we find the probability for the project duration not taking longer than 18 days to be:

$$P\left\{z \leq \frac{x-\mu}{s}\right\} = P\left\{z \leq \frac{18-16}{\sqrt{5}}\right\} = P\{z \leq 0.8944\} = 0.8144$$

If the CPM calculations result in more than one critical path the variance of the project duration distribution is taken to be the maximum of the variances of the critical paths. The reason for that choice is that the larger variance results in a more conservative estimate of the probability of a particular event occurring on or before a prespecified date.

6.4 Shortcomings of the PERT Method

A drawback of PERT is that it results in an optimistic estimate for the mean project duration. In other words, the PERT estimate is an underestimate of the actual mean project duration in general (see, e.g., Fulkerson 1962; Mitchell and Klastorin 2007; Acebes et al. 2014).

The true mean project duration is of the form $E_r[[\max PD]_r]$, where PD is any path duration between the initial and end nodes of the project network and r is any realization and the expectation is taken over all realizations. A realization is obtained by assigning feasible durations to all the activities of the project network. The mean project duration value obtained by PERT is the maximum of all the paths' expected durations and can be expressed as max E[PD]. Hence, we can write the following relation between the true mean project duration and its PERT estimate as follows:

$$\text{True mean} = E_r[[\max PD]_r] \geq \max E[PD] = \text{PERT mean}$$

This inequality implies that the PERT estimate of the mean project duration will always underestimate the actual value. To illustrate this result, consider four paths with the path length distributions shown in Fig. 6.3, where the dotted vertical lines correspond to the expected value of the corresponding path length distribution. We see that there is considerable overlap across each path's length distributions. Let P_4 correspond to the path with the maximum expected value. Consider the cross-hatched area under the path length distribution P_4. This area represents the probability that the mean project duration for path P_3 will be larger than that for P_4, which corresponds to the PERT project duration. The accuracy of the PERT estimate will

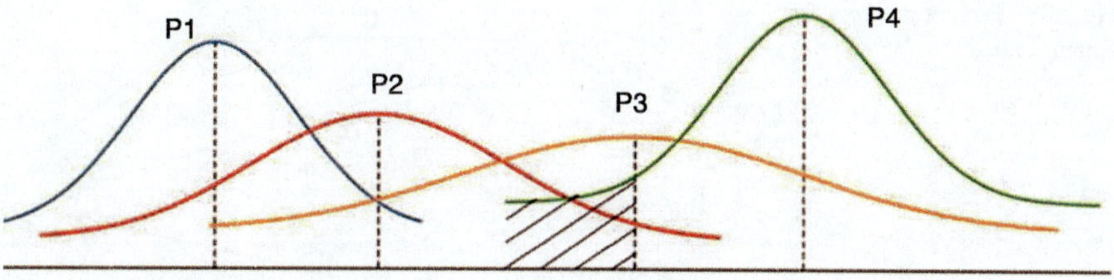

Fig. 6.3 Path length distributions for four paths on the project network

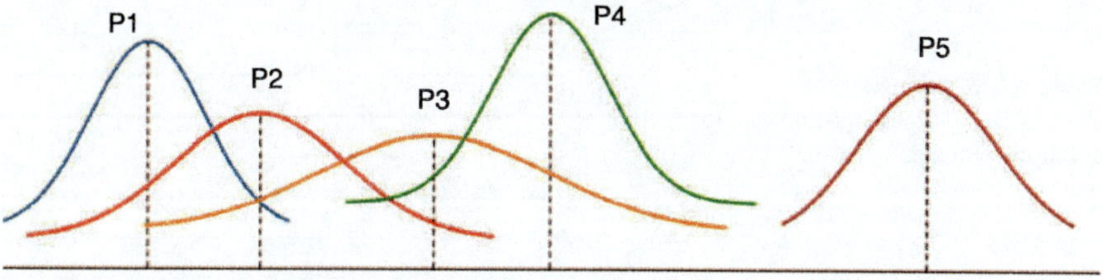

Fig. 6.4 Example for a disassociated path length distribution

improve as a path stands out from the other paths as illustrated by the rightmost distribution P_S in Fig. 6.4. In such an extreme case, the true mean is approximately equal to the PERT mean.

Another drawback of PERT is that it does not distinguish between the critical activities. This could be a drawback in the implementation. Depending on its location on the network, a critical activity might contain more risks than the other critical ones. Consider the project network in Fig. 6.2. The mean activity durations for each arc are indicated in Table 6.2. Based on these values, the critical path consists of the activities {B, F, I}. One can intuitively argue that B is more critical than F since B would also be critical if {B, E, H} turns out to be the longest path in the implementation. This point will be demonstrated in Example 6.2 through the use of a problem with a finite number of *instances* and integer activity durations for the sake of simplicity. A project network with all its activities assigned with a feasible duration is called an *instance* of that project network.

Example 6.2 Consider the following network in Fig. 6.5. On that network, we will enumerate all possible instances using the estimates for the optimistic (a), most likely (m), and pessimistic durations (b) given in Table 6.3 and each occurring with probability 1/3.

Solution

The PERT critical path is calculated as {A-C-G}. But the total enumeration of the instances reveals that {A-C-G} will only be the longest path with a *ratio of occurrence* of 0.27. This value is the ratio of the number of instances with

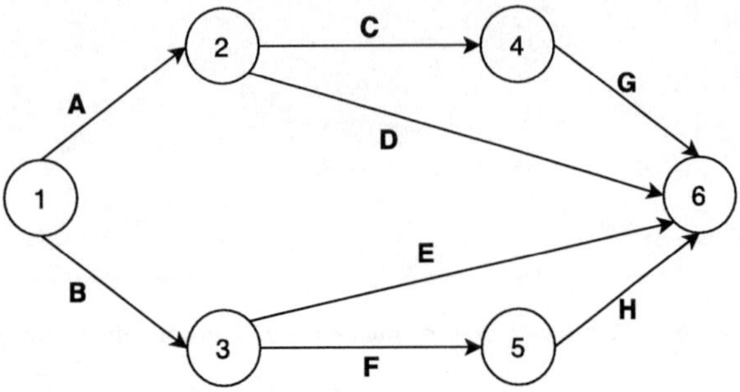

Fig. 6.5 Project network for Example 6.2

Table 6.3 Optimistic, most likely, and pessimistic duration estimates

Activity	a	m	b
A	7	9	11
B	7	8	10
C	1	3	5
D	1	3	7
E	1	3	7
F	1	3	5
G	1	1	1
H	1	1	1

Table 6.4 The criticality indices of the activities

Activity	A	B	C	D	E	F	G	H
Criticality Index	0.58	0.42	0.27	0.30	0.24	0.19	0.27	0.19

{A-C-G} as the longest path to the total number of instances. When we evaluate all longest paths, we observe that the path with the highest number of occurrences is the path {A–D}. The ratio of occurrence it will become the critical path is 0.30, which is larger than that of the PERT critical path {A-C-G}. We can perform a similar criticality analysis for each activity and find out its *criticality index* (CI). CI is defined as the ratio of the number of times a particular activity is on a critical path to the total number of instances, which for that activity corresponds to the *probability of being critical*. The results are reported in Table 6.4.

Although activity C was found to be critical in PERT analysis, its criticality index is 0.27 which is less than those for activities B and D, which are non-critical according to PERT analysis. In other words, activities B and D both have higher probabilities of being critical than activity C.

All these findings indicate that PERT results can indeed mislead the PM, in identifying both the most likely critical path and the criticality of individual activities.

6.5 A Model for Managing Uncertainty in Activity Costs

Case (1972) used the three-estimate approach of PERT to estimate activity costs. This adaptation will be designated here as PERT-Costing. It is a relatively naïve approach and has not been as widely accepted as PERT. Still, in many project planning exercises, the resource requirements and costs are taken as deterministic. PERT-Costing is based on the assumption that the duration and cost of activities are independent of each other, which does not necessarily reflect the reality. Actually, the cost and duration of an activity are dependent on each other (see the time/cost trade-off analysis in Chap. 5).

The PERT-Costing uses the same probability distribution assumptions as PERT. Activity costs are assumed to follow a Beta distribution. The estimates for the average and standard deviation of the activity cost distribution are obtained by using the following three estimates about the cost:

Optimistic cost estimate (c_{opt}).
Pessimistic cost estimate (c_{pess}).
Most likely cost estimate (c_{mode}).

The average and the standard deviation of the Beta distribution are taken to be equal to $[(c_{opt} + 4\, c_{mode} + c_{pess})/6]$ and $[(c_{pess} - c_{opt})/6]$, respectively. The project cost is then estimated by the sum of all activity costs. By the Central Limit Theorem, total project cost will be approximately normally distributed with a mean cost as the sum of the mean cost of each activity and variance as the sum of the variances around the mean cost of each activity. However, recall the assumptions of the Central Limit Theorem: the project has a relatively large number of activities ($n \geq 30$) and the activity cost distributions are independent of each other. Since the requirement is for the total number of activities, not solely those on the critical path, it is much less restrictive than it would be for the duration of the critical activities. We can safely state that a great majority of the projects in the real life would satisfy the above assumption on the number of activities.

We will demonstrate PERT-Costing through an example.

Example 6.3 Consider the following project consisting of 6 work packages. The optimistic, pessimistic, and most likely cost estimates for each work package are given in Table 6.5. Estimate the project mean cost and the probability, under normality assumption, that the project cost will not exceed $1500.

Solution
The mean and variance for the work packages are calculated and reported in Table 6.5.

Summing the mean values for the work packages and the variances result in the project mean cost and project cost variance, respectively:

Table 6.5 Estimates for activity costs and Beta distribution parameters

Work Packages	c_{opt}	c_{mode}	c_{pess}	Mean	Variance
1	80	100	120	100	44.44
2	400	500	900	550	6944.44
3	150	180	180	175	25.00
4	120	150	240	160	400.00
5	70	100	130	100	100.00
6	100	200	600	250	6944.44
Parameters for project cost distribution				1335	14458.32

Project mean cost = \$1335,
Project cost variance = $\$^2$14458.32.

Since the project cost is assumed to be normally distributed, we claim that the project cost is below \$1335 with a probability of 0.50. Let us find the probability under normality assumption that the project cost will not exceed \$1500. Using the Normal Table, we get:

$$P\left\{z \le \frac{1500 - 1335}{\sqrt{14458.32}}\right\} = P\{z \le 1.3723\} = 0.9150.$$

The probability that the project cost will not exceed \$1500 is 0.9150. That the project cost will exceed \$1500, on the other hand, is then the complement, i.e., 1–0.9150 = 0.085.

6.6 Monte Carlo Simulation Approach to Handle Uncertainty

Monte-Carlo simulation is a well-known technique based on random sampling (see, e.g., Law 2014). In project management, Monte-Carlo simulation is applied to investigate problems arising from uncertainty issues; particularly, from uncertainties around activity duration and cost and the assessment of risks. Monte-Carlo simulation is useful for performing schedule risk analysis and planning scenarios for projects exceptionally critical in terms of project time and risks.

In this section, we will investigate via simulation the problems for which we sought answers earlier using PERT analysis. Further use of Monte Carlo simulation in project management is dealt with in Sect. 12.4.3.

In Monte Carlo simulation, the choice of the underlying probability distribution is a critical step. In PERT applications, the lack of sufficient observations for the application of standard statistical techniques to estimate the probability distribution of the activity duration forces us to represent it using a robust approximation. For that purpose, we will employ PERT distribution to represent the uncertainty in activity durations. The PERT distribution is a continuous probability distribution defined over the range [a, b] and has two parameters α and β. According to Vose (2008), "The PERT distribution gets its name because it uses the same assumption about the mean as the PERT networks. It is a version of the Beta distribution and requires the same three parameters as the triangular distribution." These parameters are the minimum (a) and the maximum (b) of the range over which the distribution is defined and the mode (*m*) of the distribution. Its probability density function is given in Eq. (6.7) (Vose 2008):

$$f(X) = \frac{(X-a)^{\alpha-1}(b-X)^{\beta-1}}{B(\alpha,\beta)(b-a)^{\alpha+\beta-1}} \quad a \leq X \leq b \tag{6.7}$$

where $\alpha > 1$, $\beta > 1$, and $B(\alpha,\beta) = \frac{(\alpha-1)!(\beta-1)!}{(\alpha+\beta-1)!}$.

The expected value and variance of the PERT distribution is given as:

$$E(X) = \mu = \frac{a+4m+b}{6} \tag{6.8}$$

$$Var(X) = \sigma^2 = \frac{(\mu-a)(b-\mu)}{7} \tag{6.9}$$

In order to generate PERT random variates, we need to calculate α and β parameters of the assumed underlying PERT distribution for each activity. The expressions for α and β parameters over the interval [a,b] are obtained by the following relationships in terms of the three estimates of the activity duration: *a*, *m*, and *b* (Vose 2008):

$$\alpha = \frac{4m+b-5a}{b-a} \tag{6.10}$$

$$\beta = \frac{5b-a-4m}{b-a} \tag{6.11}$$

Let us consider now the case for random activity durations. Using the expressions above we calculate the parameters α and β for each activity, which we will employ in the simulation runs for generating random activity durations.

At each run, using the PERT pseudo-random number generator we create an instance by randomly generating activity durations for all activities of the project network. Then we apply CPM and determine the critical path(s), the critical activities, and the project duration.

After a pre-specified number of runs have been completed, we calculate the following statistics:

(i) The mean and the variance of the project duration.
(ii) For all paths and over all the runs, the number of times a path is observed as critical.
(iii) For all activities and over all the runs, the number of times an activity appears on one or more critical paths.

Table 6.6 The α and β parameters of the activities

Activity	A	B	C	D	F	G	H	I
α	3	1.5	3	3	1.8	3	1.8	3
β	3	4.5	3	3	4.2	3	4.2	3

Example 6.4
Solve the problem stated in Example 6.1 using Monte Carlo simulation.

(a) Determine the mean and variance of the project duration, the ratio of occurrences for the critical paths, and the criticality index for each activity.
(b) Find the probability that the project will not take longer than 18 days.

Solution
(a) The estimates for the three parameters *a*, *m*, *b* and the mean duration for each activity are as given in Table 6.1. The variance, on the other hand, is calculated using expression (6.9).

We calculate the parameters α and β of the PERT distribution for each activity using expressions (6.10) and (6.11), respectively, which we will employ in the runs for generating random activity durations. The resulting α and β parameters are reported in Table 6.6.

Using these α and β parameters, we take 240 runs, each time generating activity durations for all activities except activity E, whose activity duration is deterministic. For each run, we determine the project duration PD and the critical path applying CPM. The results of a sample of runs together with PD and the critical path (CP) are given in Table 6.7.

Based on the 240 PD values obtained in this simulation, the expected PD is found to be 16.80 days and the variance as 3.94 days^2. Comparing these to the mean and variance of PD obtained in Example 6.1, which were determined as 16.00 and 5.00, respectively, we observe that PERT has resulted in an optimistic expected PD compared to the simulation experiment, and the simulation solution has a lower variance.

Among the critical paths generated through the simulation experiment we observe 4 different critical paths: {A,C,G}, {A,D,H}, {B,E,H}, and {B,F,I}. Their respective ratios of occurrence are reported in Table 6.8. {B,F,I}, which is also the critical path obtained by PERT, has the highest ratio of occurrence with 0.6064. Note that the ratio of occurrence of {A,D,H} indicates that {A,D,H} has roughly one third chance of becoming the critical path. The PERT analysis misses this possibility completely. This would have been a valuable information for the PM.

The CI value for each activity is reported in Table 6.9. Although the CI values are highest for the activities B, F, and I, which are designated as the critical activities in the PERT solution, the activities A, D, and H also have relatively high CI values. Missing this possibility points to another deficiency of the PERT method.

(b) Find the probability that the project will not take longer than 18 days.

Table 6.7 Sample run values for the activity durations, PD, and CP

Run Number	Activity Durations									
	A	B	C	D	F	G	H	I	PD	CP
1	4,60	4,45	2,53	4,85	10,11	3,39	7,47	2,99	17,56	BFI
2	5,04	6,85	2,21	4,20	6,16	3,20	7,15	2,52	16,38	ADH
3	6,20	6,24	1,38	3,42	7,05	2,40	8,87	1,85	18,49	ADH
4	6,23	6,98	3,39	4,42	8,67	2,68	10,69	3,41	21,33	ADH
5	4,62	4,63	2,72	3,50	7,23	2,61	7,00	3,12	15,13	ADH
6	4,40	6,47	2,44	4,12	10,23	3,85	7,51	3,20	19,90	BFI
7	5,82	6,12	2,71	3,88	5,30	2,64	3,26	2,25	13,68	BFI
8	3,38	4,64	2,39	3,51	8,37	2,38	5,38	3,95	16,96	BFI
9	4,35	5,86	4,23	3,80	10,70	3,80	6,09	3,79	20,35	BFI
10	5,69	7,48	3,25	3,79	8,78	2,40	7,87	2,53	18,79	BFI

Table 6.8 Ratios of occurrence for the critical paths

CP	ACG	ADH	BEH	BFI
Ratio of occurrence	0.0008	0.3380	0.0548	0.6064

Table 6.9 The CI values for the activities

Activity	A	B	C	D	E	F	G	H	I
CI	0.34	0.66	0.08	0.33	0.05	0.60	0.08	0.39	0.60

To obtain this probability we will calculate the ratio of simulation runs with PD value of 18 days or less to 240, which is the total number of simulation runs:

$$\text{Prob}\{\text{PD} \leq 18\} = 192/240 = 0.8.$$

Compared to the results of Example 6.1, although the expected PD is slightly longer in the simulation analysis, the probability that PD does not exceed 18 days is slightly less. This result is due to the smaller variance resulting from the simulation analysis.

Example 6.5 Solve the problem stated in Example 6.3 using Monte Carlo simulation. Estimate the project mean cost and the probability that the project cost will not exceed $1500.

Solution

We apply the same methodology implemented in Example 6.4. The estimates for the three parameters and the mean for each WP are as given in Table 6.5. The variance for the WPs is calculated using the expression (6.9). Using the expressions (6.9) and (6.11), the α and β parameters of the PERT distribution are calculated for each WP and reported in Table 6.10.

Table 6.10 The α and β parameters of the WPs

WP	1	2	3	4	5	6
α	3	1.8	5	2	3	1.8
β	3	4.2	1	4	3	4.2

Table 6.11 Sample run values for the WP costs and the total cost

	Work Package Costs						
Run Number	WP1	WP2	WP3	WP4	WP5	WP6	Total Cost
1	107,63	563,04	178,73	162,63	116,97	298,36	1427,36
2	109,07	527,94	176,86	161,23	126,81	258,93	1360,85
3	90,39	638,13	178,92	145,75	114,34	178,29	1345,83
4	93,75	646,67	178,82	191,90	76,10	442,09	1629,33
5	98,22	671,24	179,21	161,89	116,47	295,20	1522,22
6	114,04	678,56	165,03	156,81	78,03	312,51	1504,98
7	99,51	496,93	168,61	162,72	107,67	226,56	1262,01
8	94,71	413,02	176,78	173,62	111,98	228,50	1198,61
9	87,30	623,11	177,90	225,83	104,39	361,79	1580,33
10	101,47	524,74	179,23	211,31	101,54	116,82	1235,10

α and β parameters are employed in generating costs for all the WPs in each run for 240 runs. For each run, the total cost is calculated. The results for 10 runs are displayed in Table 6.11.

The mean total cost is found as \$1336.01 with a variance of $\$^2$15760.28. Comparing the PERT result (\$1335.00, $\$^2$14458.32) with the simulation result we observe that the mean total cost of PERT is almost the same, and its variance is slightly less than that of the simulation.

The probability that the project cost will not exceed \$1500 is calculated using the frequency data and found as:

$$P\{TC \leq 1500\} = 211/240 = 0.8792.$$

The probability value resulting from the PERT calculations was 0.9150. The lower probability value obtained here is an expected result, since the mean values are rather close for the PERT and the simulation whereas the variance is relatively larger for the simulation leading to a smaller probability.

In this section, we presented the Monte Carlo simulation and solved two problems, which we solved earlier using the PERT method. These two applications show the versatility of the Monte Carlo simulation and its possible extensions for further use – particularly in scenario analysis, which is treated in detail in Chap. 12.

6.7 Conclusions, Recent Developments, and Some Future Research Directions

As we mentioned in Sect. 6.2, according to Littlefield Jr. and Randolph (1987), "the early workers in PERT reasoned that the standard deviation is about one sixth of the range". The set of assumptions on which this reasoning rested are stated in Sect. 6.2 as well. One of these assumptions is that the distribution of PERT times is Beta. Experts are asked for their estimates for m, a, and b, where a and b correspond to the endpoints of the range for the activity duration. Another approach to acquire the estimates for a and b is to ask for the 5 and 95 percentile estimates. According to a study by Moder and Rodgers (1968), when a and b defined as the 5 and 95 percentile values, the difference $(b - a)$ varies from 3.1 to 3.3 standard deviations for a variety of distributions including exponential, normal, rectangular, triangular, and Beta distributions. The average is 3.2. But for the same set of distributions, there is no similar convergence around an average multiple of standard deviations to represent the whole range given by the difference between 0 and 100 percentiles. Indeed, such a multiplier varies from 3.5 to 6.0. It is comforting to observe that defining the estimates for a and b as 5 and 95 percentiles and using 3.2 in the denominator rather than 6 leads to an estimator of the standard deviation that is rather robust to the actual underlying distribution eliminating the need for assuming a distribution *a priori* (Moder and Phillips 1970).

Project managers should consider the impacts of uncertainty and use appropriate tools to predict unexpected deviations in project completion time and project cost. PERT and PERT-Costing have been used for project planning under uncertainty. These approaches are based on stochastic modeling, which is an effective approach when we can accurately identify uncertainty with probability distributions. For this, there is an absolute need for data containing reliable and sufficient observations. However, due to the unique, one-time nature of many projects, historical data might not be available, or the existing data might not be representative of the actual conditions a specific project may encounter. There is a need for studies on other methods, such as robust optimization. The comparison of the actual results using the planning approaches based on stochastic modeling and other optimization methods would be highly valuable.

On the other hand, even if the probability distribution can be determined correctly, an analytical solution of stochastic models is difficult in many cases. For this reason, instead of optimization approaches, simulation techniques, mostly Monte- Carlo simulation, have been used in the analysis of the systems that address dynamic environments. However, according to Hall (2016), many companies are unwilling to implement simulation, mainly due to the unfamiliarity of managers with the statistical concepts it requires. In this regard, there is a need to embed the simulation approaches in project management software and present managers user friendly interfaces (Hazır 2015).

The validity of the results of the models introduced in this chapter depends on the assumptions on the probability distributions for activity/project time and cost variables. The researchers need "real" data for validating the theoretical models. In this regard, the study of Batselier and Vanhoucke (2015) is very valuable. There is a need for additional "real" databases on project management.

6.8 Case Study: Constructing an Earthquake Resistant Residential House

BW Company is a specialized construction company building high quality residential homes in the suburbs of Istanbul. It has developed expertise in designing and building earthquake resistant residential buildings, as researchers have warned that Istanbul could be hit at any time by a massive earthquake with devastating effects. For this reason, BW Company has received considerable attention over the past few years.

BW Company has recently been awarded a contract for building a residential house in the Emirgan district in Istanbul. The design of the house resembles traditional Japanese houses, which are built by erecting wooden columns on top of a flat foundation made of packed earth or stones. Unique architectural and interior features of traditional Japanese houses will be incorporated by hiring specialized subcontractors for interior design work.

The project consists of the following work packages, all of which will be executed by subcontractors:

- Foundation Work
- Plumbing Work
- Electrical Work,
- Interior and Exterior Woodwork
- Flooring
- Kitchen
- Bathrooms
- Interior Design Work
- Wallpaper
- Tatami mat
- Fusuma partition
- Interior and Exterior Plaster and Paint Work

After consultation and negotiation with all the sub-contractors, the project manager has determined the job definitions, precedence relations, and job duration estimates in working days as provided in Table 6.12. All parties agreed on the following working conditions: Saturday and Sunday are not regular working days and daily working hours are 8:00 to 18:00 with a lunch break during 12:00–13:00. With the approval of the management and prior knowledge of the sub-contactors, the project manager announced the project start date as **June 1, 2017, at 8:00 h**.

Some of the precedence relationships are of a generalized precedence type:

- Interior woodwork must start 2 working days after wiring is finished.
- Foundation 2 can finish 4 working days after foundation 1 is finished.
- Tatami mat must start 1 working day after cleaning is finished.
- Exterior painting must start 1 working day after drying is finished.

We determine the mean durations for the activities employing the relationship (6.5). The mean durations are reported in Table 6.12 as integer working days rounded to the nearest integer.

Table 6.12 Activity definitions, precedence relations and durations[a]

Activity No	Activity Description	Predecessors	Duration (Working Days)				
			Optimistic	Most likely	Pessimistic	Mean Duration	Variance
1	Placement	--	1	1	1	1	0.0000
2	Temp. plumbing	--	1	1	1	1	0.0000
3	Temp. electricity	--	1	1	1	1	0.0000
4	Foundation 1	1,2,3	2	3	10	4	6.2500
5	Hardening 1	4	2	3	4	3	0.3906
6	Foundation 2	5	1	1	1	1	0.0000
7	Rough plumbing 1	6	1	1	1	1	0.0000
8	Foundation 3	7	1	1	1	1	0.0000
9	Rough plumbing 2	7	3	6	7	6	1.5625
10	Hardening 2	8	1	1	2	1	0.0977
11	Mounting scaffolding	9,10	1	2	2	2	0.0977
12	Framing	11	4	6	10	6	3.5156
13	Exterior woodwork	12	3	6	9	6	3.5156
14	Roofing	12	1	3	8	4	4.7852
15	Plumbing 1 (gas)	12	1	1	1	1	0.0000
16	Plumbing 2	12	1	2	3	2	0.3906
17	Wiring	12	2	2	2	2	0.0000
18	Interior woodwork	14,15,16,17	11	14	18	14	4.7852
19	Sheet metal work	13	1	2	3	2	0.3906
20	Wire mesh	19	1	5	7	5	3.5156
21	Exterior plaster	20	1	3	7	3	3.5156
22	Drying of exterior plaster	21	3	5	9	5	3.5156
23	Exterior painting	22	2	2	4	2	0.3906

(continued)

Table 6.12 (continued)

Activity No	Activity Description	Predecessors	Duration (Working Days)				
			Optimistic	Most likely	Pessimistic	Mean Duration	Variance
24	Interior plaster	18,23	2	4	5	4	0.8789
25	Doors & windows	18	3	5	8	5	2.4414
26	Interior painting	24	4	5	8	5	1.5625
27	Gutter	23	1	2	2	2	0.0977
28	Tiles	26	4	5	6	5	0.3906
29	Wall paper	26	3	5	7	5	1.5625
30	Demounting scaffolding	27	1	1	2	1	0.0977
31	Fusuma partition	25,28,29	2	2	2	2	0.0000
32	Carpet	28,29	1	2	3	2	0.3906
33	Electric appliance	28,29	1	3	5	3	1.5625
34	Miscellaneous parts	32	1	1	1	1	0.0000
35	Mounting boiler	32	1	1	2	1	0.0977
36	Kitchen	34	1	2	2	2	0.0977
37	Cleaning	30,31,33,35,36	1	2	3	2	0.3906
38	Screen	37	1	1	1	1	0.0000
39	Storm windows	37	1	2	2	2	0.0977
40	Tatami mat	37	1	2	2	2	0.0977

[a]The activity descriptions and project network are based on Tsubakitani and Deckro (1990)

(1) The project network is represented employing a project management software and PERT is applied to obtain the expected values for the critical path and the project duration. The project is estimated to finish on **September 08, 2017**. The critical path consists of activities {S, 1, 2, 3, 4, 5, 6, 7, 9, 11, 12, 13, 19, 20, 21, 22, 23, 24, 26, 28, 29, 32, 34, 36, 37, 40, T} and its length is 72 working days. Project duration is normally distributed with a mean of 72 days and a variance of 31.6406 days^2 (N ~ (72, 31.6406)).

(2) The management of BW Company wishes to complete and deliver the building to the owner earlier in order to receive the remaining payments earlier. To plan the cash flow for this project, the management would like to know the probabilities of finishing the project 5, 10, and 15 working days earlier. The required probabilities are calculated as follows:

(a) Probability that the project will not take longer than 67 working days.

$$P\left\{z \le \frac{x-\mu}{s}\right\} = P\left\{z \le \frac{67-72}{\sqrt{31.6406}}\right\} = P\{z \le -0.8889\} = 0.1870$$

(b) Probability that the project will not take longer than 62 working days.

$$P\left\{z \le \frac{x-\mu}{s}\right\} = P\left\{z \le \frac{62-72}{\sqrt{31.6406}}\right\} = P\{z \le -1.7778\} = 0.0377$$

(c) Probability that the project will not take longer than 57 working days.

$$P\left\{z \le \frac{x-\mu}{s}\right\} = P\left\{z \le \frac{57-72}{\sqrt{32.6406}}\right\} = P\{z \le -2.6667\} = 0.0038$$

(3) A key event is the start of the activity 18 – Interior woodwork, as it takes 14 days to finish. According to the PERT result it finishes on July 28, 2017. The management would like to know the impact of finishing this activity 7 working days earlier on the project duration. Note that this activity is not on the critical path, so deploying extra resources to finish this activity earlier will not have an impact on the overall finish date of the project. However, if the duration of this activity exceeds 20 working days it will be on the critical path, and any additional delay will also delay the estimated project finish date.

(4) Another key event is exterior woodwork and this activity is on the critical path. The estimated finish date of this activity is July 12, 2017. If the management decides to spend extra resources on this activity and make it finish 2 days earlier, i.e., July 10, 2017, then the finish date is also updated accordingly, and the new finish date would become **September 06, 2017**.

Exercises

6.1 What is the basic assumption underlying the PERT approach?

6.2 Explain why PERT is said to be optimistic. (You can use a figure to illustrate your answer.)

6.3 What would you consider as the shortcomings of the PERT approach? Given these shortcomings why would one prefer PERT to CPM?

6.4 You are in the process of estimating the mean duration value for an activity. You use the PERT approach and ask for the optimistic, most likely, and pessimistic duration estimates for the activity to 7 different experts. Having obtained these values, you calculate the mean activity duration estimated by each expert as reported in the table below. Now you need to reconcile these values to obtain a single estimate for the mean activity duration. Calculate such a single estimate using both the geometric mean (the *n*th root of the product of all estimates) and the arithmetic mean. Use 2-digit accuracy.

Expert 1	Expert 2	Expert 3	Expert 4	Expert 5	Expert 6	Expert 7
7	32	5	11	8	9	25

Which one of the estimates would you use and why? (Hint: The geometric mean is always less than the arithmetic mean.)

6.5 How do you estimate the probability of a path between the starting node and the terminal node to be the longest path using the activity criticality indices (CIs)? Apply your suggestion to the following example and the CIs given in the table below.

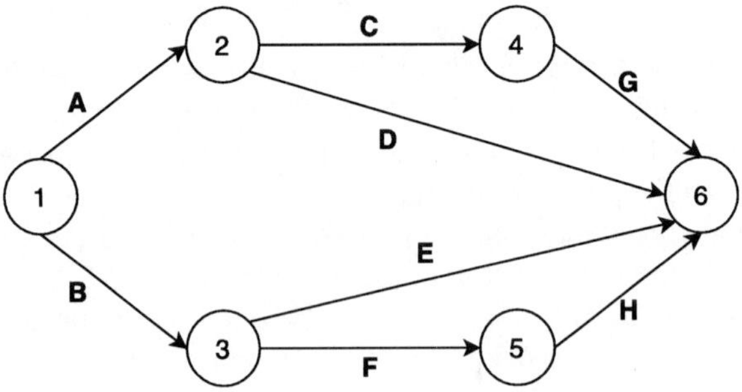

Activity	A	B	C	D	E	F	G	H
CI	0.58	0.42	0.27	0.30	0.24	0.19	0.27	0.19

6.6 For the planning of a project, we have estimated the optimistic, most likely, and pessimistic durations of each activity. The durations are given in days. Keep 2 significant digits in all your computations.

		Duration		
Activity definitions	Predecessors	a	m	b
A	*	2	6	10
B	*	3	5	7
C	A	2	4	10
D	B	1	3	8
E	B	4	6	10
F	C	3	7	9

(continued)

G	D	5	6	8
H	E	7	9	10
I	H	5	9	11

(a) Determine the expected project duration.
(b) What is the probability that the completion of activity H will not exceed 18 days?

6.7 In the project shown below all the activity durations have normal distributions with means and standard deviation as given. What is the probability that the path A – C – E – G will be the longest path in the network?

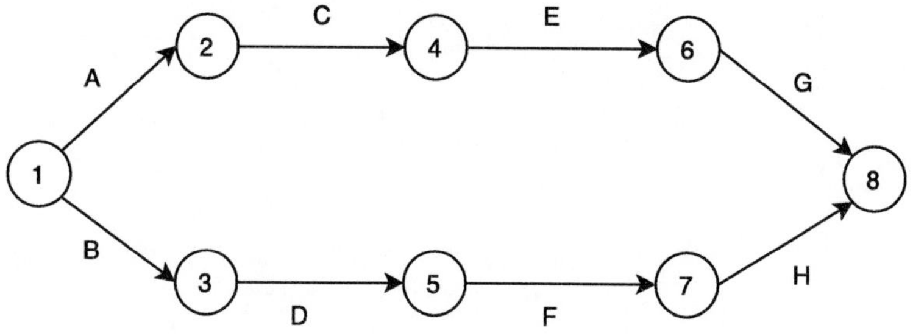

Activity	μ	σ
A	30	4
B	50	4
C	40	3
D	20	2
E	80	8
F	60	6
G	40	5
H	40	4

6.8 Consider the following project network, where the durations are probabilistic. The optimistic, most likely, and pessimistic estimations for the durations are provided by the experts in the company as follows.

Activity	Immediate predecessor	Estimates for durations (Days)		
		a	m	b
A	–	4	5	6
B	–	1	3	5
C	A	1	2	3
D	B	1	2	3
E	A	3	6	9
F	C, D	1	2	3

(continued)

G	B	9	12	15
H	E	3	5	7
I	F, H	3	5	7

(a) Draw the AOA project network and determine the PERT estimate for the project duration and the corresponding critical path.
(b) Calculate an estimate of the probability that the critical path found in part (a) will indeed be the critical path. Is your estimate an overestimate or an underestimate of the actual project duration? Explain.

6.9 Consider the following project consisting of 6 work packages. The optimistic, pessimistic, and most likely cost estimates are given.

	Cost (TL)		
Work packages	a	m	b
A	200	400	600
B	300	500	900
C	400	500	700
D	300	600	750
E	400	500	800
F	500	900	1000

(a) Calculate project mean cost and project cost standard deviation.
(b) What is the probability that the project cost will not exceed 3500TL?
(c) Assume that at some point in the project, work packages A, B, C and D are completed with their most likely cost. For the cost of the remaining two work packages, the manager estimates these subjective probabilities:

- E will cost 400 with probability 10%.
  500 with probability 70%.
  800 with probability 20%.
- F will cost 500 with probability 15%.
  900 with probability 60%
  1000 with probability 25%.

Estimate the total project costs with their occurrence probabilities.
What is the probability that the project cost will not exceed 3500TL?

6.10 Consider the following stochastic project network. Use the following relationship for calculating the standard deviation: s = [(b-a)/3.2].
(a) What is the probability that the project will be finished on or before day 58?
(b) The project manager has a budget of 300,000TL. S/he can improve the duration estimates for the following activities by investing a certain amount of additional resources as given in the following table. Find the

best improvement combination without exceeding the budget provided. (Unnecessary expenditures will be penalized by your boss. Thus, be careful!)

Activity	a	b	m	Cost
F	16	28	20	100
G	12	24	16	110
L	14	29	22	180

(c) What is the probability that activity H will finish earlier than activity F?
(d) What is the time on or before which the event "activities D and E have finished" will occur with 95% probability?

Activity	Predecessors	a	b	m
A	–	4	9	6
B	–	5	13	7
C	–	11	22	16
D	A	9	17	12
E	B	3	9	5
F	B	22	36	28
G	C	17	29	22
H	G	9	21	17
I	D, E	2	9	6
J	F, H	11	19	13
K	G	8	22	13
L	K	23	36	27
M	I	4	9	7

6.11 Consider the following project network, where the three estimates in weeks (pessimistic, most likely, optimistic) for the durations are indicated on each arc.

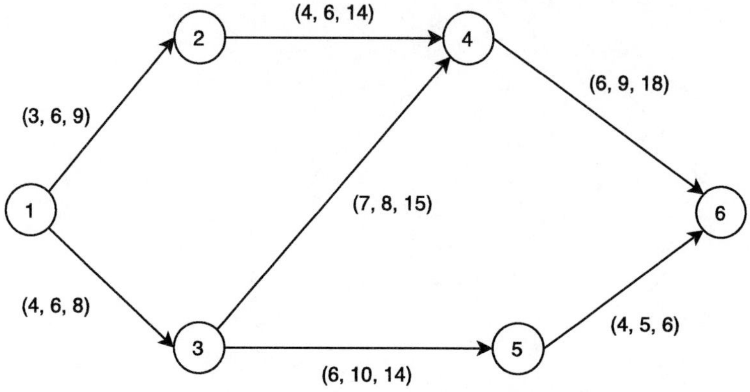

(a) Find the expected duration for the project network given above.

(b) Calculate the probability that the project duration will exceed 27 weeks.

(c) The engineers have calculated the saving in terms of the reduction in the expected project duration and the reduction in the variance of the critical path as follows:
Saving = 5650 × (reduction in the expected project duration) + 10,890 × (reduction in the variance of the critical path).

The engineers inform you of a new diagnostic technology, which, when applied, reduces the variance of that activity's duration by 80% but the expected duration stays the same. Due to the high demand, the company can rent the associated equipment to be used only once and at only one activity. Determine the activity at which you will employ the equipment and the value of the information you will obtain as a result.

(d) The contractor will receive 10,000 TL at the completion of activity (1,2) and 25,000 TL at the completion of activity (1,3). Assume these activities start simultaneously at $t = 0$ and the execution of these activities to be independent of each other. What would the contractor's expected return be at $t = 6$?

6.12 Consider the following activity network where the (optimistic, most likely, pessimistic) estimates for the corresponding activity duration are indicated on each arc. Assuming Beta distribution for each activity's duration and the independence of these Beta random variables, find the following.

(a) Expected critical path length.

(b) The probability that the project will be completed in no more than 16 time units.

(c) Will the solution change if the duration estimates for the activity (2,5) are changed to (3, 6, 9)?

(d) Will the results change if the activity (2,5) has the duration estimates given as (8, 10, 18)?

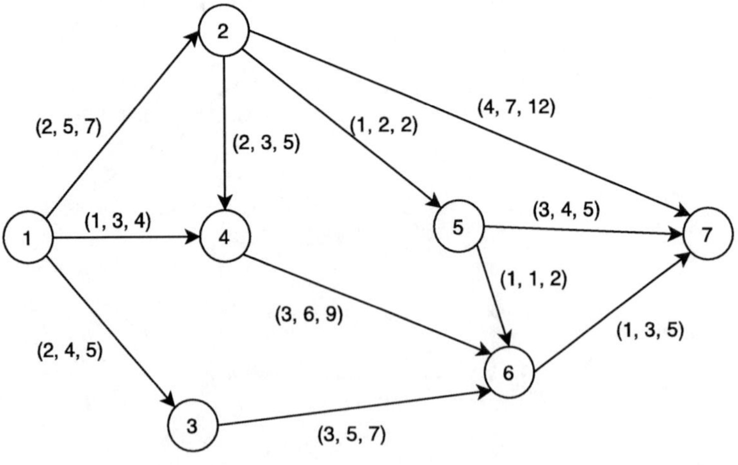

6.13 The manager of a software design project has been given the following data on activity durations by the experts consulted by the project crew. The AOA project network is given below where the dotted arcs indicate dummy activities.

Activity definitions	a	m	b
A – Specify functional features	3	4	5
B – Specify user interface features	8	9	13
C – Design and code functional component	10	12	18
D – Test and debug functional component	11	12	13
E – Internal audit of functional test	7	8	9
F – Design and code graphical user interface	13	15	16
G – Integrate functional component and interface	8	10	14
H – Train accounting personnel on interface	5	6	8
I – Train personnel on the integrated system	8	9	10

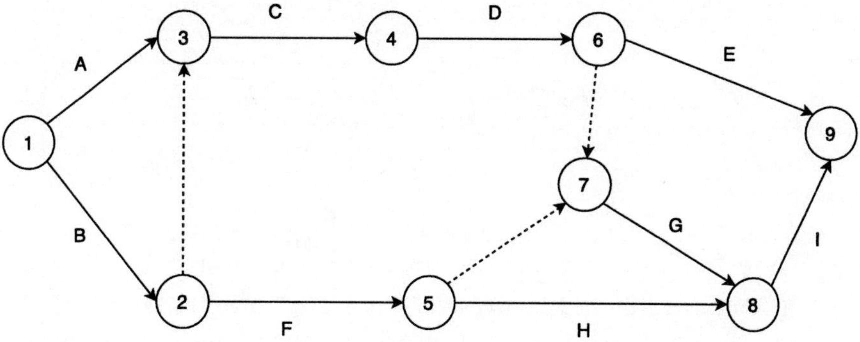

(a) Considering the duration estimation data available and the precedence relations described by the project network above, determine the expected critical path and the project duration.

(b) At time t = 25, the following progress report about the activities was handed to the project manager.

Activity	Progress level	Actual duration	a	m	b
A	100%	4	–	–	–
B	100%	10	–	–	–
C	100%	9	–	–	–
D	100%	12	–	–	–
E	0%	–	5	6	7
F	100%	15	–	–	–

(continued)

G	0%	–	9	11	13
H	0%	–	7	8	10
I	0%	–	8	9	10

Recalculate the expected critical path and project duration according to the progress information handed in.

6.14 Consider the following two projects. Project A consists of 3 work packages. The optimistic, pessimistic, and most likely cost estimates are given. Project B consists of 2 work packages. For each work package, estimated costs and its probabilities are given as follows.

		Cost (TL)		
	Work packages	a	m	b
Project A	A	300	800	900
	B	600	700	1100
	C	350	500	650

	Work packages	**Cost (TL)**	**Prob. (%)**
Project B	A_1	1000	60
	A_2	1500	40
	B_1	800	20
	B_2	1000	50
	B_3	1200	30

(a) Calculate estimated mean costs of both projects. If you had known that the company can manage only one of these projects due to resource constraints, which project should be chosen according to their mean cost?

(b) What is the probability that Project A's cost will not exceed Project B's mean cost?

6.15 Happy Ships and Pleasant Ships compete against each other by following two different routes both starting from Barcelona and ending in Istanbul. They have departed from Barcelona on the same date. The ships' routes represented as AOA project networks consist of activities resulting in paths A = {A,B,C} and D = {D,E,F} for the Happy Ships and Pleasant Ships, respectively. The durations of the activities are uncertain. Optimistic (a), most likely (m), and pessimistic (b) estimates in days for these activities are given in the table below.

Activity	a	m	b
A	8	11	24
B	7	9	11

(continued)

C	11	13	18
D	10	16	19
E	13	15	16
F	6	8	10

People are betting heavily on who will finish first. What is the probability that Happy Ships will be the winner, i.e., it will reach Istanbul first?

6.16 The planning of a project consisting of 10 activities is under way and you are a member of a project team doing the planning. The project will be implemented in an uncertain environment. The project team has agreed that the activity durations are uncertain and the planning process has to take this into account. Consulting several experts familiar with the activities in a project, the PM has obtained the following three estimates (optimistic, most likely, pessimistic) for each activity of the project in days.

Activity	**Immediate Predecessor**	**a**	**m**	**b**
A	-	4	6	8
B	-	2	7	12
C	-	6	9	14
D	A	2	3	4
E	B	2	7	10
F	C	1	3	5
G	C	3	6	15
H	E,F	5	9	19
I	G	5	9	11
J	D	7	11	13

The PM has asked you to conduct a Monte Carlo simulation analysis using these three estimates for each activity duration and report the following output obtained.

(a) Expected project duration and the variance. Report the 95% confidence interval around the expected project duration.
(b) The criticality index CI of each activity.
(c) The ratio of occurrence that a particular path will be critical.
(d) The probability that the project will be completed in 28 or less number of days.
(e) The project duration distribution.

Appendix 6A: Area Under the Normal Curve

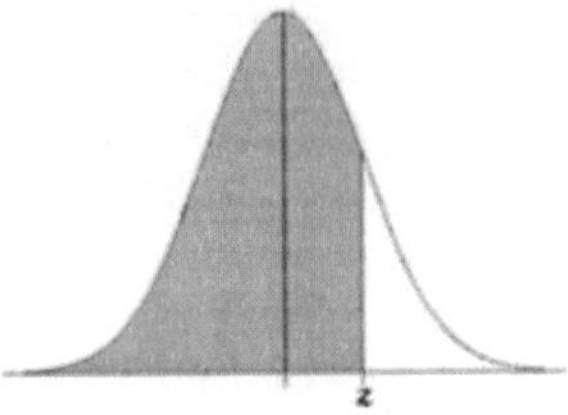

z	.00	.01	.02	.03	.04	.05	.06	.07	.08	.09
−3.4	0.0003	0.0003	0.0003	0.0003	0.0003	0.0003	0.0003	0.0003	0.0003	0.0002
−3.3	0.0005	0.0005	0.0005	0.0004	0.0004	0.0004	0.0004	0.0004	0.0004	0.0003
−3.2	0.0007	0.0007	0.0006	0.0006	0.0006	0.0006	0.0006	0.0005	0.0005	0.0005
−3.1	0.0010	0.0009	0.0009	0.0009	0.0008	0.0008	0.0008	0.0008	0.0007	0.0007
−3.0	0.0013	0.0013	0.0013	0.0012	0.0012	0.0011	0.0011	0.0011	0.0010	0.0010
−2.9	0.0019	0.0018	0.0018	0.0017	0.0016	0.0016	0.0015	0.0015	0.0014	0.0014
−2.8	0.0026	0.0025	0.0024	0.0023	0.0023	0.0022	0.0021	0.0021	0.0020	0.0019
−2.7	0.0035	0.0034	0.0033	0.0032	0.0031	0.0030	0.0029	0.0028	0.0027	0.0026
−2.6	0.0047	0.0045	0.0044	0.0043	0.0041	0.0040	0.0039	0.0038	0.0037	0.0036
−2.5	0.0062	0.0060	0.0059	0.0057	0.0055	0.0054	0.0052	0.0051	0.0049	0.0048
−2.4	0.0082	0.0080	0.0078	0.0075	0.0073	0.0071	0.0069	0.0068	0.0066	0.0064
−2.3	0.0107	0.0104	0.0102	0.0099	0.0096	0.0094	0.0091	0.0089	0.0087	0.0084
−2.2	0.0139	0.0136	0.0132	0.0129	0.0125	0.0122	0.0119	0.0116	0.0113	0.0110
−2.1	0.0179	0.0174	0.0170	0.0166	0.0162	0.0158	0.0154	0.0150	0.0146	0.0143
−2.0	0.0228	0.0222	0.0217	0.0212	0.0207	0.0202	0.0197	0.0192	0.0188	0.0183
−1.9	0.0287	0.0281	0.0274	0.0268	0.0262	0.0256	0.0250	0.0244	0.0239	0.0233
−1.8	0.0359	0.0351	0.0344	0.0336	0.0329	0.0322	0.0314	0.0307	0.0301	0.0294
−1.7	0.0446	0.0436	0.0427	0.0418	0.0409	0.0401	0.0392	0.0384	0.0375	0.0367
−1.6	0.0548	0.0537	0.0526	0.0516	0.0505	0.0495	0.0485	0.0475	0.0465	0.0455
−1.5	0.0668	0.0655	0.0643	0.0630	0.0618	0.0606	0.0594	0.0582	0.0571	0.0559
−1.4	0.0808	0.0793	0.0778	0.0764	0.0749	0.0735	0.0721	0.0708	0.0694	0.0681
−1.3	0.0968	0.0951	0.0934	0.0918	0.0901	0.0885	0.0869	0.0853	0.0838	0.0823
−1.2	0.1151	0.1131	0.1112	0.1093	0.1075	0.1056	0.1038	0.1020	0.1003	0.0985
−1.1	0.1357	0.1335	0.1314	0.1292	0.1271	0.1251	0.1230	0.1210	0.1190	0.1170
−1.0	0.1587	0.1562	0.1539	0.1515	0.1492	0.1469	0.1446	0.1423	0.1401	0.1379
−0.9	0.1841	0.1814	0.1788	0.1762	0.1736	0.1711	0.1685	0.1660	0.1635	0.1611
−0.8	0.2119	0.2090	0.2061	0.2033	0.2005	0.1977	0.1949	0.1922	0.1894	0.1867
−0.7	0.2420	0.2389	0.2358	0.2327	0.2296	0.2266	0.2236	0.2206	0.2177	0.2148
−0.6	0.2743	0.2709	0.2676	0.2643	0.2611	0.2578	0.2546	0.2514	0.2483	0.2451
−0.5	0.3085	0.3050	0.3015	0.2981	0.2946	0.2912	0.2877	0.2843	0.2810	0.2776
−0.4	0.3446	0.3409	0.3372	0.3336	0.3300	0.3264	0.3228	0.3192	0.3156	0.3121
−0.3	0.3821	0.3783	0.3745	0.3707	0.3669	0.3632	0.3594	0.3557	0.3520	0.3483
−0.2	0.4207	0.4168	0.4129	0.4090	0.4052	0.4013	0.3974	0.3936	0.3897	0.3859
−0.1	0.4602	0.4562	0.4522	0.4483	0.4443	0.4404	0.4364	0.4325	0.4286	0.4247
−0.0	0.5000	0.4960	0.4920	0.4880	0.4840	0.4801	0.4761	0.4721	0.4681	0.4641

z	.00	.01	.02	.03	.04	.05	.06	.07	.08	.09
0.0	0.5000	0.5040	0.5080	0.5120	0.5160	0.5199	0.5239	0.5279	0.5319	0.5359
0.1	0.5398	0.5438	0.5478	0.5517	0.5557	0.5596	0.5636	0.5675	0.5714	0.5753
0.2	0.5793	0.5832	0.5871	0.5910	0.5948	0.5987	0.6026	0.6064	0.6103	0.6141
0.3	0.6179	0.6217	0.6255	0.6293	0.6331	0.6368	0.6406	0.6443	0.6480	0.6517
0.4	0.6554	0.6591	0.6628	0.6664	0.6700	0.6736	0.6772	0.6808	0.6844	0.6879
0.5	0.6915	0.6950	0.6985	0.7019	0.7054	0.7088	0.7123	0.7157	0.7190	0.7224
0.6	0.7257	0.7291	0.7324	0.7357	0.7389	0.7422	0.7454	0.7486	0.7517	0.7549
0.7	0.7580	0.7611	0.7642	0.7673	0.7704	0.7734	0.7764	0.7794	0.7823	0.7852
0.8	0.7881	0.7910	0.7939	0.7967	0.7995	0.8023	0.8051	0.8078	0.8106	0.8133
0.9	0.8159	0.8186	0.8212	0.8238	0.8264	0.8289	0.8315	0.8340	0.8365	0.8389
1.0	0.8413	0.8438	0.8461	0.8485	0.8508	0.8531	0.8554	0.8577	0.8599	0.8621
1.1	0.8643	0.8665	0.8686	0.8708	0.8729	0.8749	0.8770	0.8790	0.8810	0.8830
1.2	0.8849	0.8869	0.8888	0.8907	0.8925	0.8944	0.8962	0.8980	0.8997	0.9015
1.3	0.9032	0.9049	0.9066	0.9082	0.9099	0.9115	0.9131	0.9147	0.9162	0.9177
1.4	0.9192	0.9207	0.9222	0.9236	0.9251	0.9265	0.9279	0.9292	0.9306	0.9319
1.5	0.9332	0.9345	0.9357	0.9370	0.9382	0.9394	0.9406	0.9418	0.9429	0.9441
1.6	0.9452	0.9463	0.9474	0.9484	0.9495	0.9505	0.9515	0.9525	0.9535	0.9545
1.7	0.9554	0.9564	0.9573	0.9582	0.9591	0.9599	0.9608	0.9616	0.9625	0.9633
1.8	0.9641	0.9649	0.9656	0.9664	0.9671	0.9678	0.9686	0.9693	0.9699	0.9706
1.9	0.9713	0.9719	0.9726	0.9732	0.9738	0.9744	0.9750	0.9756	0.9761	0.9767
2.0	0.9772	0.9778	0.9783	0.9788	0.9793	0.9798	0.9803	0.9808	0.9812	0.9817
2.1	0.9821	0.9826	0.9830	0.9834	0.9838	0.9842	0.9846	0.9850	0.9854	0.9857
2.2	0.9861	0.9864	0.9868	0.9871	0.9875	0.9878	0.9881	0.9884	0.9887	0.9890
2.3	0.9893	0.9896	0.9898	0.9901	0.9904	0.9906	0.9909	0.9911	0.9913	0.9916
2.4	0.9918	0.9920	0.9922	0.9925	0.9927	0.9929	0.9931	0.9932	0.9934	0.9936
2.5	0.9938	0.9940	0.9941	0.9943	0.9945	0.9946	0.9948	0.9949	0.9951	0.9952
2.6	0.9953	0.9955	0.9956	0.9957	0.9959	0.9960	0.9961	0.9962	0.9963	0.9964
2.7	0.9965	0.9966	0.9967	0.9968	0.9969	0.9970	0.9971	0.9972	0.9973	0.9974
2.8	0.9974	0.9975	0.9976	0.9977	0.9977	0.9978	0.9979	0.9979	0.9980	0.9981
2.9	0.9981	0.9982	0.9982	0.9983	0.9984	0.9984	0.9985	0.9985	0.9986	0.9986
3.0	0.9987	0.9987	0.9987	0.9988	0.9988	0.9989	0.9989	0.9989	0.9990	0.9990
3.1	0.9990	0.9991	0.9991	0.9991	0.9992	0.9992	0.9992	0.9992	0.9993	0.9993
3.2	0.9993	0.9993	0.9994	0.9994	0.9994	0.9994	0.9994	0.9995	0.9995	0.9995
3.3	0.9995	0.9995	0.9995	0.9996	0.9996	0.9996	0.9996	0.9996	0.9996	0.9997
3.4	0.9997	0.9997	0.9997	0.9997	0.9997	0.9997	0.9997	0.9997	0.9997	0.9998

Reproduced from Table A.3 pp. 755–756 in R.E. Walpole, R.H. Myers, S.L. Myers, K. Ye, *Probability & Statistics for Engineers & Scientists*, Pearson Education Limited, Essex, England, 2016

References

Acebes, F., Pajares, J., Galán, J. M., & López-Paredes, A. (2014). A new approach for project control under uncertainty. Going back to the basics. *International Journal of Project Management, 32*, 423–434.

Adlakha, V. G., & Kulkarni, V. G. (1989). A classified bibliography of research on stochastic PERT networks. *Infor, 27*(3), 272–296.

Batselier, J., & Vanhoucke, M. (2015). Construction and evaluation framework for a real-life project database. *International Journal of Project Management, 33*(3), 697–710.

Buss, A., & Rosenblatt, M. (1997). Activity delay in stochastic project networks. *Operations Research, 45*, 126–139.

Case, K. E. (1972). On the consideration of variability in cost estimating. *IEEE Transactions on Engineering Management, 19*, 114–118.

Dodin, B. H. W. S., & Elmaghraby, S. E. E. (1985). Approximating the criticality indices of the activities in PERT networks. *Management Science, 31*, 207–223.

Drake, A. W. (1967). *Fundamentals of applied probability*. New York: McGraw Hill.

Fulkerson, D. R. (1962). Expected critical path lengths in PERT networks. *Operations Research, 10* (6), 808–817.

Golenko-Ginzburg, D. (1989). PERT assumptions revisited. *Omega, 17*, 393–396.

Hall, N. G. (2016). Research and teaching opportunities in project management. In *Optimization challenges in complex, networked and risky systems* (pp. 329–388) INFORMS.

Hazır, Ö. (2015). A review of analytical models, approaches and decision support tools in project monitoring and control. *International Journal of Project Management, 33*(4), 808–815.

Iida, T. (2000). Computing bounds on project duration distributions for stochastic PERT networks. *Naval Research Logistics, 47*, 559–580.

Kamburowski, J. (1996). New validations of PERT times. *Omega, 25*(3), 323–328.

Keefer, D. L., & Verdini, W. A. (1993). Better estimation of PERT activity time parameters. *Management Science, 39*(9), 1086–1091.

Kleindorfer, P. (1971). Stochastic bounds on job completion times. *Operations Research, 19*, 1586–1601.

Law, A. M. (2014). *Simulation modeling* (Series in Industrial Engineering and Management) (5th ed.). New York: McGraw Hill.

Littlefield, T. K., Jr., & Randolph, P. H. (1987). An answer to Sasieni's question on PERT times. *Management Science, 33*(10), 1357–1359.

Malcolm, D. G., Roseboom, G. H., Clark, C. E., & Fazar, W. (1959). Applications of a technique for research and development program evaluation. *Operations Research, 7*, 646–669.

Mitchell, G., & Klastorin, T. (2007). An effective methodology for the stochastic project compression problem. *IIE Transactions, 39*(10), 957–969.

Moder, J. J., & Rodgers, E. G. (1968). Judgement estimates of the moments of PERT type distributions. *Management Science, 15*(2), 76–83.

Moder, J. J., & Phillips, C. R. (1970). *Project management with CPM and PERT* (2nd ed.). New York: Van Nostrand Reinhols Company.

Prez, J. G., Martin, M., Garca, C., & Granero, M. (2016). Project management under uncertainty beyond beta: The generalized bicubic distribution. *Operations Research Perspectives, 3*, 67–76.

Schonberger, R. J. (1981). Why projects are always late: A rationale based on manual simulation of a PERT/CPM network. *Interfaces, 11*, 66–70.

Sculli, D. (1989). A historical note on PERT times. *Omega, 17*(2), 195–196.

Tsubakitani, S., & Deckro, R. F. (1990). A heuristic for multi-project scheduling with limited resources in the housing industry. *European Journal of Operational Research, 49*, 80–91.

Van Slyke, R. M. (1963). Monte Carlo methods and PERT problem. *Operations Research, 11*, 839–860.

Vose, D. (2008). *Risk analysis. A quantitative guide* (3rd ed.). Chichester: Wiley.

Walpole, R. E., Myers, R. H., & Myers, S. L. (1998). *Probability and statistics for engineers and scientists* (6th ed.). Prentice Hall International.

Zhu, G., Bard, J. F., & Yu, G. (2005). Disruption management for resource-constrained project scheduling. *Journal of the Operational Research Society, 56*, 365–381.

Resource Constrained Project Scheduling 7

Learning Outcomes
Upon successful completion of this Chapter, the reader will be able to:

1. Define and compare different categories of resources used in projects.
2. Formulate mathematical models to address the Resource Constrained Project Scheduling Problem (RCPSP).
3. Formulate RCPSP with different objective functions.
4. Generate solutions for the RCPSP using different heuristics.
5. Apply an exact algorithm to obtain an optimal solution for the RCPSP.
6. Formulate multi-mode RCPSP (MRCPSP) with different objective functions.

7.1 Introduction

In this chapter we address the allocation of limited resources to project activities, referred as Resource Constrained Project Scheduling Problem (RCPSP). In the most common version of this problem, project completion time is minimized subject to precedence relationships among the activities, deterministic activity durations, and renewable and non-renewable resource availability constraints. The RCPSP has been extensively studied in many different versions (Özdamar and Ulusoy 1995; Herroelen et al. 1998; Kolisch and Padman 2001; Hartmann and Briskorn 2010), and several of its extensions are closely related to production planning problems for products such as ships, yachts, locomotives, and railway cars.

Since the RCPSP was introduced to the operations research literature by Kelley Jr and Walker (1959), this problem and its extensions have attracted the attention of practitioners and researchers for two principal reasons. The RCPSP is a version of the resource allocation and optimization problem, a fundamental problem in management and economics. It is also a challenging and difficult problem to solve from both theoretical and computational aspects, belonging to the class of provably difficult, hard problems. It is shown to be NP-hard in the strong sense (Blazewicz

G. Ulusoy, Ö. Hazır, *An Introduction to Project Modeling and Planning*, Springer Texts in Business and Economics, https://doi.org/10.1007/978-3-030-61423-2_7

et al. 1983). Further detail on computational complexity is given in *Appendix 7A Computational Complexity*.

The most commonly studied objective functions for RCPSPs are the project duration (C_{max}) and the Net Present Value (NPV) of cash flows, discussed in Chap. 4. To set the stage for this chapter, we begin by discussing the different categories of resources that may be encountered in the project scheduling context and introduce the resource parameters upon which the various mathematical programming formulations and analysis of the RCPSP and its extensions are based.

7.1.1 Categories of Resources

A major task in project planning is to allocate the limited resources available among the activities over time. Resources encountered in project management can be grouped into three main categories (Blazewicz et al. 1986): Renewable, non-renewable, and doubly-constrained. *Renewable* resources such as construction machinery, equipment, and personnel are non-consumable; they do not accumulate if not used but are available in a constant quantity over time. *Non-renewable* resources, on the other hand, are resources that are consumed during the execution of an activity such as chemicals, cement, and cash. Non-renewable resources are limited both over any time period and/or over the whole project. A non-renewable resource whose availability is limited both over each planning period and over the whole project is called *doubly constrained*. The most common example of such a resource is cash when both the total expenditure for the project and per period expenditures cannot exceed a specified budget.

Renewable resources that are only available at certain periods of time are called *partially renewable resources*. Similarly, non-renewable resources that are available only over certain time periods are called *partially non-renewable resources* (Böttcher et al. 1996).

In some applications, a renewable resource can be assigned to only one activity at a time. Such renewable resources are called *dedicated resources* and are represented as a resource with one-unit availability per period (Bianco et al. 1998). Beşikci et al. (2012) give a different definition of dedicated resources in multi-project settings as a set of renewable resources, such as dedicated teams of specialists, who are allocated to a certain project for its entire duration and only become available to other projects when that project is completed or terminated for some reason.

Another classification of resources is based on how their requirements are determined. In *activity related resources*, nature and work content of activity determines the level and intensity of usage for such resources. For example, machinery would be an activity related resource. The usage of *resource related resources* depends on the usage of other resources, as when one foreman is required for each team of four workers. The consumption of *project related resources*, such as supervisory management, is determined by the project duration, as they are used at constant or varying levels throughout the whole project.

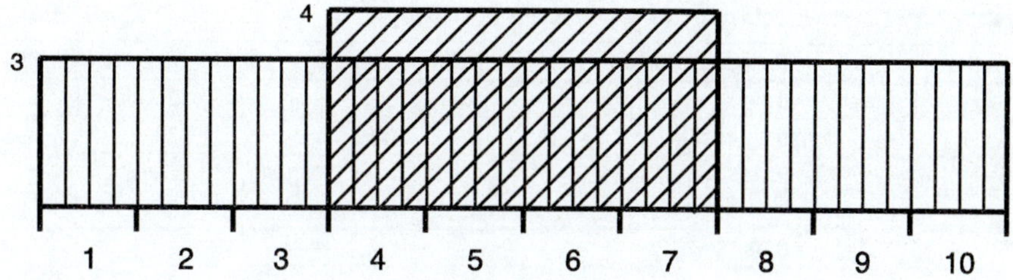

Fig. 7.1a Activity A and the resource demand histogram

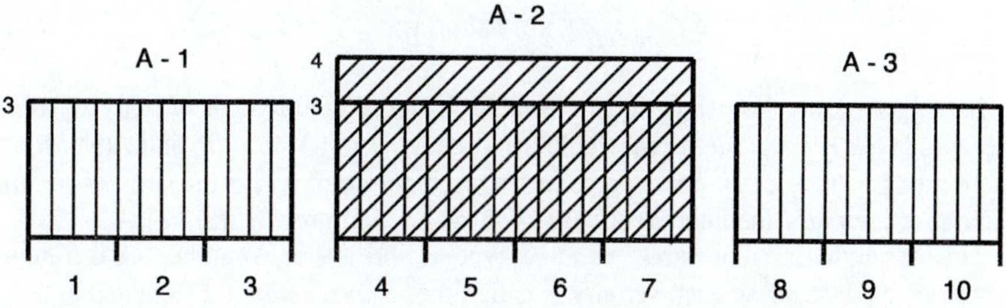

Fig. 7.1b Activity A split into three sub-arcs A-1, A-2, and A-3 each with uniform resource usage

In practice, resource usage requirements might vary over the duration of the activity. For example, a resource might only be required during certain portions of an activity's duration instead of the full duration. In this situation, the activity can be split into sub-activities to facilitate the mathematical programming formulation.

Consider, for example, an activity A of duration 10 periods using two different renewable resources. 3 units of renewable resource 1 are required throughout all 10 periods, while 4 units of renewable resource 2 are needed only during the interval [4,7] as shown in Fig. 7.1a.

Figure 7.1b illustrates how such an activity can be split into three sub-activities A-1, A-2, and A-3 to obtain uniform resource usage over each sub-activity:

Care needs to be taken to schedule these sub-activities consecutively without any break in between. Activity splitting is usually not desirable, since it increases the number of activities and can significantly complicate both data management and scheduling.

7.1.2 Resource Parameters

For a particular renewable resource type r, let K_r denote the maximum permissible usage of r per period. The ratio of K_r to the average usage of r per activity per period is defined as the resource strength RS_r given by

Table 7.1 The renewable resource requirements of activities

Activity	Resource type 1	Resource type 2	Resource type 3
1	3	5	2
2	0	4	2
3	2	4	1
4	3	3	0
5	0	4	2
6	4	0	3

$$RS_r = K_r / \left(\sum_j k_{jr} / |J| \right) for\ r \in R \tag{7.1}$$

where k_{jr} denotes the per period renewable resource requirement of activity j for resource type r, R the set of renewable resource types r = 1, . . ., | R| and J the set of activities j = 1, . . ., | J|. RS_r represents the average use of K_r for the case where all activities are performed consecutively, with no parallel processing.

For a particular renewable resource type r, the average number of different resource types used by each activity is called the *resource factor,* calculated as

$$RF = \sum_j \sum_r \delta_{jr} / (|J||R|) \tag{7.2}$$

where $\delta_{jr} = 1$ if $k_{jr} > 0$; and $\delta_{jr} = 0$ otherwise. RF indicates the homogeneity of the usage of the different resource types over the activities. If all types of resources are used by all activities, then $RF = 1$. $RF = 0$, on the other hand, represents the case of unconstrained project scheduling.

Example 7.1 Consider a project with 6 activities and 3 renewable resources whose maximum usage per period is $K_1 = 6$, $K_2 = 8$, $K_3 = 5$. The resource requirements of the activities for each resource type are given in Table 7.1. Calculate the RS values for each resource type and the RF value.

Solution

$$RS_1 = \frac{6}{12/6} = 3$$

$$RS_2 = \frac{10}{20/6} = 3$$

$$RS_3 = \frac{5}{10/6} = 3$$

The results indicate that the average usage of all resource types is 3 units per period. Although all RS values are the same, RS_1 and RS_3 represent a tighter resource availability for resources 1 and 3 than for resource 2.

$$RF = (3 + 2 + 3 + 2 + 2 + 2)/6x3 = 0.778$$

The *RF* value indicates that all resource types are required by a relatively high number of activities. If, for example, activity 3 does not require resource type 2, then RF becomes 0.722.

7.2 Objective Functions

7.2.1 Regular and Non-regular Objective Functions

As discussed in Chap. 4, the most frequently treated optimization criteria in project management are the minimization of project duration, the maximization of profit or revenue, and the minimization of cost, along with quality related objectives. We now discuss another classification of objective functions encountered in project management problems that has considerable impact on the solution algorithms we consider. A *regular performance measure* is one that is non-decreasing in the completion times of activities or tasks (French 1982). Hence, as the completion time of an individual activity increases, the value of a regular performance measure either stays the same or increases.

Definition 1 A performance measure *R* is said to be regular if the following two conditions hold:

$$C_i \leq C'_i \; i = 1\ldots, n \tag{7.3}$$

$$R(C_1, \ldots, C_n) \leq R(C'_1, \ldots, C'_n) \tag{7.4}$$

where C_i denotes the completion time of activity $i = 1\ldots, n$ in a schedule *S* and C'_i the completion time of activity $i = 1\ldots, n$ in the schedule S'.

Let us check whether the two performance measures we most frequently employ, C_{max} and NPV, are regular or not. We will consider three cases for C_{max}: *(i)* We know from the discussion of CPM in Chap. 4 that any increase of magnitude Δ in the duration of a critical activity increases C_{max} by Δ. *(ii)* Any increase in the duration of a non-critical activity of magnitude Δ≤slack of that activity leaves C_{max} unchanged. *(iii)* If Δ exceeds the slack of that activity, then C_{max} will increase by the difference between Δ and the slack of that activity. In all cases, Δ satisfies (7.4): increasing the completion time of any individual activity either leaves C_{max} unchanged or increases it.

NPV, on the other hand, is a non-regular measure because increasing the completion time of one or more activities in a schedule may cause NPV to decrease depending on the values of the cash inflows and outflows. Example 7.2 provides numerical illustrations for both C_{max} and NPV.

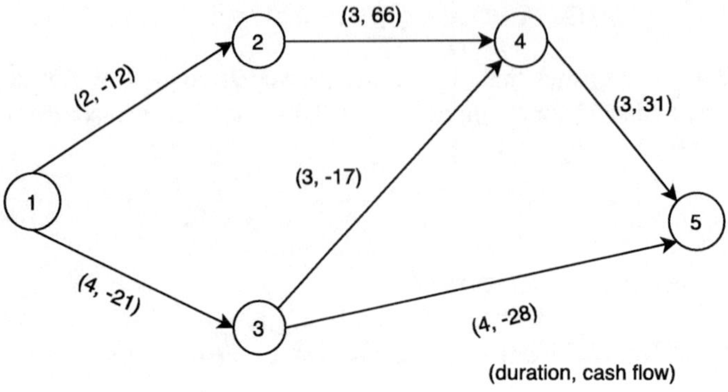

Fig. 7.2 Project network for Example 7.2

Example 7.2 Consider the AOA project network depicted in Fig. 7.2, with the (duration, cash flow value) pair shown on each arc. Both cash inflows and outflows take place at the completion of the activities, and cash outflows are indicated by a negative sign. The interest rate α is given as 10% per period.

Solution

Applying CPM to the example problem we obtain $C_{max} = 10$. If we increase the duration of activity (2,4) from 3 to 6 periods, the makespan C'_{max} increases to 11 periods. Hence, $C_{2,4} < C'_{2,4} \rightarrow C_{max} < C'_{max}$ satisfying both conditions (7.3) and (7.4).

We will demonstrate by a counterexample that *NPV* is a non-regular measure. Applying early start to activities with positive cash flow and late start to activities with negative cash flow, we calculate *NPV* as follows:

$$\begin{aligned}\mathrm{NPV} &= -21(1.1)^{-4} - 12(1.1)^{-4} + 66(1.1)^{-7} - 17(1.1)^{-7} - 28(1.1)^{-10} + 31(1.1)^{-10} \\ &= 3.762.\end{aligned}$$

However, when the duration of activity (2,4) is increased from 3 to 6 periods we obtain $\mathrm{NPV}' = 0.954$. Hence, $C_{2,4} < C'_{2,4} \rightarrow \mathrm{NPV} > \mathrm{NPV}'$, violating condition (7.4). Thus, NPV is not a regular measure.

We now present examples of the principal time-based, financial, quality, and resource-based objectives.

7.2.2 Time-Based Objectives

The makespan or project duration C_{max} is by far the most frequently considered time-based objective for a single project. Other time-based objectives include the lateness $L = (C_{max} - D)$, where D is the due date of the project. Depending on the relative magnitudes of C_{max} and D, L can be zero, positive or negative. When $L = 0$, the project is said to be on time. When $L > 0$, the project is said to be tardy, with tardiness $T = \max\{0, L\}$. When $L < 0$, then the project is early. Earliness is denoted by $E = \max\{0, -L\}$. L and T are both regular measures whereas E is not (see Exercise 7.23). These performance measures become particularly relevant in decision environments where a penalty or bonus based on project completion by a due date are involved.

7.2.2.1 The Case of Multiple Projects

In the case of multiple projects, the overall project completion time $C_{max}^{overall}$ is determined by the length of the time interval between the initiation of the first project and the completion of the last one. In order to complete the set of projects as quickly as possible, the decision-maker should try to schedule these projects to minimize $C_{max}^{overall} = max_{i=1}^{M}(C_{max}^{i})$, where M is the number of projects. This performance measure implies that the completion times of individual projects do not play a significant role in decision-making.

Another time-based objective is the average project completion time over all projects, $\overline{C}_{max} = (1/M)\sum_{i=1}^{M} C_{max}^{i}$, where C_{max}^{i} denotes the completion time for project i, $i = 1,\ldots,M$. This performance measure is directly related to customer satisfaction since it represents the average time the project owners wait for their deliverables.

The flow time F_i of project i is defined as the length of the time interval between the starting and the completion of the project. It is the project duration and corresponds to the period the project is ongoing and must be managed. Note that $C_{max}^{i} = F_i + r_i$, where r_i is the ready time for project i defined as the period from time zero to the start time of project i. The mean flow time $\overline{F}$ is defined as $\overline{F} = (1/M)\sum_{i=1}^{M} F_i$, where F_i is the flow time of project i, $i = 1,\ldots,M$. The decision-maker would be interested in minimizing $\overline{F}$ since it is expected to lead to a reduction in work-in-progress (Herroelen and Leus 2001), i.e., a reduction in the number of active projects in any given time period and hence, in operating capital. Minimizing $\overline{F}$ also helps reduce overhead costs associated with managing an ongoing project.

Consider a decision environment with multiple projects where some or all of the projects have due dates D_i for project i and let the lateness, tardiness and earliness of project i be denoted by L_i, T_i, and E_i, respectively. Relevant objectives for this case are average lateness $\overline{L}$, average tardiness $\overline{T}$, and average earliness $\overline{E}$. Another set of measures represent the extreme values: maximum lateness $L_{max} = max_i(L_i)$,maximum tardiness $T_{max} = max_i(T_i)$, and maximum earliness $E_{max} = max_i(E_i)$. As suggested above, these performance measures become particularly relevant in decision environments where penalties for late completion or bonuses for early completion are involved.

Decision-makers can assign different importance, indicated by weights w_i, to different projects in the system for various reasons, such as different project owners or financial importance. Weighted tardiness, for example, is then defined as $T_w = (1/n)\sum_{i=1}^{n} T_i$. Weighted earliness and lateness are defined similarly.

The performance measures such as lateness, tardiness, and earliness that we have defined for multiple projects can also be applied to individual *activities* of a single project. An activity i might be assigned a due date D_i if it represents work to be performed by a subcontractor who will be charged a penalty if this due date is exceeded. Tardiness would be an appropriate performance measure in this case. The decision-makers might also distinguish the importance of completion times of activities i by assigning them weights w_i.

7.2.3 Financial Objectives

Financial performance measures deal with the cash flows associated with the activities and/or events. The objective of maximizing the NPV of cash inflows and outflows in a project without resource constraints is treated in Sect. 4.4. The introduction of resource constraints makes the problem NP-hard (Baroum 1992) and will be treated in Sect. 7.4.

The trade-off between the cost and the project duration is a common problem in project management and scheduling, which is treated in Chap. 5 for the case without resource constraints. The objective of the time/cost trade-off problem is the minimization of the activity compression cost. A financial objective for a resource constrained problem would be the minimization of the cost of the resources required to complete the project by a pre-specified due date. This objective is only relevant for project scheduling problems with multiple modes, since this provides the project manager with several alternative combinations of duration and cost for each activity, from among which the most advantageous combination can be selected.

To execute a project without delays due to lack of cash, the availability of cash or operating capital is a crucial financial necessity (Goldratt 1997). Hence another financial performance measure of interest is the *maximum cash balance*, defined as the maximal gap between cumulative cash inflow and outflow in any period, representing the maximum amount of cash (operating capital) required in any period over the project duration. An objective reflecting this necessity is the minimization of the maximal cumulative gap between the project's cash inflows and outflows (Ning et al. 2017; Kucuksayacigil and Ulusoy 2020).

7.2.4 Quality Based Objectives

A major concern for PMs is how to manage project quality, which requires the formulation and implementation of appropriate quality measures. A widely employed measure of quality in both industry and services is conformance to customer's requirements. Considering the three dimensions of duration, cost, and quality in project management, PMs seek to exceed customer requirements while completing the project on time and within budget. Hence an appropriate objective for maximizing quality would be minimizing the cost and time required for rework to remediate quality problems arising from failure to meet customer specifications

(Icmeli-Tukel and Rom 1997). The crashing of activities can cause quality problems, as the reduction of activity durations through crashing often increases the risk of non-conformance with customer specifications. An objective function minimizing the sum of the direct cost of the time-cost trade-off and the non-conformance cost of rework and modification is presented by Kim et al. (2012).

7.2.5 Resource Based Objectives

Resource based objectives are typically associated with the resource levels employed. One such objective, *resource leveling*, seeks to minimize the fluctuations in resource usage over time. Another variant of the problem, *the Resource Availability Cost Problem* (RACP), seeks a feasible project schedule that does not exceed the deadline and minimizes the total cost, which is expressed as a function of the peak resource demand. Both resource leveling and RACP are treated in Chap. 8.

7.3 A Mathematical Programming Formulation of the Resource Constrained Project Scheduling Problem

There are several mathematical programming formulations for RCPSP (see, e.g., Demeulemeester and Herroelen 2002). The formulation presented in this section is a zero–one programming formulation (Pritsker et al. 1969; Patterson and Huber 1974; Patterson and Roth 1976).

The objective of this well-known problem is minimizing the project duration, i.e., the makespan. The activity precedence relations can be represented on an AON network with one source and one sink node. A dummy source is added if there is more than one activity with no predecessor, and a dummy sink if there is more than one activity with no successor. A dummy node has zero duration and no resource requirements. Precedence relations among activities are finish-to-start with zero time lags. Activities are assumed to be non-preemptive, i.e., once started they will be finished without any break in-between. Otherwise, the activity is said to be preemptive.

Data Pre-processing Given the RCPSP data, it is necessary to check for feasibility before attempting to solve the problem. (*i*) None of the renewable resource requirements for an activity should exceed the corresponding resource limit per period. (*ii*) The sum of the non-renewable resource requirements over the activities should not exceed the corresponding non-renewable resource limit. This pre-processing allows non-renewable resources to be incorporated into the formulation implicitly. In the following formulation, the constraint set limiting the non-renewable resources is added for the sake of completeness.

The zero–one decision variables are defined over the time interval $[EFT_j, LFT_j]$, where EFT_j denotes the early finish time and LFT_j the late finish time for activity j.

The number of zero-one variables is determined by the length of these time intervals. EFT_j is obtained from the forward pass of CPM for the unconstrained resource case. Calculation of LFT_j requires an estimate for the project horizon denoted here by T. To reduce the number of zero-one variables T must be as small as possible without losing feasibility. Such an estimate might be obtained by an effective heuristic procedure, if no other value is provided *a priori*. Given a value of T, the backward pass of CPM can be applied to obtain the LFT_j values.

J = number of activities including dummy activities, if any.
j = activity index.
R = set of renewable resource types.
r = renewable resource index
N = set of non-renewable resource types.
n = non-renewable resource index.
t = time index.
d_j =duration of activity j
P_j = set of immediate predecessors to activity j.
k_{jr} = per period renewable resource requirement of resource type r for activity j
k_{jn}= non-renewable resource requirement of resource type n for activity j
K_r = resource limit per period for renewable resource type r
K_n = amount of non-renewable resource type n available

The mathematical model for RCPSP can now be formulated as follows:

$$\min \sum_{t=EFT_J}^{LFT_J} t x_{Jt} \tag{7.5}$$

$$\sum_{t=EFT_j}^{LFT_j} x_{jt} = 1 \quad j = 1, 2, \ldots, J \tag{7.6}$$

$$\sum_{t=EFT_i}^{LFT_i} t x_{it} \leq \sum_{t=EFT_j}^{LFT_j} (t - d_j) x_{jt} \quad j = 2, \ldots, J, \quad i \in P_j \tag{7.7}$$

$$\sum_{j=1}^{J} k_{jr} \sum_{\tau=t}^{t+d_j-1} x_{j\tau} \leq K_r \quad t = 1, 2, \ldots, T, \quad r \in R \tag{7.8}$$

$$\sum_{j=1}^{J} k_{jn} \sum_{t=EFT_j}^{LFT_j} x_{jt} \leq K_n \quad n \in N \tag{7.9}$$

$$x_{jt} = \begin{cases} 1, \text{if activity } j \text{ is finished at the end of period } t \\ 0, \text{otherwise} \quad \textit{for all } j, t \end{cases} \tag{7.10}$$

The objective (7.5) is the minimization of the makespan. The constraint set (7.6) is the assignment constraint, enforcing the completion of each activity exactly once in the interval $[EFT_j, LFT_j]$, while (7.7) enforces the precedence constraints. The right-hand side of the constraint set (7.7) represents the starting period of activity j. Constraint set (7.8) represents the limited capacity of the renewable resources in each period for each resource. For any period t, any activity j that completes in period $[t, (t + d_j - 1)]$ will be in progress in period t and hence will consume k_{jr} units of renewable resource r in period t. An upper bound on the availability of non-renewable resource type n is given in constraint set (7.9). Constraint set (7.10) defines the zero-one variables.

If we have time dependent renewable resource profiles, the right-hand side of (7.8) must be replaced by $K_r(t)$, which represents the limit on the renewable resource r as a function of time t.

As stated earlier, RCPSP is an NP-hard in the strong sense problem (Blazewicz et al. 1983). RCPSP is a generalization of the *job-shop problem*, which itself is already an NP-hard problem for three machines and unit processing times (Lenstra and Rinnooy Kan 1979).

7.4 Exact Solution Procedures Using Branch and Bound

A branch and bound (B&B) algorithm is an implicit enumeration algorithm to solve combinatorial optimization problems to optimality (see e.g., Hillier and Liebermann 2005; Brucker and Knust 2006). A basic introduction to the B&B algorithm for a discrete minimization problem is provided in *Appendix 7B*. This will be further extended to a B&B algorithm for RCPSP in the following.

Various B&B algorithms, which differ mainly in their branching strategy, have been developed to solve the binary optimization formulation of the RCPSP in the previous section (Demeulemeester and Herroelen 2002). We present a B&B algorithm based on conflict sets, which can be viewed as a simpler version of the minimal forbidden sets approach proposed by Igelmund and Radermacher (1983a, b).

In the B&B tree each node represents a *partial solution* with some activities already scheduled and completed and others still in progress. At each node of the B&B tree we define several attributes in addition to those defined in *Appendix 7B*:

Occurrence time: The actual time $t_j = min_k \{EFT_k | k$: activities in progress at node j$\}$ that an unscheduled activity $k \in U_j$ can start, where U_j denotes the set of unscheduled activities at node j whose predecessors have been completed at or before t_j.

Lower Bound (*LB*): To obtain the lower bound LB_j at node j CPM is applied starting at t_j to the project network comprised of activities already completed and/or in progress assuming all as yet unscheduled activities will start at their early start time and ignoring resource constraints.

Branching strategy: Each branch descending from a parent node on the B&B tree represents a set of activities in progress. These activities are selected from the candidate activities available at that parent node. Candidate activities are precedence and resource feasible activities that have all their predecessors completed and whose resource requirements can be met with the currently available resources. The activity or activities assigned to each branch generated is determined by the branching strategy, which is based on the *conflict sets*. A conflict set is a set of activities whose total resource requirement per unit time exceeds the corresponding resource limit per unit time for at least one resource type. A conflict set cannot include pairs of activities with precedence relationships, i.e., connected by a path on the project network, since they cannot be scheduled simultaneously. A *minimal conflict set* is a conflict set such that it is no longer a conflict set if any one of its activities is removed.

Having identified a conflict set Q, we now identify all possible unordered pairs consisting of a discarded activity and the minimal conflict set obtained by deleting it from Q. By the definition of a minimal conflict set, the resource conflict of the original conflict set Q is automatically resolved by forcing one activity in the set to be completed before another, delaying at least one element of the set so as not to be scheduled concurrently with the remaining elements. Hence, if activities A and B form a minimal conflict set, our branching strategy will generate two branches, one in which A precedes B and another in which B precedes A. Activities A and B are said to define a *pair of disjunctive activities*. Hence, each branch represents one of the two possible resolutions of a pair of disjunctive activities, a decision uniquely specifying how that specific conflict set is resolved. Thus, each new node generated by branching specifies a unique feasible partial schedule. Once a selection i is made for a conflict set from among the set of conflict sets, say F_i, no other selections involving F_i are made on all the descending branches from this branch.

To facilitate the calculation of *LB*s and the correct propagation of the partial solution through the remainder of the search tree, each selection between a pair of disjunctive activities is represented by a dummy precedence arc added between this pair of activities in the project network. While backtracking to another node, the corresponding dummy precedence arc on the branch is removed. At each conflict set resolution, or level of branching, all possible pair selections of the next conflict set are generated if:

(i) The added precedence arc is not redundant for the pair, or,
(ii) The next conflict set represents a resource conflict under the current early start schedule, or,
(iii) The pair is not included in any previously considered conflict set.

In order to reduce the number of partial schedules evaluated, *dominance rules* are applied to the nodes generated before calculating their *LB*s. When one node dominates another, all partial schedules descending from the dominated node are fathomed, eliminating the need to generate any additional nodes.

Dominance Rules: The following dominance relationships can be shown to hold among two partial schedules (nodes) *j* and *k* of the B&B tree, neither of which is a descendant of the other.

Result 1: If the following four conditions hold simultaneously, the partial schedule *j* dominates partial schedule *k*, allowing the partial schedule *k* to be fathomed:

(i) $t_j \leq t_k$.
(ii) Any activity not yet started in partial schedule *k* is either not started, continuing or completed in partial schedule *j*.
(iii) Each activity completed in partial schedule *k* is also completed in partial schedule *j*.
(iv) Each activity *h* that is started in partial schedule *k* has progressed at least as far in partial schedule *j*. In other words, $\left(t_j - t_h^j\right) \geq \left(t_k - t_h^k\right)$, where t_h^j and t_h^k are the start times for activity *h* in partial schedules *j* and *k*, respectively.

The above dominance result is intuitive in the sense that a partial schedule is dominated if some other partial schedule has *both* less unscheduled work content and a smaller makespan, allowing more time to complete this lower amount of remaining work. Result 1 can generally be applied when the *minimum LB branch selection strategy* is utilized since the two partial schedules to be compared must be the offspring of completely different branches. The following result, on the other hand, is applicable to *both* the depth first and minimum LB branching strategies defined in *Appendix 7B*.

Result 2: If there is any activity with an already assigned start time in some partial schedule *k*, which can be left-shifted to an earlier start time without violating either a precedence or resource constraint, that partial schedule *k* is dominated and can be fathomed.

In Result 2, the partial schedule *k* can be improved by left-shifting that particular activity to an earlier start time resulting in a new partial schedule, which dominates the partial schedule *k*. Hence, the partial schedule *k* cannot be part of an optimal schedule and can be fathomed without further developing it.

Example 7.3 Consider the AOA project network with seven activities given in Fig. 7.3. The durations are in weeks. There is a single renewable resource with an availability of 5 units per period. Use the B&B approach with depth-first branch selection strategy and conflict sets to obtain the optimal solution for this RCPSP.

Solution

The depth-first branch selection strategy is employed. Node 1 is the root of the B&B tree with occurrence time $t_1 = 0$. To obtain a CUB value at the root, the shortest activity duration (SPT) priority rule (Sect. 7.5) is used, yielding a CUB value of 22 weeks. The schedule and resource profile for the initial solution are displayed in Fig. 7.4. The complete B&B tree for this example is given in Fig. 7.5. Due to space limitations, the indices for *t* and *LB* are omitted.

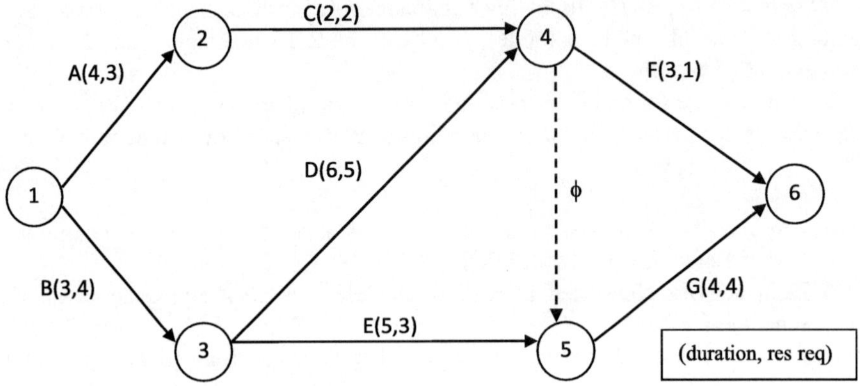

Fig. 7.3 Activity-on-arc project network for Example 7.3

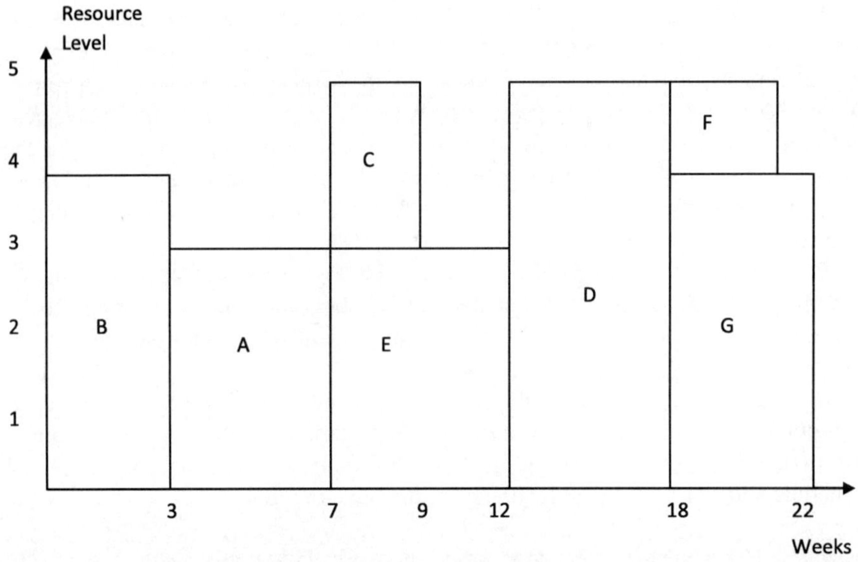

Fig. 7.4 The schedule and resource profile for the initial solution

The branching strategy is based on minimal conflict sets. The minimal conflict sets F_i ($i = 1\ldots,6$) at the root node are determined as follows: {A, B}, {B, C}, {A, D}, {A, E}, {D, E}, {C, D}. By the definition of a conflict set, the simultaneous processing of the activities in any of the minimal conflict sets F_i would violate the resource limit. Consider the set of activities {C, D, E}, which also constitutes a conflict set. By removing either C or E, we obtain two distinct minimal conflict sets {D, E} and {C, D}. The only F_i satisfying the precedence requirements at $t_1 = 0$ is $F_1 = \{A, B\}$. Hence we start branching with F_1. With the *depth-first strategy*, one branch descends from node 1 ending at node 2 with the disjunctive arc going from A

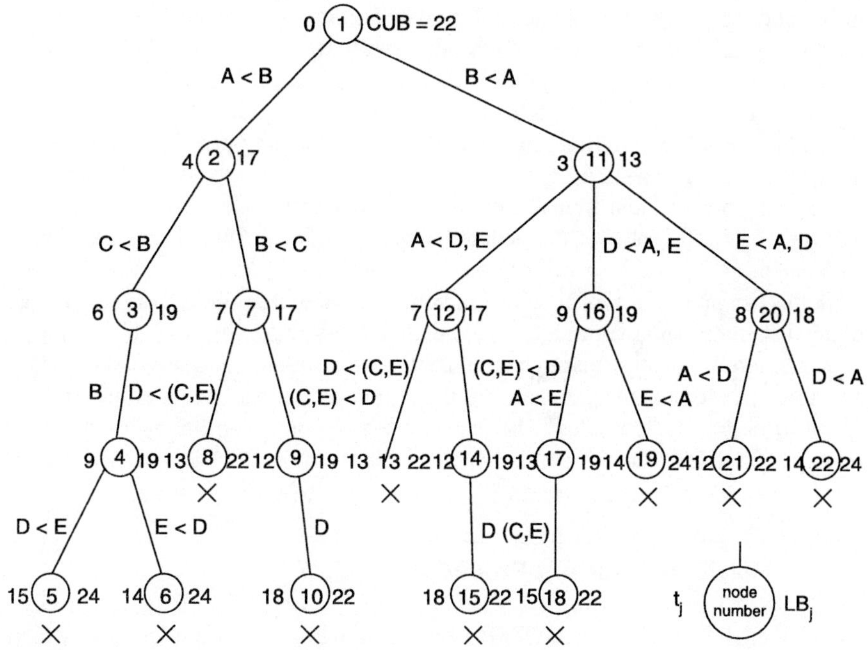

Fig. 7.5 The complete B&B tree for Example 7.3

to B, indicating that A precedes B; A$\prec$B. Assigning A$\prec$B to the leftmost branch is arbitrary; we could just as well assign B$\prec$A. At node 2, $t_2 = 4$ and $LB_2 = 17$ as explained in *Appendix 7B*. Since $LB_2 < CBU$, we do not fathom node 2.

Following the depth-first strategy, we will continue branching from node 2. Since activity A is completed at node 2, precedence relations imply that the branching from node 2 will be based on $F_2 = \{B, C\}$. On the branch emanating from node 2 the disjunctive arc goes from C to B (i.e., C$\prec$B) and ends at node 3 with $t_3 = 6$ and $LB_3 = 19$. We don't fathom node 3 since $LB_3 < CBU = 22$. Branching continues in this fashion until node 5 is fathomed. The new node to branch from must now be determined by backtracking from node 5 to node 4, which is a live node. At node 4 we resolve the conflict set {D, E}, creating a new branch representing E$\prec$D ending at node 6 with the disjunctive arc going from E to D. Since $LB_6 = 24 > CBU = 22$, we fathom node 6 and backtrack to node 4. Since we cannot branch from node 4 anymore, we backtrack to node 3. Here again we cannot branch further, and we backtrack to node 2, where we branch with B$\prec$C ending at node 7. The disjunctive arc goes from B to C.

The branches of the B&B tree where two or more activities are assigned simultaneously can be observed with the activity pair (C, E). Activities C and E have unequal durations and C terminates earlier than E. In all the branches with (C, E) pair on the B&B tree, we created one node to represent the termination of both activities. The reason for this is that even if we had created a node when the shorter activity of the two (activity C) had terminated and activity E continues, we would not be able to assign any additional activities to be processed while E was ongoing without violating the resource limit. The only exception is node 18. If we had continued with node 18, then on the branch from node 18 activities E and F would be processed simultaneously with activity E continuing with its remaining part. In general, if at a node we can assign one or more activities to be processed together with the activities already in progress, the branch coming out from that node will end at a node representing the event that the activity with the shortest (remaining) duration is completed.

In this example, since all nodes are fathomed, the initial solution turns out to be an optimal solution with the minimum project duration being 22 weeks. This exhibits a phenomenon frequently encountered in B&B algorithms: an optimal solution might actually be encountered quite early in the search, but considerable effort must be spent to confirm its optimality. The schedule and resource distribution are the same as displayed in Fig. 7.4.

7.5 Heuristic Solution Procedures

Due to the NP-hard nature of RCPSP, exact solutions are likely to require very high computation times. Hence a wide variety of heuristic algorithms have been proposed for solving this problem in reduced time at the cost of some deviation from optimality. We will present examples of constructive and improvement heuristics. In constructive heuristics, a schedule is built by adding one activity at a time to a partial schedule that is initially empty. As examples of constructive heuristics, we will introduce serial and parallel schedule generation schemes. In improvement heuristics, on the other hand, a given feasible schedule (usually generated by a constructive heuristic) is improved through local search.

7.5.1 Types of Schedules

In this section, we will define three types of feasible schedules by means of an example problem whose AOA project network is displayed in Fig. 7.6.

A feasible schedule generated from this project network is displayed in Fig. 7.7. Additional feasible schedules can be obtained from this project network by changing the starting times ST_j of the activities. A reduction of ST_j is called a *left shift*. Since activity 6 in Fig. 7.7 does not start at its earliest feasible starting point, ST_6 can be decreased by one period without changing the processing sequence or violating any precedence relations. Such a shift for an activity is called a *one-period left shift*. A *local left shift* of an activity is a left shift for that activity obtained through one or more successively applied one-period left shifts. A *global left shift* of an activity, on

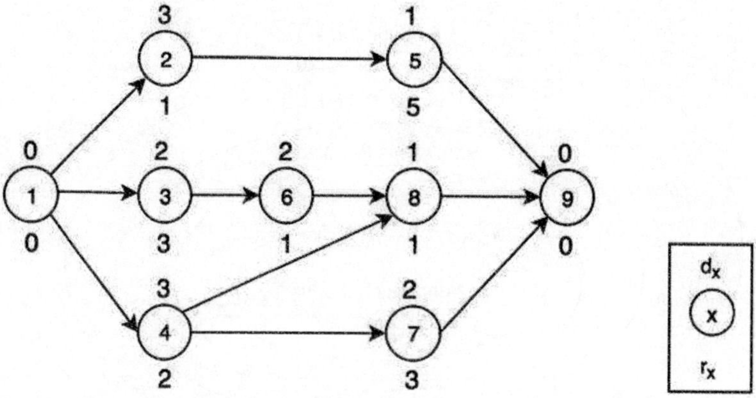

Fig. 7.6 Project network for the example project

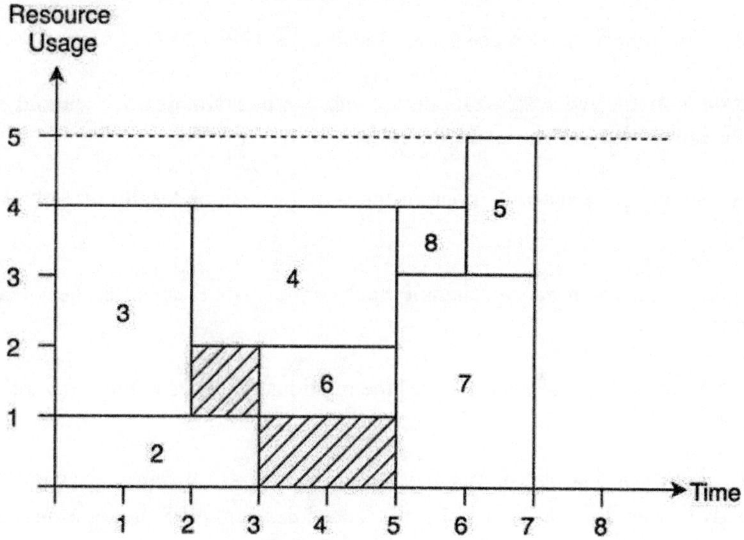

Fig. 7.7 A feasible schedule for the example project

the other hand, is a left shift, which cannot be obtained by a local left shift, and hence requires changes in the start times of some other activities.

Definition 2 In a *semiactive* schedule no activity can be started earlier without altering the processing sequence or violating feasibility.

This definition implies that none of the activities in a semiactive schedule can be locally left shifted.

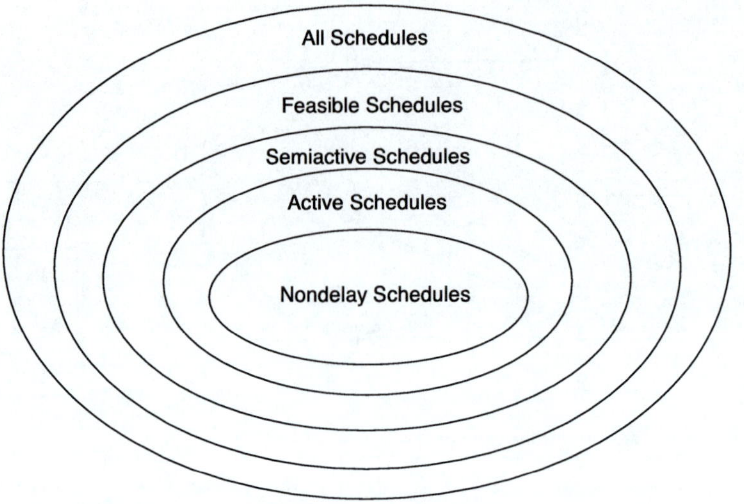

Fig. 7.8 Venn diagram representation of the schedule types (Pinedo 2008)

Definition 3 In an *active* schedule no activity can be started earlier without either delaying some other activity or violating feasibility.

Hence in an active schedule none of the activities can be locally or globally left shifted.

Definition 4 In a *non-delay* schedule each activity starts at its earliest feasible start time.

The following theorem characterizes the relationship between the different types of schedules.

Theorem (Sprecher et al. 1995) Let S denote the set of schedules, SAS the set of semi-active schedules, AS the set of active schedules, and NDS the set of non-delay schedules. Then the following holds:

$$NDS \subseteq AS \subseteq SAS \subseteq S.$$

This result is graphically shown in the Venn diagram in Fig. 7.8.

The following two results are also presented by Sprecher et al. (1995).

Result 1: For every RCPSP instance there exists at least one optimal schedule that is active.

Result 2: The set of all non-delay schedules does not necessarily include an optimal solution.

Hence a heuristic solution procedure generating active schedules might generate an optimal solution, whereas heuristic solution procedures generating non-delay schedules cannot be guaranteed to do so.

7.5.2 Priority Rules

Priority rules are employed in the design of heuristic procedures and are defined to reflect the objective(s) of the decision-maker. In the course of a heuristic procedure, when a choice must be made between two or more alternatives it is based on the priorities determined for each of the alternatives. Priority rules are also referred to as *heuristic rules* or *decision rules* or, in the context of job shop scheduling, *dispatching rules*. For example, consider the scheduling of a project under a single resource constraint employing the *shortest processing time* (SPT) priority rule. Assume that at some time *t* in the schedule two activities are precedence and resource feasible. These two activities form the *eligible activity set EAS* at time *t*. Further assume that the available resource level allows only one of these activities in *EAS* to be scheduled. Then, using the SPT priority rule the activity with the shorter activity duration is scheduled. In case of a tie, then either one of the activities is scheduled randomly or another priority rule is used as a tiebreaker.

7.5.2.1 A Representative List of Priority Rules

In the scheduling literature a large number of priority rules are suggested and employed. In this Section we will introduce only a representative list. Each priority rule described below is used to make a choice for an activity to schedule from the set of precedence and resource feasible activities *EAS*. Hence, the activity referred to in the statement is always an element of *EAS*.

Shortest processing time - SPT. The activity with the minimum duration is scheduled.

Late finish time - LFT. The activity with the minimum late finish time is scheduled.

Late start time - LST. The activity with the minimum late start time is scheduled.

Minimum slack – MSLK: For activity *j*, the slack value is obtained as $(LST_j - EST_j)$, where EST_j is the precedence and resource feasible early start time of activity *j*. The activity with the minimum slack value is scheduled.

Most immediate successors - MIS. The activity with the maximum number of immediate successors is scheduled.

Most total successors - MTS. The activity with the maximum number of successors is scheduled.

Greatest resource demand - GRD. For activity *j*, the resource demand is calculated as the product of its duration and the sum of all resources employed by activity *j*. The activity with the maximum resource demand is scheduled.

Resource scheduling method – RSM: For each pair of activities *i* and *j* in *EAS*, it is calculated how many periods activity *i* is delayed beyond its unconstrained late start time (*LST*), if scheduled after activity *j*. The activity *j* that induces the smallest delay among all activities in *EAS* is scheduled. *RSM* is applicable only to parallel schedule generation scheme (see Sect. 7.5.4).

Greatest ranked positional weight – GRPW: The ranked positional weight for an activity j is obtained by summing the activity *j*'s total resource product (sum of the products of the resource(s) with the duration of the activity) together with the total resource products of the activities that follow it in all the paths leading from the end node of activity j to the terminal node. The activity with the greatest ranked positional weight is scheduled.

Priority rules are classified as *static* if the relative assignment of priorities remains the same throughout the scheduling of all activities and *dynamic* otherwise. In the above list, *SPT, MIS, MTS, GRD*, and *GRPW* are static. Another classification is based on the information used to compute the priority index. Those rules requiring information only for the activity under consideration are called *local* rules, whereas those requiring information about other activities are called *global* rules. Thus, *RSM* and *GRPW* are global rules. A further class of priority rules are *composite priority rules*, where a priority rule is generated by the weighted sum of two or more priority rules. An example would be the *Weighted Resource Utilization Ratio and Precedence* (*WRUP*), which is a weighted combination of the *MIS* and the relative resource usage priority rules (Ulusoy and Özdamar 1989).

When a schedule is obtained using a priority rule, then this process is called a *single pass*. There are also *multi-pass* procedures, such as generating several schedules using the same schedule generation scheme but with different priority rules. This approach is called *multi-priority rule* approach. Another multi-pass approach is based on biased and random sampling of priority rules in consecutive passes (Kolish 1995).

7.5.3 Serial Schedule Generation

In a serial schedule generation scheme (SSGS), one activity is scheduled at each iteration following a precedence feasible order specified in the priority list *A*, which contains all *n* activities ordered according to a priority rule adopted for this purpose. When an activity is scheduled it is eliminated from *A* and the procedure terminates after *n* iterations when *A* is empty. Defining *J* as the set of all activities; *S* as the set of scheduled activities; and *U* the set of activities not yet scheduled, the SSGS can be stated as follows:

Step 1. Obtain the priority list *A* using the selected priority rule. Let $J = \{1, 2..., J\}$, $S = \{\phi\}$, $U = J$.

Step 2. Select the first activity, say *a*, in *A*, and start it at its earliest precedence and resource feasible starting time.

Step 3. Remove activity *a* from *A* and *U* and add it to *S*, i.e., $A = A \setminus \{a\}$; $U = U \setminus \{a\}$; and $S = S \cup \{a\}$. If *A* is empty, terminate SSGS and report the schedule. Otherwise, go to Step 2.

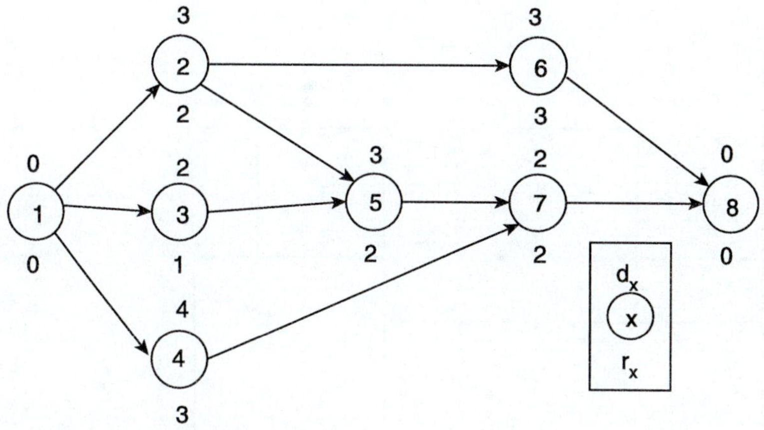

Fig. 7.9 Project network for Example 7.4

Example 7.4 Consider the AOA project network displayed in Fig. 7.9. The resource limit for the renewable resource A employed by the activities is specified as $R_A = 5$ units. Obtain a feasible schedule using SSGS with the priority list $A = \{1,2,3,4,6,5,7,8\}$.

Solution

Step 1: $A = \{1,2,3,4,6,5,7,8\}$, $J = \{1,2,3,4,5,6,7,8\}$, $S = \{\phi\}$, $U = J$.
Step 2: Schedule activity 1 at t = 0.
Step 3: $A = \{2,3,4,6,5,7,8\}$, $S = \{1\}$, $U = \{2,3,4,5,6,7,8\}$.
Step 2: Schedule activity 2 at t = 0.
Step 3: $A = \{3,4,6,5,7,8\}$, $S = \{1,2\}$, $U = \{3,4,5,6,7,8\}$.
Step 2: Schedule activity 3 at t = 0.
Step 3: $A = \{4,6,5,7,8\}$, $S = \{1,2,3\}$, $U = \{4,5,6,7,8\}$.
Step 2: Schedule activity 4 at t = 2.
Step 3: $A = \{6,5,7,8\}$, $S = \{1,2,3,4\}$, $U = \{5,6,7,8\}$.
Step 2: Schedule activity 6 at t = 6.
Step 3: $A = \{5,7,8\}$, $S = \{1,2,3,4,6\}$, $U = \{5,7,8\}$.
Step 2: Schedule activity 5 at t = 3.
Step 3: $A = \{7,8\}$, $S = \{1,2,3,4,6,5\}$, $U = \{7,8\}$.
Step 2: Schedule activity 7 at t = 6.
Step 3: $A = \{8\}$, $S = \{1,2,3,4,6,7\}$, $U = \{8\}$.
Step 2: Schedule activity 8 at t = 9.
Step 3: $A = \{\phi\}$, $S = \{1,2,3,4,6,7,8\}$, $U = \{\phi\}$. Since $A = \{\phi\}$, SSGS is terminated.

The project is completed at t = 9, i.e., in 9 time periods. The resulting schedule is given in Fig. 7.10. The reader can verify that the schedule is an active one.

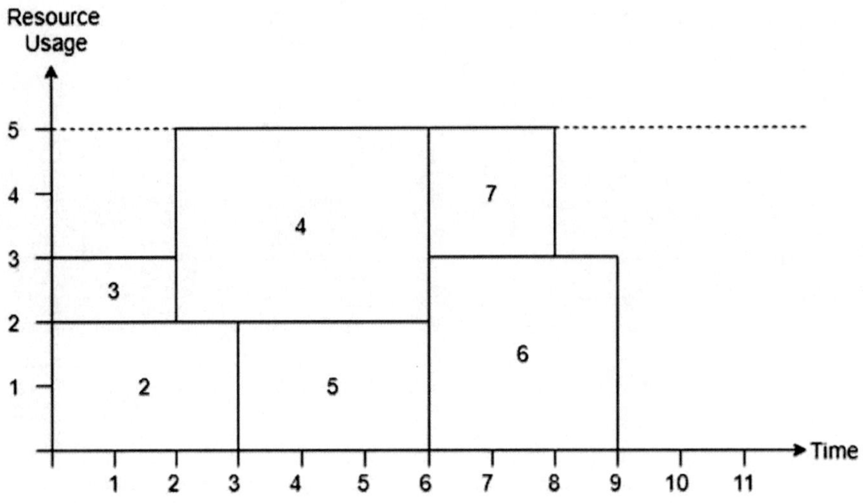

Fig. 7.10 SSGS schedule for Example 7.4

7.5.4 Parallel Schedule Generation Scheme

In a parallel schedule generation scheme (PSGS), one or more precedence and resource feasible activities are scheduled at specific times t where one or more activities terminate and release their renewable resources until all activities are scheduled.

At each time point t, we must consider two sets of activities. The first is the set of eligible activities *EAS* containing all activities that are precedence and resource feasible at time t. The activities in *EAS* are sequenced in non-increasing order of their priority level to obtain the ordered scheduling set *OSS*. If more than one activity has the same priority level, another priority rule is employed as a tiebreaker, if one has been specified. If no such tiebreaker has been specified or the tiebreaker does not resolve the conflict, then the choice is arbitrary.

Let us now state the algorithm.

Step 1. Obtain the priority list A using the priority rule adopted. Let $J = \{1...,N\}$, $S = \{\phi\}$, $U \equiv J$.

Step 2: Determine the earliest time t where one or more ongoing activities terminate. Determine their succeeding activities which are precedence and resource feasible, and include them in *EAS*. Order the activities in *EAS* according to the priority list A to obtain *OSS*.

Step 3: Schedule the first resource feasible activity a of *OSS*. Remove activity a from *OSS* and U and add it to S, i.e., $OSS = OSS \setminus \{a\}$; $U = U \setminus \{a\}$; and $S = S \text{U} \{a\}$. If $OSS = \phi$, go to Step 4. If not, continue scheduling the activities in the order of

OSS until $OSS = \phi$ or the activity next on the *OSS* list is not resource feasible. If *OSS* is empty, go to Step 4; otherwise, go to Step 2.

Step 4: If U is empty, then terminate PSGS and report the schedule. Otherwise, go to Step 2.

Example 7.5 Given the AOA project network in Fig. 7.9, obtain a feasible schedule employing PSGS. There are only 5 units of the renewable resource A employed by the activities available, i.e., $R_A = 5$ units. The priority list is given as $A = \{2,3,4,6,5,7\}$.

Solution

Step 1: $S = \{\phi\}$, $U = \{2,3,4,5,6,7\}$.

Step 2: $t = 0$. $EAS = \{2,3,4\} = OSS$.

Step 3: $RR_A = 5$. Schedule activity 2. $S = \{2\}$, $U = \{3,4,5,6,7\}$, $OSS = \{3,4\}$, $RR_A = 3$.

Schedule activity 3. $S = \{2,3\}$, $U = \{4,5,6,7\}$, $OSS = \{4\}$, $RR_A = 2$. Activity 4 requires 3 units of resource A; hence, it is resource infeasible. Go to step 2.

Step 2: $t = 2$. $EAS = \{4,5,6\}$, $OSS = \{4,6,5\}$.

Step 3: $RR_A = 3$. Schedule activity 4. $S = \{2,3,4\}$, $U = \{5,6,7\}$, $OSS = \{6,5\}$, $RR_A = 0$. Activity 6 requires 3 units of resource A; hence, it is resource infeasible. Go to step 2.

Step 2: $t = 6$. $EAS = \{6,5\} = \text{OSS}$.

Step 3: $RR_A = 5$. Schedule activity 6. $S = \{2,3,4,6\}$, $U = \{5,7\}$, $OSS = \{5\}$, $RR_A = 2$.

Schedule activity 5. $S = \{2,3,4,6,5\}$, $U = \{7\}$, $OSS = \{\phi\}$, $RR_A = 0$. Since $OSS = \{\phi\}$, to step 4.

Step 4: Since $U \neq \{\phi\}$, go to Step 2.

Step 2: $t = 9$. $EAS = \{7\} = OSS$.

Step 3: $RR_A = 5$. Schedule activity 7. $S = \{2,3,4,6,5,7\}$, $U = \{\phi\}$, $OSS = \{\phi\}$, $RR_A = 2$. Since $U = \{\phi\}$ and $OSS = \{\phi\}$, to step 2.

Step 4: Since $U = \{\phi\}$, terminate the algorithm.

The project completes at $t = 11$. Hence, the project duration is 10 time periods. The schedule obtained is given in Fig. 7.11. It can easily be verified that the schedule is semi-active.

7.5.5 A Brief Assessment of Serial and Parallel Schedule Generation Schemes

Among the various schedule generation schemes, SSGS and PSGS are preferred to the others and both SSGS and PSGS demonstrated the same computational complexity for the same feasible list (Demeulemeester and Herroelen 2002). Kolisch (1996a, b) proves the following results:

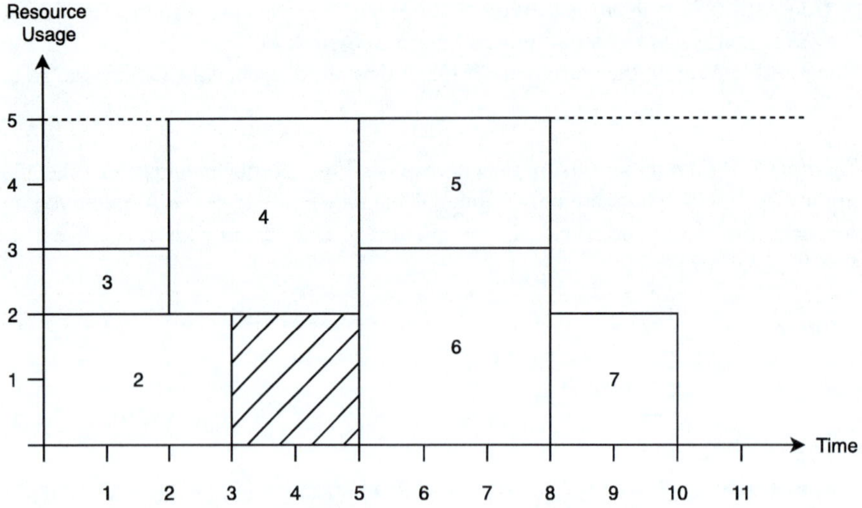

Fig. 7.11 The schedule obtained by PSGS in Example 7.5

Result 3: A schedule generated by SSGS belongs to the set of *active* schedules.
Result 4: A schedule generated by PSGS belongs to the set of *non-delay* schedules.

7.6 A Mathematical Programming Formulation with Multiple Modes

In this section we present a mathematical programming formulation of the RCPSP with both renewable and non-renewable resources and multiple modes. This problem is called the multi-mode RCPSP and denoted by MRCPSP (Demeulemeester and Herroelen 2002). Being a generalization of RCPSP, MRCPSP is NP-hard.

A *mode* associated with an activity is defined by the usage of a combination of resources at specific levels resulting in a specific activity duration. Let M_j denote the number of modes for each activity j. For example, an excavation activity might be performed in three different modes: (*i*) by 3 workers in 20 days; (*ii*) by 10 workers in 5 days; (*iii*) by 1 worker and 1 excavator in 5 days. The fact that the duration of an activity can be decreased at the expense of a larger resource demand (e.g., money) is referred to as a *time/resource trade-off* discussed in Chap. 5. The possibility of altering the resource types requested while keeping the duration constant is referred to as a *resource/resource trade-off*. Hence, switching from mode (*i*) above to mode (*ii*) is a time/resource trade-off, while changing from mode (*ii*) to mode (*iii*) is a resource/resource trade-off, since the activity duration remains the same.

Pre-processing of Data Given the data for an MRCPSP, there are several conditions to check for feasibility before attempting to solve the problem.

Table 7.2 A set of activities with multiple modes

Activity	Mode	Duration	Renewable Resource	Non-renewable Resource
1	1	5	5	6
	2	7	3	6
2	1	8	3	4
	2	11	2	3
	3	11	1	5
3	1	3	7	5
	2	5	3	8
4	1	9	8	8
	2	13	6	4
	3	15	8	4

(i) None of the renewable resource requirements for a mode should exceed the corresponding resource limit per period. In such a case, the corresponding mode is eliminated from further consideration.

(ii) The sum of the non-renewable resource requirements taken over each activity's minimum non-renewable resource consumption should not exceed the corresponding non-renewable resource limit. In such a case, the problem is infeasible.

(iii) A mode i of an activity j is called *inefficient* if there exists another mode j^* for activity i with $d_{ij} \geq d_{ij^*}$ and $k_{ijr} \geq k_{ij^*r}$ for each renewable resource $r \in R$ and $k_{ijn} \geq k_{ij^*n}$ for each non-renewable resource $n \in N$, i.e., mode j* uses fewer resources and gets the job done in less time. Inefficient modes are removed from further consideration.

We will adopt the convention that the modes are ordered in non-decreasing order of their resulting activity duration. Consider the set of multiple-mode activities in Table 7.2.

There is a resource/resource trade-off between the second and third modes of activity 2 with the duration remaining the same in both modes. Although the third mode of activity 4 uses more resources compared to the second mode still its duration is larger than that of mode 2. Hence, it is said to be inefficient and can be eliminated from further consideration. Considering the amounts of the renewable and non-renewable resources used by different modes we conclude that the renewable resource limit has to be at least 8, and the non-renewable resource limit at least 18. A renewable resource limit of at least 8 makes all the modes available for selection. A non-renewable resource limit of at least 18, on the other hand, is obtained as the sum of the minimum non-renewable resource requirement for each activity so that any combination of modes over the activities becomes feasible.

As in the single mode formulation we need to determine the limits of the time window [EFT_j, LFT_j]. EFT_j values are obtained by applying the forward pass of CPM for the unconstrained resource case using the mode with the shortest activity duration, i.e., mode 1. Once an estimate of the project duration is obtained, generally using a heuristic solution procedure, the CPM backward pass is applied to obtain LFT_j values using again the mode with the shortest activity duration.

The following mathematical programming formulation is given for an AON network representation with one source and one sink node. In the following, M_j denotes the number of modes associated with activity j and m the index for the modes.

$$\min Z = \sum_{t=EFT_J}^{LFT_J} t x_{J1t} \tag{7.11}$$

s.t.

$$\sum_{m=1}^{M_j} \sum_{t=EFT_j}^{LFT_j} x_{jmt} = 1 \quad j = 1, \ldots, J \tag{7.12}$$

$$\sum_{m=1}^{M_i} \sum_{t=EFT_i}^{LFT_i} t x_{imt} \leq \sum_{m=1}^{M_j} \sum_{t=EFT_j}^{LFT_j} \left(t - d_{jm}\right) x_{jmt} \quad j = 2, \ldots, J, \quad i \in P_j \tag{7.13}$$

$$\sum_{j=1}^{J} \sum_{m=1}^{M_j} k_{jmr} \sum_{\tau=t}^{t+d_{jm}-1} x_{jm\tau} \leq K_r \;\; r = 1, \ldots, |R|, \quad t = 1, \ldots, T \tag{7.14}$$

$$\sum_{j=1}^{J} \sum_{m=1}^{M_j} k_{jmn} \sum_{t=EFT_j}^{LFT_j} x_{jm\tau} \leq K_n \quad n = 1, \ldots, N \tag{7.15}$$

$$x_{jmt} \in \{0, 1\} \quad j = 1, \ldots, J, m = 1, \ldots, M_j, t = EFT_j, \ldots, LFT_j \tag{7.16}$$

This formulation above differs from the single mode formulation (7.5)–(7.10) in the inclusion of the indices m for the modes.

Another objective beyond the minimization of makespan employed above (7.11) would be the minimization of the cost or amount of non-renewable resources employed throughout the project. This would imply the selection of the modes from the perspective of minimum cost/usage of non-renewable resources.

7.7 A Mathematical Programming Formulation for Maximizing the Net Present Value of a Project Under No Budgetary Constraint

Since the calculation of NPV is based on the concept of time value of money, not only the magnitudes of receipts and expenditures but also their timing are of interest. Usually in projects, when an activity is in progress, the timing of cash outflows and inflows is related to the progress of that activity. In this problem we seek to determine the starting times of the activities, i.e., a schedule, to maximize the NPV of all cash flows subject to the per period resource constraints, if any. The presence of resource constraints makes finding the optimal schedule with respect to NPV considerably more difficult than minimizing the makespan, especially since NPV is a non-regular performance measure as mentioned in Sect. 7.2.1. Delaying an activity with a negative net cash flow might lead to resource bottlenecks as the project progresses, preventing the early start of a later activity with a net positive cash flow and thus adversely affecting the NPV objective.

A special class of problems is one where a negative cash flow occurs at each activity, and the only positive cash flow is realized at the completion of the project. Smith-Daniels and Aquilano (1987) showed that a right shifted (late start) schedule for all non-critical activities yields a higher NPV than schedules based on early start of all activities. Unfortunately, we cannot generalize this observation to cases where net positive cash flows occur in other activities as well, since delaying an activity with a negative cash flow might delay another with a positive cash flow due to precedence relations, reducing the overall NPV.

A mathematical programming formulation for the maximization of the NPV in the context of the RCPSP with renewable resources was given by Russell (1986). The problem of maximizing NPV under resource constraints was shown to be NP-hard by Baroum (1992).

The formulation below considers only net cash flows and is defined over an AON project network with one initial and one terminal node and no budget constraints. The net cash flow for each activity is assumed to occur at the termination of the activity. This assumption can be satisfied easily by bringing every cash flow associated with the activity to the completion time of the activity through a future worth (FW) transformation. Since the definition of a feasible schedule is the same as that already considered for the minimization of the makespan, we employ the same constraint sets (7.6)–(7.10) to represent the feasible region.

$$\max \sum_{j=1}^{J} \sum_{t=EFT_j}^{LFT_j} f_t^a CF_j x_{jt} \tag{7.17}$$

$$\sum_{t=EFT_j}^{LFT_j} x_{jt} = 1 \quad j = 1, 2, \ldots, J \tag{7.18}$$

$$-\sum_{t=EFT_i}^{LFT_i} t x_{it} + \sum_{t=EFT_j}^{LFT_j} (t - d_j) x_{jt} \geq 0 \quad j = 2, \ldots, J, \quad i \in P_j \tag{7.19}$$

$$\sum_{j=1}^{J} \sum_{q=t}^{t+d_j-1} k_{jr} x_{jq} \leq K_r \quad t = 1, 2, \ldots, T, \quad r = 1, 2, \ldots, |R| \tag{7.20}$$

$$\sum_{j=1}^{J} k_{jn} \sum_{t=EFT_j}^{LFT_j} x_{jt} \leq K_n \quad n = 1, \ldots, N \tag{7.21}$$

$$x_{jt} = \begin{cases} 1, \text{if activity } j \text{ is finished at the end of period } t \\ 0, \text{otherwise} \quad \textit{for all } j, t \end{cases} \tag{7.22}$$

where:

J = number of activities including any dummy activities.
$|R|$ = number of renewable resource types.
j = activity index.
r = renewable resource index.
t = time index in periods.
T = time horizon in periods.
d_j = duration of activity j.
P_j = set of immediate predecessors to activity j.
k_{jr} = per period renewable resource requirement of resource type r for activity j.
K_r = renewable resource limit per period for renewable resource type r in period t.
f_t^{α} = the discount factor for the cash flow in period t at an interest rate α.
CF_j = net cash flow amount for activity j.
CF_j is calculated as a net cash flow amount including both the expenditures and receipts. If receipts exceed expenditures, then CF_j is positive.

If the cash flow for an activity occurs at its start time, the corresponding discount factor is $f_{t-d_j}^{\alpha}$, where α stands for the interest rate and $(t - d_j)$ represents the start time of activity j.

The extension of the above formulation to the multi-mode case can be written as in Sect. 7.6.

In another extension of this problem, the project manager is assumed to have a limited budget of C_t^{av} in each period t that they can use towards direct costs. Unused funds remaining at the end of a period will not be available for use in the following periods. For each activity j, a direct cost of C_j is incurred in each period during the execution of activity j, i.e., costs are uniformly distributed over the duration of the activity. The mathematical programming formulation for this particular case is obtained by adding the following constraint set to the formulation given in (7.17)–(7.12).

$$\sum\nolimits_{j=1}^{J}\sum\nolimits_{q=t}^{t+d_j-1} C_j x_{jq} \leq C_t^{av} \quad t = 1, \ldots, T \tag{7.21}$$

Another case arises when, in addition to the direct cost of activities and limited budget per period, a constant overhead cost O per period is incurred at the end of each period throughout the project duration. The only change to the formulation (7.17)–(7.23) occurs in the objective function. For the sake of exposition, continuous compounding is applied, yielding the objective function:

$$\max NPV = \sum_{j=1}^{J}\sum_{t=EFT_j}^{LFT_j} e^{-\alpha t} D_j x_{jt} - \sum_{t=EFT_J}^{LFT_J} \frac{O(e^{\alpha t} - 1)}{e^{\alpha t}(e^{\alpha} - 1)} x_{Jt} \tag{7.22}$$

In (7.24), the overhead cost O in each period is discounted to the end of the last period in which the project is completed using the continuous version of the compound amount factor, and then discounted as a lump sum to the present using the continuous version of the present worth factor. After simplification of its second term (7.24) reduces to

$$\max NPV = \sum_{j=1}^{J} \sum_{t=EFT_j}^{LFT_j} e^{-\alpha t} D_j x_{jt} - O \sum_{t=EFT_J}^{LFT_J} \frac{(1 - e^{-\alpha t})}{(e^{\alpha} - 1)} x_{Jt} \tag{7.23}$$

7.8 Conclusions, Recent Developments, and Some Future Research Directions

Modeling and solving RCPSPs is imperative to reflect the requirements of the project management practice. Real life cases require solving large scale instances with thousands of activities and multiple resource types. Even though RCPSP, which is an NP-hard problem, has been extensively studied, there is still a need for developing effective and efficient problem specific heuristics to solve the various versions of RCPSP. Another important necessity in practice is managing uncertainty. However, stochastic or robust versions of RCPSP become even more difficult to solve, and their formulation is often not straightforward. Regarding uncertainty, we note that the project management literature has mainly focused on the randomness in the duration of the activities. However, the uncertainty due to the randomness in resource requirements and resource availability should also be modeled. Hall (2016) emphasizes the variability of resource availabilities over time specifically in IT projects, but research emphasizing this uncertainty is scarce.

Slack analysis is important in robust project planning, as it supports project managers to identify and focus on critical activities. However, the identification of the slacks in resource constrained projects is ambiguous because the outcome is not unique but depends on the resource allocation rules utilized in scheduling (Wiest 1964). Late start/finish and early start/finish times of the activities must be determined in order to find the slacks. When we apply classical CPM forward and backward passes, we may get a resource infeasible schedule. In resource constrained settings, the early and late times are calculated through establishing left and right justified schedules. A schedule generation mechanism and a priority rule are required to construct these schedules. Left and right justified schedules, thus the early and late times of activities depend on the schedule generation rule applied. Even in a simple resource constrained project network, alternative resource allocations are often possible resulting in a choice of schedules with identical project durations but different activity slacks.

In resource-constrained projects, critical tasks, which form the *critical chain*, need to be determined by considering both the precedence and resource constraints. There are several algorithms in the literature to calculate the slacks and identify the critical activities. These algorithms and their limitations are discussed by Lim et al. (2011). There is a need for further research to overcome these limitations and develop reliable efficient algorithms. Regarding critical chain management, Hall

(2016) stresses the importance of buffer sizing and points out the need for comparison of the algorithms through extensive computational studies.

Exercises

7.1 Show that lateness L and tardiness T are both regular measures whereas earliness E is a non-regular performance measure.

7.2 In a project scheduling environment with limited resources show that a doubly constrained resource can be represented by one renewable resource constraint and one non-renewable resource constraint.

7.3 Consider the following two projects with AOA representation. On each arc the activity duration and the requirements for renewable resources 1 and 2 are given. Determine the resource strength and resource factor values for the two projects given and compare them.

Project 1:

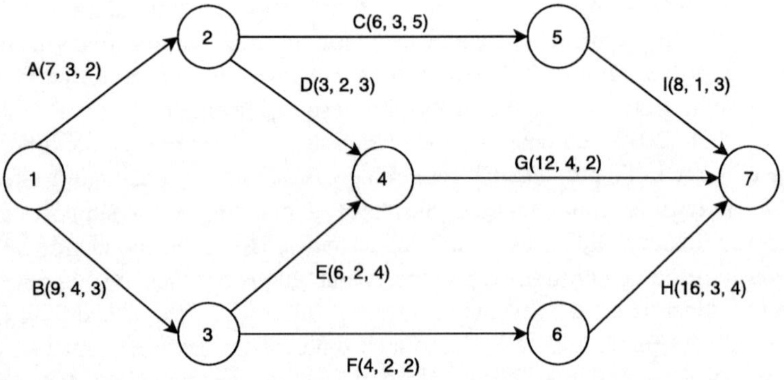

Project 2:

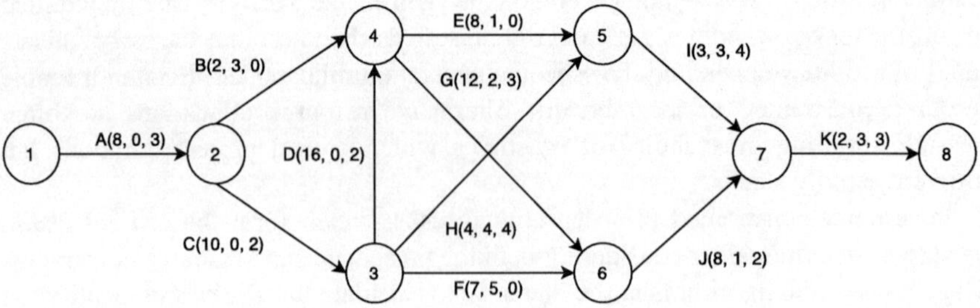

7.4 Consider the following mathematical programming formulation for an RCPSP represented with an AON project network. The notation is as defined in Sect. 7.3.

$$\min Z = \sum_{t=EFT_J}^{LFT_J} t x_{Jt}$$

subject to

$$\sum_{t=EFT_j}^{LFT_j} x_{jt} = 1 \quad \mathrm{j} = 1, \ldots, \mathrm{J}$$

$$\sum_{t=EFT_i}^{LFT_i} t x_{it} \leq \sum_{t=EFT_j}^{LFT_j} (t - d_j) x_{jt} \quad \mathrm{j} = 2, \ldots, \mathrm{J} \ \mathrm{i} \in \mathrm{P_j}$$

$$\sum_{j=1}^{J} k_{jr} \sum_{\tau=t}^{t+d_j-1} x_{j\tau} \leq K_r \quad \mathrm{r} = 1, \ldots, |\mathrm{R}|, \quad \mathrm{t} = 1, \ldots, \mathrm{T}$$

$$x_{jt} = \begin{cases} 1, \text{if activity j is finished at the end of period t} \\ 0, \text{otherwise} \qquad \text{for all j, t} \end{cases}$$

Rewrite the formulation for the case where a given deadline DL must be observed and additional resources can be employed in each period at a cost of c_r per period for resource r, but this additional usage of resource r cannot exceed $o_r K_r$, where $0 \leq o_r \leq 1$, is a factor limiting the additional amount of the renewable resource type r. Let O_{rt} denote the number of additional renewable resources of type r used in period t. The objective is to minimize the cost incurred due to the usage of additional resources.

7.5 You are in charge of managing 3 projects with deterministic activity durations. There are 2 renewable resources R1 and R2 with limits 11 and 15, respectively. The first project must start at least 15 days after the second project has been started. The third project must start at least 15 days after the second project has been completed. All projects have their individual deadlines.

(a) Draw the project networks as a single project network employing AON convention.

(b) Suggest an objective function of relevance to the project manager. (Hint: It is neither min makespan nor max (or min) NPV.)

Project 1			Project 2			Project 3		
Act.	Imm. Pred.	Dur.	Act.	Imm. Pred.	Dur.	Act.	Imm. Pred.	Dur.
A1	–	11	**A2**	–	7	**A3**	–	17
B1	–	9	**B2**	–	9	**B3**	A3	19
C1	A1	13	**C2**	–	6	**C3**	B3	9
D1	B1	16	**D2**	A2	15	**D3**	B3	10
E1	C1	15	**E2**	B2	13	**E3**	C3	22
F1	D1	8	**F2**	C2	8	**F3**	D3	8
			G2	D2, E2, F2	11			
			H2	G2	16			

7.6 Consider the following project with 6 activities. The precedence relations among activities, activity durations and resource requirements for both renewable and non-renewable resources are given in the table below.

Activity	Precedence	Duration	Resources	
			Renewable	Non-renewable
1	–	6	3	5
2	1	4	4	8
3	1	8	2	3
4	1	3	3	1
5	2,3,4	7	7	6
6	5	5	6	2

The availabilities for the renewable and non-renewable resources are given as 7 and 30 units, respectively.

(a) Check for the feasibility of the problem with respect to the non-renewable resource.
(b) Determine an estimate for the time horizon T using the parallel schedule generation scheme with the minimum slack (MSLK) priority rule, breaking ties with SPT.
(c) Determine the EFT_j and LFT_j values for all activities j to be used in the mathematical programming formulation.
(d) Write down the availability constraint of the renewable resource for $t = 8$.

7.7 The production manager of a machine shop has an order for a machine they must manufacture and deliver as soon as possible. There are 12 activities to be completed that can all be processed on the same type of machine. Precedence relations among these activities are given in the table below. The activity durations are deterministic and also given in the same table. Employ the AOA network representation.

Activity	Immediate predecessors	Duration
1	–	7
2	1	9
3	1	4
4	2	11
5	3	7
6	3	10
7	4	14
8	5,6	5
9	5,6	8
10	7	12
11	8,9	7
12	10,11	9

(a) What is the minimum number of parallel identical machines needed to achieve the project duration achievable with unlimited resources?
(b) If the production manager wants to save one machine from that calculated in part (a) to use it for other currently processed orders, how would this affect the makespan in part (a)?

7.8 Consider the following project network. The duration of each activity is indicated on the corresponding arc together with its usage of the two resources A and B in parentheses. Use the late finish time (LFT) priority rule to obtain a feasible schedule if the resource limits are given as $R_A = 5$ and $R_B = 4$.

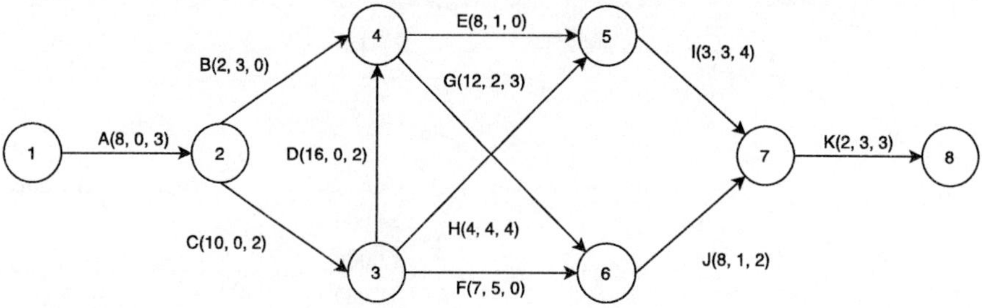

7.9 Consider the project data below.

Activity	Immediate predecessor	Duration (days)	Resource #1 usage	Resource #2 usage	Resource #3 usage
A	–	3	3	2	1
B	–	5	2	4	2
C	A	6	3	1	2
D	A	2	4	3	1
E	D	3	2	0	3
F	D	3	1	1	1
G	B	4	3	1	1
H	F, G	5	2	2	2
I	C	4	3	2	3
J	E, I	2	4	1	0
K	J	3	5	4	2

Resource limits are $R_1 = 6$; $R_2 = 7$; $R_3 = 6$.
(a) Draw the AON project network introducing appropriate dummy nodes as necessary.
(b) Obtain a resource constrained schedule for this problem using the resource scheduling method (RSM) heuristic, breaking any ties with SPT.

7.10 Consider the following project network for a project consisting of 8 activities. The activity names are indicated on each corresponding arc in the network. Each activity needs a certain number of teams to be executed, and all teams are identical in their costs and capabilities. The cost, duration and team requirement for each activity are given below.

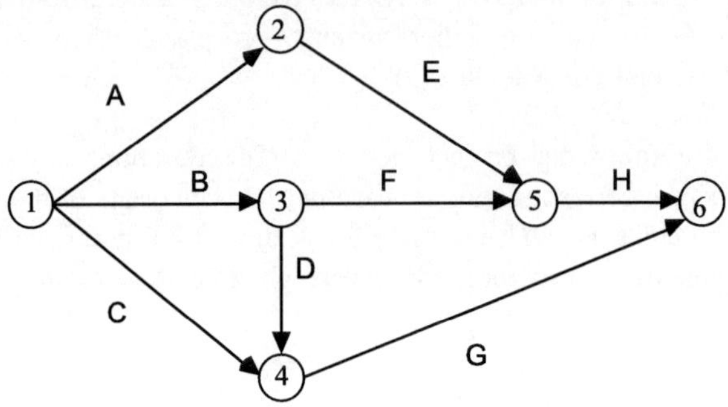

Activity	Duration	Cost	No of teams
A	4	24,000	2
B	3	27,000	3
C	3	21,000	2
D	4	36,000	3
E	5	10,000	1
F	7	35,000	2
G	9	54,000	2
H	3	32,000	2

(a) Given an unlimited number of teams, draw the resource requirement profile using early start and late start for the activities.
(b) Given only 4 teams are available, schedule the project activities to minimize the project duration using the SPT priority rule.
(c) Draw the cost profile of the project for the resource constrained case indicating alternatives resulting from the use of slacks, if any.
(d) We have the option of reducing the duration of an activity by assigning it an additional team. The marginal cost of assigning an additional team to an activity and the new duration are given below.

Activity	Duration	Marginal cost
A	3	3000
B	3	9000
C	2	0
D	3	0
E	3	2000
F	4	5000
G	6	0

Determine the project schedule resulting in the least duration. Draw the cost profile. Compare it to that obtained in part (c).

7.11 Data for a project network is given below. For administrative reasons, crews can include only 3, 4 or 5 members. The durations are in days and any non-integer duration is rounded up to an integer (i.e., 2.3 rounded up to 3). Once a crew is formed, its size cannot be changed throughout the execution of the activity. Hence, as the activity is executed, one or more team members may become idle during part of the activity duration. New crews can be formed for each activity. There are 10 workers available. Determine the shortest possible makespan for the project by employing a heuristic procedure. When scheduling, order activities in non-increasing order of man-day requirement. Break ties considering lexicographic order.

Activity description	Immediate predecessor	Man-day requirement to accomplish activity
A	–	50
B	–	60
C	A	20
D	B	30
E	B	18
F	C, D	32
G	E	52
H	F	64

7.12 Consider the following 3 projects with different ready times (TA = 12, TB = 20, and TC = 0), before which the corresponding project cannot be started.

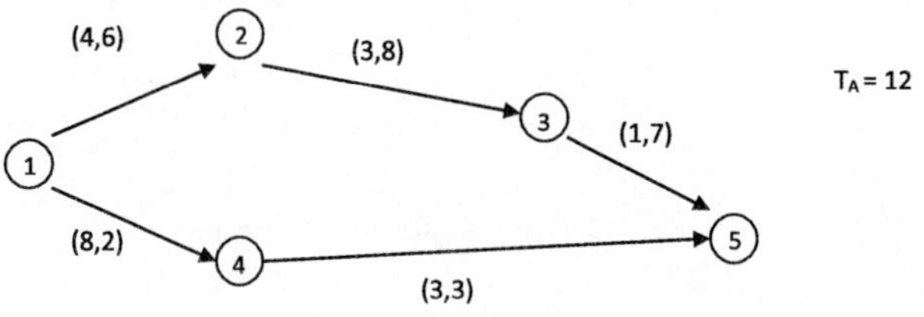

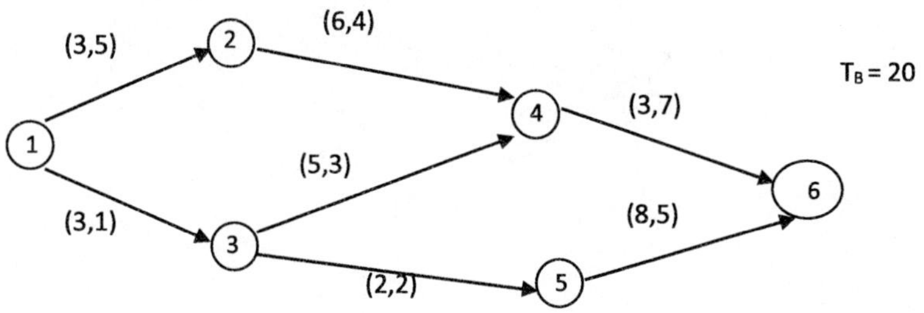

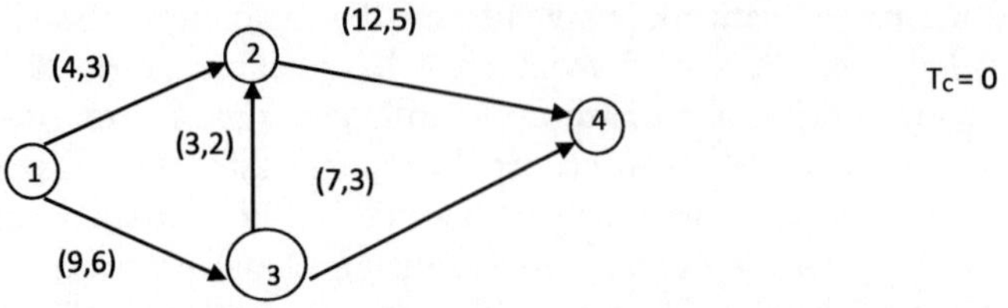

There is a single resource employed by all activities of all projects whose availability is 10 units per period. Schedule the activities of these projects to minimize the overall makespan using the SPT priority rule and break any ties arbitrarily. To limit the computational effort, when scheduling the activities, generate the EAS 6 times and make the assignments and then stop.

7.13 Consider the following two projects with AOA type representation. The durations are indicated on each arc together with the renewable resource requirement given in brackets. The renewable resource availability is 4 units.

Project 1:

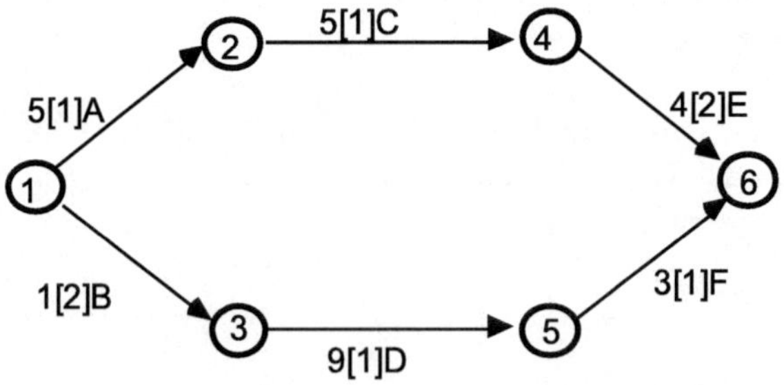

Project 2:

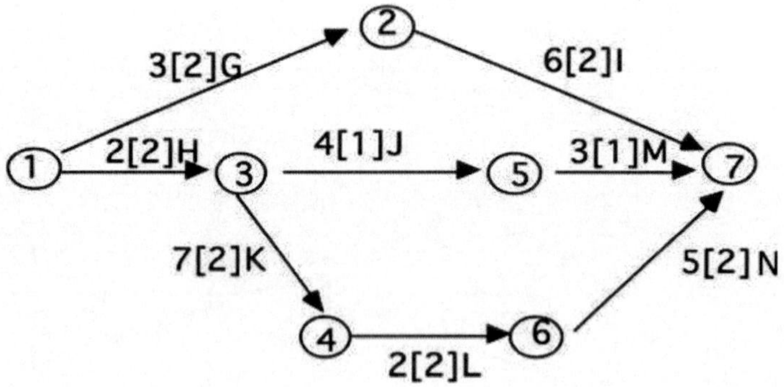

(a) Let the objective be the minimization of the throughput time $\left(= \max\left\{C^1_{max}, C^2_{max}\right\}\right)$. Find the individual project durations. Use the Shortest Operation First (SOF) as the priority rule, breaking the ties with the First Come First Serve (FCFS) priority rule. Apply parallel schedule generation scheme.
(b) Using the objective function of part (a), employ the minimum slack (MSLK) as the priority rule. Break the ties using the FCFS priority rule. Compare the solution with that of part (a).
(c) Solve the problem under the same conditions to minimize the average tardiness. Which priority rule would you employ? The due dates for projects I and II are given as 18 and 24 periods, respectively.
(d) Evaluate the average tardiness values for the schedules in parts (a) and (b) and compare them with the one you obtained in part (c).

7.14 Consider the following data for an AOA project network. The activity durations are indicated on each arc. Use late finish time (LFT) to schedule the project activities with the resource limits being RA = 5, RB = 4 for resources A and B, respectively. Use SPT as a tiebreaker.

Activity	Immediate predecessor	Resource requirement (A)	Resource requirement (B)
A	–	2	0
B	–	3	2
C	A	1	1
D	B	4	4
E	A	2	1
F	C, D	1	0
G	B	1	3
H	E, F	2	1

7.15 A project consisting of 8 activities makes use of a certain resource, whose availability is limited to 5 units. There is also an upper limit on spending per day. It is given as 300 MU/day (MU stands for monetary unit). The indirect cost is reported over the whole project on a daily basis and is estimated to be 120 MU/day. The direct cost for each unit of resource usage is 25 MU/day and this cost is only incurred when the corresponding resource is employed. The predecessors, resource requirements, and duration are given in the table below.

Activity	Predecessors	Duration (Days)	Resource requirement
A	–	5	3
B	–	7	1
C	–	3	3
D	C	4	2
E	A	6	1
F	A	10	3
G	B, E	6	1
H	D	4	3

You can hire additional resource units for certain durations in order to speed up the project as long as you stay within your budget. These additional resource units are not meant to crash the activities, but rather allow you to schedule activities by overcoming the bottleneck resulting from the resource limit. The hiring cost for each additional resource is 40 MU/day. Hence, the direct cost of hired units/day will be the sum of the hiring cost/day and the direct cost of the regular resources/day. Determine the least cost schedule for the activities. Use the LFT priority rule breaking ties with SPT.

7.16 Consider the following project network, with the activity duration and resource requirement shown in parentheses on each arc. The resource is of renewable type and the scheduling is non-preemptive.

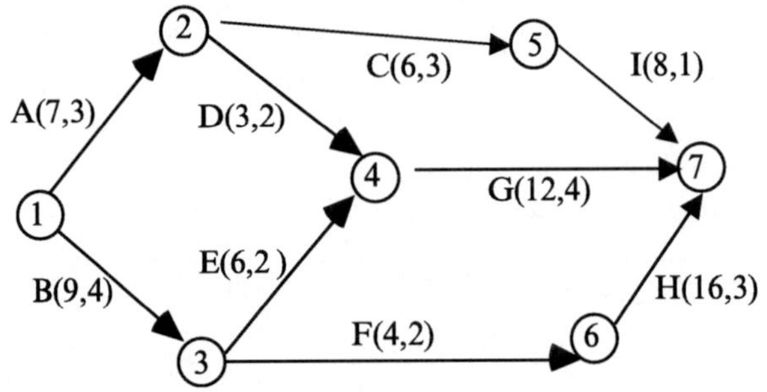

Event	1	2	3	4	5	6	7
E_i	0	7	9	15	13	13	29
L_i	0	14	9	17	21	13	29

Activity	D	ES	EF	LS	LF	TS
A	7	0	7	7	14	7
B	9	0	9	0	9	0
C	6	7	13	15	21	8
D	3	7	10	14	17	7
E	6	9	15	11	17	2
F	4	9	13	9	13	0
G	12	15	27	17	29	2
H	16	13	29	13	29	0
I	8	13	21	21	29	8

The results of the necessary calculations concerning event occurrence times, activity times, and total slack are reported in the Tables above.

(a) For the project network above find a schedule to minimize the project makespan where the resource is constrained to 6 units throughout the project duration. Employ the parallel schedule generation scheme using the greatest resource demand (GRD) heuristic, breaking ties with SPT.

(b) Apply the B&B approach of Sect. 7.4 to the project network given initially. Assume the resource to be constrained to 6 units throughout the project duration. Employ the result of the (a) part to improve the effectiveness of the B&B process. At each node specify the following: Schedulable set (SS), LB, current time, resource remaining, and activity(ies) in progress, if any. Apply depth first strategy. Provide the above specified information for the first 8 nodes including the root or "all solutions" node.

7.17 Consider the following project network, where each arc is labeled with its name, its activity duration in days, and its resource requirements for resources A and B, in that order. The resource A is of renewable type with the availability of 6 units, while the second resource B is non-renewable type such that 1 unit of this resource arrives at the beginning of each day. An activity cannot be initiated unless its full non-renewable resource requirement is on-hand. Assume no renewable resource is on hand at the start of the schedule. Since the non-renewable resource B is not perishable, it can be stored if not consumed. The scheduling is non-preemptive.

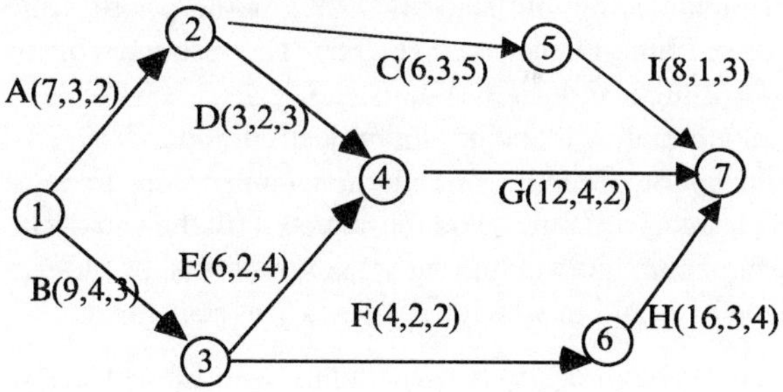

(a) Applying the parallel schedule generation scheme determine the project duration using greatest resource demand (GRD). Break ties, if any, using SPT.
(b) Draw the resource diagram for both resources A and B separately

7.18 Consider the following project data.

Activity	Immediate predecessor	Duration (days)	Resource requirement
A	–	3	2
B	–	3	6
C	A	2	2
D	B	2	4
E	A	3	7
F	C, D	3	6
G	B	7	4
H	E	5	3
I	F, H	5	3

Solve the RCPSP using the Branch & Bound approach. Generate 10 nodes (including the initial all-solutions node) using the depth first approach and calculate their LBs.

7.19 Consider the following project network where activity A precedes activity C; B precedes D and E; C and D precede F. All these activities use two different types of renewable resources: resource A and resource B. The duration and the resource requirements for each activity are given in the following table.

Activity	Duration (weeks)	Resource requirement A	Resource requirement B
A	7	2	2
B	3	3	3
C	2	1	2
D	5	1	1
E	6	2	2
F	4	2	3

(a) Apply the parallel schedule generation scheme to obtain the project duration. Employ minimum slack (MSLK) as the priority rule breaking ties with First-Come-First-Served (FCFS). The resource constraints are given as $R_A = 4$ units and $R_B = 4$ units.
(b) The saving due to reducing the project duration is 10,000TL/week. You have the option of renting one additional unit of resource B starting at some period t. Once you rent it you must keep it till the end of the project. What is the maximum amount of rent/week you should be willing to pay for one resource of B and in which period will you start the rent?

7.20 Consider the problem of project scheduling with limited renewable resources and with the objective of minimizing the project duration. Your boss tells you that she is not interested in minimizing the project duration but rather wants to minimize the sum of idle times of the resources throughout the project. How would you change the heuristic procedure to accommodate this new type of objective function? Do it only for two renewable resources A and B.

7.21 Apply preprocessing to the following multi-mode project scheduling data.

Activity	Mode	Duration	Ren Res* 1	Ren Res 2	N Ren Res* 1	N Ren Res 2
A	1	9	4	3	6	9
	2	8	4	3	5	8
	3	6	3	5	9	13
B	1	11	6	7	3	5
	2	7	9	10	5	8
C	1	18	7	8	8	4
	2	14	9	10	9	5
	3	14	7	12	10	6
D	1	13	7	3	13	7
	2	11	9	4	14	9
	3	10	11	4	15	10

*Ren Res: Renewable resource; N Ren Res: Non-renewable resource

Ren Res 1 is limited by 10 units; Ren Res 2 by 12; N Ren Res 1 by 35 units; and N Ren Res 2 by 40 units.

(a) Determine whether any two modes display a resource-resource trade-off.
(b) Are any of the modes inefficient?
(c) Check feasibility with respect to renewable and non-renewable resources. How will you remove the infeasibilities?

7.22 A company has taken on a reconstruction and remodeling project consisting of 7 activities. The precedence relationships, durations and the resource requirements of the activities are given in the table below.

Activity	Predecessors	Duration	Number of workers
A	–	6	5
B	–	4	4
C	A	5	3
D	B	6	3
E	B	4	3
F	C, D	3	3
G	E	4	5

(a) Assume that there are 9 regular workers of the company. However, some workers may take some days off counting towards their annual leave. The available number of workers over the next 20 days is given below. After day 20 all the workers return to work. Use the parallel schedule generation scheme with the minimum slack (MSLK) priority rule to determine the project duration in this case.

Days	1–4	5–6	7–9	10–20
Available number of workers	9	6	9	6

(b) Assume in addition to the above case, the company can hire additional workers when necessary at a cost that depends on what day the worker is hired. The cost of hiring for the next 14 days is given below, where x is the number of hired workers.

Days	1–5	6–11	12–14
Cost of hiring	200TL per worker	400TL per worker	$150x^2$

The indirect cost per unit day of the project is 750 TL. Determine the least cost schedule, if and when to hire additional workers and how many workers the company should hire.

Appendixes

Appendix 7A: Computational Complexity

Throughout this book we frequently refer to problems being NP-hard, implying that these problems are hard to solve – specifically, that the computation time for any algorithm yielding a provably optimal solution increases exponentially in terms of the input size of the problem. With this brief introduction to complexity theory we would like to present a more formal discussion of the basis for these statements. The basic question posed is: "*Are some problems inherently easier/harder to solve than others?*" The answer to this question can guide us in developing a solution strategy to that problem such as "*should we seek an exact solution procedure?*" or "*should we try to develop good heuristic solution procedures?*"

We use the word "problem" rather freely in daily speech, but in the context of computational complexity a *problem* denotes a general question to be answered, usually requiring some parameters whose values must be specified. A example of a problem is the deterministic project scheduling problem with no resource constraints investigated in Chap. 4, whose parameters are the number of tasks, their durations and the precedence relations between them. Let us designate this problem by *PN*. Assigning specific numerical values to all the problem's parameters yields an instance of the problem, such as the instance of *PN* depicted in Fig. 7A.1. The instance is one with 6 activities, each with an activity number, duration, and a set of predecessors.

The size of an instance depends on the *encoding* scheme employed. Separators such as ',' or '/' are used between data items. Examples of encodings of the instance in Fig. 7A.1 are given below. For each activity, its activity number, duration, number of predecessor(s), a dn the predecessor(s) themselves are listed.

1,2,0,2,4,0,3,3,1,1,4,1,1,2,5,4,1,2,6,3,2,3,4

1,10,0,10,100,0,11,11,1,1,100,1,1,10,101,100,1,10,110,101,10,101,100 (*binary encoding*)

1,11,0,11,1111,0,111,111,1,1,1111,1,1,11,11111,1111,1,11,111111,111,11,111,1-111 (*unary encoding*)

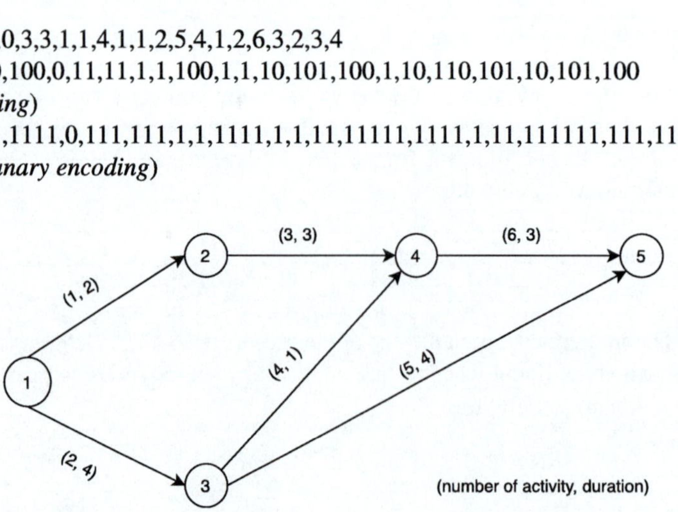

Fig. 7A.1 Project network for problem *PN*

There are also other encodings with integer base greater than 2. The most preferred one is the binary encoding - particularly for computer algorithms. The size π of an instance is defined by the length of the data string necessary to specify the instance.

A question is: "*Given a data string of size π is input to an algorithm for a solution, how long will the algorithm take to reach a solution in terms of the number of operations?*" Rather than an average value for a given size π, the worst case value is sought for. The larger the size of the instance, the larger the number of operations we would expect to be required. This relationship is expressed as *the time complexity function* $f(\pi)$, defined as an upper bound on the number of elementary operations (addition, multiplication, subtraction, division, and comparison) required for the algorithm to reach a solution given an input of size π.

A central question in complexity theory is how $f(\pi)$ increases with instance size π. Instead of an exact expression for $f(\pi)$ an asymptotic order is used. We say, $f(\pi)$ is $O(g(\pi))$, i.e., $f(\pi)$ is of order $g(\pi)$, if for some constant c, $f(\pi) \leq c\, g(\pi)$ as π approaches infinity. For example, $f(\pi) = 3\pi^3 + 2\pi + 12$ is $O(\pi^3)$; $f(\pi) = 2^\pi + 6\pi^4 + 2\pi^2$ is $O(2^\pi)$.

If the time complexity function of an algorithm is bounded by a polynomial function of the input size π, the algorithm is said to be of *polynomial time complexity*. Otherwise, the algorithm is of *exponential time complexity*. Algorithms with polynomial time complexity are also called *efficient* algorithms. Note that there is a certain arbitrariness to this definition of efficiency; it may well be that a polynomial-time algorithm has such a high time complexity to render it impractical, such as $O(\pi^{250})$.

A small numerical exercise will clarify the difficulty of solving algorithms of exponential time complexity (Garey and Johnson 1979). Let us compare two algorithms of $O(\pi^3)$ and $O(2^\pi)$ running them on a computer executing 10^6 operations per second. As reported in Figure 1.2 of Garey and Johnson (1979), for π=30, the algorithm of $O(\pi^3)$ will take 0.027 seconds and for π=60, it will take 0.216 seconds. The algorithm of $O(2^\pi)$, on the other hand, will take 17.9 minutes and 366 centuries, respectively. Of interest is not only the difference in computation times of the two algorithms but also the magnitude of the increase in the computation time of the exponential time algorithm when the size of the input doubled. Garey and Johnson further observed that with the $O(2^\pi)$ algorithm, a thousand-fold increase in computation speed increases the size of the largest problem instance one solves in 1 hour by 10, whereas with the $O(\pi^3)$ algorithm this size increases tenfold.

The observations above tell us clearly why problems with algorithms of polynomial time complexity are designated as "*easy*" problems. Unfortunately, a large number of scheduling problems are "*hard*" problems, i.e., the algorithms applied are of exponential time complexity assuming $\mathcal{P} \neq \mathcal{NP}$. Hard problems are said to be *intractable*. This distinction between algorithms of polynomial and exponential time complexity leads to the definition of the two classes of problems $\mathcal{P}$ and $\mathcal{NP}$. Class $\mathcal{P}$ of problems consists of problems solvable with algorithms of polynomial time complexity. The other class of problems designated as $\mathcal{NP}$ is the set of nondeterministically polynomially solvable problems. The classes $\mathcal{P}$ and $\mathcal{NP}$ are defined in terms of a *decision* (*recognition*) *problem* with a "*yes*" or "*no*" answer.

We will refer to an instance I of the decision problem with a yes-answer as a *yes-instance* and with a no-answer as a *no-instance.*

If a particular instance I is a yes-instance of the decision problem, then we require that there exists a *concise certificate* for I, whose length is bounded by a polynomial in the input size of I and which can be checked in polynomial time for validity. Hence, the fact that this particular instance I is a yes-instance can be verified in polynomial time. We say that a decision problem is in $\mathcal{NP}$ if a concise certificate satisfying these properties exists for each yes-instance of this decision problem. Let us now consider the following decision problem.

In an AOA project network with activities i, $i = 1,\ldots,n$, each with a duration d_i and a set of finish to start precedence relations $(i,j) \in A$ with time lag zero, does there exist a feasible schedule with makespan $C_{max} \leq z$, where z is a given threshold value.

A concise certificate for this decision problem is given by the completion times C_i, $i = 1, \ldots, n$, of the activities. It is polynomially bounded in the input size. We check the following:

$$C_i + d_i \leq C_j \quad \text{for each} \quad (i,j) \in A$$

$$\max_{i=1,\ldots n}\{C_i\} \leq z.$$

There are as many inequalities as the number of precedence relations, i.e., $|A|$. We can check the certificate through a certificate checking algorithm in polynomial time $O(|A|)$. Hence, the decision problem *PN* belongs to the class $\mathcal{NP}$∎.

Let the problem *P* be in $\mathcal{P}$. Thus, *P* is solvable by an algorithm of polynomial time complexity. This algorithm can then be used to verify the certificate for the associated decision problem. This implies that *every problem in $\mathcal{P}$ is also in $\mathcal{NP}$.* Thus, $\mathcal{P}$ is a subset of $\mathcal{NP}$. However, there is as yet no rigorous proof that there exist problems in $\mathcal{NP}$ for which no polynomial time algorithm exists, i.e., that $\mathcal{P} \neq \mathcal{NP}$. This question remains one of the most important open questions in mathematics today.

A concept of practical relevance is the reducibility between decision problems. We say a decision problem P_1 reduces in polynomial time to another decision problem P_2 if for every instance I_1 of P_1 a polynomial time function $f(|I_1|)$ exists transforming I_1 to an instance I_2 of P_2 so that I_1 is a yes-instance of P_1 if and only if I_2 is a yes-instance of P_2. P_1 reduces to P_2 is denoted by $P_1 \propto P_2$.

If P_1 reduces to P_2, then this implies that P_1 is *at least as hard as* P_2. If P_2 is an easy problem in $\mathcal{P}$, i.e., solvable in polynomial time and P_1 is reducible to P_2 in polynomial time, then P_1 can be solved in polynomial time as well. If, on the other hand, $P_1 \propto P_2$, and P_1 is not polynomially solvable, then P_2 is also not polynomially solvable unless $\mathcal{P} = \mathcal{NP}$. Despite the lack of a formal proof that $\mathcal{P} = \mathcal{NP}$, there is broad consensus in the scientific community that polynomial-time algorithms for some problems in $\mathcal{NP}$ are extremely unlikely to exist.

We will now introduce these NP-complete problems.

If P_1, P_2, P_3 are decision problems, if $P_1 \propto P_2$ and $P_2 \propto P_3$, then this implies $P_1 \propto P_3$. Hence, *reducibility* ($\propto$) *is transitive*. A decision problem *P* in $\mathcal{NP}$ is said to be *NP-complete* if every other problem in $\mathcal{NP}$ is reducible to P_1 in polynomial time. Given P is NP-complete, then if one can find a polynomial-time algorithm that solves P_1, then all problems in $\mathcal{NP}$ can be solved in polynomial time and hence, $\mathcal{P} = \mathcal{NP}$. But so far such a case has not been found, and the conjecture $\mathcal{P} \neq \mathcal{NP}$ still prevails. If a decision problem P is NP-complete, then the corresponding *optimization problem* is said to be *NP-hard*. The NP-complete problems thus represent the hardest class of problems in NP, and there is broad agreement that it is extremely unlikely that polynomial-time algorithms exist for any NP-complete problem.

If the time complexity function of an algorithm is bounded by a polynomial function of the size of the input and an upper bound on the magnitude of each data item, then the algorithm is said to be of *pseudo-polynomial time complexity*. In pseudo-polynomial time complexity, the complexity of an algorithm is polynomially bounded by the size of an instance in *unary* encoding but *not in binary encoding*. Thus, if unary encoding is employed, the algorithm is polynomially bounded and the associated problem is easy. Considering that the usual encodings employed are other than unary, pseudo-polynomial time algorithms are not really of polynomial time complexity. But still there is hope! Pseudo-polynomial algorithms may exist for so-called *number problems*. A decision problem P is called a number problem, if no polynomial f exists such that max(I) $\leq f(|\,\mathrm{I}\,|)$ for each instance I of P, where max(I) is the upper bound on the magnitude of the data in the instance I and | I | is the size of the instance I.

Let P be a decision problem and $f(\pi)$ a polynomial of problem size π. A sub-problem of P - called P_p - is defined by restricting P to those instances for which all data are bounded by $f(\pi)$, i.e., max(I) $\leq f(|\,\mathrm{I}\,|)$. Hence, P_p is *not* a number problem. If a pseudo-polynomial algorithm exists for the decision problem P, then P_p is solvable in polynomial time.

A decision problem P in $\mathcal{NP}$ is *NP-complete in the strong sense* or *strongly NP-complete*, if for some polynomial $f(\pi)$, P_p is NP-complete. Assuming $\mathcal{P} \neq \mathcal{NP}$, if P is NP-complete and is not a number problem, i.e., cannot be solved in pseudo-polynomial time, then it is NP-complete in the strong sense, and the corresponding optimization problem is said to be NP-hard in the strong sense.

In order to prove that a decision problem P_1 is NP-complete we go through two steps: (*i*) Show that the decision problem P_1 is in $\mathcal{NP}$. (*ii*) Choose another decision problem P_2 already proven to be NP-complete. If P_2 is polynomially reducible to P_1 ($P_2 \propto P_1$), then P_1 is NP-complete.

The first decision problem shown to be NP-complete was the *satisfiability* problem from Boolean logic (Cook 1971). Cook used reduction to show that other decision problems are NP-complete. Later, Karp (1972) proved several other decision problems are NP-complete. The books by Garey and Johnson (1979), Papadimitriou (1994), and Papadimitriou and Steiglitz (1998) contain a large number of NP-complete decision problems.

In the domain of project scheduling, the first complexity result was reported by Krishnamoorty and Deo (1979), who showed that the minimum dummy activity problem is NP-hard. The corresponding decision problem is stated as follows: *Given a set of activities, their precedence relations of finish-start with a time lag of zero and a positive integer k, is there an AOA network that satisfies the given precedence relations and which contains k or fewer dummy activities?* They proved the NP-hardness of the minimum dummy activity problem by showing the reducibility of the vertex (node) cover problem in graphs of degree 2 or 3 to the minimum dummy activity decision problem.

Another NP-hard optimization problem is the celebrated RCPSP. Blazewicz et al. (1983) showed that the optimization problem (P2|res111, chain, $p_j = 1|C_{max}$) is NP-hard in the strong sense. The problem is a machine scheduling problem on two parallel identical machines with jobs of unit processing times and chain-like precedence relations. There is a single renewable resource type with availability of one. The objective is to minimize the makespan. The authors employed the 3-PARTI TION decision problem for proving the NP-hardness in the strong sense for the RCPSP: *We are given a set of 3 m elements, a positive integer bound B, and a positive integer s(a) for each* $a \in A$ *such that* $B/4 < s(a) < B/2$ *and such that* $\sum_{a \in A} s(a) = mB$*. Can A be partitioned into m disjoint sets* $A_1, A_2, \ldots, A_m$ *such that, for* $1 \leq i \leq m$, $\sum_{a \in A_i} s(a) = B$ *(note that each* A_i *must therefore contain exactly 3 elements from A)?* This decision problem is the first number problem proven to be NP-complete in the strong sense (Garey and Johnson 1975).

Hence, minimizing the makespan for an RCPSP is NP-hard in the strong sense. This implies that all generalizations of RCPSP and its versions with different objective functions are hard to solve. Things become even more complicated for the multi-mode RCPSP. Kolisch and Drexl (1997) verified that the associated feasibility problem for two or more non-renewable resources was NP-complete. As suggested above, RCPSP with unit processing times, on the other hand, is an easy problem.

So how does all this help us in developing solution strategies for these problems, especially project scheduling problems? We know that if a decision problem is NP-complete, then the corresponding optimization problem is NP-hard. In developing solution algorithms for NP-hard optimization problems, we have two choices. If we insist on a guaranteed optimal solution, we know that the worst-case time complexity of the algorithm producing it will be exponential (unless $\mathcal{P} = \mathcal{NP}$, which is very unlikely). If the application motivating our optimization problem is one that will produce very small instances of the problem, and we have plenty of time to produce a solution, this may be a perfectly viable option. If, on the other hand, the application is such that solutions are required in a short amount of time, a solution algorithm must have low-order polynomial time complexity to be viable. In this case, our only option is heuristic procedures that run in polynomial time, but cannot guarantee a provably optimal solution. It is also possible that special cases of difficult problems might lead to easy solutions. An example would be the RCPSP with unit processing times. Hence, before we attempt to solve an optimization problem for which we do not know whether it is NP-hard or not we need to check whether the corresponding decision problem is NP-complete. The concept of polynomial-time reductions is central to making this determination.

Further Reading

Ahuja RK, Magnanti TL, Orlin JB (1993) Network Flows. Theory, Algorithms, and Applications. Prentice Hall, NJ (Appendix B)

Brucker P (2007) Scheduling Algorithms. Fifth Edition, Springer Verlag, Berlin

Brucker P, Knust S (2006) Complex Scheduling. Springer Verlag, Berlin

Garey MR, Johnson DS (1979) Computers and Intractability. A Guide to Theory of NP-Completeness. WH Freeman and Company, San Francisco

Papadimitriou CH (1994) Computational Complexity. Addison-Wesley Educational Publishers Inc., NY

Papadimitriou CH, Steiglitz K (1998) Combinatorial Optimization. Algorithms and Complexity. Dover Publications Inc., NY

Pinedo ML (2008) Scheduling. Theory, Algorithms, and Systems. Third Edition, Springer Science + BusinessMedia, NY (Appendix D)

Complexity results for scheduling problems: http://www.informatik.uni-osnabrueck.de/knust/class/

Appendix 7B: Branch and Bound

In this appendix, we introduce the Branch and Bound (B&B) algorithm for reaching an exact solution for a combinatorial optimization problem, specifically a *minimization problem*. The B&B algorithm is an enumerative procedure that proceeds by implicitly enumerating portions of the solution space in an intelligent way such that the algorithm does not need to examine every solution individually, reducing the computational time required to identify an optimal solution.

The B&B algorithm can be graphically illustrated as a B&B tree consisting of nodes and branches. The root node of the B&B tree corresponds to the set of all feasible solutions to the problem at hand, say, $\mathcal{S}$. A node represents a subset of feasible solutions. A node is called a *parent* if it has one or more descendants, each of which is called a *child*.

Branching: The branching process divides the solution set $\mathcal{S}_i$ at node i into $k > 1$ subsets $\mathcal{S}_{ij}$ whose union results in $\mathcal{S}_i$; i.e., $\bigcup_{i=1}^{k} \mathcal{S}_{ij} = \mathcal{S}_i$. Branching is recursive in the sense that each subset created corresponds to a node of the B&B tree which, in turn, is a candidate for further branching. Each branch produces a child node representing a reduced subset of solutions compared to that represented by the parent node. In the context of project scheduling, for example, branching might restrict the subset of solutions of the child to contain only solutions where activity A precedes activity B. Hence, the solutions at the nodes are referred to as *partial solutions*.

The bounding process is applied through *upper bounding* and *lower bounding* of the sub-sets of the solution space.

Upper Bounding: Initially, an upper bound on the optimal objective function value of the minimization problem, referred to as the *current upper bound* (*CUB*) is assigned at *the root node*. This might correspond to an already known feasible solution -such as one obtained through a heuristic solution procedure- or if no such solution is available, to a sufficiently large number. As the B&B tree evolves, whenever a feasible solution is obtained at some node i of the tree its objective function value UB_i constitutes an upper bound on the minimum value of the problem over the set of solutions represented by node i. If $UB_i < CUB$, then CUB is set equal to UB_i, reducing the upper bound on the minimum value of the problem obtained so far, rendering this feasible solution the best solution identified so far. Otherwise, the solution associated with CUB is retained as the best available solution.

Lower Bounding: The idea of the B&B algorithm is to obtain an optimal solution to a difficult combinatorial problem through the solution of an easier problem, specifically a *relaxation* of the original problem, at each node of the tree. Relaxation of an optimization problem can be intuitively defined as another optimization problem whose feasible solution set contains all feasible solutions to the original problem. Any feasible solution to the original problem is also a feasible solution to the relaxation, but the converse is not true. The most common way of obtaining a relaxation to a difficult integer programming problem is to eliminate (relax) a particularly problematic set of constraints, whose relaxation will result in a problem that can be solved efficiently. At each node i of the B&B tree, the algorithm generates a lower bound LB_i by solving a relaxation of the original problem over the set of solutions represented by node i. Thus, the minimum objective function value among all feasible solutions in the sub-set described by node i cannot be less than LB_i. In addition, the LB_j value of all nodes j descending from node i cannot be lower than LB_i.

Fathoming (pruning): The implicit enumeration is realized through the *fathoming (pruning)* process. A node i is fathomed, i.e., no further branching will be made from that node, when one of the following conditions hold: (*i*) No further branching is possible from node i, i.e., all possible subsets have already been generated and the node i represents a single solution; (*ii*) $LB_i \geq CUB$, implying that no solution in the feasible solution subset S_i at node i can have a smaller objective function value than the best available feasible solution associated with CUB. Hence, there is no need to search further in S_i. In order to identify multiple optima if they exist, the fathoming decision can be implemented with the strict inequality $LB_i > CUB$, although this may possibly result in significantly higher computation times. (*iii*) When LB_i represents a feasible solution. Due to its feasibility, we set $UB_i = LB_i$ and update CUB, if necessary, as explained in the *Upper Bounding* above. Any node not yet fathomed is said to be *alive* and is called a *live* or *active node*.

As the above inequalities imply, the fathoming process becomes more effective as smaller CUB values (tighter upper bounds) are obtained. This indicates the importance of obtaining a tight CUB at the root node. The additional effort invested in obtaining a tighter CUB before starting the branching process might pay off later in the process through more effective pruning.

The fathoming process is also more effective when larger LB values (tighter lower bounds) are obtained at each node. Tight LBs generally require the solution of a more complicated problem at the nodes, and hence additional computational time. Hence there is a trade-off between the tightness of the LBs and the resulting computational burden. A tighter LB can be expected to reduce the number of live nodes but may lead to a higher computational burden at each node. A less tight *LB*, on the other hand, might lead to a larger number of nodes being examined but with less computational burden at each node.

The implicit enumeration procedure above can be further improved through problem dependent *dominance rules* that reduce the size of the solution space by showing that certain solutions cannot lead to a better objective function value than that already obtained (Jouglet and Carlier 2011). In the context of the B&B algorithm covered here, dominance rules are employed to compare the partial solutions at two nodes *i* and *j* and show that the best solution obtainable from one node, say *i*, cannot be better than that obtained from the other. Hence there is no need to consider further the descendants of the dominated node, allowing it to be fathomed. An example of a set of dominance rules will be stated in the following section when we present the B&B algorithm for solving the RCPSP.

Branch Selection Strategy: The branch selection strategy selects one of the live nodes to branch from. There are three principal branch selection strategies: (*i*) *depth-first search,* (*ii*) *minimum LB search* (also called the *frontier search strategy* or *restricted flooding*), and (*iii*) *breadth first search.*

In *depth-first search*, the most recently created live node is selected for branching and only one new branch is created, leading to one new node extending the partial solution on hand. This strategy seeks to reach a feasible solution as rapidly as possible in order to improve the *CUB* and facilitate more effective fathoming. When the most recently created node is fathomed, branching is resumed from the most recently created live node. The process of reaching this live node is called *backtracking*. For backtracking, one moves from the current fathomed node back up the B&B tree to its parent node. This step is repeated until a live node is reached, which becomes the next node to branch from.

Under the *minimum LB* strategy, on the other hand, the live node with the minimum *LB* is selected, since it is considered to be the most promising subset of solutions to contain an optimal solution. At the selected node for branching all possible branches are created. A drawback of this strategy is its relatively larger memory requirement compared with the depth-first strategy resulting from the relatively larger number of live nodes at any step of the process. Since the data associated with a live node has to be kept in memory, it follows that the larger the number of live nodes at any step of the process, the larger will be the memory requirement.

The *breadth first search* strategy is an exhaustive search strategy searching the tree breadthwise exhausting the nodes at a given level– opposite to depth first strategy. Once a level of the tree is exhausted, the search continues with the next layer. The breadth first search strategy is widely employed in artificial intelligence in the exploration of data structures.

Termination: The B&B algorithm *terminates* when all nodes have been fathomed either by reaching a feasible solution or as a result of their *LB*s exceeding the *CUB*. If the decision-maker opts for early termination, s/he can impose a termination criterion (or criteria), such as a limit on the computation time or the number of nodes to be examined.

The final *CUB* and its associated schedule correspond to the *optimal* objective function value and schedule, respectively.

A pseudocode for the B&B algorithm using depth first as the branching strategy is provided in Fig. 7B.1.

We have described the B&B algorithm for a minimization problem. For a *maximization problem*, upper bounding and lower bounding are interchanged in the bounding process. The rest of the algorithm remains unchanged.

```
Create an empty stack S (set of active nodes)
Push parent (node 1) to S
CUB:= value of a known feasible solution, value of a heuristic solution or a very large value
CBS:= solution corresponding to CUB
while S not empty do
        peek at the top j of S
        while child i can be generated from parent j do
                calculate LBj
                if LBj >= CUB then
                        fathom j
                        pop S
                else
                        if i is a feasible solution then
                                CUB := LBi
                                CBS := solution corresponding to child i
                        else
                                push i to S
                        end if
                end if
        end while
end while
```

Fig. 7B.1 A pseudocode for the B&B algorithm using depth first as the branching strategy

References

Ahuja, R. K., Magnanti, T. L., & Orlin, J. B. (1993). *Network flows. Theory, algorithms, and applications*. Englewood Cliffs: Prentice Hall.

Alvarez-Valdes, R., & Tamarit, J. M. (1989). Heuristic algorithms for resource constrained project scheduling: A review and empirical analysis. In R. Slowinski & J. Weglarz (Eds.), *Advances in project scheduling* (pp. 113–134). Amsterdam: Elsevier.

Baroum, S. M. (1992). An exact solution procedure for maximizing the net present value of resource-constrained projects, Unpublished PhD Dissertation, Indiana University.

Beşikci, U., Bilge, Ü., & Ulusoy, G. (2012). Resource dedication problem in a multi-project environment.

Bianco, L., Dell'Olmo, P., & Speranza, M. G. (1998). Heuristics for multi-mode scheduling problems with dedicated resources. *European Journal of Operational Research, 107*, 260–271.

Blazewicz, J., Lenstra, J., & Rinnooy Kan, A. (1983). Scheduling subject to resource constraints: Classification and complexity. *Discrete Applied Mathematics, 5*, 11–24.

Blazewicz, J., Cellary, W., Slowinski, R., & Weglarz, J. (1986). *Scheduling under resource constraints-deterministic models*. Basel: Baltzer.

Böttcher, J., Drexl, A., & Kolisch, R. (1996). A branch-and-bound procedure for project scheduling with partially renewable resource constraints. Proceedings of the Fifth Workshop on Project Management and Scheduling, Poznan, Poland, pp. 48–51.

Brucker, P. (2007). *Scheduling algorithms* (5th ed.). Berlin: Springer.

Brucker, P., & Knust, S. (2006). *Complex scheduling*. Berlin: Springer.

Cook, S. A. (1971). The complexity of theorem-proving procedures. In *Proceedings of the 3rd annual ACM symposium on theory of computing*, 151–158.

Cooper, D. F. (1976). Heuristics for scheduling resource-constrained projects: An experimental investigation. *Management Science, 22*(11), 1186–1194.

Demeulemeester, E., & Herroelen, W. (2002). *Project scheduling. A research handbook*. Dordrecht: Kluwer Academic Publishers.

Doersch, R. H., & Patterson, J. H. (1977). Scheduling a project to maximize its present value: A zero-one programming approach. *Management Science, 23*(8), 882–889.

French, S. (1982). *Sequencing and scheduling*. Chichester: Ellis Horwood Limited.

Garey, M. R., & Johnson, D. S. (1975). Complexity results for multiprocessor scheduling under resource constraints. *SIAM Journal on Computing, 4*, 397–411.

Garey, M. R., & Johnson, D. S. (1979). *Computers and intractability. A guide to theory of NP-completeness*. San Francisco: WH Freeman and Company.

Goldratt, E. M. (1997). *Critical chain*. Great Barrington: North River Press.

Hall, N G (2016) Research and teaching opportunities in project management. In *Optimization challenges in complex, networked and risky systems* (pp. 329–388). INFORMS.

Hartmann, S., & Briskorn, D. (2010). A survey of variants and extensions of the resource-constrained project scheduling problem. *European Journal of Operational Research, 207*(1), 1–14.

Herroelen, W., & Leus, R. (2001). On the merits and pitfalls of critical chain scheduling. *Journal of Operations Management, 19*(5), 559–577.

Herroelen, W., De Reyck, B., & Demeulemeester, E. (1998). Resource-constrained project scheduling: A survey of recent developments. *Computers and Operations Research, 25*(4), 279–302.

Hillier, F. S., Liebermann, G. J. (2005). *Introduction to operations research* (International Edition. 8th ed.). New York: McGraw-Hill.

Icmeli-Tukel, O., & Rom, W. O. (1997). Ensuring quality in resource constrained project scheduling. *European Journal of Operational Research, 103*, 483–496.

Igelmund, G., & Radermacher, F. J. (1983a). Preselective strategies for the optimization of stochastic project networks under resource constraints. *Networks, 13*, 1–28.

Igelmund, G., & Radermacher, F. J. (1983b). Algorithmic approaches to preselective strategies for stochastic scheduling problems. *Networks, 13*, 29–48.

Jouglet, A., & Carlier, J. (2011). Dominance rules in combinatorial problems. *European Journal of Operational Research, 212*(3), 433–444.

Kaplan, L. A. (1988). Resource-constrained project scheduling with preemption of jobs, Unpublished PhD Dissertation, University of Michigan, MI.

Karp, R. M. (1972). Reducibility among combinatorial problems. In: *Complexity of computer computations Proceedings of the Symposium*, IBM Thomas J. Watson Research Center, Yorktown Heights, NY 1972 85–103, Plenum, New York.

Kelley, J. E., Jr., & Walker, M. R. (1959). Critical path planning and scheduling. *Proceedings of Eastern Joint Computer Conference, 16*, 160–172.

Kim, J., Kang, C., & Hwang, I. (2012). A practical approach to project scheduling: Considering the potential quality loss cost in the time–cost tradeoff problem. *International Journal of Project Management, 30*(2), 264–272.

Kolish, R. (1995). *Project scheduling under resource constraints – Efficient heuristics for several problem classes*. Heidelberg: Physica – Verlag.

Kolisch, R. (1996a). Efficient priority rules for the resource-constrained project scheduling problem. *Journal of Operations Management, 14*(3), 179–192.

Kolisch, R. (1996b). Serial and parallel resource-constrained project scheduling methods revisited: Theory and computation. *European Journal of Operational Research, 90*(2), 320–333.

Kolisch, R., & Drexl, A. (1997). Local search for non-preemptive multi-mode resource constrained project scheduling. *IIE Transactions, 29*(11), 987–999.

Kolisch, R., & Padman, R. (2001). An integrated survey of deterministic project scheduling. *Omega, 29*(3), 249–272.

Krishnamoorty, M. S. & Deo, N. (1979). Complexity of the minimum-dummy-activities problem in a Pert network. *Networks, 9,* 189–194.

Kucuksayacigil, F., & Ulusoy, G. (2020). A hybrid genetic algorithm application for a bi-objective, multi-project, multi-mode, resource-constrained project scheduling. *Frontiers of Engineering Management, 7*(3), 426–446.

Lenstra, J. K., & Rinnooy Kan, A. H. G. (1979). Computational complexity of discrete optimization problems. *Annals of Discrete Mathematics, 4*, 121–140.

Lim, A., Ma, H., Rodrigues, B., Tan, S. T., & Xiao, F. (2011). New concepts for activity float in resource-constrained project management. *Computers & Operations Research, 38*(6), 917–930.

Ning, M., Zhengwen, H., Jia, T., & Wang, N. (2017). Metaheuristics for multi-mode cash flow balanced project scheduling with stochastic duration of activities. *Automation in Construction, 81*, 224–233.

Özdamar, L., & Ulusoy, G. (1995). A survey on the resource-constrained project scheduling problem. *IIE Transactions, 27*, 574–586.

Papadimitriou, C. H., & Steiglitz, K. (1998). *Combinatorial optimization. Algorithms and complexity*. New York: Dover Publications.

Patterson, J. H., & Huber, W. D. (1974). A horizon-varying, zero-one approach to project scheduling. *Management Science, 20*, 990–998.

Patterson, J. H., & Roth, G. W. (1976). Scheduling a project under multiple resource constraints: A zero-one programming approach. *AIIE Transactions, 8*, 449–455.

Pinedo, M. L. (2008). *Scheduling. Theory, algorithms, and systems* (3rd ed.). New York: Springer.

Pritsker, A. A. B., Watters, L. J., & Wolfe, P. M. (1969). Multi-project scheduling with limited resources: A zero-one programming approach. *Management Science, 16*, 93–108.

Russel, R. A. (1986). A comparison of heuristics for scheduling project with cash flows and resource restrictions, *Management Science, 32,* 1291–1300.

Smith-Daniels, D. E., & Aquilano, N. J. (1987). Using a late-start resource-constrained project schedule to improve project net present value. *Decision Sciences, 18*, 617–630.

Smith-Daniels, D. E., Padman, R., & Smith-Daniels, V. L. (1987). Maximizing the net present value of a project subject to materials and capital constraints. *Journal of Operations Management, 7* (1), 33–45.

Smith-Daniels, D. E., Padman, R., & Smith-Daniels, V. L. (1996). Heuristic scheduling of capital constrained projects. *Journal of Operations Management, 14*(3), 241–254.

Sprecher, A., Kolisch, R., & Drexl, A. (1995). Semi-active, active, and non-delay schedules for the resource-constrained project scheduling problem. *European Journal of Operational Research, 80*, 94–102.

Ulusoy, G., & Özdamar, L. (1989). Heuristic performance and network/resource characteristics in resource-constrained project scheduling. *Journal of Operational Research Society, 40*, 1145–1152.

Wiest, J. D. (1964). Some properties of schedules for large projects with limited resources. *Operations Research, 12*(3), 395–418.

8 Resource Leveling and Other Resource Management Problems

Learning Outcomes
Upon successful completion of this Chapter, the reader will be able to:

1. Formulate resource leveling problems with different objective functions.
2. Obtain a solution to the resource leveling problem and evaluate the results obtained.
3. Apply heuristic solution procedures to efficiently obtain near-optimal solutions.
4. Have an understanding of different resource management models.
5. Suggest project resource management policies.

8.1 Introduction

In this chapter, we will treat problems related to resource management. The first one is the *resource leveling problem*, which deals with projects where renewable resources are not limited for practical purposes but the distribution of their usages over time is not smooth, i.e., there are relatively large fluctuations in resource usage over time, which make resource management more difficult, inefficient, and costly. Thus, a project manager would prefer to have uniform renewable resource usage over time. Several different objective functions are formulated in the literature so as to reflect these fluctuations in renewable resource requirements from period to period over the project duration. We will introduce these formulations in this Chapter.

An extension of the resource leveling problem is the *total adjustment cost problem*. In this problem, the fluctuations are penalized like in the leveling problem but this time not as a function of fluctuations of the resource utilization levels but as a function of the cost of these fluctuations. Hence, the total adjustment cost problem differs from the resource leveling problem by its objective, which is the minimization of the total cost of positive and negative fluctuations (jumps).

Under the more general title *resource management problems,* we will study the problems that address the determination of resource requirements of projects. We will cover several such problems, one being the *resource availability cost problem*

G. Ulusoy, Ö. Hazır, *An Introduction to Project Modeling and Planning*, Springer Texts in Business and Economics, https://doi.org/10.1007/978-3-030-61423-2_8

(RACP). RACP is similar to the resource leveling problem as it also addresses renewable resource requirements throughout the project implementation. However, unlike the leveling problem, RACP assumes constant renewable resource availabilities over the project duration. The required resource levels are searched to minimize the total resource availability cost over all resource types given the project deadline.

In multi-project settings, in general, resources constitute pools, from where they are shared by all projects. Renewable resources once employed return to their pools for further usages by other activities. This resource management policy will be referred to as the *resource sharing policy* (RSP). RSP has been the conventional approach used in resource optimization problems. In some cases, a subset of the resources might be dedicated to some specific projects over the whole duration of these projects. Such an assignment policy will be called here the *resource dedication policy* (RDP). These policies are needed for example, for geographically distributed or fixed location renewable resources. In some other cases such as software projects, developers might not transfer to other projects before the current project is completed, because of the learning concerns. We will examine also the time and/or costs of transferring resources within or between projects.

Resource portfolio management problem, which addresses the issue of allocating a resource budget among renewable and non-renewable resources in a *multi-project environment* such that an objective or objectives are satisfied, will be introduced. Two versions of the problem will be covered. One version will be the resource portfolio management problem under RDP. The other version will be solved under the *relaxed resource dedication policy* (RRD), where a resource is allowed to be transferred to a project not yet started once the last project at which it was employed is completed.

Finally, a case study on the resource leveling problem is presented. It addresses the project planning of an annual audit process. MS Project software is employed for planning purposes.

8.2 Resource Leveling

Since the resource leveling problem deals with renewable resources, in the sequel, the term "resource" will be used to mean a renewable resource. As the decision environment is deterministic, so we treat this problem as an extension of the unconstrained deterministic project scheduling problem. The resource leveling problem can be stated as of the problem of minimizing the variance of resource usage over time without exceeding the pre-specified project deadline.

The objective function to be minimized is given in (8.1).

$$\min \sum_{r=1}^{|R|} \sum_{t=1}^{PD} C_r(u_{rt}) \tag{8.1}$$

where

r = resource type index (1, . . ., $|R|$)
t = period index (1, . . ., PD)
PD = project duration
u_{rt} = level of resource usage of type r in period t
$C_r(u_{rt})$ = performance measure for resource type r in period t at a level u_{rt}

The ideal solution would be where the resources are uniformly distributed over the project duration. However, in most of the problem instances, this solution could not be reached. To distinguish between alternative solutions, we need to define a performance measure that reflects the variance of the resource usage distribution over the project duration. For this purpose, several different functions, represented as $C_r(u_{rt})$, can be formulated. The classical function, the classical performance measure $C_r(u_{rt})$ is the square of the usage of resource r in period t multiplied by a nonnegative weight factor C_r for each resource type r over all resources and periods (8.2).

$$C_r(u_{rt}) = c_r \cdot u_{rt}^2 \tag{8.2}$$

In this function, the resource usage value is squared to accentuate the impact of increasing values of resource usage. The project manager might also attach more importance to some specific resource types over the others. This can be done by defining a relative weight factor $c_r \geq 0$ for each resource type r, where a larger c_r value corresponds to a higher importance attached to resource type r. This performance measure will be used in the schedule generation algorithm by Burgess and Killebrew (1962).

Another function can be formulated to penalize the fluctuation in both directions between two consecutive periods. It is given in (8.3).

$$C_r(u_{rt}) = c_r \cdot (u_{rt} - u_{r,t-1})^2 \tag{8.3}$$

A similar performance measure would be: $C_r(u_{rt}) = c_r \mid u_{rt} - u_{r,\,t-1} \mid$.

On the other hand, if only the increase from one period to the next is to be penalized, then the performance measure can be formulated as in (8.4).

$$C_r(u_{rt}) = c_r \cdot (\max(0, u_{rt} - u_{r,t-1}))^2 \tag{8.4}$$

In some projects, a target level of usage, say a_k, can be explicitly specified for each resource type r. In this case, the fluctuations of the resource usages around a_k could be measured and penalized using the function (8.5).

$$C_r(u_{rt}) = c_r \cdot (u_{rt} - a_r)^2 \tag{8.5}$$

Similarly, one can employ $C_r(u_{rt}) = c_{kr} \mid u_{rt} - a_r|$.

Sometimes project managers focus on avoiding fluctuations above a_k, i.e., interested in overload only (Easa 1989). For these cases the performance measure would be as in (8.6).

$$C_r(u_{rt}) = c_r \cdot max\,(0, u_{rt} - a_r) \tag{8.6}$$

Next we present a mathematical programming formulation for the resource leveling problem, (8.7) through (8.11). The expression (8.7) employs one of the possible objective functions. Depending on the project requirements, one of the above given functions can replace this one.

The project network is AOA-type. In the following, T_i denotes the occurrence time of event i, where $i = 1\ldots, m$. S_t, the set of activities active in period t, and k_{ir} the consumption by activity i of resource type k.

$$min \sum_{r=1}^{|R|} c_r \sum_{t=1}^{PD} u_{rt}^2 \tag{8.7}$$

subject to

$$T_j - T_i \geq d_{ij} \quad \text{for all } (i,j) \in A \tag{8.8}$$

$$T_1 = 0 \tag{8.9}$$

$$T_m \leq PD \tag{8.10}$$

$$\sum_{i \in S_t} k_{ir} \leq u_{rt} \quad \text{for } r = 1, .., |R| \text{ and } t = 1, .., PD \tag{8.11}$$

The resource leveling problem is *NP*-hard in the strong sense and hence, is difficult to solve to optimality in polynomial time (Neumann et al. 2003, Section 3.4). In order to be able to solve problems of large sizes, heuristic algorithms have been proposed in the literature.

In Sect. 8.2.1, a heuristic algorithm proposed by Burgess and Killebrew (1962) is presented together with an example problem. It is a simple algorithm to demonstrate the basics of the resource leveling problem and is easy to implement. Other solution procedures for the resource leveling problem are given in Sect. 8.2.2.

8.2.1 The Burgess and Killebrew Algorithm

We will consider here the heuristic algorithm proposed by Burgess and Killebrew (1962) utilizing the *sum of squares performance measure* (8.2) but it is also applicable using the other performance measures introduced earlier in (8.3)–(8.6).

The Burgess and Killebrew algorithm can be called a non-gradient algorithm, as the consecutive iterations might lead to deteriorating objective function values. It is neither constructive nor improvement type heuristic. The algorithm considers activities, which have been scheduled according to a priority list, and shifts the activities to the best place choosing the start time, which minimizes the objective function under consideration.

Before introducing a pseudo-code for the Burgess and Killebrew algorithm, let us briefly summarize how the algorithm operates. Initially, randomly $|M|$ distinct priority lists are generated, which satisfy the precedence relationships of the activities. Starting with the first priority list in the set of distinct priority lists, M, the start times of all activities are set to their early start times. The algorithm has two loops; an outer loop and an inner loop. In the outer loop, the algorithm iterates over the activities of the selected priority list in reverse order. In the inner loop, the algorithm shifts the activity being processed to the best start time possible to minimize the objective function under consideration. In case there are multiple best start times possible, then the algorithm assigns the activity to the latest of all these possible start times. The rationale behind this choice is to increase the assignment flexibility of activities yet to be assigned. Once all activities of the selected priority list are processed by both loops, then the schedule obtained is stored together with its objective function value. Starting with the outer loop, the algorithm iterates over the activities of the last stored schedule in reverse order, and the inner loop seeks the best start time for the activity under consideration each time. The loops are repeated as long as there is a change in the start time of at least one activity from the previous loop's stored schedule. When there is no change, the iteration is terminated with the stored schedule and its objective function value. This objective function value is compared with the current best objective function value of all priority lists processed so far and replaces the current best objective function value as the best value obtained so far if it is smaller. The algorithm terminates when all priority lists in M are exhausted declaring the best objective function value and the associated schedule as the heuristic solution to the problem. Otherwise, the algorithm continues with the next priority list in M and the application of both loops in an iterative manner.

The pseudocode for this heuristic is given below. For this purpose, we introduce the following notation.

M = set of distinct priority lists that satisfy the precedence order of activities,
k = iteration number,
SS_{it} = sum of square of resource utilization of activity i starting in period t (performance measure),
EST_i = early start time of activity i,
LST_i = late start time of activity i,
ST_i = start time of activity i,
CB = the current best value within an iteration
BST_i = the start time of activity i corresponding to the current best value
CBS = tuple dictionary containing schedule corresponding to CB as (activity, best start time)
OB = the overall best value

OBS = tuple dictionary containing schedule corresponding to *OB* as (activity, best start time)
$A_k = k^{th}$ priority list among |*M*| distinct priority lists.

```
k = 1
OB: = inf
change = True
while k ≤ |M| do
    // select A_k among the priority lists
    for each activity i in A_k do
        ST_i = EST_i
    end for
    CB: = inf
    while change do
        change := False
        for each activity i in A_k in reverse order do
            for each possible starting time t in [EST_i,LST_i] do
                calculate SS_it
                if SS_it ≤ CB then
                        CB: = SS_it
                        BST_i := t
                end if
            end for
            if BST_i ≠ ST_i then
                ST_i := BST_i
                    change := True
            end if
        end for
    end while
    for each activity i in A_k do
        CBS.add(i, BST_i)
    end for
    if CB < OB
        OB := CB
        OBS := CBS
    end if
    k = k + 1
end while
```

Example 8.1 A software development project requires 7 work packages to be prepared. The pre-specified deadline *PD* is 24 weeks. The only resource type is the software programmer. Management needs a project plan with an even distribution of the software programmers over time. The project manager has decided to apply the Burgess and Killebrew resource leveling algorithm using the sum of

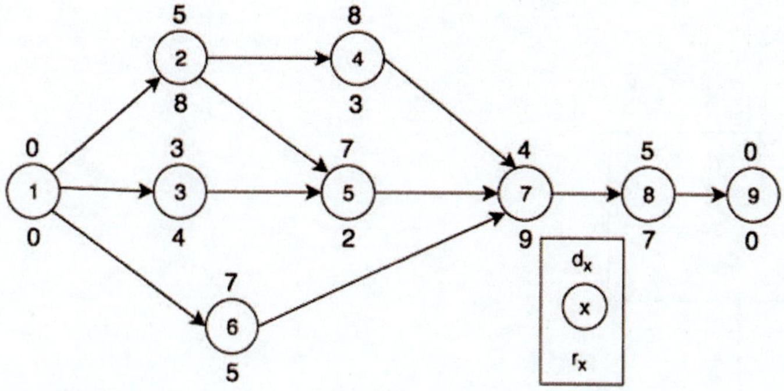

Fig. 8.1 AON project network

squares of resource usages as the performance measure. The precedence relations between the work packages, their expected durations in weeks, and the required number of software programmers for each work package are all indicated on an AON project network (Fig. 8.1). The activity nodes 1 and 9 are the dummy initial and terminal nodes, respectively. The unit cost c_l of the software programmers is given as 1 per week.

Solution

The First Loop of the First Iteration. We start iteration number $k = 1$ by considering the priority list <1, 2, 3, 4, 5, 6, 7, 8, 9>. We can neglect the dummy activities 1 and 9, since they have no impact on the resulting schedule. The early-start schedule for the problem instance is represented in Fig. 8.2. The performance measure value for the early-start schedule is determined to be 2108.

- The first activity to be considered is activity 8 with a current early start time of 17 and a late start time of 19 obtained considering *PD*. The values of SS_{8t} for each possible starting time t are as follows. (The values of SS_{8t} are indicated in brackets): 19 (2108), 18 (2108) and 17 (2108). The minimum value of the performance measure SS_{8t} is 2108 and is obtained at starting times 17, 18, and 19. The Burgess and Killebrew procedure, however, picks the latest of these possible starting times, delaying activity 8 to start at $t = 19$.
- The next activity to be considered is activity 7 with the possible starting times in the interval [13,15]. The values of SS_{7t} for these starting times are: 15 (2108), 14 The next activity to be considered is activity 7 with the possible starting times in the interval [13,15]. The values of SS_{7t} for these starting times are: 15 (2108), 14 (2108) and 13 (2108). Activity 7 is therefore delayed starting at time instant 15.
- The next activity to be considered in the priority list is activity 6 with the following results: 8 (1718), 7 (1768), 6 (1818), 5 (1838), 4 (1868), 3 (1898), 2 (1968), 1 (2038) and 0 (2108). Activity 6 is thus delayed starting at time instant 8.

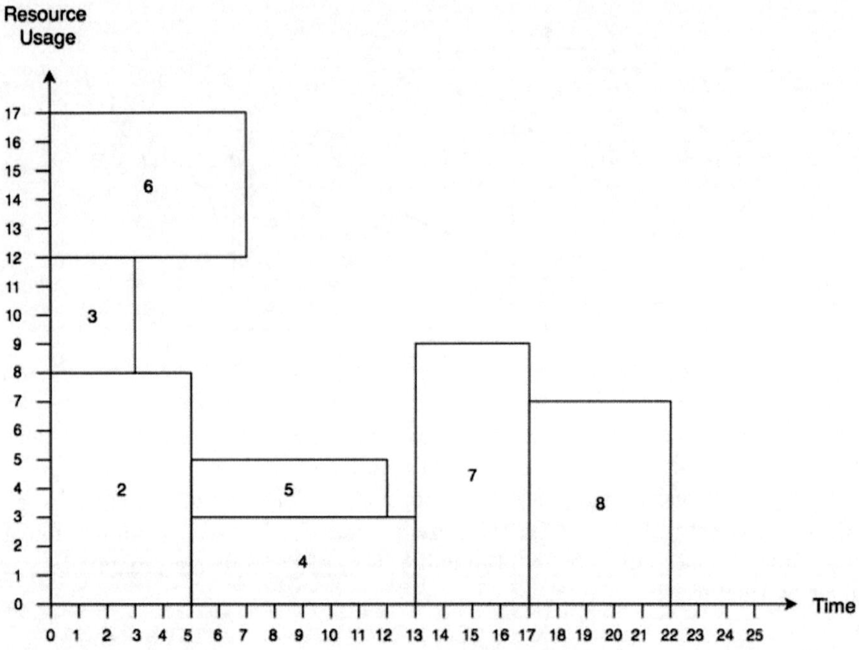

Fig. 8.2 Resource usage in the early-start schedule for Example 8.1

- Activity 5 is the next activity to be considered in the priority list. The resulting SS_{5t} values are: 8 (1754), 7 (1746), 6 (1738), 5 (1718). As a result, the starting time of activity 5 remains unchanged at time instant 5.
- The calculations for activity 4 are: 7 (1754), 6 (1736) and 5 (1718). So, for activity 4 the starting time remains at time instant 5.
- Activity 3 can start between time instants 0 and 2. The SS_{3t}values for all these time instants are equal to 1718. Hence, the starting time of activity 3 is delayed to time instant 2.
- Activity 2 has only one possible starting time, namely time instant 0 with the corresponding performance measure value of 1718.

In this way, we have completed one loop for the priority list <1, 2, 3, 4, 5, 6, 7, 8, 9 > decreasing the objective function value from 2108 to 1718. The resulting schedule is presented in Fig. 8.3.

The Second Loop of the First Iteration During the first loop, activities 8, 7, 6 and 3 have been delayed. The Burgess and Killebrew procedure requires us to repeat the loop over all activities in the priority list.

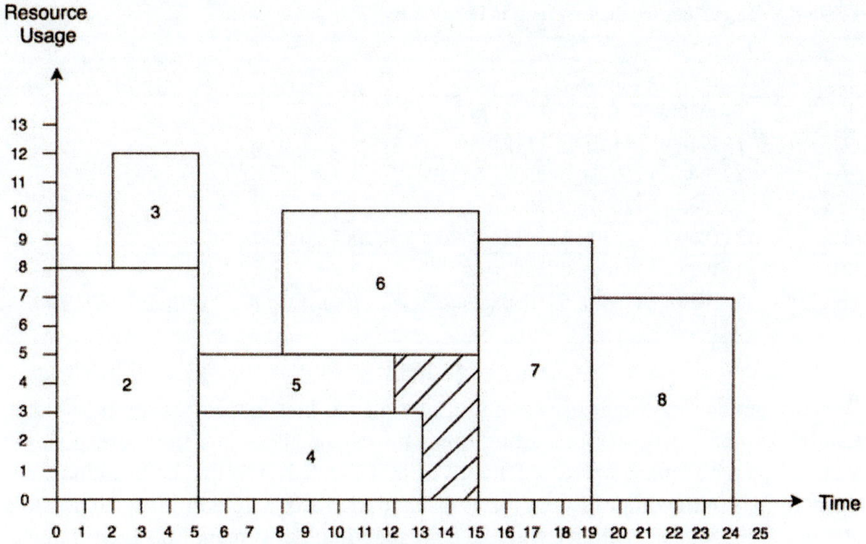

Fig. 8.3 Improved schedule after the first loop of the first iteration

- A second loop through the priority list starts again with activity 8. Since activities 8 and 7 are scheduled at their latest starting times in the previous iteration, they cannot be delayed any further.
- Activity 6 has possible starting times in the interval [5, 8] and the SS_{6t} values are: 8 (1718), 7 (1768), 6 (1818) and 5 (1838). The starting time of activity 6 remains as $t = 8$.
- The current early start time of activity 5 is 5 and the late starting time is 8. The calculations for this activity give the following results: 8 (1754), 7 (1746), 6 (1738), 5 (1718). Activity 5 is therefore not delayed.
- Next in the priority list is activity 4 with possible starting times in the interval [5,7] and the SS_{4t} values are: 7 (1754), 6 (1736) and 5 (1718). The starting time of activity 4 remains at $t = 5$.
- Activities 3 and 2 are again scheduled at their latest starting time; hence, their starting times will not change.

The second loop did not change the solution of the first loop. Hence, the first iteration is terminated and the best solution is stored.

The Second Iteration Let the iteration number $k = 2$ and select a different priority list from the priority list pool, e.g., priority list <1, 6, 2, 3, 4, 5, 7, 8, 9>. The early-start scheduling for this priority list results in the same schedule as in the previous iteration (Fig. 8.2).

Table 8.1 SS_{it} values for the first loop in iteration $k = 2$

Activity i	Starting time (SS_{it})
8	**19 (2108)**, 18 (2108), 17 (2108)
7	**15 (2108)**, 14 (2108), 13 (2108)
5	**8 (2044)**, 7 (2056), 6 (2088), 5 (2108)
4	**7 (2008)**, 6 (2026), 5 (2044)
3	**5 (1800)**, 4 (1880), 3 (1944), 2 (2008), 1 (2008), 0 (2008)
2	2 (1928), 1 (1864), **0 (1800)**
6	**8 (1670),** 7 (1690), 6 (1680), **5 (1670),** 4 (1700), 3 (1730), 2 (1760), 1 (1790) and 0 (1800)

For demonstration purposes, we will present the first loop of iteration k = 2 in tabular format. The possible starting times for each activity and the corresponding value of its performance measure are given in Table 8.1. The first column indicates the activities in the order of the priority list and the second column lists the possible starting times for each activity with the corresponding performance measure value in brackets. The best starting time for each activity and its corresponding performance measure value are highlighted both in bold.

The resulting schedule given in Fig. 8.4 has an objective function value of 1670. Since the new objective function value is better than the best one obtained so far, we update our current best solution *CB*.

- A second loop through the same priority list starts with activity 8. Activities 8, 7, and 5, however, are scheduled at their late start times and thus cannot be delayed anymore.
- The next activity in the priority list is activity 4, which is also scheduled at its latest starting time with 7 (1670). Activity 4 can start in the interval [5,6] and the corresponding SS_{4t} are 6 (1652) and 5 (1634), respectively. Note that if we start activity 4 at its current early start time, our objective function value improves. So, activity 4 is scheduled to start at t = 5.
- Activities 3, 2, and 6 are all already scheduled at their latest starting times and therefore cannot be delayed.

Since the new objective function value of 1634 < $CB = 1670$, we update our current best solution to $CB = 1634$. The resulting new schedule is given in Fig. 8.5. We observe that all the activities except activity 4 are at their late start values and activity 4 is at its early start time and consequently, looping through the current priority list will no longer improve the objective function value. Thus, we terminate the second iteration.

If the number of distinct priority lists is larger than 2, then we increase the number of iterations by one and let $k = 3$. The algorithm continues with a different priority list from the priority list pool and terminates when the pre-specified number of priority lists are considered.

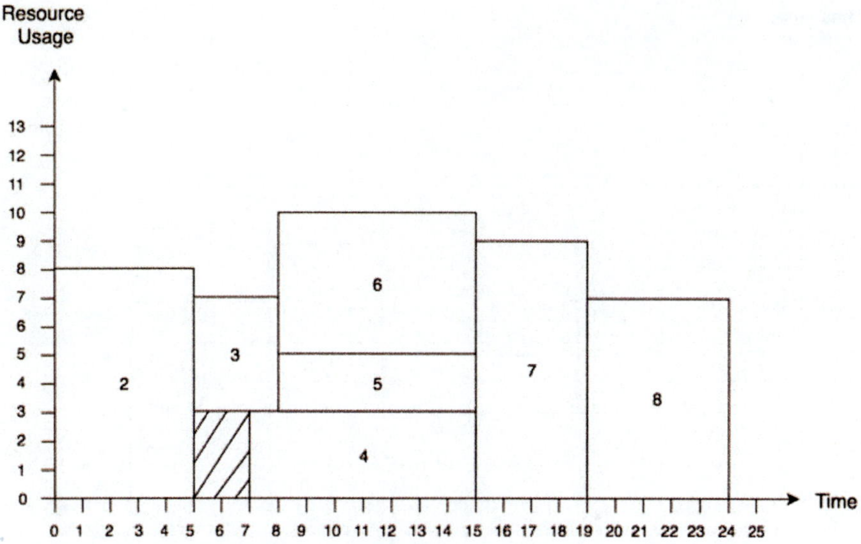

Fig. 8.4 Improved schedule after the first loop of the second iteration

8.2.2 Other Solution Approaches

Let us first briefly cover a sample of exact solution procedures. The first exact solution approach was introduced by Petrovic (1969). It was based on dynamic programming with the objective of minimizing the total squared utilization cost. Easa (1989) dealt with the case of minimizing the absolute deviations between the resource requirements and a uniform resource level for a single resource using an integer linear optimization model. His work was extended by Rieck et al. (2012) in order to deal with general temporal constraints and a set of renewable resources. An integer programming-based method with the objective of minimizing the sum of absolute deviations of the resource utilization from the desired resource profile was proposed by Younis and Saad (1996). Mattila and Abraham (1998) presented an integer linear programming model minimizing the absolute deviation of daily resource usage from an average resource usage rate. A branch and bound procedure with general temporal constraints given by minimum and maximum time lags between activities was introduced by Neumann and Zimmermann (2000). Minimization of several different objective functions was considered such as the deviation from a resource level, fluctuations in consecutive time periods, and costs due to resource level fluctuations. They also introduced heuristic solution procedures. A tree-based enumeration scheme was proposed by Gather et al. (2011). They considered the minimization of two objective functions: The total squared utilization cost and the total positive deviation from the desired resource utilizations.

Due to the computational complexity of the resource leveling problem, the emphasis has been on the development of heuristic solution procedures. One of the earlier studies on resource leveling is by Burgess and Killebrew (1962), which we have presented in Sect. 8.2 in detail. A combination of this methodology together

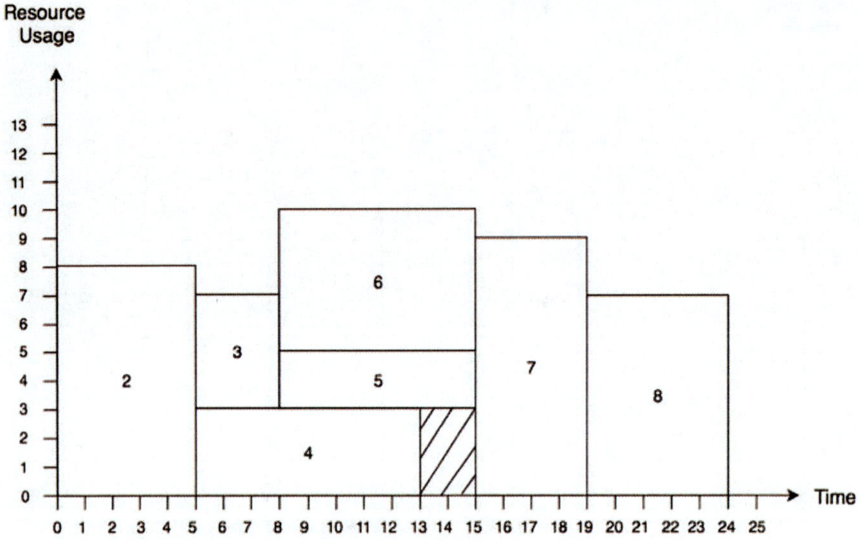

Fig. 8.5 Improved schedule after the second loop of second iteration

with PERT was presented by Woodworth and Willy (1975) particularly aiming at multiple project and multiple resource problems. Harris (1978) proposed a method based on minimizing the moment of resource histogram called the minimum moment method. Harris (1990) later presented a variation of this method called the packing method (PACK). The case of resource leveling problems with minimum and maximum time lags between the activities was treated first by Brinkmann and Neumann (1996) who presented two heuristic procedures. Later Neumann and Zimmermann (1999) introduced polynomial heuristic procedures for different types of resource leveling problems.

Metaheuristics, particularly genetic algorithms (GA)s, have also been proposed for solving resource leveling problems. GA-based methods have been introduced by several researchers. Leu and Yang (1999) presented a multi-criteria approach. Leu et al. (2000) suggested a decision-support system application. Hegazy (1999) considered resource allocation and leveling simultaneously. Senouchi and Eldin (2004) devised a solution methodology for considering resource leveling and resource constrained scheduling simultaneously. Li and Demeulemeester (2016) presented a GA for the robust resource leveling problem. Li et al. (2018) studied the resource leveling problem with generalized precedence relations.

Savin et al. (1996, 1997) employed artificial neural networks to solve resource leveling problems. Christodoulou et al. (2010) revisit the minimum moment method and restate it as an entropy maximization problem. Geng et al. (2011) presented a directional ant colony optimization approach with the objective of minimizing the squared deviation of the daily resource consumption intensity from the mean resource consumption strength. A further metaheuristic approach was proposed by Ranjbar (2013), which is a path-relinking metaheuristic algorithm generating paths between as well as beyond related solutions in the neighborhood space defined.

Li et al. (2015) studied the stochastic resource leveling problem subject to activity duration uncertainty. The objective considered was the minimization of the expected sum of the weighted coefficient of variation of the resource usage. Scheduling policies were generated using two heuristics. The first heuristic was designed as a modified version of the Burgess and Killebrew algorithm to obtain a scheduling policy by solving the deterministic equivalent of the stochastic resource leveling problem. For the second heuristic, on the other hand, a simulation-based tabu search procedure was proposed, which directly works with the stochastic resource leveling problem.

8.2.3 The Total Adjustment Cost Problem

In a specific resource leveling problem, the resource utilization cost is formulated as the sum of the costs of increase or decrease in the level of resource levels. This problem is called the *total adjustment cost problem*. We refer to Kreter et al. (2014) for an extensive review of applications, models, and solution algorithms of this problem.

We note that the unit costs of increases and decreases can be different. For example, the hiring and firing costs of personnel differ significantly. Next, we formulate this interesting problem.

Let $\Delta^+ u_{rt} \geq 0$ and $\Delta^- u_{rt} \geq 0$ denote the magnitude of positive jumps (increase in resource level) and negative jumps (decrease in resource level) of resource type r in period t, respectively, and $c_r^+ \geq 0$ and $c_r^- \geq 0$ the unit costs of positive and negative changes in the utilization of resource type r. The total adjustment cost C is then given by:

$$C = \sum_r \sum_t \left(c_r^+ \, \Delta^+ u_{rt} + c_r^- \, \Delta^- u_{rt} \right) \tag{8.12}$$

Note that Expression (8.12) does not take into account the time value of money. For relatively short projects this might be a valid assumption but for longer projects, time value should be integrated into the solution procedure (see Exercise 8.9).

We start the project with a number of resources - itself a positive jump from zero. We end the project by a negative jump reducing the final number of resources to zero. Thus, starting from zero and ending at zero resource level, the sum of all the positive jumps equals the sum of all the negative jumps. Based on this observation, the total adjustment cost (8.12) can be written as:

$$C = \sum_r \sum_t \left(c_r^+ + c_r^- \right) \Delta^+ u_{rt} = \sum_r \sum_t \overline{c}_r \, \Delta^+ u_{rt} \tag{8.13}$$

where $\overline{c}_r = \left(c_r^+ + c_r^- \right)$.

Note that $\Delta^+ u_{rt} = \max\,(0, u_{r,\,t+1} - u_{rt})$. Hence, a mathematical programming formulation for the total cost adjustment problem can be written as follows:

$$\min \sum_{r=1}^{|R|} \sum_{t=0}^{PD} \overline{c}_r \{ \max (0, u_{r,t+1} - u_{rt}) \} \tag{8.14}$$

subject to

$$\sum_{t=EFT_j}^{LFT_j} x_{jt} = 1 \quad j = 1, \ldots J \tag{8.15}$$

$$\sum_{t=EFT_i}^{LFT_i} t x_{it} \le \sum_{t=EFT_j}^{LFT_j} (t - d_j) x_{jt} \quad j = 2, \ldots J \quad i \epsilon P_j \tag{8.16}$$

$$\sum_{j=1}^{J} k_{rj} \sum_{\tau=t}^{t+d_j-1} x_{jt} = u_{rt} \qquad r \in R, \quad t = 1, \ldots, PD \tag{8.17}$$

$$u_{rt} \ge 0 \qquad r \in R, \quad t = 1, \ldots, PD \tag{8.18}$$

$$u_{r,0} = u_{r,PD+1} = 0 \qquad r \in R \tag{8.19}$$

$$x_{jt} = \begin{cases} 1, & \text{if activity } j \text{ finishes at period } t \\ 0, & \text{otherwise} \end{cases} \qquad j = 1, \ldots, J, \quad t = 1, \ldots, PD \tag{8.20}$$

(8.14) is the objective considering only the positive deviations as expressed in (8.13). The constraint set (8.15) is the assignment constraint ensuring the completion of all activities. (8.16) represents the immediate precedence relations among the activities. The constraint set (8.17) defines the decision variable u_{rt} as the total usage of resource r in period t. Constraint sets (8.18) through (8.20) are the non-negativity constraints for the decision variables u_{rt} and x_{jt}.

The function $\overline{c}_r\{ \max (0, u_{r,t+1} - u_{rt})\}$ is convex and since the sum of convex functions is convex, it follows that the objective function is also convex. It can be linearized by defining a new non-negative decision variable, v_{rt}, $r = 1, \ldots, |R|$; $t = 0, \ldots, (PD - 1)$ as follows:

$$\min \sum_{r=1}^{|R|} \sum_{t=0}^{PD-1} \overline{c}_r v_{rt} \tag{8.21}$$

subject to

$$v_{rt} \ge u_{r,t+1} - u_{r,t} \qquad r \in R, \ t = 0, \ldots, PD - 1 \tag{8.22}$$

$$v_{rt} \ge 0 \qquad r \in R, \ t = 0, \ldots, PD - 1 \tag{8.23}$$

(8.15) through (8.20).

Example 8.2 Let us consider the resource leveling problem stated in Example 8.1 as a total cost adjustment problem with unit cost components $c^{+}= 10$ and $c^{-} = 3$, respectively.

Solution

We solve the linearized version of the total cost adjustment model using CPLEX. Since there is only one resource, the formulation is simplified by dropping the subscript r indicating the resource type. The optimal schedule obtained is the same as that displayed in Fig. 8.5 with an objective function value of 169. The positive increase in resource utilization occurs in periods $t = 0$, $v_0 = 8$, $t = 8$, $v_8 = 3$ and $t = 20$, $v_{20} = 2$. Thus, the optimal objective function value is $13 \times (8 + 3 + 2) = 169$.

From the managerial point of view, it is interesting to note that the same schedule would have been obtained if the values of c^{+} and c^{-} were reversed. This observation follows from the result that the sums of positive and negative jumps cancel each other, and time value of money is ignored.

In the remainder of this chapter we will deal with the resource management problems introduced in Sect. 8.1.

8.3 Resource Availability Cost Problem

The *resource availability cost problem* (RACP) is similar to the resource leveling problem as it also addresses resource requirements throughout the project execution. However, unlike the resource leveling problem, which aims to minimize the variation in resource utilization over time; in the RACP the maximum resource demand for each resource type r is considered, and the total resource cost is minimized, which is expressed as a non-decreasing resource cost function $C_r(u_r)$ of the resource demand u_r. u_r is the amount of resource of type r acquired and is determined as the peak level of the resource profile for resource type r.

The motivation behind investigating the maximum demand has its roots in the relation between peak resource demand and resource cost. Consider resources that are used only for a limited duration. It is a common practice in projects to lease or subcontract these resources. In that case, the number of resources to be leased and hence, the resource cost is determined by the peak level in the resource loading profiles. It is assumed that once a resource is acquired, then it is assigned to the project for the whole project duration. This dependence of resource cost on peak resource demand is addressed in RACP, which is called also the *resource investment problem* in the literature.

Like RCPSP and the resource leveling problem, RACP is an *NP*-hard problem (shown by Möhring, 1984). Real size instances could only be solved approximately. The RACP can be used to optimize the resource allocation in projects carried out in various industries. The readers are referred to Möhring (1984) for an application to bridge construction.

RACP seeks a feasible project schedule that does not exceed the deadline and minimizes the total resource cost. The RACP is defined below on AON project networks with a one start and one end activity. In the following, FT_i denotes the finish time of activity i, where $i = 1\ldots, n$, S_t the set of activities active in period t, H the set of precedence relations, and r_{ik} the consumption by activity i of resource type r, $r = 1\ldots, |R|$. Like in resource leveling problem, the project duration is not allowed to exceed a prespecified project deadline PD.

We present below a mathematical programming formulation for the RACP (Demeulemeester 1995):

$$\min \sum_{r \in R} C_r(u_r) \tag{8.24}$$

subject to

$$FT_j - FT_i \geq d_j \qquad \text{for all } (i,j) \in H \tag{8.25}$$

$$FT_1 = 0 \tag{8.26}$$

$$FT_n \leq PD \tag{8.27}$$

$$\sum_{i \in S_t} k_{ir} \leq u_r \qquad \textit{for } r \in R \textit{ and } t = 1, .., PD \tag{8.28}$$

The RACP has basically the same constraint set as the resource leveling problem differing in the last constraint set. The *rhs*, u_{rt} in the resource leveling problem varies over time t, whereas the *rhs* of the RACP, u_r, is constant over the project duration.

To solve this problem, Demeulemeester (1995) proposed a methodology based on the iterative solution of decision problems with different resource availabilities. A search strategy is proposed leading to an optimal solution. The decision problem requiring a yes or no answer is stated as follows: "Given the project data, the resource availabilities and a deadline for the project, does there exist a solution with a project length that does not exceed the deadline of the project and for which none of the precedence or resource constraints is violated?" The solution of the decision problem is based on the depth-first branch-and-bound algorithm introduced by Demeulemeester (1992) and Demeulemeester and Herroelen (1992). The solution procedure is applied to obtain exact solutions for some small size problems.

Drexl and Kimms (2001) developed lower bounds using Lagrangean relaxation and column generation techniques. Their procedures generate feasible solutions; hence, upper bounds for the problem leading to two optimization guided heuristics. Yamashita et al. (2007) formulated RACP with uncertain activity durations within a robust optimization framework. They modeled uncertainty through scenarios. Rodrigues and Yamashita (2010) improved significantly Demeulemeester's algorithm by limiting the search space. Zhu et al. (2017) divided the problem into a sequencing sub-problem and a resource decision sub-problem and applied an effective multi-start iterative search heuristic to solve it. Further exact and heuristic solution methods can be found in Rodrigues and Yamashita (2015) and Van Peteghem and Vanhoucke (2015), respectively.

8.4 Resource Management Policies

As mentioned in Sect. 8.1, a *resource sharing policy* (RSP) refers to multi-project settings. Renewable resources are taken from their respective pools for use in performing an activity and once the activity is completed, they are returned to their pools for possible usage by another activity. Usually, the assumption is that the resources do not break down or deteriorate. The resource sharing policy has been the conventional approach employed in limited resource problems. Depending on the decision environment and the nature of the resources some or all resources might be dedicated to individual projects for the whole duration of these projects. Once a resource is dedicated to a project, then it is not allowed to be shared by other projects. Such a policy is called the *resource dedication policy* (RDP) (Beşikci et al. 2012). Dedication of resources would be the case, for example, for geographically distributed or fixed location renewable resources.

A mathematical programming formulation for RDP in a multi-project setting is provided by Beşikci et al. (2012). They consider multi-mode version of RCPSP (MRCPSP) for individual projects. The resources are limited by pre-specified general resource capacities and the problem is to dedicate resources to individual projects such that the general resource capacities are not exceeded and the total weighted tardiness over all projects is minimized. Two different solution procedures are suggested. The first solution approach is a GA employing an improvement move called combinatorial auction for resource dedication, which is based on preferences of projects for resources. Two approaches are proposed for calculating the projects' preferences: One approach is based on linear relaxation and the other one on Lagrangian relaxation. The second solution approach is a Lagrangian relaxation-based heuristic employing sub-gradient optimization.

Beşikci et al. (2019) extended the RDP to the case where a resource is allowed to be transferred to a project not yet started once the last project at which it was employed is completed. This policy is called the *relaxed resource dedication policy* (RRD).

Further resource management policies are provided by Krüger and Scholl (2009) for multi-project resource constrained scheduling problem (RCMPSP), where resources are subject to sequence- and resource-dependent transfer times for transfers within the individual projects as well as for transfers between projects. They define the RCMPSP with transfer times: RCMPSPTT. Since RCMPSPTT is a generalization of RCPSP, it is an *NP*-hard problem. Hence, they propose heuristic solution frameworks combining time and resource scheduling.

Krüger and Scholl (2010) define three different approaches for dealing with resource transfers in an RCMPSP setting: (*i*) The transfer-neglecting approach assumes RSP, where there is no restriction on resource flow between projects and transfer times are assumed to be negligible. (*ii*) In the transfer-reducing approach, with the cost of resource transfers being prohibitive RSP is not allowed and resources are dedicated to projects. A resource is allowed to be transferred from one project to another only when the project it is dedicated to is completed. (*iii*) In the transfer-using approach, on the other hand, resources can be transferred between projects with time requirements and explicit costs. This is said to be the most general approach since it includes all the previous ones. In their paper, Krüger and Scholl focus on the transfer-using approach and provide a mixed-integer programming model for each version of the problem. They recommend the use of specialized solution procedures including heuristic approaches rather than standard solvers due to excessive computational burden.

8.5 The Resource Portfolio Management Problem

A resource management problem of interest is the *resource portfolio management* problem. The resource portfolio management problem addresses the issue of allocating a resource budget among renewable and non-renewable resources in a *multi-project environment* such that an objective or objectives are satisfied. The resource portfolio management problem is one step higher in the decision hierarchy than RCPSP and extensions. When dealing with RCPSP and extensions, resource levels are already allocated to the problem under consideration and are expressed in terms of upper bounds.

The resource portfolio management problem under RDP is investigated by Beşikci et al. (2015) with the objective function being the minimization of the total weighted tardiness over all projects. The resource portfolio management problem is indeed a higher-level problem than RCPSP and extensions in the sense that, given a budget, the problem involves the determination of the general resource capacities, which are then to be allocated to individual projects to be employed for resource constrained scheduling of the activities. In other words, in addition to resource dedication space, the whole solution space of RPP has another dimension of general resource capacity. This implies that a search algorithm needs to explore: (*i*) The resource capacities space, i.e., different general resource capacity instances, and (*ii*) the resource dedication space, i.e., different resource dedication instances constrained by a general resource capacity instance. Furthermore, for each general resource capacity and corresponding resource dedication, an MRCPSP is to be set up and solved for each project. For searching through this search space, a two-phase GA is proposed. For comparative evaluation of the results of the two-phase GA, a monolithic GA is developed, which applies different search algorithms simultaneously. Multi-project problems of 6 projects each with 22 activities and 32 activities are generated as test problems for different combinations of *Network Complexity* (NC) and *Maximum Utilization Factor* (MUF). NC is defined as the number of arcs by the total number of nodes and MUF is defined as the ratio of the no-delay schedule resource requirement and the available resource. On the overall, two-phase GA performed better than the monolithic solution procedure.

As stated earlier, an extension of RDP, namely RRD represents the case where resources are dedicated to individual projects can be transferred to other projects not yet started once the project, they are dedicated to is completed (Beşikci et al. 2019). Besides this relaxation in RDP, the problem setting is the same as the one in the paper by Beşikci et al. (2015). For solving this problem, they propose a modified Branch and Cut (B&C) procedure based on CPLEX. The B&C procedure of CPLEX is modified through different branching strategies, valid inequalities, and feasible solution approaches. A test problem set consisting of 6 projects with 22 activities and 32 activities and with MUF values 1.4–1.5–1.6 is generated. The solution procedure suggested performs better than the unmodified CPLEX for all test problems and MUF values on all four performance measures: Average run time, average weighted tardiness, average gap measured by the average distance between the objective function value and lower bound, number of feasible solutions.

8.6 Conclusions, Recent Developments, and Some Future Research Directions

In this chapter, we studied resource leveling and cost availability problems and introduced some resource management policies and problems. We presented several alternative mathematical models and algorithms to schedule the projects. We listed the assumptions and discussed the application areas of these models and algorithms.

Resource leveling has an important role in both industry and service projects because fluctuations in resource requirements are costly. The majority of the studies in the literature assumed deterministic project settings. Differently, Li and Demeulemeester (2016) considered the impacts of uncertain activity durations on resource leveling and formulated the robust leveling problem. Different than the deterministic case, they minimize the expected positive deviation of resource utilizations and activity starting times. They developed a genetic algorithm for solving the robust problem.

To construct the robust RACP, Yamashita et al. (2007) developed two alternative models: The first model minimizes the sum of the mean and variance of the costs, whereas the second one minimizes the maximum regret function. To model uncertainty, they adopted a scenario-based approach, where a scenario represents a realization of the duration of the activities. For solution, they used scatter search heuristic.

There is a need for further studies in robust and reactive versions of resource leveling and cost availability problems. The existing literature is not rich in this regard. New scheduling algorithms are needed to minimize effect of disruptions on project performance. These algorithms will determine how to develop the baselines and when to react and how to react in uncertain environments.

Another interesting research direction is examining the effect of resource leveling on schedule stability and robustness. Shariatmadari and Nahavandib (2020) have recently developed several surrogate measures to evaluate schedule robustness. Some of these measures integrate the variability of resource usages. New measures could be developed and tested in different project settings. Testing with real data would be an interesting contribution.

A future research direction regarding resource management policies is testing new policies such as allowing transfers between projects with a time and/or value cost. Testing different objective functions such as cost/revenue related ones can also lead to interesting results supporting practical applications.

8.7 Case Study: Conducting Annual Audit for Celik Company

Celik Co. is a durable household goods manufacturer registered company with the tax authorities and as per the government regulations, Celik Co. must submit an annual audit report conducted by reputable third-party auditors. In this regard, Celik Co. has hired ECOVIS Audit Company to conduct a complete internal audit for the fiscal year 2016–2017. As per government regulations, the audit report must be submitted no later than December 31, 2017, i.e., the last day of the current fiscal year.

After the initial meeting with the legal department of Celik Co., auditors from ECOVIS would start the planning of the audit based on the requirements communicated to them by the legal department of Celik Co. The planning phase is estimated to take in 3 days. ECOVIS appointed Ms. Gizem Ermis as the lead auditor. She will hold a preliminary meeting with the representatives from Celik Co to communicate the final audit plan as well as formalize the remaining modalities before actual audit can start.

As per the agreed upon plan, the initial activities of the auditors include a review of internal control procedures, compliance tests, year-end procedures, and general audit procedures. They will observe the inventory and collect all documents related to inventory pricing. Afterwards, they will start detailed individual audits as listed below, each corresponding to a work package.

WP1. Audit cash
WP2. Audit receivables
WP3. Audit other current assets
WP4. Audit fix assets
WP5. Audit liabilities
WP6. Audit capital stock and revenues and expenses
WP7. Audit sales
WP8. Audit cost of goods sold
WP9. Audit other revenues and expenses.

As this time of the year is the busiest for ECOVIS due to the closing of the fiscal year, efficient utilization of available resources is of critical importance. For this purpose, Ms. Ermis decided to use a project management software to ensure that all project activities are controlled effectively while utilizing all resources in the most efficient manner. There are two types of resources designated as staff members and senior staff members. She tabulated all job definitions, their precedence relations, durations, and resource requirements as in Table 8.2. Saturday and Sunday are not regular working days. Daily working hours are 8:00 to 18:00 with a lunch break during 12:00–13:00. The project will start on October 16, 2017, at 8:00 hrs.

The estimated finish time of the project is December 13, 2017, and hence well within the specified deadline. To analyze resource usage across the project duration, Ms. Ermis constructed resource profiles for both staff members (Fig. 8.6) and senior staff members (Fig. 8.7). As per resource profiles, this project would require 12 staff members and 8 senior staff members. However, allocating these many resources to this project is not feasible due to other ongoing projects as well as resource requirements for projects already planned. Therefore, Ms. Ermis was asked to revise resource requirements accordingly.

Table 8.2 List of tasks, precedence, durations, and resources

Job No	Job	Predecessor	Duration (hrs)	Resources	
				Staff	Senior
1	Planning 2017 audit	–	21	2	1
2	Preliminary discussions with the company	1	2	0	3
3	Review internal control	2	8	2	1
4	Compliance tests	3	5	0	1
5	Yearend procedures	4	13	2	2
6	General audit procedures	9	15	1	0
7	Audit cash	9	20	4	2
8	Audit receivables	7	10	2	1
9	Observation of inventory	5	26	2	2
10	Inventory pricing	9	151	5	4
11	Audit other current assets	6	11	3	2
12	Audit fix assets	11	21	3	1
13	Audit liabilities	10	2	4	4
14	Audit capital stock and revenues and expenses	11	1	1	1
15	Audit sales	8	6	1	2
16	Audit cost of goods sold	10	25	5	1
17	Audit other revenues and expenses	15, 16	9	1	3
18	Lawyer's letter	27,32	2	2	2
19	Management's letter	12, 14	4	1	3
20	Subsequent review	12, 14	19	2	1
21	Prepare financial statements	18, 19, 20	15	0	2
22	Prepare tax returns	18, 19, 20	12	3	1
23	Partner / manager review	21,22	6	2	1
24	Mail confirmations	13	12	2	0
25	Test pension plan	13	4	3	1
26	Vouch selected liabilities	13	61	1	1
27	Test accruals and amortization	26	4	2	1
28	Process confirmations	24	40	2	0
29	Reconcile interest expense to debt	25,26	8	3	1
30	Verify debt restriction compliance	29	7	2	0
31	Investigate debit balances	28	5	3	2
32	Review subsequent payments	30,31	11	4	1

She conducted resource leveling within the available free slack to ensure that the project finish date does not exceed December 13, 2017. As per new resource profiles generated after resource leveling (Fig. 8.8 and 8.9), this project can be completed with 9 staff members and 5 senior staff members. However, top management wanted to reduce these requirements further and therefore asked Ms. Ermis to negotiate the finish date for this project with Celik Co. so that resource requirements might be further reduced.

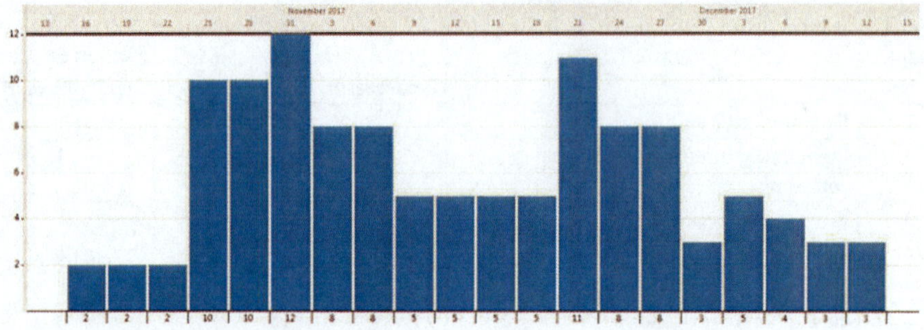

Fig. 8.6 Resource profile for staff members

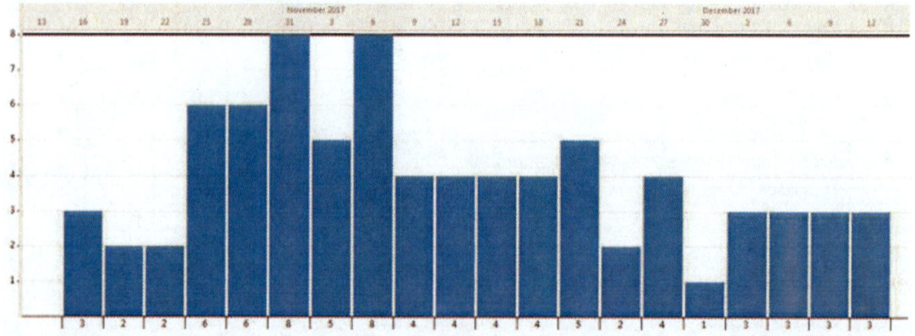

Fig. 8.7 Resource profile for senior staff members

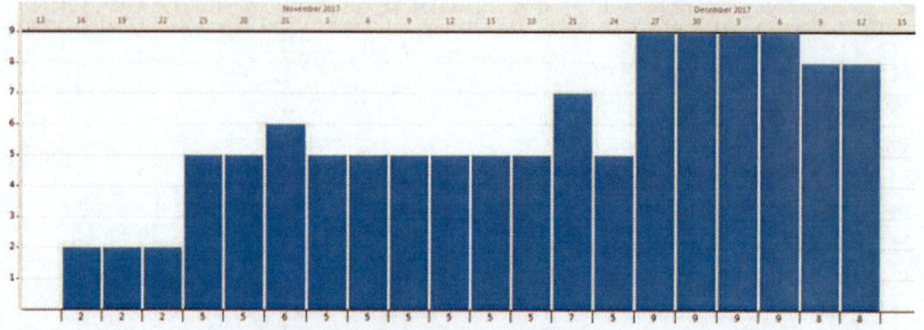

Fig. 8.8 Resource profile for staff members after resource leveling

Ms. Ermis constructed various scenarios with different resource requirements and finish dates. Celik Co. wants this audit to be finished no later than December 22, 2017, so that it has sufficient time to review and submit it to the authorities. Based on various scenarios, Ms. Ermis proposed December 20, 2017 as the finish date for this project, to which Celik Co. agreed. As per the new plan (Fig. 8.10, 8.11, and 8.12), project can be completed with seven staff members and four senior staff

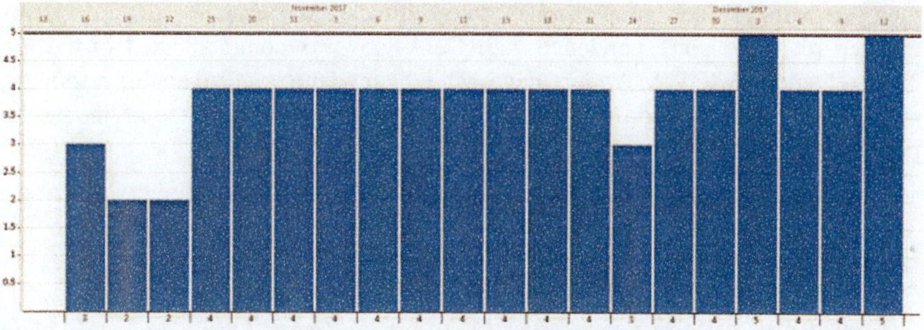

Fig. 8.9 Resource profile for senior staff members after resource leveling

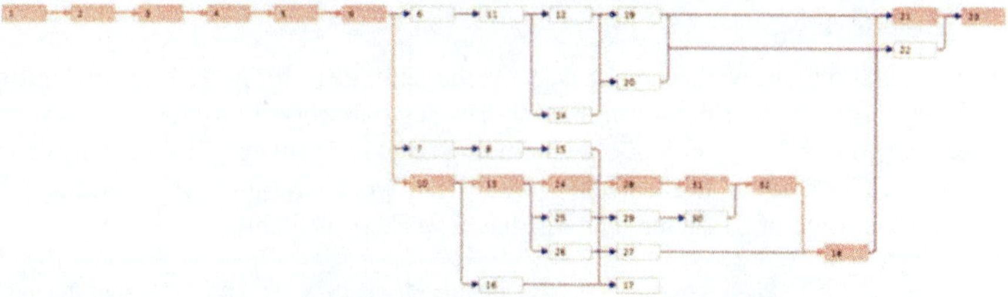

Fig. 8.10 Project network diagram and the critical path

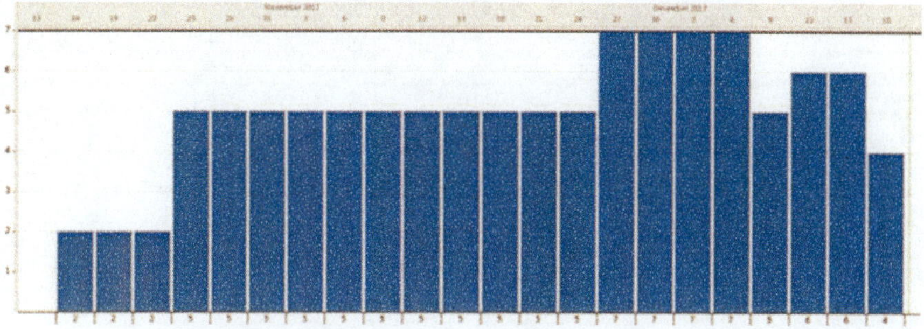

Fig. 8.11 Final resource profile for staff members

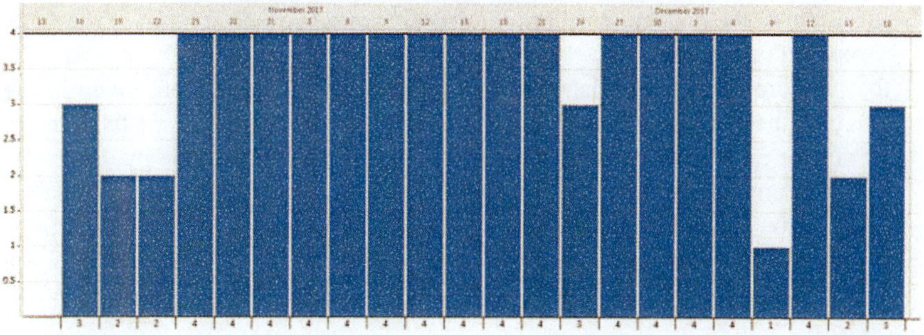

Fig. 8.12 Final resource profile for senior staff members

members with the new project finish date estimated as December 20, 2017. This updated plan and correspondence with Celik Co. was communicated to the top management of ECOVIS. After reviewing all the modalities and available resources, the updated plan was approved and Ms. Gizem Ermis was instructed to initiate the project.

Exercises

1. Consider the performance measure $C_k\ (u_{kt}) = c_k \cdot$ (max $(0, u_{kt} - u_{k,t-1})$. Explain its relationship to the Resource Availability Cost Problem (RACP).
2. Write the set of constraints in expression (8.11) in explicit form without making use of the set S_t.
3. The duration and work content data for the activities of a project consisting of 8 activities given in the Table below. The regular work hours for the workers employed is 8 hours per workday including breaks. The wage for a regular work hour is \$28/hour. If needed, the workers can work overtime for an additional overtime wage of \$14/hour. The PM has 5 workers available.

Activities	Predecessors	Duration (Days)	Work Content (hours)
A	–	5	27
B	–	7	9
C	–	3	25
D	C	4	19
E	A	6	9
F	A	10	26
G	B, E	6	9
H	D	4	31

   (a) What should be the optimum schedule for the project so as to minimize the overtime cost? Use any solver available
   (b) Is the optimal solution feasible if the regulations allow each worker to do overtime at most 2 hours per workday?
   (c) In case the optimal solution violates the regulations, how would you proceed?
4. (a) Rewrite the mathematical programming formulation (8.7) – (8.11) for AON project networks.
   (b) The project network data and the associated CPM results are given in the table below. There is only one renewable resource with the relative weight factor $c_1 = 1$. Obtain the optimal resource profile solving the mathematical programming formulation you obtained in part (a) using any solver available. Do you have alternative optima?

Activity	Immediate Predecessor	Duration	Resource Requirement
A1	–	2	3
A2	A1	2	2
A3	A1	3	5
A4	A3	2	6
A5	A2, A4	4	4
A6	A2, A4	3	3
A7	A5	1	4

(c) Solve the same problem penalizing this time the fluctuations in both directions between two consecutive periods. Compare with the solution in part (b).

5. In the following project environment, it is desired to minimize the hiring and firing costs of workers. The hiring cost is given as $c_H = 3$ and the firing cost as $c_F = 9$.

Activity	Immediate Predecessor	Duration (Days)	Worker Req'ment
A	–	2	3
B	–	3	3
C	A	3	2
D	A	3	4
E	B	4	4
F	D	4	3
G	C	2	1

Obtain the optimal schedule and the resulting total hiring and firing cost.

6. Obtain the optimal solution for the following RACP using any solver available.

Activity	Immediate Predecessor	Duration	Resource Req'ment A	Resource Req'ment B
A1	–	2	3	2
A2	A1	2	2	5
A3	A1	3	5	2
A4	A3	2	6	2
A5	A2, A4	4	4	6
A6	A2, A4	3	3	5
A7	A5	1	4	3

The cost coefficients for the resources are given as $c_A = 4$ and $c_B = 8$.

7. Johnson's Contractors has taken on an office remodeling project. The activities and the precedence relationships for the project are given in the table below, along with the number of workers required for each activity

Activities	Predecessor(s)	Duration	# of Workers Needed
A	–	5	6
B	–	3	5
C	A	6	6
D	B	3	5
E	B	4	2
F	C, D	8	7
G	E	4	2

Assume that the company needs to hire all the workers at the start of the project, for the whole duration of the project (i.e., they cannot hire or fire workers during the execution of the project). Using Burgess and Killebrew resource leveling algorithm find out the schedule of activities that will require the least number of workers without increasing the project duration. Draw the resulting resource profile.

8. Apply resource leveling to the following project network. On each arc, the arc designation is given together with the activity duration and resource requirements for resources A and B, all in parenthesis and in that order. The resources are of renewable type and the scheduling is non-preemptive.

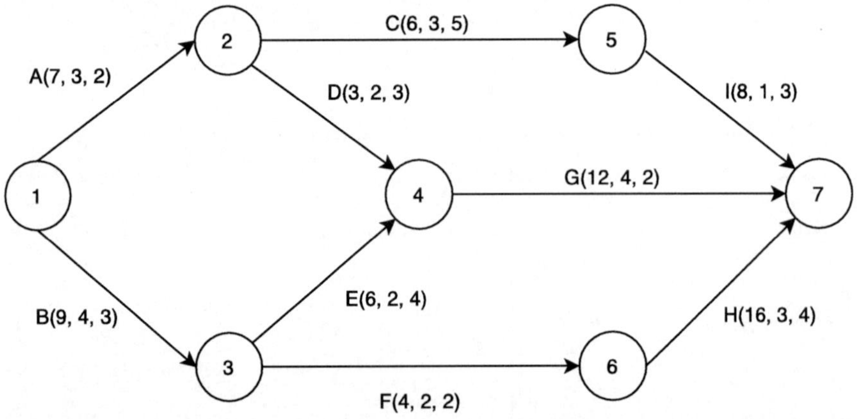

9. Consider the following project network data with the durations, precedence relationships, and resource requirements for each activity.

Activity	A	B	C	D	E	F	G	H
Immediate predecessor	–	–	A	A	B	C	D, E	B
Duration	3	5	6	4	5	2	6	3
Resource X	5	2	4	3	5	6	7	2
Resource Y	3	1	0	3	1	4	3	6

(a) Draw resource profiles for the early and late schedules and evaluate them using the sum of squares performance measure.
(b) Implement the Burgess and Killebrew resource leveling algorithm to obtain an initial solution.
(c) Search around the initial solution for further improvements over the solution obtained in part (a).

10. Consider the following project network with the mean duration of each activity indicated on the corresponding arc.

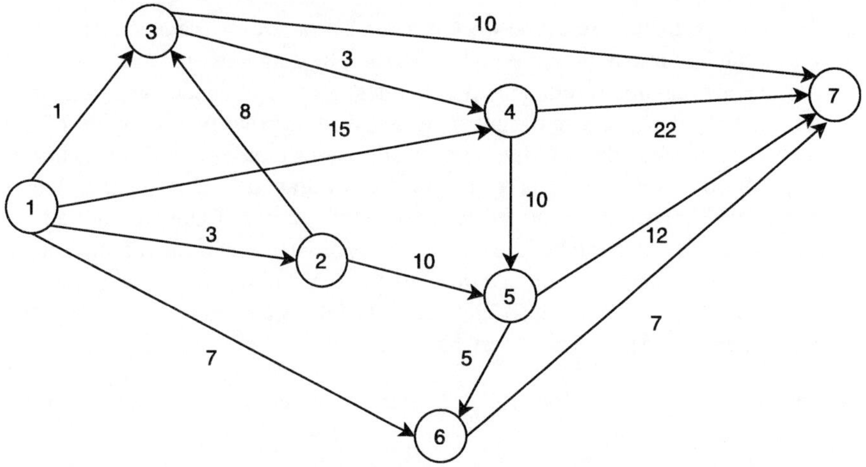

Activity	No of Men	Activity	No of Men
(1,2)	1	(3,7)	9
(1,3)	2	(4,5)	8
(1,4)	5	(4,7)	7
(1,6)	3	(5,6)	2
(2,3)	1	(5,7)	5
(2,5)	4	(6,7)	3
(3,4)	10		

Find the early and late scheduling manpower requirements. Indicate the minimum number of men required at any time and distribute the noncritical activity manpower requirements over these in such a way that the fluctuations for manpower are reduced.

11. In a labor-intensive project, each activity has multiple modes with two types of resources: low skill labor and high skill labor. Scheduling has been performed and a schedule is obtained. Keeping the mode assignments and the project duration fixed the management wants to reduce the cost further by rearranging the starting times of the activities. The costs that will make a difference are the

hiring and firing costs of the respective labor types. Suggest a performance measure based on which the management will search for improvement in cost.

Skill Level	Hiring Cost	Firing Cost
High	4	6
Low	1	2

12. When applying the Burgess and Killebrew algorithm, what bias does first adding the resource requirements of all the resources and then squaring the sum and using this value as the performance measure on the solution?
13. Data for a particular project network is given below. Due to managerial reasons, crews can be formed only of size 3–4-5 men. The durations are in days and any non-integer duration is rounded up to an integer (i.e., 2.3 rounded up to 3). Once a crew is formed, its size cannot be changed throughout the execution of the activity. Hence, as the activity is executed, one or more of the team members might become idle during part of the activity duration because of the integer crew size. Crews can be formed for each activity anew. There are 10 workers available. Try to determine the shortest possible makespan for the project by employing a heuristic procedure. When scheduling, order activities in non-increasing order of man-day requirement. Break ties considering the lexicographic order of the activity names.

Activity Name	Imm. Pred.	Man-day Requirement to Accomplish Activity
A	–	50
B	–	60
C	A	20
D	B	30
E	B	18
F	C,D	32
G	E	52
H	F	64

References

Beşikci, U, Bilge, Ü, Ulusoy, G (2012) *Resource dedication problem in a multi-project environment.*

Beşikci, U., Bilge, Ü., & Ulusoy, G. (2015). Multi-mode resource constrained multi-project scheduling and resource portfolio problem. *European Journal of Operational Research, 240* (1), 22–31.

Beşikci, U., Bilge, Ü., & Ulusoy, G. (2019). Resource portfolio problem under relaxed resource dedication policy in multi-mode multi-project scheduling. *International Journal of Mathematics in Operational Research, 14*(4), 541–567.

Brinkmann, K., & Neumann, K. (1996). Heuristic procedures for resource-constrained project scheduling with minimal and maximal time lags: The minimum-project duration and resource levelling problem. *Journal of Decision Systems, 5*, 129–156.

Burgess, A.R., Killebrew, J.B. (1962) *Variation in activity level on a cyclic arrow diagram.* Industrial Engineering, March–April: 76–83.

Christodoulou, S. E., Ellinas, G., & Michaelidou-Kamenou, A. (2010). Minimum moment method for resource leveling using entropy maximization. *Journal of Construction Engineering and Management, 136*(5), 518–527.

Demeulemeester, E.L. (1992) *Optimal algorithms for various classes of multiple resource-constrained project scheduling problems.* Ph.D. Thesis, Department of Applied Economics, Katholieke Universiteit Leuven, Belgium.

Demeulemeester, E. (1995). Minimizing resource availability costs in time-limited project networks. *Management Science, 41*(10), 1590–1598.

Demeulemeester, E. L., & Herroelen, W. S. (1992). A branch-and-bound procedure for the multiple resource-constrained project scheduling problem. *Management Science, 38*, 1803–1818.

Drexl, A., & Kimms, A. (2001). Optimization guided lower and upper bounds for the resource investment problem. *Journal of the Operational Research Society, 52*(3), 340–351.

Easa, S. M. (1989). Resource leveling in construction by optimization. *Journal of Construction Engineering Management, 115*, 302–316.

Gather, T., Zimmermann, J., & Bartels, J. H. (2011). Exact methods for the resource levelling problem. *Journal of Scheduling, 14*(6), 557–569.

Geng, J., Weng, L., & Liu, S. (2011). An improved ant colony optimization algorithm for nonlinear resource-leveling problems. *Computers and Mathematics with Applications, 61*, 2300–2305.

Harris, R. (1978). *Precedence and arrow networking techniques for construction.* New York: Wiley.

Harris, R. B. (1990). Packing methods for resource leveling. *Journal of Construction Engineering Management, 116*, 39–43.

Hegazy, T. (1999). Optimization of resource allocation and leveling using genetic algorithms. *Journal of Construction Engineering and Management, 125*(3), 167–175.

Kreter, S., Rieck, J., & Zimmermann, J. (2014). The total adjustment cost problem: Applications, models, and solution algorithms. *Journal of Scheduling, 17*, 145–160.

Krüger, F., & Scholl, A. (2009). A heuristic solution framework for the resource constrained (multi-)project scheduling problem with sequence dependent transfer times. *European Journal of Operations Research, 197*, 492–508.

Krüger, F., & Scholl, A. (2010). Managing and modeling general resource transfers in (multi-) project scheduling. *OR Spectrum, 32*, 369–394.

Leu, S. S., & Yang, C. H. (1999). GA-based multicriteria optimal model for construction scheduling. *Journal of Construction Engineering and Management, 125*(6), 420–427.

Leu, S. S., Yang, C. H., & Huang, J. C. (2000). Resource leveling in construction by genetic algorithm-based optimization and its decision support system application. *Automation in Construction, 10*(1), 27–41.

Li, H., & Demeulemeester, E. (2016). A genetic algorithm for the robust resource leveling problem. *Journal of Scheduling, 19*, 43–60.

Li, H., Xu, Z., & Demeulemeester, E. (2015). Scheduling policies for the stochastic resource leveling problem. *Journal of Construction Engineering and Management, 141*(2). https://doi.org/10.1061/(ASCE)CO.1943-7862.0000936.

Li, H., Xiong, L., Liu, Y., & Li, H. (2018). An effective genetic algorithm for the resource levelling problem with generalised precedence relations. *International Journal of Production Research, 56*(5), 2054–2075.

Mattila, K. G., & Abraham, D. M. (1998). Resource leveling of linear schedules using integer linear programming. *Journal of Construction Engineering and Management, 124*(3), 232–244.

Möhring, R. H. (1984). Minimizing costs of resource requirements in project networks subject to a fixed completion time. *Operations Research, 32*(1), 89–120.

Neumann, K., & Zimmermann, J. (1999). Resource levelling for projects with schedule-dependent time windows. *European Journal of Operational Research, 117*(3), 591–605.

Neumann, K., & Zimmermann, J. (2000). Procedures for resource leveling and net present value problems in project scheduling with general temporal and resource constraints. *European Journal of Operational Research, 127*(2), 425–443.

Neumann, K., Schwidt, C., & Zimmermann, J. (2003). *Project scheduling with time windows and scarce resources*. Berlin: Springer.

Petrovic, R. (1969). On optimization of resource leveling in project plans. In *Project planning by network analysis: Proceedings of the 2nd international congress* (Vol. 268, p. 273).

Ranjbar, M. (2013). A path-relinking metaheuristic for the resource levelling problem. *Journal of the Operational Research Society, 64*(7), 1071–1078.

Rieck, J., Zimmermann, J., & Gather, T. (2012). Mixed-integer linear programming for resource leveling problems. *European Journal of Operational Research, 221*(1), 27–37.

Rodrigues, S. B., & Yamashita, D. S. (2010). An exact algorithm for minimizing resource availability costs in project scheduling. *European Journal of Operational Research, 206*(3), 562–565.

Rodrigues, S. B., & Yamashita, D. S. (2015). Exact methods for the resource availability cost problem. In C. Schwindt & H. Jürgenmann (Eds.), *Handbook on Project Management and scheduling Vol.1* (pp. 319–338). Berlin: Springer.

Savin, D., Alkass, S., & Fazio, P. (1996). Construction resource leveling using neural networks. *Canadian Journal of Civil Engineering, 23*, 917–923.

Savin, D., Alkass, S., & Fazio, P. (1997). A procedure for calculating the weight-matrix of a neural network for resource leveling. *Advances in Engineering Software, 28*, 277–283.

Senouci, A. B., & Eldin, N. N. (2004). Use of genetic algorithms in resource scheduling of construction projects. *Journal of Construction Engineering and Management, 130*(6), 869–877.

Shariatmadari, M., & Nahavandi, N. (2020). A new resource buffer insertion approach for proactive resource investment problem. *Computers & Industrial Engineering, 106582*.

Van Peteghem, V., & Vanhoucke, M. (2015). Heuristic methods for the resource availability cost problem. In C. Schwindt & H. Jürgenmann (Eds.), *Handbook on Project Management and scheduling Vol.1* (pp. 339–359). Berlin: Springer.

Woodworth, B. M., & Willie, C. J. (1975). A heuristic algorithm for resource leveling in multi-project, multi-resource scheduling. *Decision Sciences, 6*(3), 525–540.

Yamashita, D. S., Armentano, V. A., & Laguna, M. (2007). Robust optimization models for project scheduling with resource availability cost. *Journal of Scheduling, 10*(1), 67–76.

Younis, M. A., & Saad, B. (1996). Optimal resource leveling of multi-resource projects. *Computers and Industrial Engineering, 31*(1–2), 1–4.

Zhu, X., Ruiz, R., Li, S., & Li, X. (2017). An effective heuristic for project scheduling with resource availability cost. *European Journal of Operational Research, 257*(3), 746–762.

Project Contract Types and Payment Schedules

9

Learning Objectives

Upon successful completion of this Chapter, the reader will be able to:

1. Distinguish between different types of contracts.
2. Formulate payment scheduling models.
3. Manage change and project control in contracts.
4. Understand the risk issues in contracts.
5. Understand the bidding issues in contracts,
6. Formulate mathematical programming models to support negotiation in project management.

9.1 Introduction

Projects are carried out to deliver specific outcomes, services or products. The delivery of these outcomes is documented through formal written clauses, which constitute contracts. Formally, a *project contract* is a legally binding document written with the aim of compensating the contractor for the work she agrees to accomplish for the client. It is prepared to formally define the relationships and obligations between the client and the contractor. The terms of the contract are agreed upon and formally approved, by both the client and the contractor organizations. Ideally, the purpose of a project contract is to create a cooperative project environment in which all parties are motivated to achieve common objectives and have their goals aligned (Turner and Simister 2001).

A project contract specifies the scope of the project, a baseline project plan with the due date and other important dates, the cost of the project to the client, and a payment schedule. However, there are further items that must be clearly stated in the project contract to prevent disagreements, disputes, litigations, and claims between the client and the contractor that might lead to project delays, cost increases, and even premature termination of the project.

G. Ulusoy, Ö. Hazır, *An Introduction to Project Modeling and Planning*, Springer Texts in Business and Economics, https://doi.org/10.1007/978-3-030-61423-2_9

The means and frequency of communication between the parties should be decided upon prior to the initiation of the project, preferably as part of the contract. In addition to communication issues, quality management and risk management are also topics of importance when writing a project contract. The quality management clauses define the quality specifications and means of quality assurance to be employed throughout the project implementation. How the identified risks will be allocated between the client and the contractor, and how unidentified risks will be managed once they occur are topics to be treated under the title of risk management.

In this chapter, we will first review different types of contracts, different payment scheduling models, and bidding. But first, we need to define a number of terms.

The *client* (*owner*) is the person, firm, or authority that is the legal owner of the project, requires the project to be completed as specified in the contract, and benefits from its completion. The client may establish a project control team to manage the progress, scope, quality, and timing of the project so as to enforce the associated clauses in the contract.

The *contractor* is the person, firm, or authority responsible for all the work to be accomplished as specified in the contract. The contractor may outsource services from some subcontractors.

A *subcontractor* is a person or firm working on a specific portion of a project such as the electrical installations in a building. The subcontractor is directly responsible to the contractor. The client may specify certain guidelines for the selection of the subcontractor but in general, the client is not involved in the financial and other contractual transactions concerning the subcontractor; all these take place between the contractor and the subcontractor. The contractor is accountable for all the work performed, even if most of the actual work is performed by the subcontractors. In spite of the risks involved, the contractor might prefer to work with subcontractors. In order to reduce the risks arising from working with a subcontractor, the contractor might take several measures. The major ones would the application of a well-designed selection process and appropriate selection criteria (see, e.g., Hartmann et al. 2009), the transfer of risks to the subcontractor through proper contract clauses, subcontractor coordination and management, and preparation of response plans for the possible replacement of a subcontractor or the addition of other subcontractors in case of failure of providing the service specified in the contract.

The *tender* is the application of a contractor to the client for the award of the contract. Either the client invites a contractor to tender for the contract or puts out a contract to tender. The *tenderer* is the person, firm, or authority that puts forward the tender.

The *bill of quantities* is an itemized list of material, parts, and labor needed for accomplishing a specified work package, especially for construction work and repair. It contains information on the item codes, item descriptions, units of measurement, quantities, unit rates for each item, labor, and total cost. The bill of quantities provides a unified framework upon which the contractors can base their tender and serves as a communication tool between the parties involved such as the client and the contractor.

The *quantity surveyor* is an employee of the client or a cost consultant who prepares bills of quantities and cost estimates for units of work to be accomplished throughout the project. Once bills of quantities and cost estimates for unit rates of work are agreed upon by the client and the contractor, the quantity surveyor assesses the work accomplished by the contractor up to a given point in time or within a period of time as agreed upon by the client and the contractor.

Mark-up is the amount added by the contractor to the estimated cost of the project in order to arrive at the price to be paid by or proposed to the client. It is usually expressed as a percentage of the estimated cost. Mark-up is equivalent to profit when overhead is incorporated into the estimated cost. If this is not the case, mark-up includes overhead plus profit. Farid and Boyer (1985) suggest that the mark-up value is expected to be fair and reasonable and should satisfy the required rate of return of the contractor.

Progress payments are payments to be made by the client to the contractor throughout the progress of the project. The *payment schedule* defines the amount and timing of the progress payments.

Retainage is a portion of the payment due held by the client until the successful termination of the project and is usually expressed as a percentage of the payment due.

Bonuses and penalties, if applicable, are incorporated in a contract as a means for recognizing the performance of the contractor as above or below expected levels, respectively. Cost, quality, and schedule items might be subject to bonuses and/or penalties. For example, the contractor might receive a bonus for finishing the project on time or by a specified deadline. A bonus or penalty scheme may be defined as a fixed amount or as a percentage of the payment. These schemes are also referred to as *incentive plans*. If the client declares the incentive plan without consulting the contractor, it is said to be a *unilateral incentive plan*. If, on the other hand, both parties agree upon it as a result of a negotiation, it is a *bilateral incentive plan*. What to incentivize depends on the aspect(s) of the project that is particularly important to the client.

This brief introduction to contracts highlights the importance of assigning the PM as early as possible so that she is involved in the writing of the contract, rather than being committed to decisions she might not agree with. Any changes at the project implementation stage will be relatively costly and may not even be possible. Furthermore, almost any change in the contract requires the involvement of the PM.

9.2 Contract Types

In this section, a classification of contracts based on how the costs are determined will be given. Concerning the classification of contracts, Herroelen et al. (1997) state the following: "Though different practices do exist throughout countries and industries, a crucial distinguishing factor is whether the contract is *fixed price* or not." Contracts not set for a fixed price will be referred to here as *variable price* contracts. The issue of how the associated payments will be made over the project duration is then discussed in Sect. 9.3 on payment schedules.

9.2.1 Fixed Price Contracts

The *lump-sum payment contract* is one of the more commonly used contract types in practice for relatively small projects of short duration where mutual trust exists between the client and the contractor. In lump-sum contracts, the client and the contractor agree upon a fixed amount covering the expenses and profit for the whole project. The client pays this amount to the contractor upon successful termination of the project. Lump-sum payment schemes are also encountered in large projects where the cost of a work package within a larger project, which has been subcontracted through a lump-sum contract, is paid as a lump sum.

The *bills of quantities contract* is the most common fixed price contract type. The quantities required are estimated by the client before the contractors make an offer for the tender. To make an offer a contractor assigns a unit price to each bill of quantity. In this type of contract, the client and the contractor agree on the fixed unit price for each bill of quantity necessary to complete the work. Once the work is completed in total or in part, both parties agree on the actual quantity of work accomplished and the payment to be made to the contractor is then based on the unit price declared in the contract and the actual quantity of work agreed upon.

Price list contract. The client declares a list of fixed unit prices for each bill of quantity in addition to the estimates of the quantities required. The contractors are then asked to compete by reducing the total estimated amount.

In a *schedule of rates contract* (also referred to as measurement or remeasurement contract), the extent of the work cannot be known in advance. Consider, for example, the construction of a water well in your backyard. You may not know for sure at what depth water will be found and hence cannot specify the extent of the work. In such a case, the contract is based on a schedule rate. For example, in the case of the water well, the client and the contractor agree on how much the contractor will charge for each meter of construction. Although not binding, depending on the nature of the work an initial estimate of the quantity of work might also be given.

In the case of fixed price contracts, the re-negotiation of the contract price might be needed if the client asks for changes in the scope of the project after the contract is awarded. Such a re-negotiation would also be needed due to unit price increases or decreases of quoted items in the contract or due to changed conditions resulting in price escalations in general.

9.2.2 Variable Price Contracts

Cost plus contracts. Under a *cost-plus percentage fee contract*, the contractor is paid the direct cost plus a specified percentage of the direct cost as a fee covering the overhead plus profit. A basic disadvantage of such a contract is that the profit of the contractor increases with the cost to the client, leaving no apparent incentive for the contractor to be efficient.

A *cost-plus fixed fee contract* is another type of cost-plus contract. In this case, the fee meant to cover the overhead and profit of the contractor is specified as a fixed amount rather than a percentage of the direct cost. The fixed fee is determined either through negotiations between the client and the contractor or as a result of a tender process.

In *target contracts*, an initial estimate of the project cost is given. Such an estimate could be directly supplied by the client or be the result of the mutual agreement of both parties. This estimate constitutes the target. Both parties agree on a fixed fee as part of the contract. Once the project is completed and the actual cost computed, the deviation from the estimate is determined. If the estimated cost is exceeded, a certain percentage of this deviation is subtracted from the fixed fee initially agreed upon. If, on the other hand, the estimated cost is not exceeded, a certain percentage of this deviation is added to the fixed fee. Hence, the target contracts involve the sharing of the cost or the profit defined as the deviation from the target specified in the contract. The target contract might be interpreted as a bonus or penalty incentive related to the project cost rather than the deadline.

Variable price contracts should be used if the scope of the work is expected to change significantly during the execution of the contract (PMI 2016).

Although the total amount of payments made by the client to the contractor for the deliverables of a project might be expected to be the same under different contract types, this is generally not the case. The total payment can differ significantly depending on the type of contract adopted. Therefore, the choice for the type of contract, the way the clauses are formulated, and the mechanisms put in place for monitoring the progress of tasks and their satisfactory completion are essential for maximizing both parties' benefit from the project.

An essential component of a contract is the *payment schedule* agreed upon by both parties. The payment schedule for a project specifies the amount, location, and timing of progress payments in that project. Different payment schedules can be applied in all types of contracts, whether fixed price or variable price. The payment schedule finally adopted is the result of negotiations between the client and the contractor.

9.2.3 Cost of Managing Contracts

There are certain costs associated with managing contracts. These costs, called *transaction costs*, arise due to planning, formulating, and adapting the contract, and task control and monitoring. A classification of transaction costs can be based on the specification and managing variations of both the product (the deliverables) and the processes (production, managerial, and administrative) associated with the project: (*i*) the cost of specifying the product, (*ii*) the cost of specifying the processes, (*iii*) the cost of managing changes to the specification of the product during project delivery, (*iv*) the cost of managing changes to the specification of the processes during project delivery (Turner and Simister 2001). Turner and Simister based their analysis on the premise that clients and contractors are motivated to achieve common objectives and their goals are aligned. The total transaction cost depends on the contract type. They summarize their conclusions as follows: "Build-only remeasurement (schedule of rates) contracts are used where the uncertainty of both product and process is low. Design and build fixed price contracts are used where the uncertainty of the product is low, but the uncertainty in the process of delivery is high. Fixed price contracts should be used where both are high."

The total cost of the project to the client is the sum of the total payments to the contractor and the transaction costs. Both components of the total cost are affected by the contract type, so the final choice of the contract is determined by the relative magnitudes of the two cost components. Usually, the component associated with the total payments to the contractor is the dominant one leading to the choice.

9.3 Payment Scheduling Models

The manner in which payments are made by the client to the contractor is crucial to the interests of both parties and must be clearly and openly stated in the contract. In addition to reducing the risk of project failure, it is in the interest of the client to offer the contractor a schedule for progress payments, allowing the contractor to undertake the project without unnecessary financial stress. This also reduces the cost to the client since the tenderers do not have to inflate their tender prices to cover financing charges, and also the possibility of having to work with a contractor with insufficient cash to cover expenses. However, a relaxed, extended payment schedule increases the total cost of the project. The client and the contractor need to strike a balance by considering their financial positions and the financial environment in which the project is executed.

Four payment scheduling models are of particular interest in practice: Lump-sum payment, payment at event occurrences, payment at equal time intervals, and progress payment (Ulusoy et al. 2001; Mika et al. 2005). The payment scheduling models will be considered here from the perspective of the contractor. The aim is to maximize the return to the contractor by scheduling the project activities in coordination with the payment schedule.

The *lump-sum payment model* has already been introduced in Sect. 9.2.1, so we shall discuss here only the remaining payment scheduling models.

Under the *payments at event occurrences*, payments are made at predetermined event nodes of the project plan. The amounts and timing of these payments are determined through negotiations between the client and the contractor. The amounts of payments are not necessarily known a priori. A model for determining the amounts of payments would be, for example, to base the payments on the contractor's expenditures since the last payment.

The *payments at equal time intervals model* differs from the payments at event occurrences model in that the client makes a predetermined number H of payments throughout the project by mutual agreement. The first (H-1) of these payments are scheduled at equal time intervals over the duration of the project, and the final payment at project completion. Hence, the time between the last two payments is uncertain.

In the *progress payment model*, the contractor receives payments from the client at regular time intervals until the project is completed. For example, the contractor might receive at the end of each month a payment for the work accomplished during that month multiplied by a mark-up agreed upon by both the client and the contractor. The difference between the payments at equal time intervals model and the progress payment model is that in the latter case, the number of payments and the payment amounts are not known in advance. In all three of these payment schedules, a retainage might be applied by the client.

In terms of project scheduling, the simplest situation is when the client and contractor agree on a lump-sum payment and the project is not subject to any resource constraints. The NPV of the contractor's net return is maximized when the non-critical activities with cost disbursements are delayed as much as possible by absorbing the slacks, and the project is completed at its critical path duration without any delay. On the other hand, as the client pays for all the expenditures as a single payment at the termination of the project, the activities can be scheduled with respect to the contractor's optimal schedule. Hence, for this problem setting, there is no conflict between the client and the contractor regarding project scheduling. On the other hand, when resource constraints are included, no simple algorithm exists for solving the scheduling problem.

The solution procedures for the payment scheduling problems other than the lump-sum payment with no resource constraints are relatively difficult. In some cases, delaying a certain payment might be to the benefit of the contractor. Hence, to determine optimal or near optimal payment schedules, new models and solution techniques are required.

As a final remark, we emphasize that there is no pricing and payment scheduling policy that is preferable to all parties under all possible conditions. Both the client and contractor should negotiate for a fair combination satisfying both parties.

9.4 Change and Control Issues in Contracts

9.4.1 Managing Change Issues in Contracts

While undertaking the projects there can be changes in the scope due to unexpected events or due to the changes in the stakeholders' requirements. In order to protect the interests of both parties, it is good practice to include a clause in the contract delineating how changes in the project will be monitored and how the resulting burden and risk will be shared.

A major change item to be followed closely by both the client and the contractor are escalations in costs during the project implementation. Various factors such as inflation, interest rate, contractual issues and taxes can be the underlying factors for escalation. The consideration of escalation factors becomes especially important in project environments with relatively high inflation rate and also for large and/or long projects. In such cases, care should be given to include clauses in the contract on how to determine and employ escalation factors.

Price indices can be employed for estimating escalations. Two such composite indices are published by the Engineering News-Record (ENR) in the USA covering construction, architecture, and engineering: The Construction Cost Index (CCI) created in 1921 and the Building Cost Index (BCI) created in 1938. They are both based only on four components: Cement, lumber, structural steel, and labor. Such indices are usually issued by related public authorities. If no such index is available, then Consumer Price Indices can be employed. Various quantitative methods have been employed for forecasting escalation factors such as the Box-Jenkins method, regression methods, neural networks, and causal methods. A qualitative method, on the other hand, would be the survey of expectations, which is easily applied and practically of no cost (Touran and Lopez 2006).

9.4.2 Control Issues in Contracts

Control issues in contracts usually take place along three dimensions: (*i*) Quality, (*ii*) schedule, and (*iii*) cost. Both the client and the contractor need to allocate resources to control the project in all these three dimensions. Besides timing and financial issues, depending on the type of contract, the contract will also include specific statements of the client‘s quality standards as well as the means of quality control and quality assurance systems. Any work performed will only be acceptable if it meets the quality standards of the client. Both client and contractor will track the progress of the project according to the project plan for obvious reasons.

In most of the contract types presented in Sect. 9.2, an estimate of the cost incurred during an interval of time such as a month is needed on which to base the payment to the contractor. Both the client and the contractor need to agree on this value. The contractor uses its own records of the amount of work accomplished during a given period and the associated costs it has incurred and reports these to the client. The client must assess the values reported by the contractor through its own means, such as employing a quantity surveyor. The client and contractor must then agree on the amount due. Not only the amount of work accomplished but also its quality needs to be ratified by the client as meeting the client's standards before the payment becomes due. The client might employ a third party for this purpose.

The client also follows the progress of work and warns the contractor if the project is behind the project schedule, asking for expediting if necessary.

9.5 Identification and Allocation of Risk in Contracts

Risks, whether realized during the implementation of the project or not, can constitute a major source of additional cost and deviation from the schedule. Examples are uncertainties associated with the weather conditions, the availability of resources such as worker absenteeism and equipment failure, over or underestimation of work content. All risks associated with these uncertainties should be reflected in the project cost, which affects both the client and the contractor. The identification of these risks would require an extensive study.

Once the risks are identified and assessed in terms of their occurrence probabilities and estimated impact, the client and contractor can apply one or more of the following risk management policies (see, e.g., PMI 2009): (*i*) Risk avoidance; (*ii*) risk mitigation; (*iii*) risk transfer; and (*iv*) risk acceptance. In risk avoidance, the associated activity causing the risk is eliminated. Risk avoidance would be more effective during the contract negotiation since once the contract is signed, risk avoidance would not be an option anymore. An exemption clause included in the contract concerning such risks can be added to avoid them. In risk mitigation, on the other hand, a certain cost is incurred to reduce the probability of occurrence, or the impact of the risk if it should occur, or both. Losing key personnel during the implementation of a project is a real risk, the probability and/or the impact of which might be reduced by offering a bonus for completing the project. Risk transfer implies the partial or total transfer of the risk to a third party through means such as insurance or outsourcing. One has to make sure that the third party chosen is the best capable one to undertake this task. If the risk is accepted with all its consequences, then this is called risk acceptance. For such risks, it is good practice to prepare a set of contingency plans (PMI 2016).

The contract has to be very clear about how the identified and agreed upon risks are allocated between the client and the contractor. An important purpose of the contract is to provide a framework for the parties to establish which one has assumed which risk. The omission or ambiguities in describing the risks involved and their allocation among the parties can lead to disagreements, disputes, litigation, and claims between the parties with undesirable consequences for both.

The following examples of risk allocation will serve to illustrate the range of possibilities. Man- made events such as strikes, wars, or riots and epidemics interrupting work are considered a force majeure (or natural hazards such as extreme weather conditions, earthquakes are considered an act of God) and do not fall on any one of the parties. The associated risk is accepted (risk retention). A fire at the project execution site is a risk to be assumed by the contractor through fire prevention measures to mitigate the risk of fire and/or risk sharing through fire insurance. The client, however, would like to avoid the risk of a project delay as a result of such a fire and the client‘s risk is covered through penalty clause(s) in the contract for late completion of the project.

All risks are a potential source of additional cost. Once a risk is included in the contract, then there is a certain cost premium to be paid by the client or the contractor or shared by both. The risks depend on the project and the project environment, but some risks can be managed differently as a result of mutual trust between the parties. Trust influences almost every aspect of the management, execution, and the cost of any project.

9.6 Bidding

A bid, more properly referred to as a proposal, is a written offer, tendered by the contracting firm to the owner, which stipulates the price for which the contractor agrees to perform the work described in the contract documents. A proposal also contains a promise that, upon its acceptance by the owner, the contractor submitting the proposal will enter into a contract with the owner for the amount of the proposal. Thus, if the owner accepts the contractor's proposal in a timely fashion, the proposal is binding on the contractor (Clough et al. 2015).

As suggested above, either the client invites a contractor to tender for the contract or puts out a contract to tender. The first case is called *negotiated bidding*. The invited contractor makes an offer and the client and the contractor negotiate the price. The second case is called *competitive bidding*. If the client's principal concern is the delivery of the project at the lowest possible price, competitive bidding is used. Within the scope of this chapter we will focus on only competitive bidding. Based on the project requirements provided by the client in its invitation for tender the bidders quote a price, and the bidder quoting the lowest price wins the bid. To win the bid the contractor might be bidding low, which might increase the risk of failure for the contractor to complete the project according to the requirements stated in the contract. When, on the other hand, the contractor proposes a bid with a relatively high mark-up, then this might reduce her chance of winning the contract. Hence, determining the right bid price is a critical and difficult problem for the contractor. In negotiated bidding, although it might not be the lowest price, the price might come out to be more realistic, since it is reached through negotiation between the client and the contractor. The supervision of the competitive bidding process can be in general a lengthy process compared to that of the negotiated bidding.

When the client considers the attributes of the proposal other than the price, such as the quoted project duration, proposed project financing means, or the work experience of the bidder, a multi-attribute selection approach must be used (see Chap. 14 Project and Portfolio Selection).

In the evaluation of the bids, the client should take into account that bids are usually prepared with limited information; the schedule and budget are usually quoted under a high level of uncertainty. As these budget and schedule promises affect the chance of the contractor to win the bid, there might be a tendency to overpromise. To control these overpromises, the contracts might impose considerable amounts of penalties for activity delays and failures to meet the milestones (Zhu et al. 2007).

The client might also impose restrictions on qualifications for bidding. If the project is of a special nature due to its size, or complexity, or security requirements, etc., the client might simply form a short list of selected contractors and invite them to bid. This will also be the case when the client prefers to work with a selected group of trusted contractors who have a proven record of success and dependability.

Bid preparation is an iterative process that includes a series of scenario analyses, what-if analyses, and sensitivity analyses. Once an estimate of the total cost is obtained, it provides the basis for the statement of the bid price together with the probability of winning the bid. The difference between the bid price and the estimated total cost is *the mark-up value* corresponding to the profit. The aim is to determine the bid price which maximizes the expected profit. The following model is provided by Friedman (1956):

$$E(B) = (B - C)\,P(B) \tag{9.1}$$

where B is the bid price, C the estimated total cost, E(B) the expected profit, and P(B) is the probability of winning the bid. The expected profit is obtained by the multiplication of the mark-up value with the probability of winning the bid, which is among other factors a function of the bid price. We observe that the uncertainty around the estimate of the total cost can have a distorting impact on the bid price. This is emphasized by King and Mercer (1985) as: "... any subtle variation in the mark-ups, calculated from the bidding strategies, are outweighed by the large variation in the cost estimate".

Chapman et al. (2000) propose the following *stages for bid preparation*:

"(i) Estimation of project cost components.
(ii) Combining project cost components to obtain C, the estimated total cost of performing the contract.
(iii) Estimation of the probability of winning a bid B, P(B).
(iv) Determination of the bid value B*, which maximizes the expected contribution to profit E(B) using expression (9.1).
(v) Determination of the final bid price, taking into account other qualitative, commercial considerations.
(vi) Iterations through the previous stages to refine the analysis where it is useful to do so."

Bid preparation can be a lengthy and relatively costly process with an appreciable probability of no financial return. Hence, the question becomes: Given an opportunity, should the contractor bid or not? A concept related to this question would *the attractiveness of a bid for the bidder*. According to King and Mercer, "In reality, the contracts vary in their attractiveness to the bidder, depending on such factors as the size of the contract, the bidder's resources available at that time, and the location of the contract, if much of the work must be done away from the bidder's base". The same applies to the competitors as well but with different weights in general. They suggest, "Thus the aim of a bidder's strategy should be to try and win those contracts which are more attractive to itself and less attractive to its competitors".

A quantity called the *bid spread* is of interest both to contractors as well as the client. Bid spread is the difference between the lowest and the second lowest bids expressed in monetary terms or in percentage difference. It represents the amount of money that the lowest bidding contractor could have saved and still be the winner of the contract. If the bid spread is larger than a few percentage points, then the client should question whether this difference is due to the winning contractor's strong operational and technical capabilities or due to a miscalculation or oversight on the part of the contractor, in which case the timely and successful termination of the project at pre-specified budget level might be at risk. Another quantity called the *total bid spread* is defined as the difference between the highest and the lowest bids and is expressed in monetary terms or in percentage difference. A larger than expected total bid spread should lead the client to question the low bidding contractor's offer (Clough et al. 2015).

9.6.1 Estimating the Total Cost of the Project

The basic aim of the bid preparation process is to reach a good estimate of the total cost. This requires the establishment of a tentative project schedule so that the cash flow profile and other financial aspects of the project can be estimated. The WBS of the project, its breakdown into work packages and activities, the ways and means of accomplishing the activities including make-or-buy analysis, estimations of activity durations and resource requirements, their associated estimated costs, support functions, and their associated indirect costs, all need to be determined, and a payment schedule specified. A further data source would be a detailed database of previous contracts the organization was awarded or failed to win.

The cost spent for generating additional and more accurate data for reducing the uncertainty around the estimate of the total cost can be defined as, the *value of information.*

An important concern in estimating the total cost of the project is the uncertainty associated with the activity durations. the uncertainty associated with the activity durations Elmaghraby (2005) maintains that activity duration uncertainty is the combined result of external and internal factors. External factors would be like weather conditions, worker absenteeism, and equipment failure. Internal factor, on the other hand, is a consequence of work content estimates. He warns the contractor against working with averages or any other statistic such as the median or the mode in estimating the expected project cost and basing the bid on them – particularly in projects where the internal factors play a dominant role. He suggests taking into account the uncertainty in the work content of an activity through resource allocation to that activity. Then, the duration of the activity becomes the consequence of the resources allocated to that activity, and not the source of uncertainty. Rather than using the expected values for resource allocation, Elmaghraby (2005) recommends the use of dynamic resource allocation throughout the implementation of the project to resolve the uncertainty in the work content of the activity. A static analysis based on the use of expected values leads to higher expected project costs compared to those obtained through dynamic resource allocation. Hence, when generating scenarios to estimate the total cost of the project, the plans need to reflect the PM's ability to take corrective actions through allocating the resources dynamically based on the status of the project results. This adaptive behavior is identified as a form of managerial flexibility by Jorgenson and Wallace (2000). They indicate that a deterministic, static model underestimates the total project cost on average no matter which scheduling method is applied. On the other hand, a stochastic and static model results in an expected cost, which is larger than that obtained through a dynamic strategy as suggested above by Elmaghraby (2005). They refer to this reduction in expected cost through the use of a dynamic strategy as the value of flexibility.

In countries with high inflation rates, another important factor in estimating the total cost for projects would be the inflation rate, particularly for projects with long durations (Sarker et al. 2012).

9.6.2 Front-loaded Bidding Strategy

When in a bid the bidder purposefully attaches higher prices to some items and lower prices to some others than expected reasonable prices in a unit price contract, then this practice would be called an *unbalanced load bidding strategy*. The unbalance is due to the unreasonably high and low prices compared to the estimations of the owner. The *front-loaded bidding strategy* is a type of unbalanced load bidding strategy, which has bigger short-term financial implications (Gates 1967). Front-loaded bidding is motivated by the fact that the proposal for many contracted projects often specifies the number of units of various items required, which must be bid on a per unit basis. The contract is awarded to the company with the total lowest bid. Hence, given a total project bid, it might be to the bidder's advantage to inflate the unit cost for work to be completed early and to deflate those to be completed late in the project without changing the total bid. Given the payment schedule imposed is based on progress payments, the front-loaded bidding strategy allows the bidder to increase the receipts in the earlier phases of the project, increasing the NPV of the project. However, there are potential disadvantages. Since the quantity of work to be performed and the costs are only estimates, if those activities underbid require more work than initially estimated and/or their costs turn out to be higher than initially estimated, the front-loaded bidding strategy can lead to substantial losses (Bey et al. 1981).

9.6.3 A Bidding Model for Determining the Bidding Price

In this section, we will present a bidding model for determining the bid price by Elmaghraby (1990). The model assumes deterministic activity durations and costs incurred at the start of the activity. The contractor‘s objective is to determine the events, represented on the AOA project network, that demand partial payment from the client and the amount of each payment. The solution approach assumes a given set of partial payment events, called *key events* (KE). For each KE, represented by node *i*, a sub-graph SG(i) having the initial node of the project network as its initial node and node *i*, i.e., the KE, as its terminal node is identified. If the client receives an initial payment from the client, the initial node is a KE and the corresponding sub-network consists of the initial node. The idea of a sub-graph is that the activities included in the sub-graph contribute to the realization of the associated KE. A KE occurs only when all activities in its sub-graph are completed.

Given the activity durations and costs, CPM is applied, and the activities are scheduled at their early start times, identifying the planned occurrence time of each KE. The corresponding sub-graph for each KE is then determined. The cost of the sub-graph for any KE is denoted by c[KE(.)]. It is *the compounded sum* of the costs of the activities in the sub-graph and is called the bid cost for the corresponding KE. A *compounding factor* α is used to transfer the activity cost from its time of occurrence to the time of occurrence of the KE it is associated with. The bid value for this KE, denoted by v[KE(.)], is then obtained by multiplying the cost of the sub-graph by $(1 + p)$, where p is the markup value. The sum of the NPVs of the individual v[KE(.)] values constitute the NPV of the total bid value.

In general, there will be activities that appear in two or more sub-graphs. The way the cost of the common activities will be allocated among the corresponding KEs can be achieved through different approaches two of which are explained below:

(i) In this approach, the activity cost is allocated in total to a single KE regardless of whether it is a common activity or not. The bid cost for a KE, c[KE(.)] is calculated as the sum of the costs of the activities, which have been completed since the previous KE has occurred. For the earliest occurring KE, c[KE(.)] is calculated as the sum of the activity costs of the activities, which have been completed since the initiation of the project.
(ii) The allocation of the cost of the common activities among the corresponding KEs is accomplished by assigning inverse proportions to their occurrence times. In other words, the KE closer to a common activity receives a higher share in the allocation of its cost.

Obviously, the contractor would like to know the most advantageous way of selecting the set of KEs and the allocation of activity costs among these KEs. This is indeed a complex problem and would require a mathematical programming formulation and solution procedure. A simplistic approach would be to try a number of combinations of KEs and select the best one.

The assumption of scheduling the activities at their early start values requires further elaboration. The contractor can reduce the total bid value by delaying all the non-critical activities, and hence their associated disbursements, by their slack values without delaying the occurrence times of the KEs unless there is a managerial reason for not doing so. But the activity durations are not really deterministic as assumed. Hence, when a non-critical activity is scheduled to start at its late start and if its duration takes longer than estimated, then the project duration might exceed the project deadline possibly resulting in a penalty depending on the contract. Therefore, the contractor has to weigh the pros and cons before deciding which non-critical activities to delay and by how long.

The model presented above is further extended to *random activity durations* by Elmaghraby (1990).

We illustrate this deterministic model for determining the bidding price by the following example.

Example 9.1 A contractor is asked to bid for a project with a project network shown in Fig. 9.1. The annual interest rate is $r = 4.2\%$ yielding a daily compounding factor of $\alpha = 1.000113411$. The contractor has a mark-up value of 20%; $p = 0.20$. The key events (KEs) are specified as 4, 6, and 7.

Activity costs are assumed to be incurred when the activity starts, and all the activities are assumed to start at their early start (ES) value. The activity costs are given in Table 9.1. Report the bid values at the KEs, the total bid value and its Net Present Value (NPV).

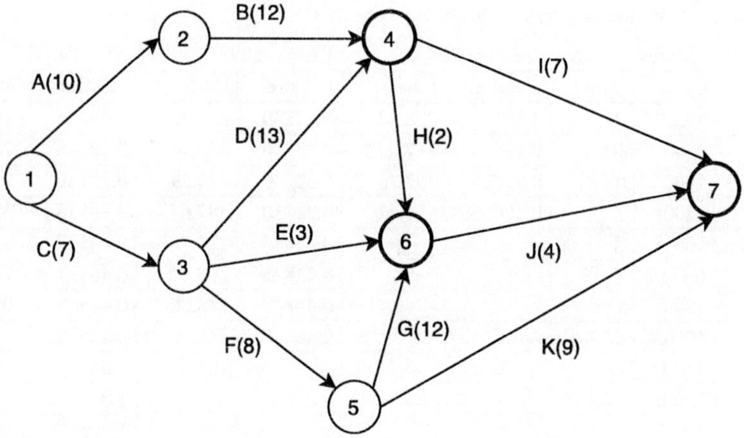

Fig. 9.1 The project network for Example 9.1

Table 9.1 The data and the ES and EF values for Example 9.1

Activity	Duration (day)	Total Cost ($)	ES	EF
A	10	700	0	10
B	12	800	10	22
C	7	1000	0	7
D	13	1200	7	20
E	3	500	7	10
F	8	600	7	15
G	12	2200	15	27
H	2	1000	22	24
I	7	1300	22	29
J	4	2000	27	31
K	9	900	15	24

Solution

The subgraphs for the key events as determined from the project network are: SG(4) = {A, B, C, D}; SG(6) = {A, B, C, D, E, F, G, H}; SG(7) includes all the arcs of the project network by definition, i.e., SG(7) = {A, B, C, D, E, F, G, H, I, J, K}. The ES and early finish (EF) values obtained by applying CPM are reported in Table 9.1.

The occurrence time of a KE is determined by the largest EF value of all activities in its sub-graph. Accordingly, the occurrence times for KE(4), KE(6), and KE(7) are obtained as 22, 27, and 31 days, respectively.

We allocate the cost of the common activities among the corresponding KEs by assigning inverse proportions to their occurrence times:

Sum of the inverses = 1/22 + 1/27 + 1/31 = 0.0455 + 0.0370 + 0.0323 = 0.1148.
Proportion of KE(4) = 0.0455 / 0.1148 = 0.39634.

Table 9.2 Allocation of activity costs to the three KEs

	Activity		KE(4) at 22		KE(6) at 27		KE(7) at 31	
Name	Cost	Time[a]	Fraction	Cost	Fraction	Cost	Fraction	Cost
A	700	0	0.39634	278.13	0.32230	226.30	0.28136	197.65
B	800	10	0.39634	317.50	0.32230	258.34	0.28136	225.62
C	1000	0	0.39634	397.33	0.32230	323.29	0.28136	282.35
D	1200	7	0.39634	476.42	0.32230	387.64	0.28136	338.55
E	500	7	–	–	0.44848	224.75	0.46577	233.52
F	600	7	–	–	0.44848	269.70	0.46577	280.22
G	2200	15	–	–	0.44848	988.00	0.46577	1026.55
H	1000	22	–	–	0.44848	448.73	0.46577	466.03
I	1300	22	–	–	–	–	1.0	1301.33
J	2000	27	–	–	–	–	1.0	2000.91
K	900	15	–	–	–	–	1.0	901.63
Cost *c*[KE(*i*)]				1469.38		3126.75		7254.38
Value *v*[KE(*i*)]				1763.26		3752.10		8705.25

[a]Indicates the time at which the cost of the activity is incurred

Proportion of KE(6) = 0.0370 / 0.1148 = 0.32230.
Proportion of KE(7) = 0.0323 / 0.1148 = 0.28136.

Consider, for instance, activity B, which is present in all subgraphs of all three KEs and whose cost of $800 is incurred at time t = 10.

Share of KE(4) of activity B cost is = $800{*}0.39634{*}\alpha^{22\text{-}10}$ = $317.50;
Share of KE(6) of activity B cost is = $800{*}0.32230{*}\alpha^{27\text{-}10}$ = $258.34;
Share of KE(7) of activity B cost is = $800{*}0.28136{*}\alpha^{31\text{-}10}$ = $225.62.

All the required calculations to obtain the costs of key events, c[KE(i)], are given in Table 9.2. The distribution of activity costs is displayed in Table 9.2.

We obtain the bid values for v[KE(.)] by making use of the markup value *p* and c [KE(.)]:

v[KE(.)] = (1 + *p*) * c[KE(.)] and *p* = 0.20
v[KE(4)] = (1 + 0.20)*c[KE(4)] = 1.20*1469.38 = $1763.26
v[KE(6)] = (1 + 0.20)*c[KE(6)] = 1.20*3126.35 = $3752.10
v[KE(8)] = (1 + 0.20)*c[KE(8)] = 1.20*7254.38 = $8705.25
Total Bid Value = 1763.26 + 3752.10 + 8705.25 = $14,220.61.

NPV for the total bid value is calculated as follows:

$$\text{NPV} = 1,763.26(1/\alpha)^{22} + 3,752.10(1/\alpha)^{27} + 8,705.25(1/\alpha)^{31} = \$14,174.19.$$

The above solution was obtained assuming that all activities begin at their early start times. As stated earlier, the contractor can reduce the total bid value by using the slacks of the non-critical activities and delaying activities that incur costs.

9.7 Mathematical Models for Negotiation in Project Management

Negotiation must be considered here as a process between the client and the contractor to reach an agreement that is accepted by both parties. The client and the contractor both have usually more than a single performance measure such as project cost and duration to consider. The project cost for the client is the budget it will provide to the contractor. For the contractor, on the other hand, the project cost represents the cost of completing the deliverables to the satisfaction of the client. Once the objectives are clearly stated, the negotiation process is expected to lead to an *efficient solution*. Assuming minimization of all objectives, a definition for an efficient solution is given as follows:

Definition: A solution is said to be efficient, if a decrease in one objective can only be obtained at the cost of an increase of one or more other objectives.

A more formal definition is given in Sect. 14.4.2.

In the following, we will introduce three mathematical programming models with two objectives each addressing certain aspects of the negotiation process in project management.

9.7.1 Preparation for a Bid

In this model, which is presented by Vanucci et al. (2012), an engineering consulting firm is preparing a bid covering the engineering design of the facilities of a new mining plant. The decision problem is a multi-mode, resource constrained project scheduling problem with two objectives. One of the objectives is minimizing the total cost of assigning individuals to tasks and the other is minimizing the project duration. In an engineering consulting firm, the employee cost is usually the major cost item. The solution methodology employed is the Non-Dominated Sorting Genetic Algorithm II (NSGA II). It produces efficient points displaying the trade-off between the two objectives in quantitative terms. This information on efficient solutions will then support the engineering consulting firm in the preparation of preferable counter offers to offers made by the mining company during the negotiation process.

9.7.2 Client-Contractor Bargaining Problem

Kavlak et al. (2009) present a model to investigate the client-contractor bargaining problem. The model is formulated as a multi-mode resource-constrained bargaining problem with progress payments and a deadline. Renewable resource constraints are considered. The objective of the client is to minimize the NPV of her payments to the contractor. The objective of the contractor, on the other hand, is to maximize the NPV of her net profit. When the solution minimizing the objective function of the client is used to evaluate the objective function of the contractor, it results in a value from which the contractor would like to move away. This value is called as an undesired value for the contractor. Similarly, the solution maximizing the objective of the contractor corresponds to a value of the objective function of the client, from which the client would like to move away and is called as an undesired value for the client. Hence, these undesired values correspond to cases resulting from one party imposing its best solution on the other one. In the bargaining process, both parties try to move away from their undesired values. The bargaining objective function reflects the objectives of both parties. The normalized distance of the party involved from the respective undesired solution becomes the fundamental metric for each party. The bargaining objective function seeks to maximize the minimum of these normalized distances raised to their bargaining power weights. In other words, it tries to improve the position of the worse-off party among the client and the contractor. The bargaining power weight is defined as a positive quantity taking on values between zero and one and adding up to 1.00. A large bargaining power weight implies a strong bargaining position.

9.7.3 Equitable Payment Schedule Problem

Developing an equitable payment schedule that specifies the amount and timing of the payments of the client to the contractor in the context of a multi-mode resource-constrained project scheduling problem subject to a deadline is studied by Ulusoy and Cebelli (2000). An equitable solution is defined as one where the client and contractor both deviate from their respective ideal solutions, which maximize their respective NPVs, by an equal percentage. The resources employed are renewable resources, and the contractor is allowed to borrow funds. The objective function value $0 \leq Z < 1$ reflects how close the solution is to an equitable solution, where lower values of Z are preferred. The negotiation process starts with the client proposing a payment distribution over the event nodes to the contractor. Based on this information, the contractor determines a project schedule that maximizes their NPV and shares this information with the client, allowing the client to calculate their NPV in turn. The objective function value is calculated, and the client suggests a new payment distribution. This process is repeated until an equitable solution is obtained. As iterations evolve, the NPVs of the client and the contractor decrease or increase, respectively, or vice versa depending on the initial solution adopted.

9.8 Conclusions, Recent Developments, and Some Future Research Directions

Delivering projects before their deadline is a principal concern of PMs. In this regard, Bordley et al. (2019) focused on contract related uncertainties, specifically the deadline uncertainties, which are usually due to changes in stakeholders' requirements. They developed an approach to incorporate deadline uncertainty into managerial decision making in projects. There is a need for further studying this uncertainty and the impacts on project contracting.

An emerging research area in contract management is the research into *smart contracts*. Smart contracts are described as “a computerized transaction protocol that executes the terms of a contract” (Christidis & Devetsikiotis 2016). The incorporation of blockchains into contract management holds great opportunities for research (Hargaden et al. 2019).

In projects, the coordination of different actors, the client, the contractor, and subcontractors has been always a challenging issue. In this regard, managers should be engaged to design effective contracts. In this perspective, Game Theory (GT) offers various tools to investigate the coordination problems and design the contracts accordingly. Contract design has been widely studied as a way of coordinating the supply chains and numerous GT models have been developed. However, applications of GT in project planning, more specifically to contract design, are scarce.

Perng et al. (2005) highlighted the potential benefits of cooperation and modeled the allocation of costs and benefits using GT. They argued that a subcontractor may even earn more profit if it collaborates with others in a coalition. They have shown that a reasonable allocation to each subcontractor is possible in a coalition using the Shapley value and nucleus. Cruel (2011) considered the case where contractor companies can cooperate to decrease the completion time of a project for which the client is willing to pay an amount. She employed cooperative GT to allocate the total payoff among the companies cooperating. Estevez-Fernandez (2012) introduced a GT approach to sharing the project penalties as well as the rewards in projects. Using GT to address multi-agent scheduling, Agnetis et al. (2015) modeled the case where the agents, for instance, the subcontractors, decide autonomously and perform specific tasks in a single project. All agents are aware of the costs of benefits of accelerating each activity. GT model characterizes the Nash equilibrium strategies, i.e., none of the agents has an interest in changing her strategy, if no other strategy is modified. This study does not consider the resource constraints, nor the multi-projects settings. To reflect the real-life cases better, there is a need for extending the analysis to resource constrained multi-project scheduling problems. Hafezalkotob et al. (2017) investigated how different subcontractors of a project can cooperate to achieve more advantages and better results, if possible, and how they can compete in the sense that they can challenge the other subcontractors to make the best decisions. The objective is to find the optimum cost to minimize the time for each of them. They analyzed the problem posed proposing two models: A model based on cooperative game theory in time/cost trade-off problem of projects and a competition model among subcontractors based on non-cooperative game theory. Hafezalkotob et al. (2018) extended this solution procedure to include multiple resources in the time/cost trade-off problem. In both papers, it was shown that a fair assignment of extra profit to the subcontractors is possible. Lately, for stochastic project scheduling problems with delay costs, where the duration of activities are unknown but their corresponding probability distributions are known, Gonçalves-Dosantos et al. (2020) proposed and characterised an allocation rule based on the Shapley value.

A risk sharing practice is the use of cost sharing contracts. A group of contractors/subcontractors might agree to share the additional costs or penalties in case of unexpected events or delays. Brânzei et al. (2002) considered a case where several subcontractors carry out specific tasks and each of them might have some responsibility in the delay of the project. They address how to evaluate these responsibilities using project scheduling and path analysis and to find out a "fair" penalty sharing mechanism using cooperative game theory.

A different cost sharing environment is associated with joint projects. Tijs and Brânzei (2004) investigated a decision problem where several enterprises contribute to realize a joint project. The agents (the enterprises) have to decide about the form of the project and the associated cost. They defined a feasible project to consist of a collection of components where the cost of the joint project is the sum of the costs of the components and the benefit is assumed to increase with increasing set of components. They proposed a formal model of this problem and a related cooperative game. A similar decision problem would be the so-called inter-firm projects. By an inter-firm project, we imply here a project in which two or more independent firms work jointly on a shared activity for a limited period of time (Jones and Lichtenstein 2008). Inter-firm projects increase in popularity in different sectors of the economy with increasing costs and risks involved such as in construction and medical research. Von Danwitz (2018) presented an extensive literature search on interfirm projects and suggested several opportunities for future research.

To summarize, we consider that, it is promising to model client, contractor, and subcontractor relationships and their decision-making options using GT. In this framework, various cooperation scenarios and games could be developed. Coordination of projects through contracts is a promising research area and an interesting application for GT models.

Exercises

9.1 Consider a payment structure agreed upon between the client and the contractor where the client pays a sum at equal time intervals reflecting the work accomplished by the contractor in the last interval. How is such a payment model called?

9.2 What would be the basic disadvantage of cost-plus contracts?

9.3 Consider the case of front-loaded bidding strategy (i.e., increasing the unit price for work to be completed early and decreasing the unit price of work to be completed late in the project without changing the project bid). Can you think of any disadvantages to this approach?

9.4 Explain target contracts and in what sense they are related to bonus and penalty applications?

9.5 In Sect. 9.3 Payment Scheduling Models, considering the lump-sum payment model, it is stated that "When, on the other hand, the resources are limited, then no simple heuristic rule exists for the optimal solution of the problem". Show why this statement is correct.

9.6 You want a well to be drilled in the backyard of your summer house. You have agreed with a contractor at a price of $1000/meter to be paid according to the schedule of rates contract. Your estimate for the contractor's profit is $100/meter.

Depth of the well	Probability of hitting water
1	0.01
2	0.01
3	0.03
4	0.03
5	0.04
6	0.05
7	0.07
8	0.10
9	0.12
10	0.14

Due to municipality regulation, you cannot dig a well of depth more than 10 meters in the backyard of your house. Hence, if the contractor cannot reach a water source in 10 meters, then he has to stop. In such a case, your agreement is you pay him $5000 for all the construction work performed and he will close the well properly and safely. What would be the expected payment based on this type of contract?

9.7 Consider the following construction project for an additional wing to Göbeklitepe Museum. A simplified network diagram for the new wing construction is given below.

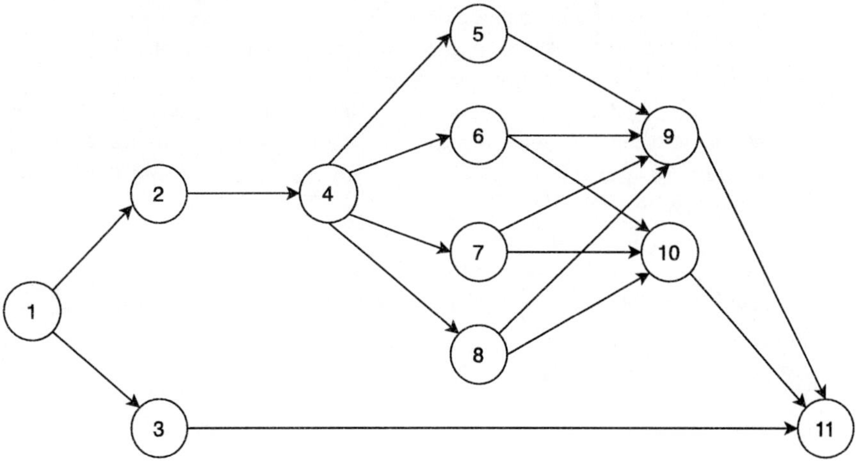

The project's associated activity durations, activity investment requirements, and monthly payments to the contractor are given in the following table where the payments are all in $1000.

Activity Number	Description	Duration (months)	Investment Required	Monthly Payments to the Contractor				
				1	**2**	**3**	**4**	**5**
1	Excavation & foundation.	2	245	10	20			
2	Erect steelwork	5	2200	10	15	15	15	20
3	Utility extension	4	690	15	20	20	25	
4	Pour floors and walls	3	1900	30	40	90		
5	Install plumbing	1	595	40				
6	Electrical contracting	2	830	20	40			
7	Heating and air conditioning	1	680	50				
8	Roof construction	1	510	30				
9	Interior finishing	3	1550	25	40	50		
10	Install security alert system	3	1130	40	40	40		
11	Install equipment and furnishing	2	1200	60	90			

According to the contract signed by the Board of the Museum (the client) and the contractor, the investment required for each activity will be paid by the client once that activity is finished. The contractor will not be involved in the financing of the activities. The fee for the contractor will be paid to the contractor by the client at the end of each month. The interest rate is given as 2% per month.

(a) Minimize the NPV of the cost to the client for building the museum wing. Provide the Gantt chart for the resulting schedule. Provide the cash flow profile for the resulting schedule.
(b) Maximize the NPV of the monthly payments for the contractor. Provide the Gantt chart for the resulting schedule.
(c) Given the schedule in part (b), calculate the NPV of the cost to the client for the whole project. Provide for this case the cash flow profile for the client considering both the monthly payment to the contractor and the activity investment in total.
(d) What would be the difference between the cost in part (c) and the cost you found in part (a)? Discuss the result and suggest arguments leading to a possible reconciliation of the client's and the contractor's perspective?

9.8 Consider the problem environment described in Problem 9.6.

Activity number	Description	Duration (months)	Activity cost
1	Excavation & foundation.	2	275
2	Erect steelwork	5	2275
3	Utility extension	4	770
4	Pour floors and walls	3	2060
5	Install plumbing	1	635
6	Electrical contracting	2	890
7	Heating and air conditioning	1	730
8	Roof construction	1	540
9	Interior finishing	3	1665
10	Install security alert system	3	1250
11	Install equipment and furnishing	2	1350

The activity costs are given in the table above. The activity costs include a fee for the contractor as well. The total project cost to the client is $12,440,000. According to the contract signed by the Board of the Museum (the client) and the contractor, the client will pay the total project cost in 5 installments. The first one will be an advance payment of the amount the $1,000,000. The remaining installments will be paid at the end of the month and by the amount given in the table below.

Month	4	8	12	17
Amount ($\$10^3$)	1500	1500	1500	6940

The contractor will incur the activity cost at the end of the activity. The discount rate is given as 2% per month.

(a) Determine the schedule to maximize the NPV of the contractor. Provide the Gantt chart for the resulting schedule. Provide the cash flow profile for the resulting schedule. What would be the NPV of the contractor?

(b) Discuss whether there is any need of financing on the part of the contractor from her own funds. If so, what is the maximum amount she has to use from her own resources?

(c) If the client agrees to make up for the loss of the contractor in NPV terms by a lump sum payment, then what would be that lump sum amount?

9.9 Consider the problem environment of Problem 9.6. The activity costs are as the ones reported in the table below. The total project cost is $12,440,000. The contractor has come to an agreement with a bank for borrowing funds at an interest rate of 2% per month if needed. The advance payment is $500,000 and there will be 4 more payments by the client at the end of months 4, 8, 12, and 17 where the payments will be the sum of activity costs of all the activities which have been completed since the last payment plus a mark-up of 4%. The contractor incurs the activity costs at the end of the respective activities.

Activity number	Description	Duration (months)	Activity cost
1	Excavation & foundation.	2	275
2	Erect steelwork	5	2275
3	Utility extension	4	770
4	Pour floors and walls	3	2060
5	Install plumbing	1	635
6	Electrical contracting	2	890
7	Heating and air conditioning	1	730
8	Roof construction	1	540
9	Interior finishing	3	1665
10	Install security alert system	3	1250
11	Install equipment and furnishing	2	1350

(a) Determine the schedule to maximize the NPV of the contractor. Provide the Gantt chart for the resulting schedule.
(b) Provide the cash flow profile for the resulting schedule. Calculate the NPV. Use 2% per month as the discount rate. Does the contractor incur any financing charges? Discuss.
(c) If the contractor does not want to incur any financing charges, what would be then the maximum amount she has to use from her own resources?

9.10 A group of fishermen (the client) has ordered a fishing boat. They made a contract with a medium sized shipyard (the contractor), which has already proven its quality through fishing boats built for other fishermen. The shipyard has determined the work packages and developed the project network. The related data are given in the table below. The payment for a work package is made by the fishermen once the work package is finished. The mark-up is agreed to be 10%. The retainage is 8%. The fishermen have agreed to pay the (cost + mark-up - retainage) for a given month with a 1-month delay; for example, the payment for the first month is paid beginning of the third month. The retainage amounts are paid with the last payment but without any interest rate applied.

The cost for a work package is incurred by the shipyard uniformly at the end of each period during the progress of the work package. For work package A, for example, the uniform cost spending rate is 6000 per month incurred at the end of the first and second months.

Work Package	Immediate Predecessor	Duration (months)	Cost (10^3)
A	–	2	12
B	–	1	16
C	-	4	8
D	A	4	20
E	B	2	6
F	C	3	18

The shipyard starts with 10,000 units of money on hand and draws from a bank at 2% interest rate per month. They pay the bank their debt at the earliest possibility. Draw the cash flow diagram over time for your solution and report the profit at the end of the project.

9.11 A manufacturer is in need of a large transformer. This transformer is to be built made-to-order, since it is a special purpose one. The manufacturer asks for a bid from the BBA Company – a manufacturer of large transformers. You are a consultant working for the BBA Company. Information on the project network and activity costs is given in the following table. Use AOA representation for the project network.

Activity	Immediate Predecessor	Duration (weeks)	$1,000
A	–	3	9
B	–	8	109
C	-	4	14
D	A	7	92
E	C	2	6
F	B	5	9
G	B	10	46
H	E,G	9	21
I	D	12	24
J	D	4	11
K	J	6	17
L	F,H,I	3	5
M	E,G	7	61
N	M	9	43

The cost for an activity is incurred once the activity is finished. The mark-up is agreed to be 10%. The retainage is 8%. The discount rate is taken to be 0.25% per week. The client has agreed to pay the (cost + markup) for the project in 3 payments. The first payment is paid in advance at $t = 0$ and consists of 15% of the total (cost + markup) of the project. An offer is expected from the BBA Company concerning the timing of the second payment, which both parties agree should coincide with the realization of an event on the project network. The (cost + markup) of all activities completed up to that point in time is to be paid by the client to the BBA Company in line with the contract. The third payment is made at the completion of the project covering the remaining (cost + markup).

(a) The BBA Company asks you to provide it with the starting times of the activities as well as nodes the timing of the second payment, which will maximize its NPV.
(b) Draw the cash flow diagram over time for your solution.

9.12 a) Solve the problem in Example 9.1 using a different policy for calculating the bid cost for a key event (KE), c[KE(.)]. c[KE(.)] is calculated as the sum of the activity costs of the activities, which have been completed since the previous KE. For the first KE, the bid cost is calculated as the sum of the activity costs of the activities, which have been completed since the initiation of the project. The annual interest rate is $r = 10\%$ yielding a daily compounding factor of $\alpha = 1.000261158$. The contractor has a mark-up value of $p = 0.15$. The KEs are specified as nodes 4, 6, and 7. The duration and cost data are given in the table below.
Report the bid values at the KEs, the total bid value, and its NPV.
b) Compare the result obtained in part (a) with that of Example 9.1.

9.13 You are asked to calculate a bid value for a project using the payments at the event occurrences model. The annual interest rate is set at 12% and the mark-up is taken as 15%. The Project Portfolio Manager wants to see a comparison of two methods where in one the activity costs are distributed over the KEs whenever there is a direct path from that activity to the KE; and where in the other method the activity costs are charged in total to the KE with the closest occurrence time to the completion time of that activity. The two KEs are the completion event of both activities C and G (event 4); and the project completion event (event 6). The durations are given in days.

Activities	Predecessor(s)	Duration	Cost
A	–	4	1600
B	–	8	2100
C	–	11	800
D	A	7	1200
E	B	12	800
F	B	13	900
G	B	18	1000
H	D,E	9	1200
I	C, G	10	700

Assume activities start at their early start and costs are incurred at the start of the activities.

9.14 A contractor is in the process of evaluating different scenarios for a project she wants to bid. She has already determined the WBS and the associated WPs and estimated the durations and the costs. She also determined the payment schedule she will propose. The data is tabulated below. The period is in weeks. The discount rate is given as 0.02% per week. Costs are incurred, and payments are made at the end of the activities. Besides the payments during the project execution, there will also be a payment of amount 300,000TL once the project is finished.

Give a mathematical programming formulation using the summation signs with the bounds explicitly stated and solve the formulation using any IP solver.

Work Package	Predecessor	Duration (weeks)	Cost (10^3 TL)	Payments (10^3 TL)
A	–	7	120	–
C	A	6	76	350
D	A	3	89	–
E	B	6	125	–
F	B	4	156	350
I	C	8	208	–
B	–	7	96	–
G	D, E	9	147	450
H	F	11	79	–

9.15 There is a bid open for the construction project for an additional wing to Göbeklitepe Museum. The client is the Board of the Göbeklitepe Museum. You are asked by your manager to prepare a tender for this bid. You have received the file of the bid from the Board's office and studied it in detail. The Board announced that they will make the offer to the bidder who satisfies all the requirements and proposes the least bid amount. A simplified network diagram for the new wing construction is given in Problem 9.7 above.

Work package	Description	Duration estimates			Cost Function	Bounds on FC
		a	m	b		
1	Excavation & foundation.	2	5	7	FC_1+25d_1	[28/32]
2	Erect steel work	5	9	11	FC_2+80d_2	[19/25]
3	Utility extension	4	5	8	FC_3+24d_3	[15/17]
4	Pour floors and walls	3	5	6	FC_4+92d_4	[15/19]
5	Install plumbing	1	4	5	FC_5+71d_5	[29/35]
6	Electrical contracting	2	4	6	FC_6+55d_6	[38/44]
7	Heating and air conditioning	2	5	7	FC_7+43d_7	[18/20]
8	Roof construction	1	3	6	FC_8+66d_8	[78/86]
9	Interior finishing	3	6	7	FC_9+76d_9	[50/52]
10	Install security alert system	3	4	7	$FC_{10}+48d_{10}$	[66/72]
11	Install equipment and furnishing	2	4	8	$FC_{11}+78d_{11}$	[25/29]

You consult the experts in your firm to obtain the optimistic, most likely, and pessimistic durations in weeks for the work packages. Analyzing the PMO's archive and working together with the quantity surveyors in your firm, you obtain the cost for each work package as a function of the duration of that work package. The costs are expressed in ($'000) and are assumed to be incurred at the completion of each work package. All this information is provided in the Table above. The fixed terms are uniformly distributed with the lower and upper bounds provided in the table above. It is your firm's standard practice to employ the PERT distribution for the work package/activity durations in construction projects. Your firm has decided to apply 10% mark-up to this project. The top management of your firm is rather risk averse and prefers to start work packages/activities at their early starts.

Apply Monte Carlo simulation for preparing the following information to be presented to the top management of your firm.

(a) The expected project duration and its variance. Determine the project duration values that will not be exceeded with 0.40 and 0.60 probability and discuss.
(b) The expected cost of the project and its variance. Determine the project cost values that will not be exceeded with 0.40 and 0.60 probability and discuss.
(c) Develop a baseline schedule. The expected duration and expected cost for the client (i.e., including 10% mark-up) of this baseline schedule you will propose in your tender as the bid values. You have decided to evaluate a scenario where there will one payment during the implementation of the project and one at the completion. What would be the expected NPV for a payment schedule where 35% of the project cost for the client for the completed work packages will be paid by the client once 50% of the project duration is reached with the remaining to be paid at the completion of the project? Assume a discount rate of 0.25% per week.
(d) Develop two further scenarios and evaluate them based on the expected NPV. Compare their results including the one in part (c).

References

Agnetis, A., Briand, C., Billaut, J. C., & ŠůCha, P. (2015). Nash equilibria for the multi-agent project scheduling problem with controllable processing times. *Journal of Scheduling, 18*(1), 15–27.

Bey, R. B., Doersch, R. H., Patterson, J. H. (1981) *The net present value criterion: Its impact on project scheduling*. Project Management Quarterly, June, 223–233.

Bordley, R. F., Keisler, J. M., & Logan, T. M. (2019). Managing projects with uncertain deadlines. *European Journal of Operational Research, 274*(1), 291–302.

Brânzei, R., Ferrari, G., Fragnelli, V., & Tijs, S. (2002). Two approaches to the problem of sharing delay costs in joint projects. *Annals of Operations Research, 109*(1-4), 359–374.

Chapman, C. B., Ward, S. C., & Bennel, J. A. (2000). Incorporating uncertainty in competitive bidding. *International Journal of Project Management, 18*, 337–347.

Christidis, K. & Devetsikiotis, M. (2016). Blockchains and smart contracts for the Internet of Things. IEEE *Access, 4*, 2292–2303.

Clough, R. H., Sears, G. A., Sears, K., Segner, R. O., & Rounds, J. L. (2015). *Construction contracting, a practical guide to company management* (8th ed.). New York: Wiley.

Cruel, I. (2011). Project management games. *International Game Theory Review, 13*(3), 281–300.

Elmaghraby, S. E. (1990). Project bidding under deterministic and probabilistic activity durations. *European Journal of Operational Research, 9*, 14–34.

Elmaghraby, S. E. (2005). On the fallacy of averages in project risk management. *European Journal of Operational Research, 165*, 307–313.

Estevez-Fernandez, A. (2012). A game theoretical approach to sharing penalties and rewards in projects. *European Journal of Operational Research., 216*(3), 647–657.

Farid, F., & Boyer, L. T. (1985). Fair and reasonable markup (FaRM) pricing model. *Journal of Construction Engineering and Management, 111*(4), 374–390.

Friedman, L. (1956). A competitive bidding strategy. *Operations Research, 4*, 104–112.

Gates, M. (1967). Bidding strategies. *Journal of the American Society of Civil Engineers, 93*, 92–97.

Gonçalves-Dosantos JC, García-Jurado I, Costa J (2020) *Sharing delay costs in stochastic scheduling problems with delays. 4OR - A Quarterly Journal of Operations Research.* https://doi.org/10.1007/s10288-019-00427-9.

Hafezalkotob, A., Houseinpour, E., & Damghani, K. K. (2017). A game theory approach for competition and cooperation among project's subcontractors with interaction activities. *Journal of Project Management, 2*, 71–86.

Hafezalkotob, A., Houseinpour, E., Moradi, M., & Damghani, K. K. (2018). Multi-resource trade-off problem of the project contractors in a cooperative environment: Highway construction case study. *International Journal of Management Science and Engineering Management, 13*(2), 129–138.

Hargaden, V., Papakostas, N., Newell, A., Khavia, A., & Scanlon, A. (2019). *The role of blockchain technologies in construction engineering project management.* In 2019 IEEE International Conference on Engineering, Technology and Innovation (ICE/ITMC): 1–6, IEEE.

Hartmann, A., Ling, F. Y. Y., & Tan, J. S. H. (2009). Relative importance of subcontractor selection criteria: Evidence from Singapore. *Journal of Construction Engineering and Management, 135* (9), 826–832.

Herroelen, W. S., Van Dommelen, P., & Demeulemeester, E. L. (1997). Project network models with discounted cash flows: A guided tour through recent developments. *European Journal of Operational Research, 100*, 97–121.

Jones, C., & Lichtenstein, B. B. (2008). Temporary inter-organizational projects: How temporal and social embeddedness enhance coordination and manage uncertainty. In S. Cropper, M. Ebers, C. Huxham, & P. S. Ring (Eds.), *The Oxford handbook of inter-organizational relations* (pp. 231–255). Oxford: Oxford Handbooks in Business and Management.

Jorgenson, T., & Wallace, S. W. (2000). Improving project cost estimation by taking into account managerial flexibility. *European Journal of Operational Research, 127*, 239–251.

Kavlak, N., Ulusoy, G., Şerifoğlu, F. S., & Birbil, Ş. İ. (2009). Client-contractor bargaining on net present value in project scheduling with limited resources. *Naval Research Logistics, 56*, 93–112.

King, M., & Mercer, A. (1985). Problems in determining bidding strategies. *Journal of Operational Research Society, 36*, 915–923.

Mika, M., Waligora, G., & Weglarz, J. (2005). Simulated annealing and tabu search for multi-mode resource-constrained project scheduling with positive discounted cash flows and different payment models. *European Journal of Operational Research, 164*(3), 639–668.

Perng, Y. H., Chen, S. J., & Lu, H. J. (2005). Potential benefits for collaborating formwork subcontractors based on co-operative game theory. *Building and Environment, 40*(2), 239–244.

PMI. (2009). *Practice standard for risk management.* Newtown Square, PA.

PMI. (2016). *Construction extension to the PMBOK guide.* Newtown Square, PA.

Sarker, B. R., Egbelu, P. J., Liao, T. W., & Yu, J. (2012). Planning and design models for construction industry: A critical survey. *Automation in Construction, 22*, 123–134.

Tijs, S. H., & Brânzei, R. (2004). Cost sharing in a joint project. In C. Carraro & V. Fragnelli (Eds.), *Game practice and the environment.* Cheltenham: Edward Elgar Publishing.

Touran, A., & Lopez, R. (2006). Modeling cost escalation in large infrastructure projects. *Journal of Construction Engineering and Management, 132*(8), 853–860.

Turner, J. R., & Simister, S. J. (2001). Project contract management and a theory of organization. *International Journal of Project Management, 19*(8), 457–464.

Ulusoy, G., & Cebelli, S. (2000). An equitable approach to the payment scheduling problem in project management. *European Journal of Operational Research, 127*(2), 262–278.

Ulusoy, G., Sivrikaya-Şerifoğlu, F., & Şahin, Ş. (2001). Four payment models for the multi-mode resource-constrained project scheduling with discounted cash flows. *Annals of Operations Research, 102*, 237–261.

Vanucci SC, Bicalho R, Carrano EG, Takahashi RHC (2012) *A modified NSGA-II for the multi-objective, multi-mode resource-constrained project scheduling problem.* WCCI 2012 IEEE World Congress on Computational Intelligence, Brisbane, Australia.

Von Danwitz, S. (2018). Managing inter-firm projects: A systematic review and directions for future research. *International Journal of Project Management, 36*(3), 525–541.

Zhu, G., Bard, J. F., & Yu, G. (2007). A two-stage stochastic programming approach for project planning with uncertain activity durations. *Journal of Scheduling, 10*, 167–180.

Progress and Cost Control 10

Learning Outcomes
Upon successful completion of this Chapter, the reader will be able to:

1. Define the project monitoring and control process and its relation to planning.
2. Describe the approaches and methods used for project control.
3. Summarize earned value management (EVM), its importance, and its role in assessing project performance.
4. Explain the EVM metrics and describe the associated graphical data display tools.
5. Make time and cost estimates using EVM.
6. Bring a critical analysis to EVM.

10.1 Introduction

Project monitoring and control is a critical responsibility of project managers (PMs). Project monitoring involves collecting and recording project realization data, whereas project control refers to analyzing the collected data to decide on corrective or preventive actions.

Monitoring and control activities have a significant impact on project success or failure (Shtub et al. 2014). For this reason, PMs try to establish a formal monitoring and control system that allows them to study the project status diligently, identify problems and their root causes, and implement appropriate actions. Responsibilities, methods, and tools must be clearly defined in order to minimize deviations from schedules and ensure achievement of project targets. As listed by (Hazır 2015), an effective formal monitoring and control system clearly defines

(a) *a monitoring policy*: what, how, where, when, and by whom to monitor, and
(b) *an intervention and control policy*: how, when, and by whom what action

G. Ulusoy, Ö. Hazır, *An Introduction to Project Modeling and Planning*, Springer Texts in Business and Economics, https://doi.org/10.1007/978-3-030-61423-2_10

should be taken to prevent and correct deviations?

However, defining such policies, designing and implementing an effective monitoring and control system are challenging tasks for organizations (Venkataraman and Pinto 2010). Adopting a user-friendly methodology is critical to support PMs in establishing a widely accepted control system. In this regard, *Earned Value Management* (EVM), developed in 1962 as part of a US Department of Defense study, has been the most widely accepted and used methodology for a wide range of industries and project sizes.

EVM is a measurement and control methodology to assess the time and cost performance of projects. It compares the planned work to the realized work on the basis of monetary units. This provides a common unit of measurement that allows deviations from the time-phased budgets to be recorded and communicated in the form of plots that managers can use to assess project performance quickly and easily. In addition to quantifying the project progress in terms of time and cost, EVM is also used to predict future performance. Trend analysis and projections made using the recorded deviations play an important role in deciding on intervention and corrective actions.

10.2 Cost/Schedule Control System Measures

EVM addresses project progress along two main performance dimensions: time and cost. For this purpose, three time-dependent values are required: planned value, earned value, and actual cost.

Planned Value (PV), also referred to as the *Budgeted Cost of Work Scheduled* (BCWS), is the value of the work to be accomplished according to project plans. As part of the project planning phase, time phased budgets are prepared, and a time scaled PV specified for the planned project duration. At the termination, the PV should be equal to the planned budget of the project.

In contrast, *Earned Value* (EV), also called the *Budgeted Cost of Work Performed* (BCWP), reflects realized progress or accomplishments as opposed to the planned ones. To be able to value the work performed, either the monetary value of the work delivered, or the completion percentages of the activities must be estimated. EV is estimated by multiplying the completion percentages (% Complete) by cost rates, which are estimated in the project budget and fixed.

The value of the work performed, EV, is usually different from the *Actual Cost* (AC) spent, also called the *Actual Cost of Work Performed* (ACWP). EV values are usually estimated by domain experts on the project team. In contrast, AC values are realized, and hence can be computed from the accounting records.

For a project of any size, before performing an EVM based evaluation, two basic planning documents must be prepared: The Work Breakdown Structure (WBS) and a time phased budget (Venkataraman and Pinto 2010). The WBS is needed to perform a complete analysis of accomplishments, while time phased budgets are necessary to calculate the PV figures. In addition to these data, as the project progresses, reliable estimates of activity completion percentages and actual spending for cost items should be recorded.

10.3 Estimating the Percent Complete

According to Brandon (1998), the main steps of the EVM are:

- Setting the work package (WP) size effectively,
- Defining the % Complete values properly,
- Setting a correct basis for costing

The classification of activities and definition of WPs is a crucial step in preparing the WBS and requires collective teamwork. At this step, the sizes of the WPs need to be discussed and carefully checked, since they affect the controlling effort during implementation. Precise cost estimation and performance analysis require smaller WP sizes. WPs whose delivery on time and within budget is perceived to be risky need more attention and control; hence, they should be divided into sub-packages and defined in more detail. On the other hand, the smaller the WPs, the more complicated the reporting system. Brandon (1998) suggests deciding on the sizes based on the frequencies of the payroll data acquisition because labor cost is the main cost driver in many projects. His suggestion is to set the maximum size of the work packages to payroll data acquisition frequency.

In addition to sizing the packages, estimating the activity realization rates (% Complete) is challenging and important for evaluating project performance correctly. In many cases, precise measurement is neither possible nor practical. In these cases, threshold levels like 25%, 50%, 75%, 100% could be set, and the corresponding progress requirements defined in concrete terms Venkataraman and Pinto (2010). Accomplishments can then be compared to the requirements of the threshold levels. Even though the estimates might not be very precise at the activity level, a practical and effective estimation and analysis can be performed at the project level.

Webb (2003) also considers the estimation of the EV as a difficult and critical task. He identifies four estimation methods:

- **Subjective Assessments:** At some specific time periods, either the % Complete or the remaining durations are estimated for each activity. In the case of time-based assessment, the activity completion time is estimated and the value per unit duration is calculated. EV is then calculated by assuming a uniform rate of value creation per time and recording the elapsed time.
- **Objective Measured Assessments:** At each reporting time, value is estimated either by comparing the achievements to the pre-assigned values for the milestone tasks that have been delivered, where milestones are defined as relatively important pre-assigned events throughout the life of the project or by assessing the

value of the number of units completed, such as the lines of code compiled in software development projects.

- **Rules of Thumb:** Simple rules such as assigning 50% of the value when activity is started and the remaining 50% when completed.
- **Indirect Assessment:** Outside factors such as the controlling or testing activities determine the evaluation process. % Complete of these activities are used to calculate the EV.

In addition to estimating the value of work accomplished, EVM also requires that actual costs be obtained. In many organizations, accounting departments may not deliver this data on time. In those cases, resource utilization figures need to be consulted and actual costs estimated by multiplying the amount of resources with the estimated unit cost of the resources (Brandon 1998).

10.4 Cost/Schedule Control Performance Indicators

To assess the project performance, PMs focus on the deviations from the plans. Using EVM metrics, they monitor two main values: the *Schedule Variance* (SV) and the *Cost Variance* (CV). SV is the difference between EV and PV and serves to evaluate the time performance. CV, the difference between EV and AC, reveals the cost performance. Note that decision-making using these variances first requires estimating the EV correctly.

$$SV = EV\text{-}PV$$

$$CV = EV\text{-}AC$$

Various graphical representations of these metrics have been widely used to assist PMs in taking control decisions. The plot of the cumulative PV over time is often called an *S*-curve due to its shape. This shape arises from the changing rate of expenditure over the project duration, which is expected to be relatively low in the initial phases, increasing during the implementation phase and decreasing later in the termination phase. AC and PV could be dynamically sketched in the *S*-curve and a snapshot view could be presented (see Fig. 10.1). Note that PVs are predetermined, while EV and AC need to be dynamically recorded so that variances can be calculated at any time and performance could be evaluated.

A positive schedule (cost) variance indicates that the project is performing ahead of schedule (budget). On the other hand, negative variances show that the project is behind schedule or over budget, signaling the PM that corrective action is required. Figure 10.1 illustrates a project with $EV < PV < AC$, and hence $0 < SV < CV$, at the control date designated as the Elapsed Time.

PMs should consult the variance data frequently, so graphs plotting only the variance values over time might support decision-making. Upper and lower control limits could be integrated into these graphs as in Fig. 10.2, based on managerial

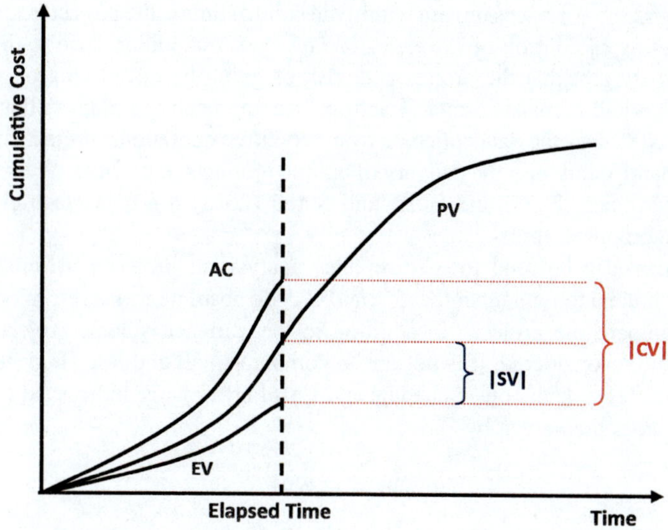

Fig. 10.1 Project progress data and *S*-curves

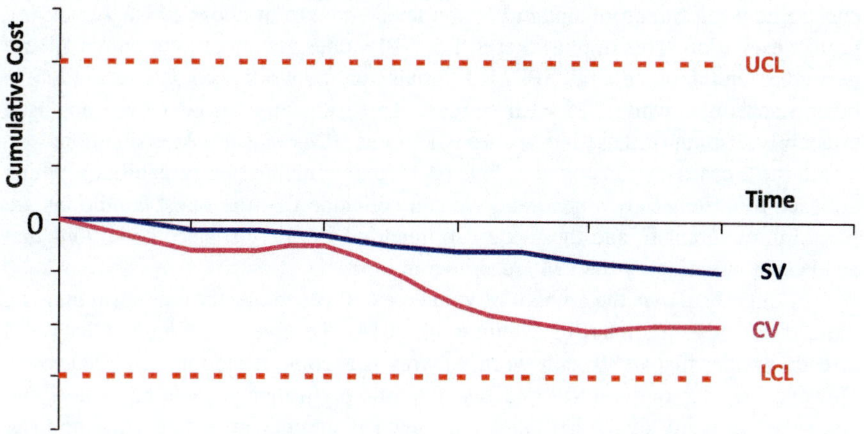

Fig. 10.2 Project variance graph and limits

concerns and tolerances. These limits could be calculated using the techniques of *Statistical Process Control* (SPC), which has been commonly used to monitor and control the quality of the processes in manufacturing and services. In quality management, these limits are usually set with an assumption of normal distribution (with mean, μ, and standard deviation, σ) for the *upper and lower control limits* (UCL = $\mu + 3\sigma$, LCL = $\mu - 3\sigma$). We refer to the project control example given by Colin and Vanhoucke (2014) for setting the limits using statistical process control.

As long as the variances remain within the control limits, the project is assumed to be progressing in control; however, variance values outside or close to the control limits may suggest that the project is at risk of going be out of control, requiring investigation and perhaps corrective action. The approach is similar to that of SPC. However, SPC uses the data collected from repetitive operations; project control, on the other hand, deals with the delivery of unique products or services. We refer to the studies of Anbari (2003) and Hazır and Shtub (2011) for EVM metrics and the illustrative graphical tools.

EVM can also be used for performance analysis of project portfolios. At this point, it is crucial to remember that SV and CV are absolute measures. In order to be able to compare the projects, in addition to the variance values, project budgets should also be considered. It is not fair to compare small and large budget projects based on absolute deviations. Schedule and Cost Performance Indices (SPI, CPI) are the two main ratios, given by.

$$\mathrm{SPI} = \mathrm{EV}/\mathrm{PV}$$

$$\mathrm{CPI} = \mathrm{EV}/\mathrm{AC}$$

In order to ensure that projects progress in line with time plans and budget constraints PMs want to maintain EV at a level equal to or above PV or AC during project execution. This implies that SPI or CPI values are equal to or above 1.0 are desirable. On the other hand, SPI <1.0 signals that the work accomplishment rate is behind schedule, while CPI <1.0 suggests that monetary resources are not used efficiently. If both of these indices are below one, PMs face a riskier situation.

All these cases are illustrated in Fig. 10.3. Analyzing the four possibilities, which are each represented by a quadrant, we can conclude that the safest conditions are placed in quadrant II, and the riskiest in quadrant IV. Performance conditions that need close attention by the PM are shown in bold.

CPI directly shows the amount of value created per monetary unit spent and is a more direct measure than SPI (Shtub et al. 2014). To give an example, CPI $= 0.8$ directly implies that an 80 cents worth of work is accomplished for each USD spent. However, relying only on SPI to assess the time performance could be misleading, since the criticality of the activities completed is another important criterion. The critical path(s) define(s) the project completion time, but the EVM metrics do not distinguish critical from noncritical activities. SPI $= 0.8$ might not be a major problem if most of the critical activities have been completed.

Another point to pay attention to while using SPI is its evolution towards the project termination. SPI $= 1$ at the end of the project by definition since EV=PV when all the tasks are delivered. At that point, the time performance cannot be evaluated; however, this is not the case for CPI, since AC usually differs from EV. Towards project termination, SPI converges to one, but it does not show the amount of delay observed regarding the project due date.

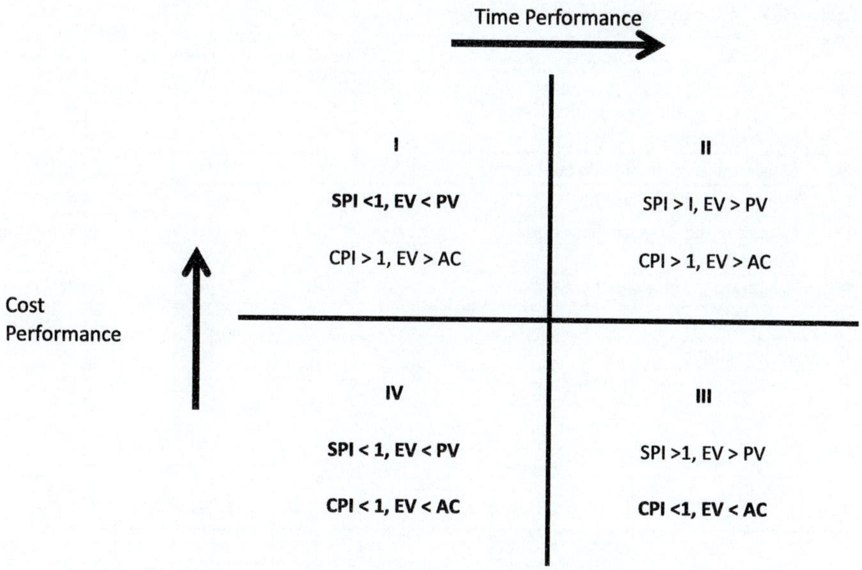

Fig. 10.3 EVM performance indices. (Adapted from Vanhoucke 2009)

Recall that durations and costs are interdependent for many activities. During project execution PMs might need to crash some activities by using additional resources, adversely affecting cost performance. Considering such interdependencies, the SPI and CPI metrics could be combined to define a composite measure such as the *Schedule and Cost Performance Index* (SCI). This takes into account both schedule and cost performance and is equal to the product of SPI and CPI. Note that if projects have both SPI and CPI values smaller than 1.0, or both are above 1.0, the resulting SCI values will be amplified as shown in the following relationships.

$$\text{SCI} = \text{SPI} * \text{CPI}$$

If $\text{SPI} < 1.0$ and $\text{CPI} < 1.0$, then $\text{SCI} < \min\{\text{SPI}, \text{CPI}\}$,
If $\text{SPI} \geq 1.0$ and $\text{CPI} \geq 1.0$, then $\text{SCI} \geq \max\{\text{SPI}, \text{CPI}\}$.

Example 10.1 Michelle has recently been appointed as the project manager for organizing an on-campus concert during the spring break. Project objectives were clearly defined, and an approximate activity budget is set. It is uniformly distributed between the ES and EF times of the activities.

The project team started planning the activities, estimated the durations and costs, and created the following WBS (Table 10.1). The project started on 1 January 2016.

Table 10.1 WBS for Example 10.1

ID	Activities	Immediate Predecessors	Duration (Week)	Cost $'000
A	Selection of the organization company and contracting	–	3	90
B	Music bands selection & contracting	A	3	150
C	Content & stage planning	B	2	6
D	Location and Layout Planning & Contracting	A	2	12
E	Procurement planning	D	2	36
F	Procurement	E	2	108
G	Stage setup	C, F	1	50
H	Concert and stage dismantling	G	1	80

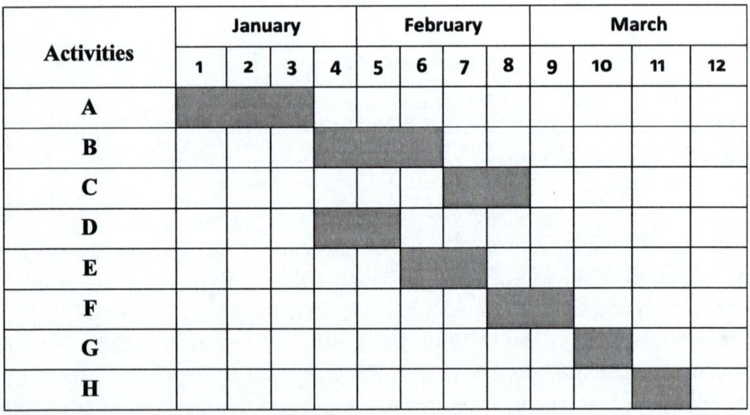

Fig. 10.4 Gantt chart of the concert project

As of February 14, an agreement was reached with the organization company, the location was chosen, and the contract was signed. Among the 4 music bands negotiated with, only one was selected and the contract was signed. A total of $150,000 was spent on all these activities. Other activities have not yet begun. Assuming that Michelle generates the schedule employing the early start approach, calculate the schedule and cost variances, and cost and schedule performance indices. Interpret these calculated values.

Solution

Let us construct the early start (ES) schedule using CPM, draw the corresponding Gantt Chart (Fig. 10.4) and prepare the time phased budget (Table 10.2).

In Table 10.2, the sum of each row and each column are calculated. The Total to Date row shows the cumulative cost values. February 14 is the end of the sixth week

Table 10.2 Time-based budget for Example 10.1

	Weeks											
Activity	1	2	3	4	5	6	7	8	9	10	11	Total
A	30	30	30									90
B				50	50	50						150
C							3	3				6
D				6	6							12
E						18	18					36
F								54	54			108
G										50		50
H											80	80
Total per week	30	30	30	56	56	68	21	57	54	50	80	
Total to date		60	90	146	202	270	291	348	402	452	532	

of the project. Checking the budget, PV = \$270,000, which is the cumulative amount to be spent at week 6. To complete the earned value analysis, activity completion percentages are estimated. Activities A and D have already been completed, but only 25% of activity B is finished. Therefore,

$$\mathrm{PV} = \$270{,}000$$

$$\mathrm{EV} = 90{,}000 + 12{,}000 + 0.25 * 150{,}000 = \$139{,}500$$

$$\mathrm{AC} = \$150{,}000$$

Given this data, variances and indices could easily be calculated.

$$\mathrm{SV} = \mathrm{EV} - \mathrm{PV} = -\$130{,}500$$

$$\mathrm{CV} = \mathrm{EV} - \mathrm{AC} = -\$10{,}500$$

$$\mathrm{SPI} = \mathrm{EV}/\mathrm{PV} = 0.52$$

$$\mathrm{CPI} = \mathrm{EV}/\mathrm{AC} = 0.93$$

$$\mathrm{SCI} = \mathrm{SPI} * \mathrm{CPI} = 0.48$$

The variance data signals Michelle that her project is behind the schedule and over budget. Carefully examining the values and data of WPs, the major problem is the pace of the project. Only 25% of the critical activity B has been completed. Since this activity requires considerable monetary resources, not enough value has been earned relative to the plans. Michelle might need to accelerate and control this activity.

10.4.1 The Impact of Inflation

The above analysis would function satisfactorily under relatively stable prices and wages. However, when the project is executed in an inflationary environment resulting in an escalation in costs, the above approach to assess project performance is no longer adequate. Note that BCWS and BCWP appear in their initially estimated constant cost values whereas ACWP must be in current cost values. In order for the analysis above to function correctly under inflation, all cost terms need to refer to the same point in time. For that purpose, we deflate ACWP using a proper *escalation factor* to obtain *Deflated Cost of Work Performed* (DCWP).

SV remains unchanged under inflation since it is based on budgeted cost values. CV, on the other hand, now consists of two components: *Inflation Variance (*IV*)* and *Deflated Cost Variance* (DCV). These two components of CV are defined as follows:

$$\mathrm{IV} = \mathrm{ACWP} - \mathrm{DCWP}$$

$$\mathrm{DCV} = \mathrm{DCWP} - \mathrm{BCWP}$$

IV represents the cost overrun as a result of cost increases due to inflationary pressures in the marketplace. Since inflation is an exogenous factor and hence, beyond the control of the PM or the contractors, it would be unfair to blame them for the IV component of CV. On the other hand, DCV is the responsibility of the PM or the contractor as is CV under relatively stable costs.

When calculating DCWP dealing with each individual cost item would create unnecessary overhead. To reduce this burden one can use escalation factors obtained from appropriate price indices. They can be specific to the line of sector dominant in the project, such as the Construction Cost Index (CCI) and the Building Cost Index (BCI) published by the Engineering News-Record (ENR) in the USA. If such indices are not available for the project, then regional or national broader indices such as Consumer Price Index (CPI) can be employed at the expense of the possibility it might lead to relatively poor results.

An escalation factor Δ_i expressed as a percentage for period i using an index would be calculated as follows:

$$\Delta_i = \left[{}^{I_i}\!/_{I_{i-1}} - 1\right]x100$$

where I_i is the index value for period i and I_{i-1} is the index value for period (i-1).

An average escalation factor r over n consecutive periods is given by the following expression:

$$r = \left[\left({}^{I_f}\!/_{I_s}\right)^{1/n} - 1\right]x100$$

Table 10.3 The consumer price indices for 7 consecutive months

Month	1	2	3	4	5	6	7
Consumer Price index	446.45	448.02	450.58	454.43	460.62	465.84	468.56

where I_s and I_f are the start and finish periods' index values, respectively. The average escalation factor *r* is synonymous with the Cumulative Average Growth Rate (CAGR).

Example 10.2 For a project, which was finished in 7 months, the management wants to assess the impact of inflation. For those 7 months the consumer price indices are reported in Table 10.3. ACWP for the project is $392,860 and BCWP is $320,110.

(a) Calculate the change in the last 2 months.
(b) Determine the average escalation factor *r* over these 7 consecutive months.
(c) Determine the Inflation Variance (IV) and Deflated Cost Variance (DCV) and assess the values obtained.

Solution

(a)
$$\Delta_7 = \left[\left(\frac{468.56}{465.84}\right) - 1\right]x100 = 0.584\%.$$

Hence, the Consumer Price Index has increased from month 6 to month 7 by 0.584%.

(b)
$$r = \left[\left(468.56\big/446.45\right)^{1/7} - 1\right]x100 = 0.693\%.$$

This result implies that the Consumer Price Index increases over these 7 months on the average by 0.693%.

(c) First, we obtain the Deflated Cost of Work Performed (DCWP):

$$\text{DCWP} = \text{ACWPx}(1\text{-}r) = 392,860\text{x}(1\text{-}0.00693) = \$390,137.48.$$

$$\text{IV} = \text{ACWP} - \text{DCWP} = 392,860\text{-}390,137.48 = \$2,772.52$$

$$\text{DCV} = \text{DCWP} - \text{BCWP} = 390,137.48\text{-}320,110 = \$70,027.48.$$

Based on the above calculations, only [2772.52/(392,860–320,110)] × 100 = 3.81% of the loss can be attributed to inflation. The remaining $70,027.48 is the responsibility of the PM.

10.5 Time and Cost Predictions Using EVM

In addition to current performance analysis, EVM is also used to predict the impact of the realized delays and cost increases on project objectives. Using time variance data, future time and cost realizations could be estimated. Based on these estimates, PMs can decide on actions to avoid delivering the projects with significant delays or cost overruns.

Project performance at completion is a function of the actual deviations and expected deviations to be observed in completing the remaining work. *Expected Total Project Cost at Completion* (EAC) will be equal to the sum of the actual cost and *Expected Spending to Complete* (CWR):

$$\text{EAC} = \text{AC} + \text{CWR}$$

where EAC and CWR respectively defines the estimated cost at completion and planned cost of work remaining. Similarly expected completion time could be expressed as the sum of *Actual Duration* (AD) and *Expected Time to Complete* (DWR):

$$\text{EAC}_t = \text{AD} + \text{DWR}$$

EAC_t and DWR respectively represent the estimated duration at completion and planned duration of the work remaining. The subscript t is used in notation to emphasize the time dimension.

Estimating CWR and DWR requires some assumptions regarding the future pace of the work remaining to be done or the efficiency in cost spending. The cost or duration of the remaining work (CWR and DWR) can be estimated by projecting the actual performance into the future. We can cite three main approaches and their corresponding assumptions (Vanhoucke 2009):

A project could be delivered;

(a) As planned, regardless of failures or achievements in the past,
(b) Continuing with the same cost performance rate (i.e., the same CPI for cost forecasting),
(c) Continuing with the same schedule and cost performance index SCI.

Under each of the above given performance assumptions total project cost could be formulated as follows:

(a) $$\text{EAC} = \text{AC} + (\text{BAC} - \text{EV})$$

(b) $$\text{EAC} = \text{AC} + (\text{BAC} - \text{EV})/\text{CPI}$$

(c) $$\text{EAC} = \text{AC} + (\text{BAC} - \text{EV})/\text{SCI}$$

In the above calculations, *Budget at Completion* (BAC), which is the sum of the budgeted costs of the activities, is used to quantify the planned value at project delivery.

Example 10.3 Predict the cost of the concert project at completion by making projections based on performance to date. Interpret the results.

Solution
BAC is the sum of the PV of all the activities, hence.

$$\text{BAC} = \$532,000$$

Coming back to the three performance assumptions given above,

(a) $\text{EAC} = \text{AC} + (\text{BAC} - \text{EV}) = 150,000 + 532,000 - 139,500 = \$542,500.$

(b) $\text{EAC} = \text{AC} + (\text{BAC} - \text{EV})/\text{CPI} = 150,000 + (532,000 - 139,500)/0.93$
$= \$572,043.$

(c) $\text{EAC} = \text{AC} + (\text{BAC} - \text{EV})/\text{SCI} = 150,000 + (532,000 - 139,500)/0.48$
$= \$966,857.$

Note that the first forecast is optimistic as it assumes that the project could be continued as planned without any further deviations. In contrast, the last forecast corresponds to a pessimistic setting. Poor schedule performance significantly lowers the SCI and as a result, current performance is rated very poorly. This frustrating performance is assumed to continue, and projection is made accordingly.

To estimate the completion time of the project, the accomplished work is projected on the time dimension and a time equivalent of the EV is formulated. For this purpose, we use the *Earned Schedule* (ES) proposed by Henderson (2005). Using temporal units, ES(t) expresses the equivalent of the work accomplished at any time t. To calculate ES(t), the earned value at time t, EV(t), is compared to the planned value PV(t). Projecting the EV on PV figure, the time t* that corresponds to that value in the plans is calculated (see Fig. 10.5).

$$\text{EV}(t) = \text{PV}(t = t*)$$

$$\text{ES}(t) = t*$$

Having calculated the ES, the project completion time can be estimated. First, the *Estimated Time to Completion* (ETC_t) is calculated. For this calculation, projections of future performance can be made based on the actual performance. For this purpose, the three performance assumptions given above will be used and the appropriate one will be implemented. ETC_t and *Estimated Time at Completion* (EAC_t) can be formulated as follows:

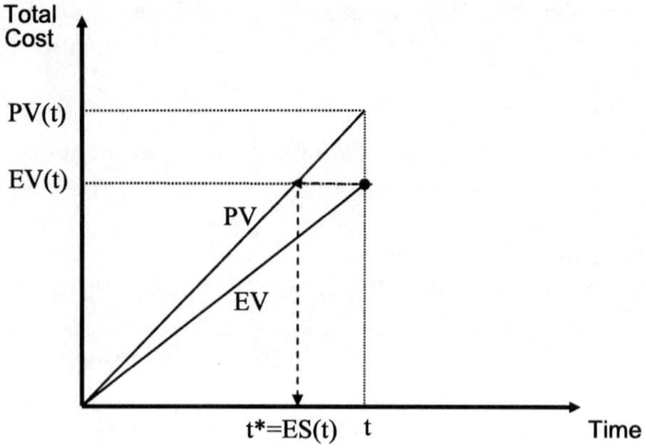

Fig. 10.5 Earned schedule and its calculation

(a) $$\mathrm{ETC_t(t) = PD - ES(t)}$$

(b) $$\mathrm{ETC_t(t) = (PD - ES(t))/SPI(t)}$$

(c) $$\mathrm{ETC_t(t) = (PD - ES(t))/SCI(t)}$$

$$\mathrm{EAC_t(t) = AD + ETC_t(t)}$$

PD and AD respectively refer to the planned and actual duration.

Example 10.4 Predict the completion time of the concert project. Assuming the current progress rate is maintained, is it possible to complete the project on March 21? Might Michelle need to allocate more resources and accelerate some of the activities?

Solution

Let us first calculate the ES.

We observe that PV $(t = 3) <$ EV $(t = 6) <$ PV $(t = 4)$.

To calculate the exact value, interpolation could be used.

$$\begin{aligned}\mathrm{ES} &= 3 + [(\mathrm{EV}(t = 6)\text{-}\mathrm{PV}(t = 3)]/[(\mathrm{PV}(t = 4)\text{-}\mathrm{PV}(t = 3)] \\ &= 3 + [139.5\text{-}90]/[146\text{-}90] = 3.88\end{aligned}$$

Returning to the three performance assumptions,

(a) $\mathrm{ETC_t(t) = PD - ES(t)} = 11 - 3.88 = 7.12$;
$\mathrm{EAC_t(t) = AD + ETC_t(t)} = 6 + 7.12 = 13.12$ weeks
(b) $\mathrm{ETC_t(t) = (PD - ES(t))/SPI(t)} = (11 - 3.88)/0.52 = 13.69$;
$\mathrm{EAC_t(t)} = 6 + 13.69 = 19.69$ weeks
(c) $\mathrm{ETC_t(t) = (PD - ES(t))/SCI(t)} = (11 - 3.88)/0.48 = 14.83$;
$\mathrm{EAC_t(t)} = 6 + 14.83 = 20.83$ weeks

All these metrics indicate that by continuing at the same pace, the project cannot be completed on time. Accelerating the critical activities is required.

10.6 Limitations of Earned Value Management

There are certain limitations of EVM which one has to consider when applying it (Hall 2012):

(i) EVM does not distinguish between critical and noncritical activities. Obviously, schedule variances for critical activities are more important than those for noncritical activities with comparable values and hence, require closer attention of PM.
(ii) EVM assumes activities are independent of each other, which often is not true in practice. Hence, a schedule variance in one activity may affect the performance of one or more dependent activities.
(iii) EVM does not reflect the impact of poor quality. Both schedule and cost variances can be satisfactory, but the quality can still be far from acceptable levels.
(iv) The precise measurement of EVM depends on the sufficiently precise determination of the project progress, which requires extensive and reliable data, particularly for projects with highly detailed WBS.

10.7 Conclusions, Recent Developments, and Some Future Research Directions

EVM has been widely used in practice to measure project performance throughout the life of a project. It has been also used to estimate expected project time and cost based on the current status of the project. For this purpose, researchers have recently been following new approaches and implementing new techniques. Artificial intelligence methods have recently been used for predicting project performance (Wauters and Vanhoucke 2016). Fuzzy logic has been applied to redefine earned value metrics considering uncertainty (Naeni et al. 2011). In addition to methodological improvements, the implementation of EVM in different project settings has always been an interesting research issue. The increasing acceptance rate of agile project management has led to the questioning of the implementation of EVM in agile projects (Manship 2018).

Other than the performance monitoring and forecasting tools, such as EVM metrics, project managers require reliable early warning systems (Martens and Vanhoucke 2018). Tools that help managers to intervene effectively and alert them to undertake corrective measures at the right time are needed. Simulation and optimal control theory, which is a field of mathematics that aims to determine optimal control policies for dynamic systems, are appropriate techniques to model the effects of the intervention. Optimal control theory has various applications in managerial decision-making (Sethi and Thompson 2002). However, applications that address intervention decisions in projects are scant. Among the limited number of studies, Schmidt and Hazır (2019) formulated and solved and an optimal control problem for a construction project. Their model addresses how to distribute the control resources among the activities optimally and their solution requires constant control effort for each activity during some specific time windows.

Besides theoretical analyses using the above-mentioned methodologies, developing decision support systems (DSS) for project control systems that model the corrective actions effectively and present the results using interactive graphical interfaces will definitely address the needs of PMs. There is a need for model driven DSSs that assist decision-makers in performing sensitivity, what/if, scenario analyses. In that sense, project control problems are a promising research area for model driven DSSs (Hazır 2015).

Finally, data analysis techniques have become a crucial tool in project management as well. Making use of big data techniques can support analyzing project performance and provide PMs with early warnings of deviations (Olsson and Bull-Berg 2015). We also stress the importance of working with real life project data. Up to now researchers have mostly used artificially generated data to test their algorithms. With the emergence of big data analytics, it has been becoming more important to create real-life project databases and making analysis and conclusions based on empirical data (Batselier and Vanhoucke 2015).

Exercises

10.1 (a) Explain the purpose of defining the Schedule and Cost Performance Index (SCI).

(b) In Sect. 10.4, the following two conditions are given:

If SPI <1.0 and CPI <1.0, then SCI $< \min\{\text{SPI, CPI}\}$,
If SPI ≥ 1.0 and CPI 1.0, then SCI $\geq \max\{\text{SPI, CPI}\}$.
Show that these conditions are correct.

10.2 One approach to estimate the forecasted budget at completion (FAC) is through the expression FAC = BAC / CPI, where CPI = BCWP/ACWP. What is the underlying assumption when adopting this approach?

10.3 When monitoring a project assume that the project is behind schedule and over cost. Referring to the metrics, specify the conditions under which the project would be under budget or over budget?

10.4 Explain how and why the results of performance evaluation measures change according to whether the cash flows are assumed to occur at the termination of an activity or uniformly over its duration?

10.5 Considering the given network and data provided, where costs are incurred upon completion of an activity:

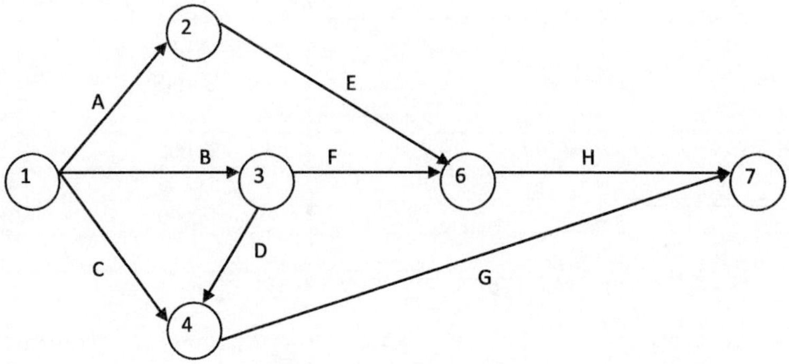

Activity	Duration	Cost ($)	Status	Duration	Cost ($)
A	4	24,000	Finished	5	27,200
B	3	27,000	Finished	3	29,400
C	3	21,000	Finished	5	21,000
D	4	36,000	Finished	6	52,000
E	5	10,000	Finished	6	12,700
F	7	35,000	Finished	9	38,300
G	9	54,000	In progress	3(20%)	12,100
H	3	32,000	–	–	–

(a) Determine the cost profiles based on ES and LS.
(b) Determine the performance measures at t = 12.
(c) Draw the actual cost profile at t = 12.

10.6 Consider the following project data. Assume costs are incurred upon completion of an activity. At the end of the project a lump sum payment of $340,000 is to be made.

Activity	Duration (periods)	Predecessors	Cost ($)
A	8	–	25,000
B	12	–	42,000
C	7	A	18,000
D	10	A	27,000
E	14	B, C	46,000
F	5	B, C	12,000
G	3	D, E	7000
H	12	D, E	42,000
I	10	F, G	36,000

(a) Find the NPV of the project at a discount rate of 3% per period.
(b) If an option is available to reduce the duration of the activities in the table below at additional cost, what would be the optimal project duration? Assume indirect costs to be negligible.

	Duration		Cost ($)	
Activity	Crash	Normal	Crash	Normal
C	5	7	22,000	18,000
E	11	14	58,000	46,000
I	8	10	44,000	36,000

10.7 Consider the following project data observed at t = 15.

Activity	Immediate predecessor	Duration	Cost ($'000)	Actual cost ($'000)	Percent complete
A	–	8	80	87	100
B	–	11	40	41	100
C	–	3	130	118	100
D	A	6	70	56	70
E	B	9	60	15	30
F	C	12	70	67	80
G	D	7	90	–	0
H	E	8	30	–	0
I	F	5	90	–	0
J	F,G,H	4	80	–	0

(a) Draw the cumulative cost curves for the ES and (LS) schedules. Assume activity cost is incurred at the completion of the activity.
(b) Draw the cumulative cost curves for the ES and LS schedules where activity cost is incurred throughout the activity uniformly.
(c) Assume the project manager has taken the early start schedule as the actual schedule and activity cost is incurred uniformly over the activity duration. Calculate the schedule variance (SV), cost variance (CV), cost performance index (CPI), schedule performance index (SPI).

10.8 Consider the following project network and the given data.
(a) Determine the cost profiles based on ES and LS. Draw the actual cost profile at t = 15. Assume costs are incurred upon finishing an activity.
(b) For the ES case, calculate the accounting variance, the schedule performance index, the cost variance, the expected performance index to evaluate the project at time t = 15. Give interpretations with regard to the status of the project using the schedule performance index, the cost variance index, and the expected performance index.

(c) Update the expected completion time and budget at completion based on original estimate approach.
(d) Update the expected completion time and budget at completion based on past performance.

Activities	Immediate predecessors	Duration (days)	Cost (1000 TL)	Percent complete	Actual Cost (1000 TL)
A	–	3	90	100	95
B	–	6	120	40	70
C	A	4	80	100	90
D	C	4	130	100	150
E	A	7	135	50	50
F	C	6	170	80	170
G	B,D	9	260	–	–
H	E,F	5	90	–	–
I	B,D	3	70	–	–
J	I	7	290	–	–
K	H,G	8	180	–	–

10.9 In industrial projects, PV and EV might not be formulated as a linear function of time. For these cases;
(a) Develop an algorithm to estimate the ES value. (**Hint**: Linear interpolation could be used for these cases.)
(b) Plot two nonlinear functions for PV and EV and show the approximation line and the ES value on the graph. (**Hint:** Consult Fig. 10.5).

10.10 Plot the EAC_t value on a graph (y-axis: Cost; x-axis: Time), given linear PV and EV functions of time and under the first and second assumptions that are described in Sect. 10.5.

10.11 In Ethical Durable Household Goods Company (EDHG Co.) a hydraulic press is needed for introducing a new series of products. The hydraulic press is to be located within an already existing part of the factory, but the construction of a special foundation is needed. The order for the hydraulic press will be placed with the Hydraulic Press and Die Company. Once the order is put, co-design activities and training of a group of personnel will start at the Hydraulic Press and Die Company and the construction of the foundation for the hydraulic press at EDHG Co. can be started to be followed by engineering work to build the infrastructure for system integration. After the completion of co-design and training, the manufacturing and testing activity can be started. Once manufactured and fully tested, the hydraulic press is to be transported to the site and assembled. Another group of personnel at EDHG Co. is to be trained. The initial plan is obtained by starting each activity at its earliest starting time. The budgeted values are given in Table 10.4. Costs are incurred uniformly over an activity.

Table 10.4 The budgeted values in Exercise 10.11

Activity	Description	Immediate Predecessors	Budgeted Values Cost (TL)	Budgeted Values Duration (Weeks)
A	Ordering of the hydraulic press	–	23,000	4
B	Preparing the site	–	28,000	4
C	Co-design and training	A	10,000	6
D	System integration	B	5000	2
F	Manufacturing and testing	C	800,000	18
G	Transportation	F	40,000	2
H	Training at EDHG Co.	D,F	5000	2
I	Assembly at site	D,G	10,000	2

Determine the schedule variance SV, cost variance CV, accounting variance AV, the Schedule Performance Index SPI and the Cost Performance Index CPI at the end of week 18, and the forecasted budget at completion FAC together with the forecasted project duration based on past information. The current situation at the end of week 18 is given in Table 10.5.

Table 10.5 The current situation at the end of week 18 in Exercise 10.11

Activity	Actual/Planned* Begin	Actual/Planned* Finish	Actual cost (TL)	Percent complete
A	0	4	28,000	100
B	2	7	39,000	100
C	4	12	12,000	100
D	9	12	6000	100
F	12	30*	320,000	33.33
G	30*	32*	0	0
H	24*	26*	0	0
I	32*	34*	0	0

10.12 In a construction project, consider the account 0800 on concrete pouring which is made up of two sub-accounts, which are further divided into three secondary sub-accounts each. Determine and report the missing values in the cells of Table 10.6 together with the Percent Complete for all levels of the accounts including the values in the row for account 0800.

Table 10.6 Account 0800 data in Exercise 10.12

Account	Sub-account	Secondary sub-account	Unit	Current quantity estimate	Total work hours to be earned	Quantities to date	Earned work hours to date	Percent complete
		0811	m^3	500	5000	500		
		0812	unit	10	1000	9		
		0813	m^3	1000	10,000	750		
	0810							
		0821	kg	550	6000	55		
		0822	unit	10	1000	2		
		0823	m^3	2500	15,000	0		
	0820							
0800			m^3					

10.13 In an expansion project for a workshop, the amounts of work for the following activities were estimated as reported in Table 10.7.

Table 10.7 The estimated amounts of work of the activities in Exercise 10.13

Activity	Unit	Current quantity estimate	Total workhours to be earned	Budgeted unit rate of work
Excavation	m^3	234	193	
Concrete pouring	m^3	94	2201	
Steel construction	Tons	2.5	119	
Mechanical work	Unit	1.1	152	
Pipe laying	m	180	470	
Electrical	m	84	220	

(a) Calculate the budgeted unit rate of work for each activity and report in the associated column in Table 10.7.
When estimates and calculations were revised just before implementation it was realized that the amounts of work were wrong. The new estimates were determined as in Table 10.8. After the initiation of the project at some point in time the amount of work performed in the activities was assessed by the quantity surveyor and reported as in Table 10.8.
(b) Determine the progress of work in each activity and the associated earned hours of work. Report the progress of the project.

Table 10.8 The revised estimates and the current realized amounts of work in Exercise 10.13

Activity	Unit	Revised quantity estimate	Total workhours to be earned	Quantities to date	Earned workhours to date	Percent complete
Excavation	m^3	257		100		
Concrete pouring	m^3	102		42		
Steel construction	Tons	2.2		0.85		
Mechanical work	Unit	1.3		0.4		
Pipe laying	m	210		35		
Electrical	m	79		22		

10.14 Fill in the empty cells in Table 10.9 and determine the Percent Complete values.

Table 10.9 The data for determining Percent Complete in 10.14

Account	Initial unit rate estimate	Current quantity estimate	Total workhours to be earned	Quantities to date	Earned workhours to date	Percent Complete
2568	1.27	1063	1350		250	
2572	1.02	100	102	70		
Total			1452		321	

10.15 The performance of a project was evaluated 12 days after its start. The relevant information concerning the project is shown in Table 10.10. Costs are incurred uniformly over the activities.
 (a) On the same Gantt chart, show the early start project plan and the actual project progress.
 (b) Calculate the schedule variance and the cost variance for the project at t = 12 days. Discuss.
 (c) Calculate the schedule performance index and the cost performance index for the project at t = 12 days. Discuss.

Table 10.10 Planned and actual values for the activities in 10.15

Activity	Predecessor	Duration (Days)		Estimated Cost	t = 12 days	
		Planned	Actual		% Complete	Actual Cost
A	–	3	3	$3000	100	$3000
B	A	7	8	$14,000	100	$15,000
C	A	4	5	$6000	100	$8000
D	C	4	13[a]	$4000	80	$4000
E	B	2	13[a]	$1000	80	$1000
F	E, D	3	16[a]	$6000	–	–

[a]Estimated Finish Times

References

Anbari, F. T. (2003). Earned value project management method and extensions. *Project Management Journal, 34*(4), 12–23.

Batselier, J., & Vanhoucke, M. (2015). Construction and evaluation framework for a real-life project database. *International Journal of Project Management, 33*(3), 697–710.

Brandon, D. M., Jr. (1998). Implementing earned value easily and effectively. *Project Management Journal, 29*(2), 11–18.

Colin, J., & Vanhoucke, M. (2014). Setting tolerance limits for statistical project control using earned value management. *Omega, 49*, 107–122.

Hall, N. G. (2012). Project management: Recent developments and research opportunities. *Journal of Systems Science and Systems Engineering, 21*(2), 129–143.

Hazır, Ö. (2015). A review of analytical models, approaches and decision support tools in project monitoring and control. *International Journal of Project Management, 33*(4), 808–815.

Hazır, Ö., & Shtub, A. (2011). Effects of the information presentation format on project control. *Journal of Operational Research Society, 62*, 2157–2161.

Henderson, K. (2005). Earned schedule in action. *The Measurable News, 8*, 23–30.
Manship, K. (2018) *Implementing earned value management on agile projects. Thesis of master of science*, Applied Information Management and the Graduate School of the University of Oregon, OR.
Martens, A., & Vanhoucke, M. (2018). An empirical validation of the performance of project control tolerance limits. *Automation in Construction, 89*, 71–85.
Naeni, L. M., Shadrokh, S., & Salehipour, A. (2011). A fuzzy approach for the earned value management. *International Journal of Project Management, 29*(6), 764–772.
Olsson, N. O., & Bull-Berg, H. (2015). Use of big data in project evaluations. *International Journal of Managing Projects in Business, 8*(3), 491–512.
Schmidt, K. W., & Hazır, Ö. (2019). Formulation and solution of an optimal control problem for industrial project control. *Annals of Operations Research, 280*(1–2), 337–350.
Sethi, S., & Thompson, G. (2002). *Optimal control theory: Applications to management science and economics* (2nd ed.). Boston, MA: Kluwer Academic Publishers.
Shtub, A., Bard, J., & Globerson, S. (2014). *Project management: Processes, methodologies, and economics. (new international edition)*. New York: Prentice Hall.
Vanhoucke, M. (2009). *Measuring time: Improving project performance using earned value management* (Vol. 136). New York: Springer Science and Business Media.
Venkataraman, R. R., & Pinto, J. K. (2010). *Cost and value management in projects*. New York: Wiley.
Wauters, M., & Vanhoucke, M. (2016). A comparative study of artificial intelligence methods for project duration forecasting. *Expert Systems with Applications, 46*, 249–261.
Webb, A. (2003). *Using earned value: A project manager's guide*. London: Routledge.

Project Risk Management 11

Learning Outcomes
Upon successful completion of this Chapter, the reader will be able to:

1. Define uncertainty and risk in projects.
2. Classify the sources of project uncertainty and give examples.
3. Appraise managing risks and its importance in projects.
4. Explore and apply project risk management and its phases.
5. Explain the risk management processes and apply qualitative and quantitative risk assessment methods.
6. Apply decision trees in project risk management.

11.1 Introduction

All projects contain a degree of uncertainty, making it a challenging task for PMs to manage projects towards project goals. Uncertain events constitute risk factors, as discussed in Chap. 1 and impact goals to various extents. The characteristics and magnitude of these impacts depend significantly on project management practices; which risk factors are identified, how they are analyzed and how the responses are managed. Hence the identification and management of risks in projects are key to achieving project success, for all sizes and types of projects, making risk management an essential dimension of project management (Krane et al. 2010).

Risk management primarily addresses the identification of risk factors ("what can go wrong"), determining responses that minimize their impact on project objectives, managing the responses, planning and running the contingency time and funds (Larson and Gray 2014). In a broader sense, Wideman (1992) defines project risk management as "the art and science of identifying, assessing and responding to project risk throughout the life of a project and in the best interests of its objectives".

Uncertainties may also create opportunities, which have also been studied as a part of risk management. Considering this, PMI (2013) defines the aim of project risk management as increasing the probability and impact of positive events

G. Ulusoy, Ö. Hazır, *An Introduction to Project Modeling and Planning*, Springer Texts in Business and Economics, https://doi.org/10.1007/978-3-030-61423-2_11

(opportunities) and reducing those of negative events (risk). All these concepts will be explained in the subsequent Chapters.

In this book, we will adopt a process-based approach to project risk management. The six risk management processes identified in PMI (2013) are:

- **Risk management planning**: Develop the risk plan.
- **Risk identification:** List the risk items.
- **Qualitative risk analysis:** Evaluate the seriousness and likelihood of each risk item.
- **Quantitative risk analysis:** Estimate the probabilities of risk items and estimate their impacts.
- **Risk response planning:** Reduce negative impacts, enhance the positive ones.
- **Risk monitoring and control:** Maintain records of the risk management processes.

Cooper et al. (2014) present a similar process-based approach and identified the key managerial questions to be addressed in these processes.

In both of the classifications, identification, analysis (qualitative or quantitative) and evaluation, response (treatments) planning, monitoring, and control are common and important steps. We now describe these steps in detail and present a case study.

11.2 Processes in Project Risk Management

11.2.1 Risk Identification

All the process steps and managerial questions summarized in Table 11.1 emphasize proactive management practices, on which project risk management is based. Anticipation and planning are preferred to simply observing and reacting (Wideman 1992). Identification of risks is thus a crucial initial step that supplies valuable risk data to planning and analysis. Due to its criticality, risk identification is not performed only by the PM or the project team; stakeholders must also engage in the process so that risk items can be listed comprehensively and studied following a

Table 11.1 Project risk management processes

Process step	Managerial questions
Risk identification	What could happen and how?
Risk analysis	What would that mean for the objectives?
Risk evaluation	What are the most important things?
Risk treatment	What are we going to do about them?
Risk monitoring	How do we keep them under control?
Risk communication	Who should be involved in the process?

Source: Cooper et al. 2014

systems approach. Depending on the number of stakeholders involved, questionnaires can be prepared, or their viewpoints noted in the meetings or brainstorming sessions.

Instead of simply listing the risk items, grouping them into specific categories helps PMs to focus on common characteristics and relationships, and facilitates a systematic risk management approach. A useful approach is focusing on the sources, instead of the effects. PMI (2009) classifies the *sources of risks* as follows:

- External but unpredictable,
- External predictable but uncertain,
- Internal, non-technical,
- Technical,
- Legal.

In this classification, external risk items are those over which PMs have minimal control. Some threats such as natural disasters are external but unpredictable. However, for some external risk items, trends can be analyzed, consequences predicted, and mitigation actions taken. For example, the source for changes in market conditions such as increases in interest rates would be external predictable but uncertain. They can be foreseen, and financial plans revised accordingly but still uncertain. In contrast, PMs are expected to manage internal risk items using appropriate managerial tools. For instance, risks arising from the departure of team members can usually be avoided by creating a pleasant and safe working environment with good benefits programs. Technical risks might arise due to various causes such as new developments and changes in product and process technology, and legal risks due to changes in laws and regulations.

Another classification for the sources of risks is given by Komendantova et al. (2012). They identified nine classes of risks: Technical, construction, operating, revenue, financial, force majeure (which includes terrorism), regulatory, environmental, and political.

The list of risks identified is called the *risk register*. In Table 11.2, an example for a risk register is given of a new product development project in a durable household goods manufacturer. For each source of risk, *risk items* from this risk source are listed and possible causes for each risk item are indicated. For example, considering the external-predictable source of risk "competitive environment" would be a risk item. Possible causes would be "new technologies developed by the competitors" and "changes in standards and regulations". To mitigate the risks associated with these possible causes the company has to closely monitor these areas and develop appropriate response plans.

After classifying the risks, PMs prepare a *risk breakdown structure* (RBS). Just as project tasks are broken into smaller work packages in the WBS; in RBS risk items are organized under a hierarchical system. While WBS plays a central role in project planning, RBS is the major source of information for planning and monitoring project risks. An example RBS for a construction project is given in Table 11.3. Hillson (2002) discusses the importance of preparing an appropriate RBS in projects.

Table 11.2 An example for risk register

Risk sources	Risk items	Risk causes
Technical	Maturity level of the technology used	Use of new-to-the-firm technology
		Use of new-to-the-world technology
	Complexity and uncertainty of the technical content	High uncertainty in technical content
		Difficulty in defining the project scope
	Inadequacy of the technical personnel	Absence of qualified technical people (persons who have the experience and knowledge about the technology)
Resource management	Inadequate resources	Inadequacy of labour units for this project because of overloading
		Inadequacy of laboratories / equipment because of overloading
		No experience with the use of the laboratories / equipment
		Equipment breakdown / lack of maintenance
		Reduction in project team size
	Changes in team members	Turnover in project team
Non-technical internal (managerial – Project Management)	Inadequate communication	Inadequacy of communication with upper management
		Inadequacy of communication within the project team
	Changes in strategy / project priorities	Changing objectives / expectations
	Inadequate project experience	Inexperienced project leader
		Lack of teamwork experience in the project team
Customer related	Uncertainty in the communication with the customer	No previous experience of working together with the customer
		Customer violating the written and oral agreements / understandings
	Uncertainty in customer requests	Frequent change requests by the customer
		Project aborted by the customer
	Project budget	Payment delays / cash flow irregularities
External- predictable	Material / service acquisition	No previous experience of working together with the supplier / consultant
		Difficulty in material procurement
		Limited service alternatives
		Interruption of provided services
		Problems in deliveries
	Competitive environment	New technologies developed by the competitors
		Changes in standards and regulations

(continued)

Table 11.2 (continued)

Risk sources	Risk items	Risk causes
External- unpredictable	Natural hazards	Earthquake, flood, etc.
	National / international economic crises	Economic crises and exchange rate fluctuations affecting the project
	International relations and legal regulations	Changes in international relations affecting the project
		Legal and bureaucratic obstructions affecting the project

He presents an example RBS for a construction design and a software development project. Hillson et al. (2006) list the main uses and benefits of RBS as follows:

- It serves as a risk identification aid
- It facilitates risk assessment. Identified risks can be shown on the RBS and classified by source.
- In bidding and tendering risks of alternatives could easily be compared. To describe the risks RBS presents a common structure and terminology.
- It aids risk reporting. Risk terminology in the reporting can be based on the RBS.
- RBS could also be used to aid in reporting the lessons learned.

During the risk identification process, a risk register, a formal detailed document that includes the list of risk items, can be prepared, containing a taxonomy of the risk items and an RBS along with a description of each risk item and information on their causes and effects.

The list of risks is dynamic in that it must be updated throughout the project life cycle. During execution, risk items are reconsidered, and the RBS revised whenever necessary.

11.2.2 Risk Analysis

As we have already discussed in Chap. 1, over the years projects have become more complex and the number and variety of stakeholders involved has been increasing. As a result, risk registers may contain thousands of risk items and managing these risks involves many organizational and environmental factors. PMs need to prioritize risks to focus on a limited set of these items. This prioritization constitutes the initial step of the risk analysis, which makes use of both qualitative and quantitative methods.

Table 11.3 An RBS of a construction project

Total risk				
1. Economic	2. Contractual	3. Political	4. Construction	5. Management
1.1 Inflation	**2.1. Failure of payment**	**3.1 Environmental**	**4.1 Uncertainty in labor**	**5.1. Productivity**
1.1.1. Labor	2.1.1. Owner	3.1.1. Air	4.1.1. Availability	5.1.1 Labor
1.1.2 Material	2.1.2 Contractor	3.1.2 Noise	4.1.2. Skills	5.1.2 Equipment
1.1.3. Equipment		3.1.3 Water		
1.2. Energy shortage	**2.2 Delay disputes**	**3.2. Public disorder**	**4.2 Uncertainty in equipment**	**5.2. Quality control**
		3.2.1. Demonstrations	4.2.1. Breakdown	**5.3. Safety**
		3.2.2 War	4.2.2. Availability	**5.4. Mistakes**
1.3. Financial uncertainty	**2.3. Coordination failure**	**3.3. Government Acts & Regulations**	**4.3. Uncertainty in material**	**5.5. Management incompetence**
1.3.1. Owner	2.3.1. Owner	3.3.1. Tax rate changes	4.3.1. Storage	
1.3.2. Contractor	2.3.2. Contractor	3.3.2. Permits	4.3.2. Protection	
1.3.3. Designer			4.3.3. Availability	
	2.4. Change orders		**4.4. Delayed site access**	**5.5. Variation in quality**
	2.4.1. Delay		4.4.1. Permit	
	2.4.2. Design change		4.4.2. Title	
1.4. Currency fluctuation	**2.5. Labor disputes**		**4.5. Quantity variation**	
			4.6. Defective construction	

Source: Kangari and Boyer 1989

Qualitative methods help to focus on individual risk items, understand their causes, estimate the likelihood of their occurrences and potential effects on project goals, and clarify their relationships with other risk items. *Quantitative methods* help PMs to study identified risks as well as to model and quantify their combined effects on goals (PMI 2009).

11.2.2.1 Qualitative Methods

The primary purpose of risk analysis is to support the PM's decision making, especially in prioritizing the risks so that appropriate resources can be allocated to mitigating and managing high priority risks. Ranking the risk items requires estimates of two principal parameters of the risks: the estimations of their probability of occurrence (also called likelihood) and the estimations of their impacts on the project goals. However, in practice, it is not always possible to quantify these parameters for all risk items. In many cases, project records are missing, or PMs might not be able to obtain the data of the previously undertaken projects. Existing data is often inaccurate, resulting in misleading findings from analysis using this data. In such cases, qualitative methods such as brainstorming sessions, the Delphi Technique, and SWOT (Strength, Weakness, Opportunities, and Threats) Analysis have been commonly used (PMI 2013). All these methods seek to collect expert opinions effectively. Especially in a dynamic project where environmental conditions are rapidly changing, integrating the information given by experts becomes crucial.

In the following, we will introduce two qualitative methods both based on an ordinal scale where we want to order the risks according to their *severity level*. The severity level is defined as a function of both the *probability of occurrence* and the *overall impact* of the risk on the project goals.

Qualitative Method 1 In this method, experts first define several categories for risk parameters; i.e., very low, low, medium, high, and very high. Note that the categories for risk parameters are defined based on ordinal scale ordering the categories among themselves but we cannot quantify the differences between the levels of the scale. In other words, we do not know how the difference, for example, between very high and high compares with the difference between medium and low. The fact that precise numerical values are not required is an advantage when data is missing or inaccurate. Table 11.4 illustrates the categories for the probability of occurrence and impact and the corresponding severity level for the risk item.

Having defined the categories and discussed the severity levels in the expert meetings, a detailed analysis is performed for each risk item. The results of the study can be presented using tools such as a risk assessment matrix.

Table 11.4 Risk severity matrix

Impact / Probability of Occurrence	**Very Low**	**Low**	**Medium**	**High**	**Very High**
Very Low	Very Low	Very Low	Low	Low	Low
Low	Very Low	Low	Low	Medium	Medium
Medium	Low	Low	Medium	High	High
High	Low	Medium	High	High	
Very High	Low	Medium	High		

Table 11.5 Risk assessment matrix

Risk description	Risk category (Impact)	Impact	Probability of occurrence	Risk evaluation (Severity Level)
Departure of a key team member	Schedule	High	Low	Medium
Design changes	Scope	Medium	Medium	Medium
Natural disasters	Schedule	High	Very low	Low
Increase in inflation	Budget	Medium	High	High

Example 11.1 Table 11.5 illustrates a case for four different risk items, which are grouped based on their impact areas: schedule, scope and budget. Given the opinions of the experts on the probability of occurrence and the impacts of the risks, the risk evaluation column of the matrix is added by consulting the risk severity matrix (Table 11.4). Risk items that are placed in high and very high grids of the risk severity matrix are considered to be severe risk items and shown in darker colors. These severities result in the prioritized list of project risks with high severity corresponding to high priority.

The *Risk Assessment Matrix*, an example of which is illustrated in Table 11.5, helps PMs to focus on high probability or high impact risks and use scarce project resources more effectively to address them. In practice, risk items with high or very high-risk assessments are monitored more frequently and specific response strategies are developed for them. In the above example, the risk of an increase in inflation needs more attention than the other items.

In practice the risk assessment process is not fully linear; there are interactions among the inherent processes (Rees 2015). For instance, having defined the risk response actions, new risk items might be identified. Working with subcontractors might create quality problems, unexpected delays due to communication problems. This might require list of risks to be updated. Another example of interaction could be observed between identification and ranking. While creating the risk registers, some of the items could be considered as not important, meaning that they have been already non-prioritized, and they might be excluded from the list with no need for ranking.

Since the above analyses require that a group of experts or project team members define the categories and match the risk items with these categories, it is by nature subjective. To reach objective results PMs must use quantitative analysis, which depends heavily on the availability of reliable data. To give an example, if the frequency of the design changes and the resulting cost increases can be estimated accurately from the records of previous projects, the impact of the design changes on the project budget can be predicted accurately.

Qualitative Method 2 In this method, generic impact scales and probability of occurrence scales given earlier in the literature were decided to be used in prioritization (Table 11.6). Using this ordinal 1–5 scale and consulting with possible experts and the project team, the PM *a probability of occurrence* and a *level of impact for each impact category*. Each risk item will be assessed with its impact on four different aspects of the project: Quality, functionality of the deliverable(s), schedule, and cost. Individual states described corresponding to scale values 1–5 for each impact category are specified in Table 11.6 (Graves, 2000; Patterson and Neailey, 2002).

Table 11.6 Scales for probability and impact estimations (Graves 2000; Patterson and Neailey 2002)

Probability of occurrence	**Scale**
Very low probability of risk to happen (0–5)%	1
The risk less likely to happen than not (6–20)%	2
The risk is just as likely to happen as not (21-50)%	3
The risk is more likely to happen than not (51–90)%	4
The risk will happen almost definitely (91–100)%	5
Quality impact (Quality is defined here as the conformance quality of the project end item with its technical specifications)	
Quality degradation barely noticeable	1
Only very demanding applications are affected	2
Quality reduction requires client approval	3
Quality reduction unacceptable to the client	4
Project end item is effectively not usable	5
Functionality impact	
Functionality decrease none or barely noticeable	1
Minor areas of functionality are affected	2
Major areas of functionality are affected	3
Functionality reduction unacceptable to the client	4
Project end item is effectively useless	5
Schedule impact	
Insignificant schedule slippage	1
Overall project slippage < 10%	2
Overall project slippage 10–20%	3
Overall project slippage 20–50%	4
Overall project slippage > 50%	5

(continued)

Table 11.6 (continued)

Cost impact	
Insignificant cost increase	1
< 5% cost increase	2
5–10% cost increase	3
10–20% cost increase	4
> 20% cost increase	5

Table 11.7 Severity score matrix

Probability of Occurrence	**5**	7	9	11	13	15
	4	6	8	10	12	14
	3	5	7	9	11	13
	2	4	6	8	10	12
	1	3	5	7	9	11
		1	**2**	**3**	**4**	**5**
	Overall Impact					

For further processing, the levels of impact of the individual risk categories are combined into an *overall impact score*. When determining the overall impact score, the highest among the impacts on quality, functionality, schedule, and cost is accepted as the overall impact score. Another way to measure the overall impact score would be to determine the weighted average of the individual impact scores. The weights are between zero and one and add up to 1. The weighted average score will be in general less than the maximum impact score among the impacts. A disadvantage resulting would be that if the difference between the maximum impact score and the weighted average is relatively high, then this might lead the PM to adopt an ineffective risk response plan.

Note that the assessments are subjective. The probability ranges given in Table 11.6 are the ones suggested by Patterson and Neailey (2002). Obviously, probability of occurrence scores can be different in different situations and for different organizations. If a numerical value can be assigned for a probability rather than a scale value, then obviously this option should be used. The same applies to the individual states described for each impact category.

The severity level for each risk is determined as a combination of its probability of occurrence and the overall impact. These combinations constitute a 5×5 matrix designated as the risk severity matrix (Table 11.7). Note that if scales of 1–6 are defined, then a 6×6 severity matrix would be employed. The cell (5,5) corresponds to the highest severity whereas the cell (1,1) to the lowest. In Table 11.7, the severity score of each cell is determined based on the relationship of Severity score = Probability of occurrence score + 2 × Overall impact score. The multiplier for the overall impact score can be less or more than "2" depending on the assessment of the organization regarding the relative importance of these two dimensions on the risk severity.

Table 11.8 Risk severity matrix (Graves 2000)

Probability of Occurrence	5	**19**	**14**	**9**	**4**	**1**
	4	**21**	**16**	**11**	**6**	**2** A
	3	**23** C	**18**	**13** B	**8**	**3**
	2	**24**	**20**	**15**	**10**	**5**
	1	**25**	**22**	**17**	**12**	**7**
		1	2	3	4	5
	Overall Impact					

A=High severity, B=Moderate severity, C=Low severity.

Using the severity scores the severity levels are specified as High, Moderate, and Low severity. We assume here that severity scores greater than 11 imply a high severity level for the risk item involved. The moderate ones are between 8 and 11 and low ones are below 8. These ranges on severity scores are subjective and reflect the choice of the organization. With the above choice of ranges there are 6 high severity, 10 moderate severity, and 9 low severity cells. Extending, for example, the range of the severity scores for the high severity level to severity scores greater than, say, 9 will increase the number of cells with high severity level to 11 and thus the probability of a risk item to be classified as a high severity risk.

The risk severity matrix is given in Table 11.8. The classes of risk severity cells A, B, and C are distinguished on the risk severity matrix by the level of darkness of their cells. The cells are numbered starting with the high severity risk cells. Whenever there is a tie, the tie is broken in favor of the cell with the higher overall impact.

If two risk items share the same cell of the risk severity matrix, then this does not give any clue with which one to deal first. For this special case as well as in general, the severity scores in Table 11.7 can be employed for the *prioritization of the risks* involved, i.e., for the *ranking* of them.

Once risks are identified, assessed, and prioritized, then they need to be further processed such as by the development of risk response plans, which are covered in Sect. 11.2.3. The number of risk items selected for further processing depends on the availability of resources and management effort allocated for that purpose.

Example 11.2 In the following we will provide an example for risk analysis from a project in a tire manufacturer. The project is on online gauge measurement and optimization of textile material used in tire manufacturing. When calculating the overall impact, all three impact coefficients are taken to be equal. The list of risks and overall impact calculations are given in Table 11.9. The risk severity assessment is then provided in Table 11.10.

Table 11.9 List of risks and overall impact calculations

Risk item	Risk description	Quality impact	Schedule impact	Cost impact	Overall impact
1	Current gauge adjustment equipment capability	3	3	4	3
2	Current gauge measurement system capability	4	3	5	4
3	Additional investment need for measuring the output gauge	3	5	4	4
4	IT infrastructure related problems	4	4	4	4
5	Budget limitations for investment	4	4	3	4
6	New measurement device delivery delays	1	5	2	2
7	Departures from the project team	2	2	2	2
8	Lack of qualified technical personnel in production	2	2	2	2
9	Sharp decrease in tire demand	1	1	1	1

Table 11.10 The risk severity assessment

Risk item	Risk description	Probability of occurrence	Overall impact	Risk severity
1	Current gauge adjustment equipment capability	2	4	Moderate
2	Current gauge measurement system capability	2	4	Moderate
3	Additional investment needs for new tool for measuring the output gauge	2	5	High
4	IT infrastructure related problems	4	4	High
5	Budget limitations for investment	3	4	Moderate
6	New measurement device delivery delays	2	5	High
7	Departures from the project team	2	2	Low
8	Lack of qualified technical personnel in production	3	2	Low
9	Sharp decrease in tire demand	1	1	Low

Based on the results reported in the risk severity column in Table 11.10, three risk items are assessed to be high level risks. The remaining risk items are of moderate and low levels.

Note that the risk item 6 is a demonstration of how the use of weighted average approach might underscore the risk level. Assuming equal weight for all impact types results in a weighted average value of 2.25, which together with the probability of occurrence implies low severity.

There are several deficiencies of the qualitative methods described above. One such weakness stems from the subjective nature of the evaluations. A measure to reduce the negative impact of subjective evaluation would be to employ group decision-making tools. Systematic checks on the consistency of judgments would also be useful in that respect. The qualitative methods evaluate each risk item by itself but do not assess the overall project risk. Also, using a scoring model imputes a degree of precision that simply does not exist. A halo effect (i.e., if a risk scores high on one criterion, it tends to score high on many of the remaining criteria) is also possible for a scoring model.

Quantitative methods, which will be discussed in the next section, can also be combined with qualitative methods. A qualitative analysis that focuses on identifying the risk items can be followed by a quantitative analysis that works on evaluating the impacts of these items (Chapman 2001).

11.2.2.2 Quantitative Methods

Unlike qualitative methods, quantitative methods allow a numerical analysis of risks. Among these methods, sensitivity analysis, Monte Carlo simulation, and decision trees are the most widely used.

Sensitivity analysis is a quantitative tool to predict how risk items affect the project goals (see Borgonovo and Plischke (2016) for the other application areas of this commonly used analysis tool). Analyzing the risk items one at a time, sensitivity analysis helps to identify those that need to be checked frequently and determine their tolerance intervals, which shows the level of changes that could be tolerated. To give an example, PMs could measure how sensitive the profitability of the project is to inflation rates. Therefore, they can specify a range of inflation rates that can be managed or that require special attention or preventive actions.

While sensitivity analysis usually addresses one variable at a time, real life problems require considering many variables simultaneously. The impacts of several variables and their interdependencies can be modeled using *Monte Carlo simulation*, allowing alternative plans of action can be tested. However, probability distributions need to be estimated correctly for uncertain variables such as inflation or interest rates.

Decision trees model a decision problem under uncertainty as a sequence of decision steps. By dividing a complex decision-making problem into components (decisions, random events, and results), helps to solve the problem consecutively. Before constructing the tree, random events (scenarios) and the decisions need to be clearly defined. Financial consequences of each decision choice for each realization of the random event are required to be estimated.

Since we will discuss sensitivity analysis and Monte Carlo simulation in Chap. 12, we focus on decision trees in this section.

Decision Trees A *decision tree* consists of decision and probability nodes, and arcs connecting these nodes. Decision nodes are indicated by squares and show the decision points. An outgoing arc is drawn for each alternative that can be chosen at this point. Probability nodes represent random events and are indicated by circles. Estimated probability values for the random events are shown on the arcs leaving these nodes. The tree is constructed starting from one or more decision points on the

Elements	Symbol	Explanation
Decision Node	■	Decision-maker chooses one of the alternatives
Chance Node	●	One of the scenarios will be realized.
End of Branch	◀	Expected gain or loss under each scenario is summarized.

Example

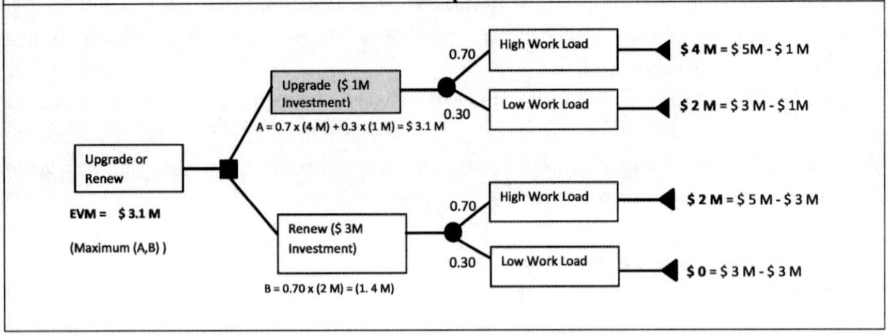

Fig. 11.1 A decision tree example

left, continuing to the right. Each decision node is followed by a decision node or probability nodes. For each decision event, expected monetary values (EMV) are calculated. To calculate the EMV, the probabilities of the occurrences are multiplied with the associated predicted monetary values and summed up considering the entire event set. Actions that maximize EMV are chosen.

An example decision tree is given in Fig. 11.1. In this example, decision-makers must choose between two alternatives: upgrade the current IT system or invest in new technologies and replace the current system. Depending on the predicted market conditions and customer orders, two scenarios are considered: low and high workload. Profitability depends on the decision taken and the workload. Maximizing the EVM, decision-makers chose to upgrade the current system in this example.

In the risk analysis application, random events represent the risk events and decision nodes the managerial actions taken to hedge against these risk events. In practice, PMs need to consider several factors and their interactions. Considering this complexity, a team discussion case of quantitative risk analysis is given in Sect. 11.3.

11.2.3 Risk Response Planning

After identifying the risks (opportunities) and assessing and ranking them, PMs need to decide on measures to take to hedge against the threats or to benefit from the opportunities. At this stage, PMs develop a proactive strategy for each risk item by considering the resources allocated, preparing action plans to mitigate the threats or exploit the opportunities.

For threats, prevention aims at controlling the sources of the risks to eliminate the risk factor completely or at least reduce their impacts, occurrence probabilities, or both. On the other hand, the PM will seek to maximize the gains or occurrence probabilities of positive opportunities. For each risk item, managers opt for one of the following strategies (PMI, 2009).

Avoid a Threat or Exploit an Opportunity A risk response plan is prepared to eliminate the negative effects of risks or to create new opportunities. The actions might include changing the scope, the methods, or replanning/rescheduling. To give an example, using new materials, a new design, or a technique might eliminate product failure risks. New investments could create new opportunities to enhance project profitability.

Transfer a Threat or Share an Opportunity This involves sharing the risk with another party in exchange for a risk premium that is paid (as in an insurance policy), or a share of the expected benefits. For many risk items, insurance has been a common practice to share losses, such as those due to natural disasters. Another risk sharing practice is using cost sharing contracts. A group of contractors/subcontractors might agree to share the additional costs or penalties in case of unexpected events or delays. Brânzei et al. (2002) consider a case where several subcontractors carry out specific tasks and each of them might have some responsibility in the delay of the project. They address how to evaluate these responsibilities using project scheduling and path analysis and to find out a "fair" penalty sharing mechanism using cooperative game theory.

Mitigate a Threat or Enhance an Opportunity This includes all managerial actions to reduce (increase) the likelihood or the impact of a threat (opportunity). In product development projects, simulation experiments, process tests, and market surveys might be conducted to minimize the risks regarding product quality and acceptance. These experiments, tests, and surveys can provide more information to predict the acceptance of the products by clients or customers. To give an example, consider the development of a new computer game. First versions of the game could be tested with the potential customers and the product design could be improved by integrating the feedback of testers. These improvements would decrease the risk, the probability that the product will not be accepted.

Accept a Threat or an Opportunity The risk is accepted, and no action is taken. This strategy is generally used in cases where other strategies are perceived to be infeasible or cost ineffective. For example, if the contractor has an equipment pool with newly acquired machines, managers might consider the probability of equipment failures very low and ignore this risk.

These strategies cover a wide range of actions, from simple ones such as accepting the risk and their consequences to preparing a comprehensive plan that requires the deployment of considerable resources (Wideman, 1992). To provide the additional resources, contingency and management reserves are allocated.

Contingency reserves refer to the additional resources added into the baseline resource levels and used to manage the identified risks that are accepted or those where mitigating actions are required. On the other hand, management reserves are held against the occurrence of unforeseen risk items that are not specifically identified but might be encountered during project execution (PMI 2013).

11.2.4 Risk Monitoring and Control

In projects, changes are inevitable. Throughout the life cycle, project scope, priorities, environmental conditions, technology may change. As a result, new risks (opportunities) may arise or the likelihood or impact of identified risks change, requiring analysis to be revised. Therefore, risks should be monitored and assessed continuously throughout the project life.

Risk monitoring facilitates learning by the project team and improves the effectiveness of their managerial practices. In this regard, documentation of the monitoring and control process is important. It includes recording how risk lists are updated, and which preventive or corrective actions have been taken. The results of these activities are monitored, recorded and their effectiveness is discussed.

Along with these regular reviews, audits should be done periodically to assess the risk management process (PMI 2009). In addition to record keeping, it is always beneficial to communicate the process and results to the team and also to stakeholders so that they can assess the current project performance and predict the future.

11.3 Project Risk Analysis: A Team Discussion Case

Alex plans to replace the project management software used in Alex & Julie Construction and Engineering Company with an integrated project/process management platform. The platform has various functions that support project managers in decision-making and can be integrated with the company's current database system. The software company has offered two sales options for the new platform. The first option covers full installation (Module 1 & 2) with a cost of $ 80,000. The second offer includes the installation of the main module (project management, business process management) with a cost of $ 50,000 and the option to buy module 2 (risk management) for $ 40,000 within 6 months.

Company managers decided to conduct a cost /benefit analysis before taking the investment decision. Having organized a brainstorming session, the risks of the project were listed in two main groups.

External Risks The future contracts of the company are dependent on public investments and economic indicators. As a result of the economic contraction, the company's business volume might decrease significantly.

Internal Risks The compatibility of the software with the existing database system and the adoption of the software by the users are crucial. In case of unsuccessful integration, data flow could be impeded, leading to significant losses.

The economic added value of the program was predicted in a long meeting with the company managers and experts. Scenarios were defined by considering the external and internal risks. Two scenarios have been emphasized for both risk factors (Economic contraction and growth, Successful and Unsuccessful Software Integration). Some decisions were taken at the meeting, and estimates were made for the scenarios.

The first decision is to install the modules gradually; the option of purchasing all the modules directly was rejected. A survey will be conducted with project managers and teams 4 months after the installation of module 1. The decision to purchase module 2 will be taken based on the survey results.

Adoption of the program and success of the installation will be assessed with the survey result. In the case of success, the present value of the economic gain of installing modules 1 and 2 is estimated as $ 70,000 if economy contracts, and $ 150,000 in case of growth. If only module 1 is installed, the expected gains are expected to be 40% less in both economic growth scenarios. On the other hand, if the modules are not successfully adopted, 30% less return is anticipated in all scenarios.

At Alex's request, several companies that have been using the module have been consulted. The probability of successful integration and adoption in the sector is estimated to be 80%. Company managers are optimistic about economic growth and an increase in their business volume. They predicted the probability of economic growth as 60%.

The company risk manager, Juliette, constructed a decision tree and made a presentation to support the decision-making process. As a result of her analysis, she suggests buying the main module, conduct the survey and buy additional modules if the results are positive.

(a) Construct a decision tree to verify Juliette's decision. Do you agree with her?
(b) Based on the recent reports on software implementation in the sector, Alex has concluded that the probability of successful implementation was overestimated. Should the investment decision be changed if the success probability is revised to 50%?
(c) What happens if the expectations of the company managers turn out to be more pessimistic; the probability of economic growth falls to 30% and the success rate decreases to 50%?

Suggestions for possible responses to the discussion points above are provided in the Solution Manual for Chap. 11.

11.4 Conclusions, Recent Developments, and Some Future Research Directions

Managing risks is considered to be a principal factor affecting project success. (Krane et al. 2010). Identifying the relevant risks and hedging against them are essential to managing all types of projects, small or large. As governments and large companies undertake ever larger projects all over the world, they are facing many types of risks, some of which could have devastating financial consequences. Identifying all the relevant risks, predicting their probabilities and impacts, and ranking them using these predictions is crucial for the viability of the projects and of the undertaking organizations. To identify and analyze the risks that might be faced in their projects, organizations can benefit from big data analytics. Since there will be more data collection channels such as internet-based systems, both the business operations and risk management practice will benefit from it (Choi et al. 2017). Big Data will also support investment decision making and evaluating the projects (Olsson and Bull-Berg 2015) through increasing the opportunities to determine relevant assessment criteria, including risk factors. To conclude, there is a need for further research to identify and improve how risks are managed in megaprojects (Sanchez-Cazorla et al. 2016) and also on how big data analytics can affect current risk management practices.

There are many studies that address risk management in large companies in the literature. However, studies on project risk management in small and medium enterprises (SMEs) and analyses focused on the management of small projects are scarce (Marcelino-Sádaba et al. 2014). SMEs need simpler tools that can be implemented without extensive experience or training and at a low cost. Hence, there is a need for developing simple tools that correspond to the needs of SMEs operating in various industries, as well.

Integration of risk analysis with project planning is also an interesting research question. In this regard, Herroelen (2014) proposed the integration of quantitative risk analysis with proactive and/or reactive project scheduling procedures. Hu et al. (2013) developed a decision-support tool for stakeholders to effectively control project risks by integrating risk analysis and planning in software development projects. Integration facilitated continuous monitoring of the risk items and dynamic revision of the plans throughout project execution. There is a need for integrated approaches in projects carried out in different industries.

Exercises

11.1 An Engineering Faculty is planning to organize a conference on renewable energy systems. The project team has come up with the following Risk Checklist and determined the level of occurrence probability and the overall impact for each risk.

RISK	Probability of occurrence	Overall impact	Risk rank (risk priority)
Not being able to find enough sponsors	High	Medium	
Not being able to reach targeted number of participants	High	Medium	
Not being able to find a suitable conference venue	Medium	High	
Bad weather conditions on the day of the conference	Medium	Medium	
Having a similar event on the same week	High	High	
Not being able to find high caliber speakers to attract participants	Low	Medium	

Determine the risk priorities using the risk severity matrix in Sect. 11.2.2.1 and report in the rightmost column of the matrix above. State at least five (in total) risk responses for the risks stated above.

11.2 In the R&D Department of Defense Technologies Inc., a weapons system is being developed. One of the work packages – development of a subsystem – is very critical and is on the critical path. It has a relatively high probability that its duration exceeds twice its mean duration. The company decides to subcontract this work package to a Sabanci University Lab in addition to its own laboratory team, such that the two teams will work in parallel and independently. How would you describe this risk policy? Explain.

11.3 Consider the four risk response planning strategies introduced in Section 11.2.3: Accept, Avoid, Mitigate, and Transfer. How would you place these strategies on the risk severity matrix?

11.4 When building an assembly line, different suppliers are employed for the purchase of various software and hardware. The engineers question the compatibility of the software package managing the movement of the assembly line with another one managing the movement of one of the overhead lines feeding the assembly line. They have gone through the probability and impact analysis to determine the severity of this particular risk of incompatibility between these two software packages. Their assessments are as follows:
The probability the risk of incompatibility will happen is estimated to be 0.35.
Quality reduction unacceptable to the client.
Functionality reduction unacceptable to the client.
An overall project slippage of 25% is expected.
A 25% increase in the associated cost is estimated.
How would you classify the severity of this particular risk?

11.5 Consider that you are editing a textbook on project management. You have two coauthors and your project covers planning the content, writing the textbook together, delivering it to the publisher, and publishing it in 2 years. Identify the major and minor risks that could be encountered in this project. Explain how you could respond to these risks.

11.6 Explain the main uses and benefits of preparing a risk breakdown structure (RBS). Discuss with examples.

11.7 Consider the construction of a large airport in a city with ten million inhabitants. Classify the external risk factors that the project can face. Give examples for each external risk group.

11.8 A European energy company plans to invest in renewable energy, install solar panels in North Africa and sell the electricity generated in the electricity market in that region. List the possible sources of project risks that the company could face. Suggest risk items for each possible risk source.

11.9 Three entrepreneurs consider establishing a startup company to develop an application to facilitate car sharing in their hometown. They might rent a small office in the town, adopt a home office system or abandon the project. Moving to an office will bring more professional customers but rents are high in the town. Taking the decision depends on the expectations regarding the market size. They have recently performed a brainstorming session to estimate their revenues and expenses.

$100,000 net income is expected if the market is favorable, but $60,000 will be lost if it is not. Working at home brings $50,000 profit in a favorable market and a $10,000 loss in an unfavorable market. Experts predict that the market will be favorable with a chance of 60%. To minimize the risk of losing money, the entrepreneurs also consider the option of buying a marketing research study and decide based on the results. The study costs $10,000. The results are reliable; there is a 0.95 probability that a positive result will be output if the market conditions are favorable. If the marketing environment is unfavorable, there is only a probability of 0.10 of getting positive results from the research. Develop a decision tree for these entrepreneurs and support their decision making.

11.10 Reconsider Problem 11.8. What could be the risk mitigation actions for this example? Are there any interactions between the risk management processes (identification, ranking, and responses planning)? Discuss with examples.

11.11 How might progress in big data analytics affect the practices in project risk management? Discuss with examples.

11.12 Ali, Ada, and Ece, three newly graduated industrial engineers, have decided to a create startup company for proposing customized software solutions to production optimization problems of small and medium sized companies operating all over the world. List the risk factors that they can face in their software development projects and prepare a risk breakdown structure.

References

Borgonovo, E., & Plischke, E. (2016). Sensitivity analysis: A review of recent advances. *European Journal of Operational Research, 248*, 869–887.

Brânzei, R., Ferrari, G., Fragnelli, V., & Tijs, S. (2002). Two approaches to the problem of sharing delay costs in joint projects. *Annals of Operations Research, 109*(1–4), 359–374.

Chapman, R. J. (2001). The controlling influences on effective risk identification and assessment for construction design management. *International Journal of Project Management, 19*(3), 147–160.

Choi, T. M., Chan, H. K., & Yue, X. (2017). Recent development in big data analytics for business operations and risk management. *IEEE Transactions on Cybernetics, 47*(1), 81–92.

Cooper, D., Bosnich, P., Grey, S., Raymond, G., Purdy, G., Walker, P., & Wood, M. (2014). *Project risk management guidelines: Managing risk with ISO 31000 and IEC 62198*. New York: Wiley.

Graves, R. (2000) Qualitative risk assessment. PM Network, October: 61–66.

Herroelen, W. (2014). A risk integrated methodology for project planning under uncertainty. In P. S. Pulat, S. Sarin, & R. Uzsoy (Eds.), *Essays in planning, scheduling and optimization: A festschrift in honor of Prof. S. E. Elmaghraby* (pp. 203–217). Boston: Springer.

Hillson, D. (2002). *Use a risk breakdown structure (RBS) to understand your risks* (Paper presented at Project Management Institute Annual Seminars and Symposium, San Antonio, TX). Newtown Square: Project Management Institute.

Hillson, D., Grimaldi, S., & Rafele, C. (2006). Managing project risks using a cross risk breakdown matrix. *Risk Management, 8*(1), 61–76.

Hu, Y., Du, J., Zhang, X., Hao, X., Ngai, E. W. T., Fan, M., & Liu, M. (2013). An integrative framework for intelligent software project risk planning. *Decision Support Systems, 55*(4), 927–937.

Kangari, R., & Boyer, L. T. (1989). Risk management by expert systems. *Project Management Journal, 20*(1), 40–45.

Komendantova, N., Patt, A., Barras, L., & Battaglini, A. (2012). Perception of risks in renewable energy projects: The case of concentrated solar power in North Africa. *Energy Policy, 40*, 103–109.

Krane, H. P., Rolstadås, A., & Olsson, N. O. (2010). Categorizing risks in seven large projects-which risks do the projects focus on? *Project Management Journal, 41*(1), 81–86.

Larson, E. W., & Gray, C. F. (2014). *Project management. The managerial process* (6th ed.). New York: McGraw Hill.

Marcelino-Sádaba, S., Pérez-Ezcurdia, A., Lazcano, A. M. E., & Villanueva, P. (2014). Project risk management methodology for small firms. *International Journal of Project Management, 32*(2), 327–340.

Olsson, N. O., & Bull-Berg, H. (2015). Use of big data in project evaluations. *International Journal of Managing Projects in Business, 8*(3), 491–512.

Patterson, F. D., & Neailey, K. (2002). A risk register database system to aid the management of project risk. *International Journal of Project Management, 20*, 365–374.

PMI. (2009). *Practice standard for project risk management*. Newtown Square: Project Management Institute. ISBN: 978-1-933890-38-8.

PMI. (2013). *A guide to the project management body of knowledge (PMBOK® guide)* (5th ed.). Newtown Square: Project Management Institute.

Rees, M. (2015). *Business risk and simulation modelling in practice: Using excel, VBA and@ RISK*. Singapore: Wiley.

Sanchez-Cazorla, A., Alfalla-Luque, R., & Irimia-Dieguez, A. I. (2016). Risk identification in megaprojects as a crucial phase of risk management: A literature review. *Project Management Journal, 47*(6), 75–93.

Wideman, R. M. (Ed.). (1992). *Project and program risk management: A guide to managing project risks and opportunities*. Newtown Square: Project Management Institute.

Project Scheduling Under Uncertainty 12

Learning Outcomes
Upon successful completion of this Chapter, the reader will be able to:

1. Interpret the relationship between planning and uncertainty.
2. Summarize the basic approaches and methods used in project scheduling under uncertainty.
3. Compare reactive, proactive, stochastic, and fuzzy scheduling methods.
4. Explain the concepts of robustness and sensitivity analysis and understand their importance for project scheduling.
5. Describe the concepts of buffer management and the critical chain.
6. Employ scenario and simulation analyses in project planning.

12.1 Introduction

As we have studied earlier in previous chapters (see, e.g. Chap. 6), projects are subject to various sources of uncertainties; various unanticipated events take place during project execution making the implementation of the baseline schedule as planned initially extremely difficult. Equipment failures, inaccurate time estimates, quality problems, and other problems frequently prevent project managers from implementing the baseline schedule without revision. Since uncertainty is an inevitable part of decision-making in projects, strategies to hedge against it should be developed and effective planning and control tools employed to manage its potentially detrimental effects. In this chapter, we focus on scheduling under uncertainty and illustrate the use of optimization models in schedule generation.

Herroelen and Leus (2005) cite five fundamental scheduling approaches to minimize the impact of unexpected events on project performance: reactive scheduling, robust (proactive) scheduling, stochastic scheduling, fuzzy scheduling and sensitivity analysis (Fig. 12.1) (see Hazır, 2008 for a more detailed review).

Reactive scheduling involves modifying or re-optimizing the schedules in the face of disruptive events that render them suboptimal or infeasible. In practice,

G. Ulusoy, Ö. Hazır, *An Introduction to Project Modeling and Planning*, Springer Texts in Business and Economics, https://doi.org/10.1007/978-3-030-61423-2_12

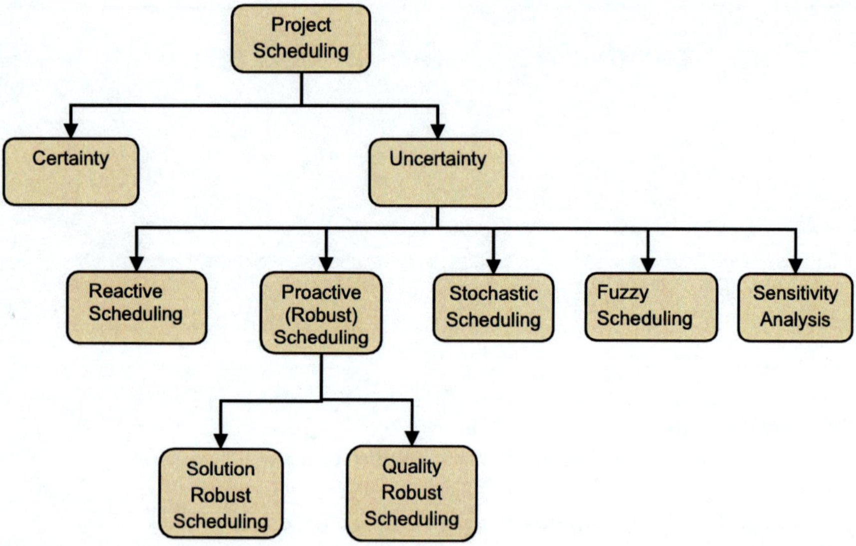

Fig. 12.1 Classification of project scheduling approaches. (Source: Herroelen and Leus 2005)

project managers either prepare a baseline schedule for the whole project lifetime, or they dynamically construct the schedule during project execution. The first approach is called predictive-reactive scheduling, and the second dynamic scheduling. Rescheduling could be performed only when an unexpected event happens (event driven approach) or at a preset time interval to check and modify the schedule (periodic approach). In both cases, either the current schedule is partially updated (partial rescheduling) or all remaining activities are rescheduled (full rescheduling) (Sabuncuoğlu and Bayız 2000; Vieira et al. 2003; Aytug et al. 2005).

In contrast to the reactive approach, *robust scheduling* takes a proactive approach. Variability is considered in the models, and schedules that are less sensitive to disruptions are sought. Herroelen and Leus (2005) study schedule robustness in two categories: solution robustness (stability) and quality robustness. *Stability* refers to the insensitivity of the activity start times to variations in the input data and *quality robustness* to the insensitivity of schedule performance measures such as project completion time to the input data. In the context of CPM, increasing total slacks improves quality robustness, while free slacks enhance stability.

In practice, these two approaches could be combined. *Proactive-reactive scheduling* requires preparing a baseline schedule considering the uncertainty and avoiding its negative impacts on project execution. For instance, a robustness measure might be maximized so that the schedule contains sufficient safety time to absorb the impacts of disruptions. Although this schedule will be less vulnerable to disruptions, all disruptive events and impacts might not be anticipated. Reactive scheduling serves as an additional protection mechanism to prevent large performance deviations. Lambrechts et al. (2008) present a nice application of this combined approach.

Stochastic project scheduling differs from the reactive and robust approaches in requiring the specification of probability distributions for all random variables. The stochastic resource-constrained project scheduling problem (SRCPSP) is an extension of the RCPSP that models activity durations as random variables. Since durations are not known at the beginning, a baseline schedule is not created. Instead, scheduling policies (or scheduling strategies) are predefined and the activities to start are chosen dynamically at appropriate decision points, usually the completion times of other activities (Stork 2001). Decisions are made based on past realizations and the characteristics of the assumed probability distribution. Stochastic dynamic programming is usually used to solve the problem (see Hazır et al. 2016 for a scheduling application under cash flow uncertainty).

Instead of using probability distributions, *fuzzy project scheduling* uses fuzzy membership functions to model the activity durations. Advocates of the fuzzy approach claim that probability distributions cannot always be defined due to lack of accurate data. On the other hand, if expert judgments are used to estimate the distributions, estimates are potentially inaccurate.

Uncertain durations are modeled with fuzzy numbers, constraints with fuzzy sets and membership functions. In this approach, some constraint violations might be allowed and the degree of satisfaction of the constraints is measured with membership functions. We recommend the book of Zimmerman (2001) and the paper by Hapke and Slowinski (1994) for further information on the theory of fuzzy programming and its applications to project scheduling.

Sensitivity analysis differs from stochastic, fuzzy, and robust scheduling in that it is reactive in nature since uncertainty is not addressed in the modeling or schedule generation phase (Borgonovo and Plischke 2016). The impact of perturbations in the input parameters, such as activity durations or resource availability, on schedule performance is investigated. In the literature, sensitivity analyses of project scheduling problems are rare; examples for machine scheduling problems are more common (Hall and Posner 2004).

In the following sections, we will focus on the three main approaches: robust scheduling, stochastic scheduling, and sensitivity analysis. We will also illustrate the use of scenario and simulation analysis in project scheduling.

12.2 Robust Scheduling

12.2.1 Critical Chain Project Management

A well-known method for achieving quality robustness in project schedules is *Critical Chain Project Management* (CCPM). Goldratt (1997) extended the basics of his Theory of Constraints (TOC), which emphasizes identifying and controlling system constraints to improve the overall system performance, to project management (Goldratt and Cox 1984) and introduced CCPM (Rand 2002).

The project management implementation of CCPM focuses on resource constraints. Therefore, unlike the critical path, a critical chain is defined as the sequence of both precedence and resource dependent activities that determine the completion time of the project. Planning and controlling buffers play a critical role, so CCPM is also referred to as Critical Chain Scheduling and Buffer Management. Buffers are protection mechanisms against uncertainty, especially in activity durations. CCPM defines three types of buffers (Rand 2002):

1. **Project buffer:** In the estimates of the activity durations usually safety factors are included to make up for the possible uncertainties involved. These safety factors are eliminated from the activities and aggregated at the end of the critical chain as a project buffer, seeking to avoid project delays.
2. **Feeding buffer:** In addition to the critical chain, paths merging with the critical chain are also considered and feeding buffers are added at the end of these paths. This allows delays on feeding paths to be absorbed without affecting the start time of critical chain activities.
3. **Resource buffer:** This buffer ensures that resources are available some time before they are needed by a critical chain activity, again ensuring that critical chain activities are not delayed by resource availability.

To calculate the size of buffers, two methods are well known: the 50% rule and the Root Square Error Method (RSEM). In the 50% rule, the duration of the chain is calculated with safe estimates, and half of it is added as a buffer. The 50% rule is also known as the "cut and paste method" (Tukel et al. 2006). This can be seen as an implementation of risk pooling; we put buffers against uncertainty in the sum of several durations instead of buffering against uncertainty in each activity duration.

The RSEM requires two estimates: the safe estimate and the average value. Their difference is assumed to be equal to two standard deviations of the activity duration. Assuming the independence of activities, the standard deviation of the sum of activity times in the chain can easily be calculated. Twice this standard deviation is used as the buffer size.

Even though inserting buffers creates some flexibility for project managers in their actions, it is not free of limitations (Herroelen and Leus 2001; Raz et al. 2003):

- The tough time estimates, which can be used by management to create pressure on the team to increase productivity, may result in quality problems.
- To identify the critical chain, CCPM does not require generating optimal schedules for RCPSP; most CCPM software packages use myopic heuristic rules for this purpose. However, Herroelen and Leus (2001) showed that the choice of scheduling and rescheduling algorithms might significantly affect the length of the critical chain.
- Resource conflicts might occur after inserting buffers into the project baseline. CCPM does not provide algorithms to solve these conflicts.

12.2.2 A Robust Optimization Model for Project Scheduling

Robust optimization is a mathematical modeling approach to identify solutions that are insensitive to data uncertainty. To achieve this, the worst-case performance of the solutions is optimized and solutions that perform well under worst-case scenarios are sought. Minmax and minmax regret objectives are commonly studied in these optimization models (Kouvelis and Yu 1997). As robust optimization is worst-case oriented, it is a relatively conservative approach. However, the use of uncertainty budgets allows the degree of conservatism to be controlled (Bertsimas and Sim 2003).

Robust optimization has been attracting attention work in a wide range of areas, from engineering to finance. In operations research, it has been used to formulate robust versions of well-known combinatorial optimization problems, such as the shortest path and traveling salesman problems (Coco et al. 2014). However, there are only a few project scheduling applications (Hazır and Ulusoy 2020). We present a robust optimization application to model a multi-mode project scheduling problem under uncertainty.

Recall that activity durations are dependent on the number of resources allocated for performing these activities. It is often possible to accelerate some activities and thus speed up the project completion by incurring additional costs (see Chap. 5). Considering this time/cost relation, we study the deadline version of the discrete time/cost trade-off problem (DTCTP). In this problem, a set of time/cost pairs (modes) for each activity, precedence relationships among activities, and the project deadline are given and project completion time is minimized. We reproduce here the mathematical formulation of the deterministic problem given in Sect. 5.4.1.

Consider a project with a set of n activities with the corresponding precedence graph in the activity-on-node (AON) representation, $G = (N, A)$, where N is the set of nodes and A is the set of arcs. The AON representation includes the n activities and two dummy nodes (0 and $n + 1$). Among these dummy nodes, 0 indicates the project start and $n + 1$ its completion. $A \subseteq N \times N$ is the arc set that defines the precedence relationships. Each activity j can be performed in exactly one of the $|M_j|$ modes, where each mode $m \in M_j$ is characterized by a processing time p_{jm} and a cost value c_{jm}. The deterministic DTCTP can be formulated as follows:

$$\text{Min} \sum_{j=1}^{n} \sum_{m \in M_j} c_{jm}\, x_{jm} \tag{12.1}$$

subject to

$$\sum_{m \in M_j} x_{jm} = 1 \quad j = 1, \ldots, n \tag{12.2}$$

$$FT_j - FT_i - \sum_{m \in M_j} p_{jm}\, x_{jm} \geq 0 \quad \forall (i,j) \in A \tag{12.3}$$

$$FT_{n+1} \leq \delta \tag{12.4}$$

$$FT_j \geq 0 \quad \forall j \in N \tag{12.5}$$

$$x_{jm} \in \{0, 1\} \quad \forall m \in M_j, \quad j = 1, \ldots, n \tag{12.6}$$

In this formulation, the non-negative continuous decision variable FT_j defines the completion time of activity j (12.5), while the binary decision variable x_{jm} chooses a mode for this activity (12.6). Total cost is minimized (12.1), while only one mode is assigned to each activity (12.2); precedence constraints must be satisfied (12.3); and the deadline should not be exceeded (12.4).

In contrast to the deterministic problem (12.1–12.6), in the robust model the durations or cost values are assumed to lie in intervals. In our application, the durations are fixed, while the costs take values in the interval, $\left[\underline{c_{jm}}, \overline{c_{jm}}\right]$. To justify fixing the durations, many PMs focus on the time objectives and control the time performance tightly, to the extent of allowing cost overruns to achieve time targets. For example, in Build-Operate-Transfer (BOT) projects, when faced with problems, PMs often opt for allocating additional resources and working overtime to complete the project on time. In BOT contracts a public service or infrastructure investment is made and operated for a (usually long) period of time by private enterprise, after which ownership is transferred to a public institution. The operating period is quite long, so that the investment is paid off and the private enterprise realizes a considerable profit. Therefore, in case of delays, BOT projects can incur large penalties or losses of operating profits.

Minmax optimization models require considering the worst-case realizations, therefore c_{jm} should be replaced with $\overline{c_{jm}}$ in the model. However, assuming these upper bound values for each activity is overly pessimistic, leading to excessively conservative solutions. To avoid this, we follow the restricted uncertainty approach of Bertsimas and Sim (2003) and assume that only a subset of the uncertain parameters can deviate from the nominal estimates to the extent of taking their worst-case values.

In particular, at most Γ activity cost values (out of a total of n) are assumed to deviate and others take their nominal values. Note that if $\Gamma = 0$ or $\Gamma = n$, nominal and maximal cost values are set for all the activities and the problem becomes the regular DTCTP. In between these values, the parameter Γ manages the pessimism/optimism level of the decision-maker. High (low) values of this parameter indicate risk averse (risk taking) decision-making behavior.

Focusing on the worst-case realizations of the uncertain parameters, let $\Delta_{jm} = \overline{c_{jm}}$ - c_{jm} denote the maximum deviation of the cost from the nominal value. Given these

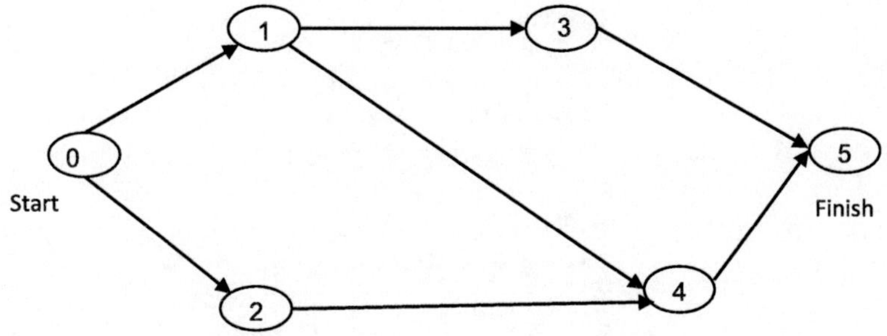

Fig. 12.2 Example project network

Table 12.1 Data for the example project

Activity	Mode	Duration	Cost $'1000 Lower bound	Upper bound
1	1	3	40	48
	2	5	20	35
2	1	4	10	12
	2	5	6	9
3	1	2	12	22
	2	4	6	8
4	1	3	6	7
	2	4	3	5

parameters and the stated assumptions, the robust problem can be modeled as follows:

$$\min\left\{\sum_{j=1}^{n}\sum_{m\in M_j} c_{jm}x_{jm} + \max\left\{\sum_{j=1}^{n}\sum_{m\in M_j} \Delta_{jm}x_{jm}u_j : \sum_{j=1}^{n} u_j \leq \Gamma, u \in B^n\right\} : x \in X\right\} \tag{12.7}$$

In this formulation, X denotes the set of feasible solutions, which is defined by the constraints (12.1) through (12.5). The set of activities that are subject to cost deviations are identified by the n-dimensional *binary vector u*; i.e., B^n denotes the set of n-dimensional binary vectors u. The indicator variable u_j takes a value of 1 if and only if the activity j achieves its upper bound in the robust approach and 0 otherwise.

We illustrate the differences between deterministic and robust problem with a simple example. The network is illustrated in AON format in Fig. 12.2 and a deadline of $\delta = 8$ is set. Each activity has two alternative execution modes, characterized by the duration (days), and lower and upper bounds on the activity cost as given in Table 12.1.

Table 12.2 Mode assignments and the deterministic and robust objectives

	Activity						$\Delta_{jm} = \overline{c_{jm}} - c_{jm}$				Robust objective: $f(\Gamma)$			
	1	2	3	4	FT_5	$f(0)$	$j = 1$	$j = 2$	$j = 3$	$j = 4$	$\Gamma = 1$	$\Gamma = 2$	$\Gamma = 3$	$f(4)$
Feasible mode combinations (m)	1	1	1	1	7	68	8	2	10	1	78	86	88	89
	1	1	1	2	8	65	8	2	10	2	75	83	85	87
	1	1	2	1	7	62	8	2	2	1	70	72	74	75
	1	1	2	2	8	59	8	2	2	2	67	69*	71*	73
	1	2	1	1	8	64	8	3	10	1	74	82	85	86
	1	2	2	1	8	58	8	3	2	1	66	69*	71*	72*
	2	1	1	1	8	48	15	2	10	1	63	73	75	76
	2	2	1	1	8	44*	15	3	10	1	59*	69*	72	73

(*) indicates optimal solutions

For this simple example, the objective function values of the deterministic (f(0) and f(4)) and robust problems (f(1), f(2), f(3)) are illustrated in Table 12.2. The rows of the table correspond to the 8 feasible mode combinations for the activities. Column 6 (FT_5) shows the completion time of the schedule. The remaining 8 combinations lead to a completion time greater than 8, and hence violate the deadline constraint.

The parameter Γ that controls the level of conservatism impacts the optimal mode assignments. Note that there are multiple optimal solutions when $\Gamma = 2$ or $\Gamma = 3$, and the optimal mode assignments when $\Gamma = 0$ and $\Gamma \geq 3$ are different. Having chosen the activity modes, CPM could be used to obtain an early start schedule illustrated using a Gantt chart. Hazır et al. (2011) give a more detailed discussion of this application, propose a decomposition-based solution algorithm, and present computational results.

12.2.3 Robustness Measures and Project Scheduling

In addition to developing a robust optimization model, baseline schedules could be generating in a simpler way by directly optimizing a robustness measure. In that sense, the measure(s) used should provide a good estimate of schedule robustness. The baseline schedules that are generating by using these robustness measures would be capable of absorbing unanticipated disruptions. However, directly optimizing performance measures such as expected or worst-case values in project networks is complex, since there may be many interdependent paths and analytical derivation of the effects of disruptions on the performance under uncertainty is difficult. To avoid this complexity, surrogate measures have been used in the literature.

Schedules that contain a smaller number of critical or potentially critical activities can absorb disruptions, avoiding delays in project completion. We also note that the criticality of an activity might be defined by its slack value. Hazır et.al (2010) focused on slack based measures and introduced surrogate metrics that can provide an accurate estimate of the schedule robustness. Hazır et al. (2015) classified these measures in six main groups:

1. Average slack.
2. Weighted slack.
3. Slack utility function.
4. Dispersion of slacks.
5. Percentage of potentially critical activities.
6. Project buffer size.

Monte Carlo simulation can be used to assess the effectiveness of these measures (Mehta and Uzsoy 1998). In the implementation, several disruption scenarios are defined; correlation between these measures and performance measures is checked such as the proportion of replicates where the project is completed before the due date (see Sect. 12.4.3, Vose 2008 for details of the simulation implementation). Having replicated several project settings and calculating the correlation, Hazır et al. (2010) found project buffer size and weighted slack, with weights equal to the number of immediate successors of the activities, to be the best predictors of robustness.

12.3 Stochastic Scheduling

In the stochastic approach, the duration of any activity is modeled as a random variable, $\widetilde{p}$. The probability distribution of the project completion time is derived and the moments, mean and variance, are calculated (Demeulemeester and Herroelen 2002).

In our analysis, we assume that activity durations are independent random variables and that their joint probability distribution function is known. If they are all independent, we only need the marginal distributions of each activity. The duration of each activity i has a probability density function (pdf), $f_i()$, and cumulative distribution function (cdf), $F_i()$.

Given the cdf's, the probability distribution of the project completion time can be derived using series-parallel reductions (Demeulemeester and Herroelen 2002). The analysis is easier for series/parallel reducible networks, which can be reduced to a single arc using successive serial and parallel merge operations (Demeulemeester and Herroelen 2002). Modeling irreducible networks may require some activities to be duplicated to obtain a reducible network (Dodin 1985).

When two serial activities are merged, the cdf of the sum of two random variables (duration of activities i and j) is given by the convolution of the individual cdf's, $F_i(t) * F_j(t) = \int_0^t F_i(t-y)dF_j(y)$.

On the other hand, for two parallel activities, the maximum of the two durations defines the time required. In this case, owing to the independence assumption, the cdf's of the variables are multiplied.

$$F_k(t) = F_1(t) = \int_0^{\infty} F(t|\widetilde{p}_1 = x)dF_1(x)$$

For illustration, let us revisit the project network given in Fig. 12.2. The AOA equivalent of this network is shown in Fig. 12.3. In this network the precedence relation between activities 1 & 4, represented with a dummy arc in Fig. 12.3, makes the lengths of the paths dependent, greatly complicating the derivation.

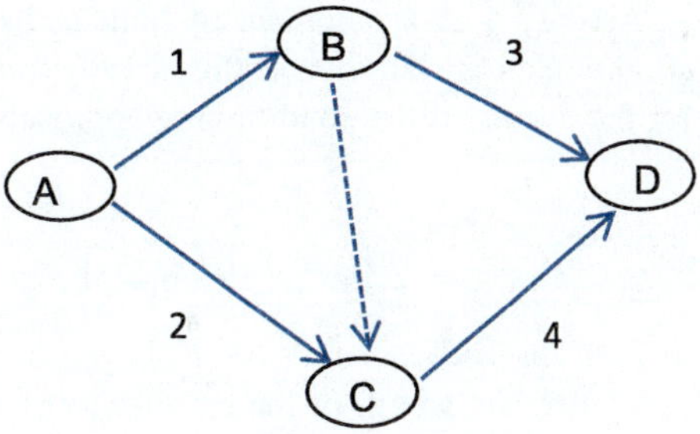

Fig. 12.3 AOA graph

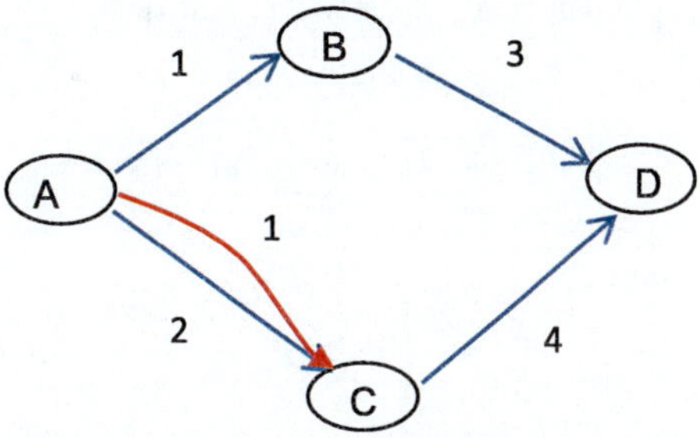

Fig. 12.4 Modified AOA graph

If this precedence relation did not exist, the resulting activity-on-arc (AOA) graph would become completely series/parallel reducible. In that case, two serial merge operations (merge activities 1 & 3 and 2 & 4); followed by a parallel merge reduce the network to a single equivalent activity. The convolutions of the cdf's are calculated for the two paths, then the resulting cdf's are multiplied:

$$F(t) = [F_1(t) * F_3(t)].[F_2(t) * F_4(t)]$$

Reintegrating the precedence relationship, three paths, i.e., 1–3, 1–4 and 2–4, need to be considered in this example to compute the project completion. The maximum of these three path lengths defines the project duration.

$$C_{max} = \max \{p_1 + p_{3;} p_{1+} p_{4;} p_{2+} p_4\}$$

Since activities 1 and 4 are common to two of the three paths, the lengths of these paths are not independent even though the individual activity durations are assumed to be so. To deal with this dependence of the paths, we first duplicate activity 1 so that the paths can be modeled separately. The modified network is illustrated in Fig. 12.4.

The project completion time can now be reformulated by considering two paths A-B-D and A-C-D separately as follows:

$$C_{max} = \max \{p_1 + p_{3;} \max \{p_{1,} p_2\} + p_4\}$$

Activity 1 is an element of both paths and to formulate the cdf of C_{max}, we condition on the duration of this activity (See Demeulemeester and Herroelen (2002) for the details of the conditioning approach).

$$F(t) = P(C_{max} \leq t\} = \int_{-\infty}^{\infty} F(t|\tilde{p}_1 = x)dF_1(x)$$

To find the length of the path A-C-D, $\max\{p_1, p_2\}$ is considered, and two cases are formulated separately: $p_2 \leq p_1$ and $p_2 > p_1$. Integration is taken over the variable $\tilde{p}_2$ considering these two conditions:

$$F(t|\tilde{p}_1 = x) = \left\{ \int_0^x F_4(t-x)dF_2(y) + \int_x^t F_4(t-y)dF_2(y) \right\} F_3(t-x)$$

$$= \left\{ F_2(x)F_4(t-x) + \int_x^t F_4(t-y)dF_2(y) \right\} F_3(t-x.)$$

In the above equations, the terms in the parentheses represent the cdf of the random variable that characterizes the length of path A-C-D. The term outside the parentheses, $F_3(t - x)$, defines the cdf of the path A-B-D, since the duration of activity 1 is fixed to x. To combine the two parallel paths, the *cdf*'s are multiplied.

The above analysis requires that probability distributions of the durations of the activities can be independently and correctly estimated, which is only possible if reliable data on the durations is available. In these cases, stochastic scheduling has the advantage of incorporating this valuable information and basing the analysis on the probability distributions of the activities. However, since each project is unique, such data may not be available for some activities. In addition to the data related problems, real life projects contain many activities with various dependencies among them, resulting in complex networks with a large number of dependent paths. This usually makes deriving the probability distribution difficult, rendering the scheduling problems complex and their solution computationally demanding.

To model this complex stochastic nature of the project scheduling problems, simulation is an appropriate modeling and analysis technique. We will make use of Monte-Carlo simulation, which has been widely used in the literature.

12.4 Sensitivity, Scenario, and Simulation Analysis

12.4.1 Sensitivity Analysis

Sensitivity analysis, which investigates the dependence of model output on input parameters, has been widely used in operations research (See the review of Borgonovo and Plischke (2016) for the application areas and recent developments). PMs have been using this methodology as a decision support tool in project selection, planning and risk analysis decisions. The basic aim is to measure the impact of changes in key parameters on project objectives and evaluate their consequences. In this regard, sensitivity analysis can be used as a quantitative risk analysis tool to predict which risk elements might affect project targets and to what extent (PMBOK Guide 2013).

Before making decisions, effective PMs usually ask "What if" questions, such as,

- If annual interest rate increases by 1%, how will the NPV of the project be affected? Up to what level of the annual interest rate does the project continue to have a positive NPV?
- How do the fluctuations in exchange rate affect the selection of the investment projects? Does 5% devaluation of the local currency against US dollar change the investment choice?

These what if questions can then be addressed by sensitivity analysis (Park 2016). Note that each question above represents a change in a single parameter. Due to the complexity of observing the impacts of changes in multiple parameters, the impact of a change in one parameter on the project objectives is observed at a time.

We also note that in project modeling and analysis, the parameters are usually interdependent; remember, for example, the time/cost trade-offs. These interdependencies make the analysis even more complicated. To overcome this limitation, i.e., changing one parameter at a time while keeping the others constant, one solution is to study a group of parameters collectively by defining scenarios, each of which corresponds to changing several parameters simultaneously while the project performance is studied regarding this combination of parameters. Generally, best case (optimistic), most likely, and worst-case (pessimistic) scenarios are formulated.

Table 12.3 Parameter estimates for the Example Project

	Activity duration (days)			Activity direct cost $'000	
Activity	Optimistic	Most likely	Pessimistic	Lower bound	Upper bound
1	3	4	5	20	48
2	4	5	7	6	12
3	2	3	5	6	22
4	3	4	7	3	7

Table 12.4 Scenarios for the Example Project

Scenario	Activity duration (days)	Activity direct cost
Best case	Optimistic	Lower bound
Most likely	Most likely	Average
Worst case	Pessimistic	Upper bound

Table 12.5 Results of the scenario analysis for the Example Project

Variable	Best case scenario	Most likely scenario	Worst case scenario
Project duration	7	9	14
Direct costs ($)	35,000	62,000	89,000
Indirect costs ($)	3500	4500	7000
Total cost ($)	38,500	66,500	96,000

12.4.2 Scenario Analysis

We now define time and cost scenarios for project activities and illustrate an application of scenario analysis to project scheduling. For this purpose, we return to the project network of Fig. 12.3 considering only a single processing mode. Five parameters are estimated for each non-dummy activity by the project team: three for the durations (optimistic, most likely, pessimistic) and two for the activity direct costs (lower and upper bounds) (Table 12.3).

Considering these time and cost estimates the PM considers three main scenarios with the activity duration and activity direct cost estimates given in Table 12.4. In addition to the direct costs, indirect costs are also considered at the rate of $500 per day. The results of the calculations are summarized in Table 12.5.

These scenarios and the corresponding project time and cost could be presented to the managers. Scenarios correspond to the pessimism level. Decision-maker might focus on some scenarios and make the project time planning and budgeting accordingly.

12.4.3 Monte Carlo Simulation

In Chap. 6, PERT is proposed as a technique to handle uncertainty in activity durations. When applying PERT, the Beta distribution has been widely used. However, some recent project scheduling studies suggest that a lognormal distribution is more appropriate to model uncertainty in project activity durations (Tavares et al. 1998; Mohan et al. 2007; Trietsch et al. 2012).

If a random variable X follows a normal distribution, the exponential function e^X will follow a lognormal distribution. Unlike the normal distribution, the lognormal distribution is not symmetric. Tavares et al. (1998) list the following reasons supporting the use of lognormal distribution:

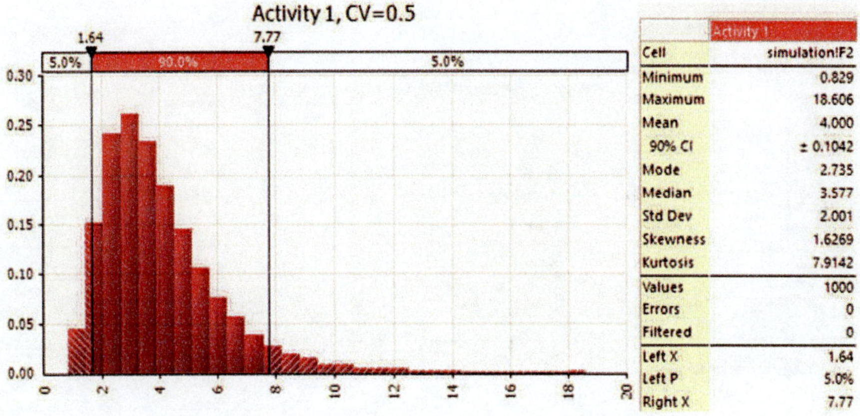

Fig. 12.5 Probability distribution of the duration of activity 1 (Lognormal, CV = 0.5) (@Risk output)

Property 1
The lower bound corresponds to a minimal feasible duration, which could be imposed by technical reasons.

Property 2
It is a right skewed distribution, which is common in project activities.

Property 3
There is no upper bound. Project activities can always be delayed extensively by unexpected events.

Let us rework the project network illustrated in Fig. 12.3 with data given in Table 12.3. In the simulation model, a lognormal distribution is assumed for the activity durations. In this example, the mean durations are assumed to be equal to the most likely estimates given in Table 12.3. Regarding the uncertainty in activity durations, two cases were considered. A coefficient of variation (CV), which is the standard deviation divided by the mean, of 0.5 and 1.0 are used to characterize different variability conditions in the activity durations.

We perform the Monte Carlo Simulation and present the outputs using Palisade @Risk Software. Figures 12.5 and 12.6 illustrate the input distribution for activity 1 (Lognormal Distributions with mean 4, standard deviation 2.0 and 4.0). These histograms show that there is a lower bound of 0 (Property 12.1), the distribution is right skewed, and there is no upper bound (Property 12.3). Even very large durations are possible, especially under high uncertainty (Fig. 12.6).

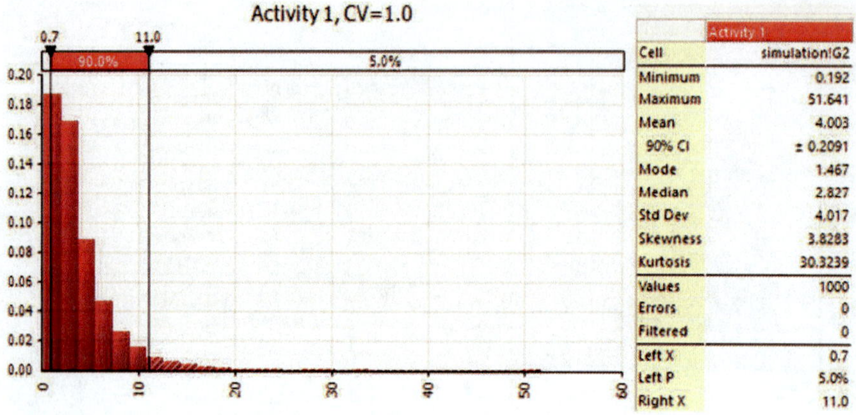

Fig. 12.6 Probability distribution of the duration of activity 1 (Lognormal, CV = 1.0) (@Risk output)

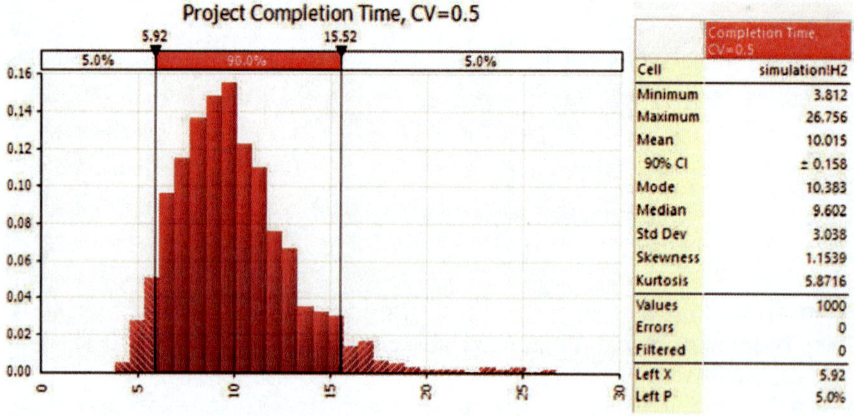

Fig. 12.7 Probability distribution of project completion time (Lognormal, CV = 0.5) (@Risk output)

If deterministic durations are equal to the most likely estimates, CPM estimates a project completion time of 9 days. However, under uncertain durations, it becomes difficult to accomplish this target. When variability of the durations increases, for instance when the CV increases from 0.5 to 1.0, the estimate of the expected project completion time increases (see Figs. 12.7 and 12.8), increasing the risk of delays. In these cases, PMs need to negotiate looser due dates to reduce the risk of delay. Figures 12.7 and 12.8 respectively show the distribution of the project completion times and confidence intervals. When CV = 0.5 a due date of 15.52 days is set to have a 95% probability of completing the project on time (Fig. 12.7), while 23.8 days are needed when CV = 1.0 (Fig. 12.8).

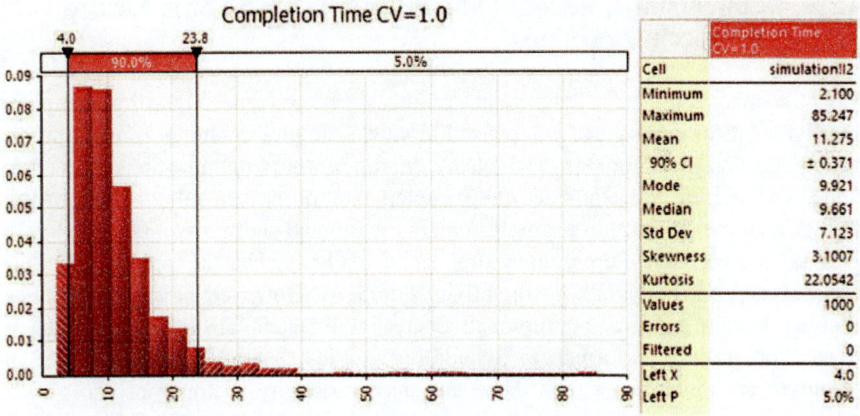

Fig. 12.8 Probability distribution of project completion time (Lognormal, CV = 0.5) (@Risk output)

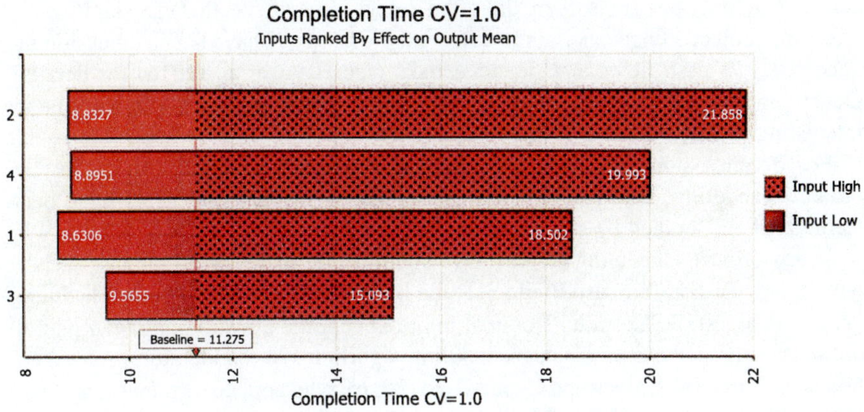

Fig. 12.9 Impact of activity durations on project completion (Lognormal, CV = 1.0) (@Risk output)

Finally, we examine which activities have the greatest impact on the distribution of the project completion time. The tornado graph in Fig. 12.9 shows that activities 2 and 4 have the largest impact on the distribution of output; activity 3 has the least. This result is intuitive, as activity 2 has the largest mean and variance, and activity 4 lies on the two paths. This also shows that in most of the replicates path 2–4 becomes the critical path. On the other hand, activity 1 has a smaller mean and is an element of only one path.

12.5 Conclusions, Recent Developments, and Some Future Research Directions

How to manage projects in the presence of uncertainty is a crucial question in project management. In this regard, we referred to stochastic programming, robust optimization and fuzzy programming, which are the fundamental optimization approaches under uncertainty. We explained their implementation in scheduling projects. We also gave some application examples on simulation and sensitivity analyses.

The project scheduling literature, specifically stochastic scheduling, has concentrated on the variability in activity durations. However, other uncertainties such as those in resource requirements or availabilities can also have a significant impact on project performance. Scheduling studies that address uncertainty in resources are much fewer than those that treat variability in durations. To give a relevant example, the project scheduling literature has addressed problems with financial objectives, mainly the optimization of NPV. A common assumption in these studies is that the amount of the cash flows is known beforehand, and cash inflows occur when the activity is finished. In practice, however, in addition to delays in activity durations, there are often delays in payments that can significantly affect the project's financial viability. Recently researchers have been addressing the uncertainty in cash flows and financial risks (see Hazır et al. (2016) for details). Developing models that address different variability sources will continue to be an interesting research area.

For decades stochastic programming has been used to model uncertainty in project scheduling. Lately fuzzy or robust versions of these problems have been formulated and studied by several researchers.

Fuzzy project scheduling studies have significantly increased in the last decade. Now fuzzy methods are also used to set the buffer sizes in critical chain scheduling (Zhang et al. 2017). Despite its critical importance, little attention has been paid to the sensitivity analysis of project scheduling algorithms in the literature. An exception is the study by, Galvez and Capuz-Rizo (2016), who applied different sensitivity analysis techniques to identify the parameters that had the largest effect on project scheduling. Regardless of the approach used to model the uncertainty, there is still a need for algorithms to solve complex large-scale project scheduling problems under uncertainty.

Among the approaches that are alternative to stochastic modelling, robust optimization has also been increasingly used. Along with the well-known minmax robust modeling, novel approaches, such as bw robustness proposed by Gabrel et al. (2013) or Lexicographic α robustness (Kalai et al. 2012), might be used to model robust project scheduling problems (Hazır et al. 2015). Combining these proactive approaches with effective reactive procedures would further improve the scheduling performance.

Scenario analysis and simulation studies have been commonly used to examine the impact of scheduling/rescheduling on project performance. These methods are appropriate to address the complexity of real life project scheduling problems under uncertainty. Monte Carlo simulation add-ins, which can be embedded in Microsoft

Excel and Project, are more frequently used as decision making support tools in many projects today (see Palisade @Risk for Project Management).

Finally, integrating theoretically tested scheduling models and algorithms as a model base of a decision support system (DSS) will be an important contribution. Embedding these DSS tools in a commercial software package such as a Microsoft Project add-in will bring substantial support to project managers. At this point, combining these scheduling algorithms with efficient project control modules that define intervening strategies in case of disruptions is a promising research and application area.

Exercises

12.1 ABC Company is working to launch a new product. The investment amount for the project is $ 15 million. The facilities including new equipment could be used for 15 years. The scrap value is assumed to be zero and straight-line depreciation is assumed. Annual fixed costs are predicted to be as $ 3 million and variable costs as approximately 70% of the sales revenue. Tax rate is 30% and opportunity cost of investment is 12%. Perform a sensitivity analysis to see how the profitability of the project will be affected by the annual sales volume Calculate the NPV values for two cases: high demand (revenue of $24 million) and low demand (revenue of $18 million).

12.2 Consider any case where a relationship is known to exist between a set of variables, dependencies between variables, but where formulating formulae for the dependencies is too complex. For this case, which uncertainty analysis tool would you consider applying for this case? Explain with examples.

12.3 A university management decided to invest in renewable energy. A solar power plant with a capacity of 500 kW (kilowatts) is to be installed in the campus. Total investment and operating cost of the plant is predicted to be $ 2 million with a useful life of 10 years. Considering weather conditions and the system efficiency, two different scenarios are considered for electricity production. In the first scenario, which is predicted to occur with a chance of 67%, total electricity production for a period of 10 years is expected to be 50 MW (Megawatt); whereas 30 MW in the second scenario, which has a likelihood of 33%. Up to 30 MW of production, the price is fixed at $0.13/kW. However, an average price of $ 0.06/kW is predicted for the sale above 30 MW. University management is looking for a preliminary contract to sell the surplus amount. One of electricity distribution companies has offered them a purchase option for $500 K. In return; the company will buy the production amount over 30 MW of electricity at the price of $ 0.13/kW. Considering the two scenarios, what would you suggest to the university management? Justify your suggestion.

12.4 Considering the probability theory and also data-driven experiences, industries might adopt some standardized probability distributions to use in their simulation applications and risk analysis. Find out some examples for these commonly used distributions and their industrial application areas. Discuss with the underlying reasons for adoption.

12.5 Real options might also be considered as hedging approaches for projects where there is high level of uncertainty. Investigate the applications of real options theory in project management and discuss how real options could serve to project management under uncertainty.

12.6 Ahmet and Mehmet are partners of DreamHouses Constructing Company. They build and sell houses in suburbs of Ankara. They acquired 15,000 m^2 of land with a price of 12,000,000 TL. They expect that the project will last for 5 years, finishing 60 houses of 100 m^2 each year and they can sell them with an average price of 250,000 TL. They predict that the project's variable expenses (material, labor, and project overheads) be equal to 40% of the sales revenue of the first year and will increase by inflation rate in the following years. Fixed costs are projected to be 1000,000TL in the first year and increase with inflation. 4,000,000TL of machinery and equipment investment is needed to realize the project; the equipment can be sold at 3,000,000 TL at the end of the project. Before starting the project engineering, insurance, etc. expenses need to be spent 1,000,000 TL. The tax rate is 20%.

The company expects a return of 5% above inflation on the project. However, Ahmet and Mehmet do not agree on the inflation forecasts. For the next 5 years, they respectively estimate the rates to be 10% and 12%. Both partners believe that they should not increase their sales prices. However, Mehmet suspects that other companies might start housing projects in that area, therefore they might need to make discounts: 10% in the 4th year and 15% in the 5th year. On the contrary Ahmet is optimistic and considers that demand will be high so that houses can easily be sold without discount.

They consulted Eda to study the financial feasibility of the project. She performed an analysis based on the assumptions of Ahmet. The project's payback period is found to be 3 years and the internal rate of return 21.24%. Ahmet also questioned about what the monetary return would be. Eda replied that, the net present values would be approximately 2,500,000TL.

(a) Are the calculations of Eda correct?

(b) If Mehmet's scenario is realized, what would be the return of the project?

12.7 Consider the robust optimization problem and its mathematical formulation (12.1–12.6) given in the text. Discuss the solution approaches. How would you reformulate the model if activity durations are also assumed to be varying in predefined intervals? How does the problem structure change in this case?

12.8 Explain the concepts of quality robustness and solution robustness with examples. Discuss the relationship of schedule robustness with total and free slacks.

12.9 An Engineering Faculty is planning to organize a conference on renewable energy systems. The project team has come up with the following Risk Checklist and determined the occurrence probability and monetary impacts for each risk.

Risk	Probability of occurrence	Expected impact ($) (monetary loss)
Not being able to find enough number of sponsors	0.200	40,000
Not being able to reach targeted number of participants	0.300	20,000
Not being able to find a suitable conference venue	0.150	50,000
Bad weather conditions on the day of the conference	0.100	20,000
Having a similar event on the same week	0.050	50,000
Not being able to find high caliber speakers to attract participants	0.200	10,000

Perform a risk analysis using the simulation approach. State your assumptions and present your findings with graphs.

12.10 A project manager who is leading a new product development project in IT business is planning to implement the critical chain project scheduling in her project and insert a project buffer in the project time schedule. However, she has some worries about the uncertainty of the activity durations and dependence of the durations. Are the following claims of her true? Why? Why not?

(a) "The uncertainty inherent in the system is not low; therefore, the lower the benefits will be as a result of risk pooling".

(b) "Aggregation benefits could be lower since there is a low correlation among the activity durations"

12.11 Interval-type of uncertainty set is assumed for the cost parameters of the robust optimization problem given in the text. Compare the interval uncertainty sets with the ellipsoidal-type uncertainty sets. Which approach would be more conservative? Why?

12.12 Suppose a critical resource is leased for a large project. There is a graduated cost associated with using the resource at a certain percentage level U. The cost is specified as $10,000 per 10% increment in the utilization level above 40%. A flat cost of $5000 is charged for utilization levels below 40%. The utilization intervals and the associated costs are presented as follows:

$U < 40\%$	$5.000
$40\% \leq U < 50\%$	$10.000
$50\% \leq U < 60\%$	$20.000
$60\% \leq U < 70\%$	$30.000
$70\% \leq U < 80\%$	$40.000
$80\% \leq U < 90\%$	$50.000
$90\% \leq U < 100\%$	$60.000

Suppose the utilization level is normally distributed random variable with mean 60% and standard deviation of 4%. Find the expected cost of using this resource.

References

Aytug, H., Lawley, M. A., McKay, K. N., Mohan, S., & Uzsoy, R. (2005). Executing production schedules in the face of uncertainty: A review and some future directions. *European Journal of Operational Research, 161*, 86–110.

Bertsimas, D., & Sim, M. (2003). Robust discrete optimization and network flows. *Mathematical Programming, 98*, 49–71.

Borgonovo, E., & Plischke, E. (2016). Sensitivity analysis: A review of recent advances. *European Journal of Operational Research, 248*, 869–887.

Coco, A. A., Solano-Charris, E. L., Santos, A. C., Prins, C., & de Noronha, T. F. (2014). *Robust optimization criteria: State-of-the-art and new issues*. Technical report UTT-LOSI-14001, ISSN: 2266-5064.

Demeulemeester, E. L., & Herroelen, W. S. (2002). *Project scheduling: A research handbook*. Dordrecht: Kluwer Academic.

Dodin, B. M. (1985). Bounding the project completion time distribution in PERT networks. *Operations Research, 33*, 862–881.

Gabrel, V., Murat, C., & Wu, L. (2013). New models for the robust shortest path problem: Complexity, resolution and generalization. *Annals of Operations Research, 207*(1), 97–120.

Gálvez, E. D., & Capuz-Rizo, S. F. (2016). Assessment of global sensitivity analysis methods for project scheduling. *Computers & Industrial Engineering, 93*, 110–120.

Goldratt, E. M. (1997). *Critical chain*. Great Barrington: The North River Press Publishing Corporation, MA.

Goldratt, E. M., & Cox, J. (1984). *The goal*. Croton-on-Hudson: North River Press.

Hall, N. G., & Posner, M. E. (2004). Sensitivity analysis for scheduling problems. *Journal of Scheduling, 7*(1), 49–83.

Hapke, M., Jaszkiewicz, A., & Slowinski, R. (1994). Fuzzy project scheduling system for software development. *Fuzzy Sets and Systems, 67*, 101–117.

Hazır, Ö. (2008). Models and algorithms for deterministic and robust discrete time/cost trade-off problems (Unpublished PhD Dissertation, Bilkent University).

Hazır, Ö., & Ulusoy, G. (2020). A classification and review of approaches and methods for modeling uncertainty in projects. *International Journal of Production Economics, 223*, 107522.

Hazır, O., Haouari, M., & Erel, E. (2010). Robust scheduling and robustness measures for the discrete time/cost trade-off problem. *European Journal of Operational Research, 207*(2), 633–664.

Hazır, O., Erel, E., & Gunalay, Y. (2011). Robust optimization models for the discrete time/cost trade-off problem. *International Journal of Production Economics, 1*, 87–95.

Hazır, Ö., Haouari, M., & Erel, E. (2015). Robust optimization for the discrete time-cost tradeoff problem with cost uncertainty. In C. Schwindt & J. Zimmermann (Eds.), *Handbook on project management and scheduling* (Vol. 2, pp. 865–874). Berlin: Springer.

Hazır, Ö., Onur, A., & Kilic, H. T. (2016). *A dynamic programming model for project scheduling under financial risks*. 15th international conference on Project Management and Scheduling (PMS2016), 239–242, Valencia, Spain, 19–23 April 2016.

Herroelen, W., & Leus, R. (2001). On the merits and pitfalls of critical chain scheduling. *Journal of Operations Management, 19*, 559–577.

Herroelen, W., & Leus, R. (2005). Project scheduling under uncertainty – Survey and research potentials. *European Journal of Operational Research, 165*, 289–306.

Kalai, R., Lamboray, C., & Vanderpooten, D. (2012). Lexicographic a robustness: An alternative to min-max criteria. *European Journal of Operational Research, 220*(3), 722–728.

Kouvelis, P., & Yu, G. (1997). *Robust discrete optimization and its applications* (Vol. 14). Dordrecht: Kluwer Academic Publishers.

Lambrechts, O., Demeulemeester, E., & Herroelen, W. (2008). Proactive and reactive strategies for resource-constrained project scheduling with uncertain resource availabilities. *Journal of Scheduling, 11*(2), 121–136.

Mehta, S. V., & Uzsoy, R. (1998). Predictable scheduling of a job shop subject to breakdowns. *IEEE Transactions on Robotics and Automation, 14*, 365–378.

Mohan, S., Gopalakrishnan, M., Balasubramanian, H., & Chandrashekar, A. (2007). A lognormal approximation of activity duration in PERT using two time estimates. *Journal of the Operational Research Society, 58*(6), 827–831.

PMI. (2013). *A guide to the Project Management Body of Knowledge (PMBOK® Guide)* (5th ed.). Newtown Square: Project Management Institute (PMI).

Raz, T., Barnes, R., & Dvir, D. (2003). A critical look at critical chain project management. *Project Management Journal, 34*(4), 24–32.

Sabuncuoğlu, I., & Bayız, M. (2000). Analysis of reactive scheduling problems in job shop environment. *European Journal of Operational Research, 126*, 567–586.

Stork, F. (2001). *Stochastic resource-constrained project scheduling*. PhD thesis, Technische Universität Berlin, Berlin, Germany.

Tavares, L. V., Ferreira, J. A., & Coelho, J. S. (1998). On the optimal management of project risk. *European Journal of Operational Research, 107*(2), 451–469.

Trietsch, D., Mazmanyan, L., Gevorgyan, L., & Baker, K. R. (2012). Modeling activity times by the Parkinson distribution with a lognormal core: Theory and validation. *European Journal of Operational Research, 216*(2), 386–396.

Tukel, O. I., Walter, O. R., & Ekşioğlu, S. D. (2006). An investigation of buffer sizing techniques in critical chain scheduling. *European Journal of Operational Research, 172*(2), 401–416.

Vieira, G. E., Herrmann, J. W., & Lin, E. (2003). Rescheduling manufacturing systems: A framework of strategies, policies, and methods. *Journal of Scheduling, 6*(1), 39–62.

Vose, D. (2008). *Quantitative risk analysis: A guide to Monte Carlo simulation modelling* (3rd ed.). Chichester: Wiley.

Zhang, J., Song, X., & Díaz, E. (2017). Critical chain project buffer sizing based on resource constraints. *International Journal of Production Research, 55*(3), 671–683.

Zimmermann, H. J. (2001). *Fuzzy set theory and its applications* (4th ed.). Boston: Kluwer Academic Publishers.

13 Planning and Scheduling of Repetitive Projects

Learning Outcomes
Upon successful completion of this Chapter, the reader will be able to:

1. Describe the basic characteristics of repetitive projects
2. Make an analogy between repetitive projects and serial production lines
3. Formulate a cost function considering direct and indirect cost items.
4. Employ mathematical models for scheduling the repetitive projects.

13.1 Introduction

An area of economic activity where project planning and scheduling can be of considerable help is in the execution of repetitive projects that involve the production or construction of similar units in batches, such as railway cars, ships, residential houses, highways and high-rise buildings. Particularly in the construction industry, repetitive projects represent a large portion of the work and hence an effective scheduling method is needed to ensure their timely and efficient completion. In repetitive projects, the workers (crews) perform identical, or very similar work as they move from one workstation to another or continue doing the identical work at a fixed location workstation. There is some similarity to assembly lines in the sense that the crews perform identical work over the units involved (Yang and Ioannou 2004). Whether the workstation is fixed, and the workers are moving from one workstation to another or similar to assembly lines the product is moving between workstations, and the workers are fixed depends on the nature of the production environment the project is realized in. Construction projects, for example, are of the first type. In make-to-order manufacturing, on the other hand, both set-ups can be observed.

As an example of repetitive projects, consider the construction of 10 independent identical residential houses on a piece of land. The construction work consists of several work packages to be accomplished consecutively with the crew, equipment, and material well defined, and the contractor has only enough resources for the

G. Ulusoy, Ö. Hazır, *An Introduction to Project Modeling and Planning*, Springer Texts in Business and Economics, https://doi.org/10.1007/978-3-030-61423-2_13

execution of one work package (WP) of each type at a time. Hence the batch size is one unit. Say the first three consecutive work packages are foundation excavation (WP_1), laying the foundation (WP_2), and first floor concrete pouring (WP_3). The work starts with WP_1 at site 1. Once WP_1 at site 1 is finished, the WP_1-crew moves to the second site to start WP_1 at site 2. The WP_2- crew moves to the first site and starts working on WP_2 at that location. Once both work packages are finished, the WP_1-crew moves to the third site, the WP_2-crew to the second site, and WP_3-crew moves to the first site to start with their respective work packages. In the next phase, WP_4-crew is introduced to the first site and the fourth site is opened for work by WP_1-crew with the other crews moving one site up. This movement of crews from site to site continues until all the work is finished at all sites. Obviously, it is important to schedule the start and finish times of these WPs (e.g., laying the foundation for all 10 houses) such that the consecutive work packages do not interfere with each other, i.e., the finish time of a work package cannot exceed the start time of the succeeding work package by at least a minimum time buffer. To meet these conditions, the PM must adjust the productivity (i.e., production rate) and work content of consecutive WPs.

The approach adopted for planning and scheduling repetitive projects is closely related to the *line of balance* (LOB) approach. The LOB approach deals also with the production of a batch of identical units. It is based on managing the production process through control points or milestones. In the *repetitive scheduling approach* adopted here control points and milestones are associated with the start and finish times of the WPs.

The WPs are assumed to continue over time without any break from start to finish. The WPs necessary for the completion of the project follow each other consecutively. They are related to each other in a sequential fashion through generalized precedence relations. A minimum time buffer between the consecutive WPs is required for the orderly execution of the jobs.

Figure 13.1 illustrates the execution of five different WPs for the production of a batch of N units over time, where ST_i denotes the start time of WP_i, FT_i its finish time, D_i its duration and B_i the minimum of the time buffers between work packages WP_i and WP_{i+1}. The project duration will be denoted by PD. The work on WP_1 starts at ST_1 and continues without a break till all N units in the batch are produced, i.e. the scheduling is non-preemptive. The line connecting ST_i and FT_i is called the *WP progress line*. The slopes of the WP progress lines in Fig. 13.1 represent the *production rate* at which the associated WP is performed, which is denoted by p_i for WP_i and corresponds to the number of units completed per unit time in WP_i. Hence, with increasing production rate the corresponding WP progress line gets steeper. Progress lines are not allowed to cross in order not to violate the precedence relation between two consecutive WPs. Physically this wouldn't make sense, since it would imply that for the units of the batch to be produced after the crossing of two consecutive WP progress lines the precedence relation is reversed.

To compensate for differing production rates between the consecutive WPs and to prevent the crossing of the progress lines, the need arises to employ time buffers. We distinguish between two such time buffers as described in the following two

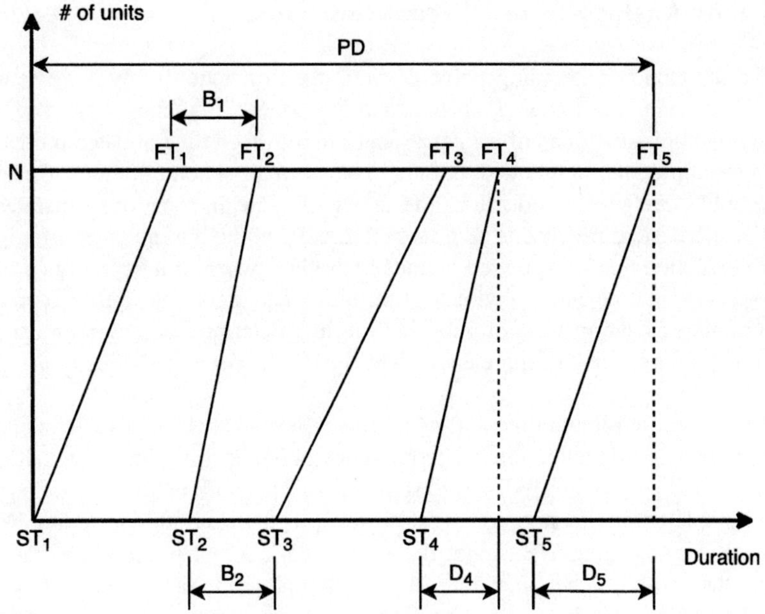

Fig. 13.1 WP progress chart and notation

Conditions. Let the preceding WP be denoted by WP_i and the succeeding one by WP_{i+1}.

Condition 1: When $p_i < p_j$, then a minimum time buffer B_i of magnitude $(FT_{i+1} - FT_i) \geq 0$ is placed between the finish times of WP_i and WP_{i+1}.
Condition 2: When $p_i \geq p_j$, then a minimum time buffer B_i of magnitude $(ST_{i+1} - ST_i) \geq 0$ is placed between the start times of WP_i and WP_{i+1}.

Examples of B_i for *Condition 1* and *Condition 2* are provided in Fig. 13.1 as B_1 and B_2, respectively. These minimum time buffers can be extended to accommodate managerial and/or technological concerns such as site preparation for the activities of the next WP. The minimum time buffer serves an analogous purpose to the generalized precedence relations. As in the definitions of the generalized precedence relations, we treat the minimum time buffer as a *finish-to-finish precedence relation*, if it is between two consecutive finish times (*Condition 1*). If, on the other hand, it is between two consecutive start times, then it corresponds to a *start-to-start precedence relation* (*Condition 2*).

A time buffer also serves as a risk mitigation measure against a possible stoppage or slowdown of the preceding/succeeding WP. When determining the length of a minimum time buffer, there is a trade-off between the length of the minimum time buffer and the project duration as can be easily deduced from Fig. 13.1.

13.1.1 An Analogy to Serial Production Lines

From a production scheduling point of view, the production of WPs depicted in Fig. 13.1 is analogous to a *serial production line*. Hence, the scheduling of WPs by employing minimum time buffers corresponds to the scheduling of batch production over a serial production line with buffers in between all or some stations. Consider, for example, the serial production line in Fig. 13.2 with 3 stations separated by 2 buffers. A basic difference is with respect to units of the buffer. The buffer in the serial production line is expressed in units of product, whereas in repetitive projects it is expressed in time units. Obviously, time buffer and units in the buffer are related to each other by the production rate. Indeed, in a deterministic environment time buffer and units in the buffer are equivalent and can be converted directly using the production rate.

The need for minimum time buffers B_1 and B_2 in Fig. 13.1 can also be explained in terms of the serial production line terminology of *blocking* and *starving*. When the operation on the preceding WP has to stop because the work on the succeeding WP is not completed due to stoppage or slowdown, then we say the preceding WP is blocked. The preceding WP cannot deliver the work completed to the succeeding WP or the crew on the preceding WP cannot proceed to the succeeding WP depending on the production environment prevailing. When, on the other hand, the operation in the succeeding WP has to stop because the work on the preceding WP is not completed due to stoppage or slowdown, then we say the succeeding WP is starving. The minimum time buffer placed between the start times of two consecutive WPs serves the purpose of accumulating enough work so as to prevent the starving of the succeeding WP.

13.1.2 Duration – Resource Relation for WPs

Specialized teams, composed of team members operating under the supervision of a team leader with the necessary tools and equipment, perform the WPs. The team size is determined for every WP based on the scope and nature of the work to be performed in the WP. A crew might consist of individuals from the same trade, such as electricians, or might include members with different but complementary skills such as a maintenance crew consisting of, e.g., electricians and mechanics. In the following, the amount of manpower available will be expressed in terms of the number of the men, to ensure compatibility with the conventional approach of estimating the work content in terms of man-days. The duration of a WP is a function of its work content *WC* in man-days and *CS* the number of men in the crew (crew size):

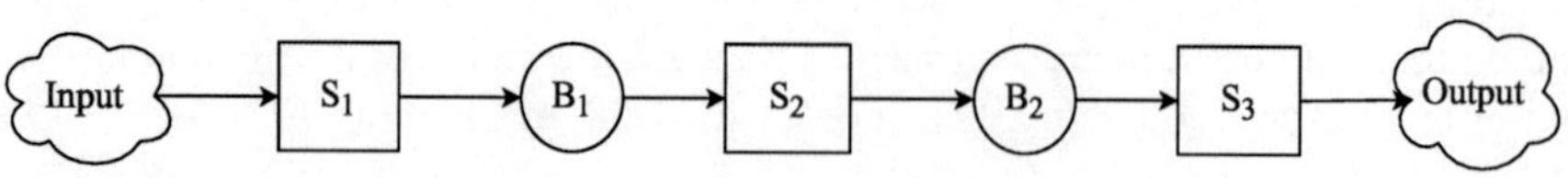

Fig. 13.2 A serial production line with 3 stations and 2 buffers

$$D_i = \frac{WC_i}{CS_i} \tag{13.1}$$

where WC_i is the work content of WP_i and CS_i is the number of men assigned to WP_i.

13.2 A Mathematical Programming Model for the Repetitive Projects

We now formulate an LP model for scheduling the WPs to minimize the total cost. An upper bound on the project duration as well as lower and upper bounds on WP durations will be integrated as constraints. The requirement that consecutive WPs do not interfere with each other will be included through another set of constraints.

To formulate the problem as an LP, the objective function and the set of constraints all need to be expressed in linear form. If the direct cost function on hand is nonlinear, then one needs to implement a linear approximation as shown in Fig. 13.3.

The total cost of a WP consists of *direct cost* and *indirect cost components*. We assume that indirect cost is directly proportional to WP duration, while the direct cost is proportional to work content *WC* and inversely proportional to WP duration.

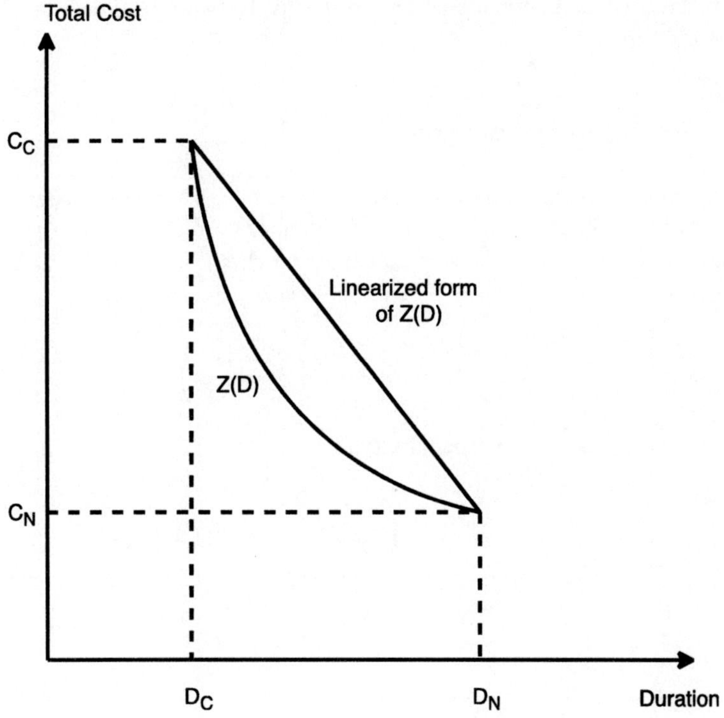

Fig. 13.3 The total cost function Z(D) for a WP

The total cost $Z_i(D_i)$ of WP_i can be expressed as

$$Z_i(D_i) = C_{1i}D_i + C_{2i}\left(\frac{WC_i}{D_i}\right) \tag{13.2}$$

where C_{1i} denotes the indirect cost per period for WP_i, and C_{2i} its direct cost per man allocated to WP_i.

Using the relation (13.1), we get the total cost CS_i as a function of the number of man assigned to WP_i as

$$Z_i(CS_i) = C_{1i}\left(\frac{WC_i}{CS_i}\right) + C_{2i}CS_i \tag{13.3}$$

(13.3) makes it clear that this problem is inherently a *time-cost trade-off problem.* The total cost is the sum of the indirect and direct cost components, where the indirect cost is a function of WP duration and the direct cost a function of the resources allocated to the WPs (see Sect. 5.2). As we increase the resources to reduce the duration of the WP at the expense of increasing the direct cost component, the indirect cost component is reduced due to the reduced duration of the WP.

Figure 13.3 illustrates the total cost of a WP_i as a function of its duration D_i, $Z_i(D_i)$. For the sake of simplicity, we have dropped the subscript i. D_C and D_N represent the crash and normal durations, respectively, as in the time/cost trade-off analysis of Chap. 5.

13.2.1 The Objective Function

We now obtain a linear approximation for the total cost function $Z_i(D_i)$ as the objective function in the mathematical programming formulation to be presented. The linear approximation of the total cost function $Z_i(D_i)$ for a duration D by $\overline{Z}(D)$ given by:

$$\overline{Z}(D) = Z_1 + (Z_2 - Z_1)\left(\frac{D - D^C}{D^N - D^C}\right) \tag{13.4}$$

We can express $(Z_2 - Z_1)$ using (13.2) as

$$\begin{aligned} Z_2 - Z_1 &= C_1D^N + C_2\left(\frac{WC}{D^N}\right) - C_1D^C - C_2\left(\frac{WC}{D^C}\right) \\ &= C_1\left(D^N - D^C\right) - C_2(WC)\left(\frac{D^N - D^C}{D^ND^C}\right) \end{aligned} \tag{13.5}$$

Dividing (13.5) by ($D^N - D^C$), substituting it into (13.4), and eliminating Z_1 and other constant terms, we obtain the following linear approximation for the total cost function, $ZL(D)$:

$$ZL(D) = C_1 D - \left(\frac{C_2 WC}{D^N D^C}\right) D = CD \tag{13.6}$$

where $C = C_1 - \frac{C_2 WC}{D^N D^C}$.

We shall use $ZL(D)$ as the objective function of the LP. Note that $ZL(D)$ is a linear approximation to the total cost function and that only up to a constant. Depending on the shape of the total cost function $Z_i(D_i)$ a better approximation can be obtained using a piecewise linear approximation.

13.2.2 The Constraints

In the decision-making environment we deal with here, an upper bound PD on project duration is imposed, which cannot be exceeded. Each WP duration is bounded from below by the crash duration D^C and from above by the normal duration D^N. The requirement that consecutive WPs do not interfere with each other and that WPs are not blocked or do not starve will be included in the mathematical programming formulation through two sets of constraints that we now develop.

The assignment of start times to the consecutive WPs determines the scheduling of WPs. For that purpose, a minimum time buffer is used in the scheduling of the WPs. As mentioned earlier, the minimum time buffer can occur between finish times or between start times. This depends on the relative productivities of the two consecutive WPs. When the preceding WP has lower productivity (i.e., slower) than the succeeding WP, the minimum time buffer is between finish times as displayed in Fig. 13.4a. The other case of the succeeding WP having lower productivity than the preceding one is shown in Fig. 13.4b. We now derive the conditions for avoiding the crossing of the consecutive WP progress lines.

Figure 13.3a depicts the case with $D_i > D_{i+1}$. The time between the completion of WP_{i+1} and the start time of WP_i is expressed as:

$$ST_i + D_i + B_i = ST_{i+1} + D_{i+1}$$

Hence, the duration between two consecutive starting times for this case is

$$ST_{i+1} - ST_i = B_i + D_i - D_{i+1}$$

Figure 13.3b shows the alternative case with $D_i < D_{i+1}$

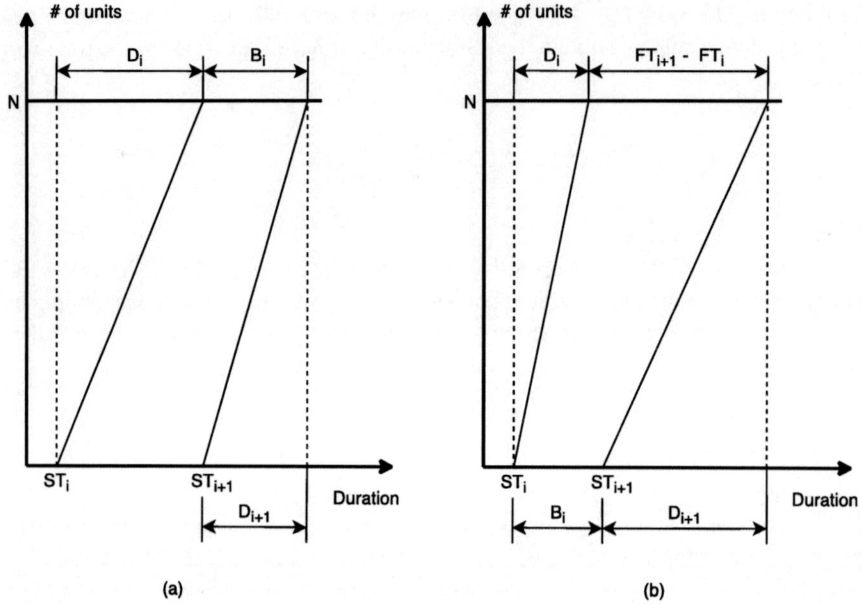

Fig. 13.4 (**a**) The case with $D_i > D_{i+1}$; (**b**) The case with $D_i < D_{i+1}$

$$D_i + FT_{i+1} - FT_i = B_i + D_{i+1}$$

Substituting ($ST_{i+1} + D_{i+1}$) and ($ST_i + D_i$) for FT_{i+1} and FT_i, respectively, we get:

$$ST_{i+1} - ST_i = B_i$$

Both cases are summarized as follows:

$$ST_{i+1} - ST_i = B_i + D_i - D_{i+1} \quad \text{if } D_i > D_{i+1}$$

$$ST_{i+1} - ST_i = B_i \quad \text{if } D_i < D_{i+1}$$

In scheduling, the above two equalities are treated as inequalities and serve as lower bounds on ($ST_{i+1} - ST_i$). Similar bounds can be derived for ($FT_{i+1} - FT_i$). These inequalities are included in the mathematical programming model as constraints.

13.2.3 The Mathematical Programming Model

We can now write the LP formulation over all WP_i to determine the minimum cost schedule for n WPs satisfying a given upper bound PD on the project duration. ST_i and D_i are the decision variables associated with the start time and duration of WP_i, respectively, while D_i^C, D_i^N, B_i and PD are input parameters.

$$\min ZL = \sum_{i=1}^{n} C_i D_i \tag{13.7}$$

s.t.

$$ST_n + D_n \leq PD \tag{13.8}$$

$$ST_{i+1} - ST_i \geq B_i + D_i - D_{i+1} \qquad i = 1, \ldots, (n-1) \tag{13.9}$$

$$ST_{i+1} - ST_i \geq B_i \qquad i = 1, \ldots, (n-1) \tag{13.10}$$

$$D_i^C \leq D_i \leq D_i^N \qquad i = 1, \ldots, n \tag{13.11}$$

$$ST_i \geq 0 \qquad i = 1, \ldots, n \tag{13.12}$$

Constraint (13.8) imposes an upper bound on the project duration *PD*. The constraint sets (13.9) and (13.10) consider both cases of the relative productivity of consecutive WPs. If $D_i > D_{i+1}$, then (13.9) is binding. On the other hand, if $D_i < D_{i+1}$, then (13.10) is binding.

In the above formulation, value of the project duration *PD* can be parameterized by assigning different values PD' to it and computing the minimum total cost for different project durations. The minimum cost project duration can be obtained by searching over the values of *PD* in the formulation [13.7–13.12] by writing (13.8) as $ST_n + D_n = PD'$.

13.3 A Model with an Alternative Total Cost Function

In the above formulation, the objective function is the sum of the direct and indirect costs for WPs. This approach is appropriate if one can easily allocate indirect costs to individual WPs. If this cannot be done, another objective function formulation can be developed where there is a single indirect cost quotation covering the whole project.

We assume here a linear direct cost function for each WP_i and an indirect cost function linearly proportional to project duration. The total cost *Z* is then the sum of the direct and indirect cost components:

$$Z = \sum_{i=1}^{n} \left[C_{2i} D_i^N + c_i \left(D_i^N - D_i \right) \right] + C_{\text{ind}} (ST_n + D_n)$$

where c_i denotes the unit direct cost of compressing WP_i, and C_{ind} the indirect cost per period for the whole project.

In the following LP formulation, the terms $C_{2i} D_i^N$ are omitted since they are constants independent of the schedule being computed. Although $c_i D_{Ni}$ terms are

Table 13.1 Data for Example 13.1

WP_1	$D_{N1} = 42$	$D_{C1} = 36$	$B_1 = 8$	$c_1 = 22$
WP_2	$D_{N2} = 81$	$D_{C2} = 71$	$B_2 = 12$	$c_2 = 61$
WP_3	$D_{N3} = 33$	$D_{C3} = 30$	$B_3 = 7$	$c_3 = 42$
WP_4	$D_{N4} = 56$	$D_{C4} = 51$	$B_4 = 21$	$c_4 = 16$
WP_5	$D_{N5} = 72$	$D_{C5} = 66$		$c_5 = 33$

also constant, we keep them in the formulation to express the trade-off between the direct cost of compression and the indirect cost component.

Since the decision environment is the same as in Sect. 13.2 except for the objective function, the constraints (13.8–13.12) remain unchanged and are reproduced below.

$$\min Z = \sum_{i=1}^{n} c_i\left(D_i^N - D_i\right) + C_{ind}(ST_n + D_n) \tag{13.13}$$

subject to

$$ST_n + D_n \leq PD \tag{13.14}$$

$$ST_{i+1} - ST_i \geq B_i + D_i - D_{i+1} \qquad i = 1, \ldots, (n-1) \tag{13.15}$$

$$ST_{i+1} - ST_i \geq B_i \qquad i = 1, \ldots, (n-1) \tag{13.16}$$

$$D_i^C \leq D_i \leq D_i^N \qquad i = 1, \ldots, n \tag{13.17}$$

$$ST_i \geq 0 \qquad i = 1, \ldots, n \tag{13.18}$$

If the decision problem on hand is to determine the minimum total cost project duration, the above formulation can be employed by parameterizing the project duration *PD*, assigning it different values *PD'* and searching for the optimal project duration as in the previous formulation.

Example 13.1 Consider a project involving the manufacturing of 50 units of a product in 160 days. Let the indirect cost coefficient C_{ind} be 50 per day. Further data is provided in Table 13.1.

Solution

Let us write the LP formulation [13.13–13.18] for this problem.

$$\begin{aligned} \min Z = {} & 22(42 - D_1) + 61(81 - D_2) + 42(33 - D_3) + 16(56 - D_4) \\ & + 33(72 - D_5) + 50(ST_5 + D_5) \end{aligned} \tag{13.19}$$

subject to

Table 13.2 The optimal durations and starting times for the WPs

D_1^*	D_2^*	D_3^*	D_4^*	D_5^*	ST_1^*	ST_2^*	ST_3^*	ST_4^*	ST_5^*
42	79	33	56	66	0	8	66	73	94

$$ST_5 + D_5 \leq 160 \tag{13.20}$$

$$ST_2 - ST_1 \geq 8 + D_1 - D_2 \tag{13.21}$$

$$ST_2 - ST_1 \geq 8 \tag{13.22}$$

$$ST_3 - ST_2 \geq 12 + D_2 - D_3 \tag{13.23}$$

$$ST_3 - ST_2 \geq 12 \tag{13.24}$$

$$ST_4 - ST_3 \geq 7 + D_3 - D_4 \tag{13.25}$$

$$ST_4 - ST_3 \geq 7 \tag{13.26}$$

$$ST_5 - ST_4 \geq 21 + D_4 - D_5 \tag{13.27}$$

$$ST_5 - ST_4 \geq 21 \tag{13.28}$$

$$36 \leq D_1 \leq 42 \tag{13.29}$$

$$71 \leq D_2 \leq 81 \tag{13.30}$$

$$30 \leq D_3 \leq 33 \tag{13.31}$$

$$51 \leq D_4 \leq 56 \tag{13.32}$$

$$66 \leq D_5 \leq 72 \tag{13.33}$$

$$ST_1, ST_2, ST_3, ST_4, ST_5 \geq 0 \tag{13.34}$$

Solving the above problem, we get the solution reported in Table 13.2 with a minimum objective function value of $Z = 8320$ and an optimal project duration of 160 days.

Constraint (13.20) concerning the project duration is satisfied at equality, which implies that the minimum cost project duration is at least 160 days. Constraints (13.22), (13.23), (13.26), and (13.28) are satisfied as equality. Hence, a minimum time buffer is located between the finishing points of the second and third WPs (based on Constraint (13.23)), and the other minimum time buffers are located between the starting points of the first and second, third and fourth, fourth and fifth WPs.

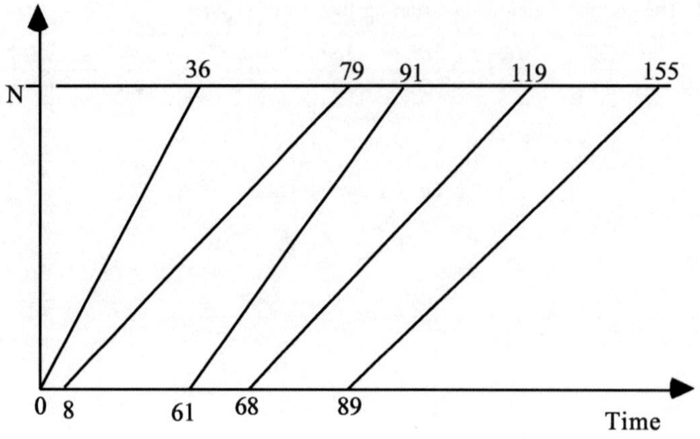

Fig. 13.5 The all-crash solution

Example 13.2 Let us re-examine Example 13.1 to understand the relations between the components of the problem. The following analysis is not meant to be an attempt for developing a heuristic approach to this problem.

Solution

The *all-crash* solution, i.e., the solution where all work packages are executed at their shortest possible duration, is represented in Fig. 13.5. The project duration is 155 days, and its total cost is (1146 + 7750) = 8896. Due to the relative productivities of consecutive work packages the minimum time buffers, B_1, B_3, and B_4 occur between the starting points of consecutive work packages. Let us try to reduce the total cost further.

Rules for expansion: WPs that complete before the starting point of a minimum time buffer can be expanded, as can WPs that start after the ending point of a minimum time buffer.

Rules for compression: WPs that complete at the starting point of a minimum time buffer can be compressed, as can WPs that start at the ending point of a minimum time buffer.

WP_3 can be expanded by 3 time units to its normal duration to reduce the project total cost to 8620 and the project duration to 152 days. The resulting schedule is displayed in Fig. 13.6. Note that expanding the duration of a WP has resulted in the compression of the project duration and that the *all-crash* solution did not result in the minimum project duration.

The compression cost of the above solution can be reduced by expanding WP_1 to its normal duration. The project duration, remains the same but the total project cost is reduced to 8620 – 6 × 22 = 8488.

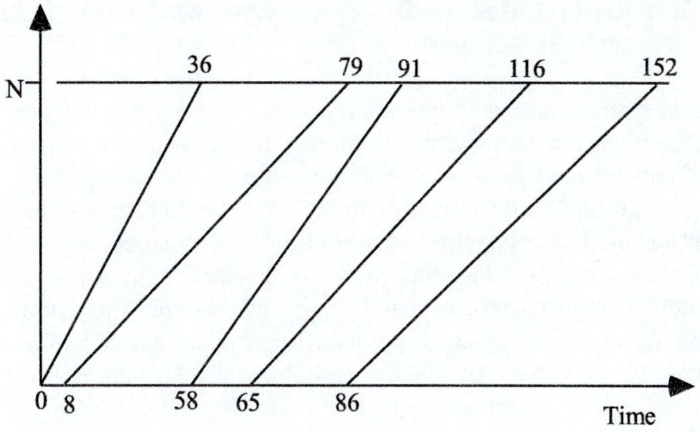

Fig. 13.6 Solution resulting from the expansion of WP_3

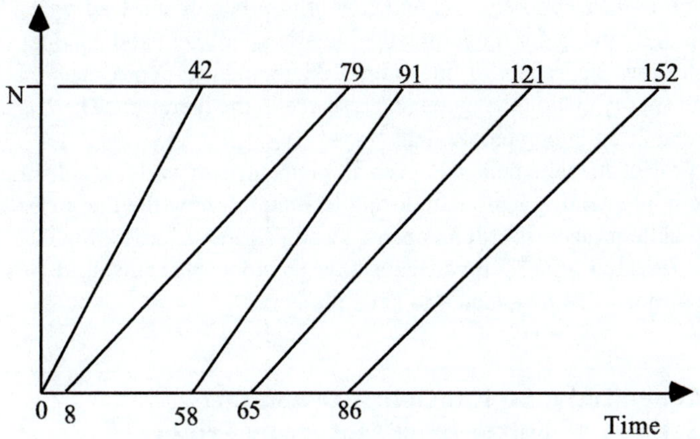

Fig. 13.7 Solution resulting from the expansion of WP_3, WP_1, and WP_4

WP_4 is expanded to its normal duration without affecting the project duration. The total project cost is reduced to (8488–5 × 16 = 8408). The result is presented in Fig. 13.7. In both cases, since the project duration does not change, the indirect cost component remains the same, while the direct cost is decreased. Expansion of either of the two as yet uncompleted WPs, WP_2 and WP_5, will be directly reflected in the project duration.

We cannot claim that the schedule presented in Fig. 13.6 is the least cost shortest duration since this solution is obtained as a result of an unsystematic search. A lower total cost of 8298 can be obtained by expanding the duration of WP_2 from 71 to 81, but this change results in the expansion of the project duration from 152 to 162, which exceeds the upper bound on project duration.

13.4 Conclusions, Recent Developments, and Some Future Research Directions

Make-to-order production is an important part of the economy covering one-of-a-kind and batch construction and manufacturing activities. In this chapter, we have developed two LP models to obtain the minimum cost schedule of the repetitive projects. The LP models differ mainly in their total cost function, which includes both direct and indirect cost components. In one case, the indirect cost is allocated to individual WPs, whereas in the other case, it is allocated to the whole project.

In repetitive projects, project teams learn and improve their progress rates. Scheduling models can integrate the productivity increases gained by the repetitions and learning. In addition to developing scheduling models, there is also a need for investigating the learning processes in projects and sectoral differences.

The contractors of repetitive projects can benefit from economies of scale, especially in procurement. In this regard, Xu et al. (2016) integrate project and supply chain management decision-making in recurrent projects. They focus on the construction industry and address uncertainty in procurement lead times. They use dynamic programming for modelling the integrated problem and conduct sensitivity analysis to show the impact of uncertainty. Mathematical approaches that combine project and supply chain management are scarce in the literature. There is a need for further research on integrated models.

Agile project management has been implemented in various industries. In the context of implementing agile monitoring techniques, how the line of balances can be used in agile projects is still an open question (Miranda & Bourque 2010). It is a promising research area to investigate how the tools that are used in managing repetitive projects can be adapted to agile projects.

13.5 Case Study: 50 Km Gebze-Orhangazi Section of Gebze-Izmir Motorway Project

The Gebze-Izmir Motorway Project is a \$6.5 billion road project whose vision is to offer safe and easy connectivity between the Marmara and Aegean regions of Turkey. The 421 km road project was inaugurated in 2010 and is estimated to finish in 7 years. The project is divided into various phases including the construction of a 3 km suspension bridge, 30 viaducts with a total length of 18.21 km, four tunnels with a total length 7.4 km, and 209 bridges.

The first phase of this project is the 53 km Gebze-Orhangazi section of the motorway. In September 2011 the firm of NOMAYG JV was awarded the contract worth \$2.3bn for the construction of this section of the project. This project includes two parts: the 3 km Izmit Bay suspension bridge and a 50 km road section. The latter is divided into four consecutive WPs, each having five constituent activities as shown in Table 13.3.

NOMAYG JV decides that for early completion of the project, the 50 km road will be divided into 10 sections each 5 km long and 25 m wide, each of which will be

Table 13.3 Main steps and constituent activities in road construction

WP No	Work packages and constituent activities
WP1	**Placement and compaction of granular materials to form the granular base course**
(a)	Load granular materials
(b)	Weigh
(c)	Transport
(d)	Dump and compact
(e)	Return
WP2	**Placement and compaction of hydrated granular materials to form the stabilized base layer**
(a)	Load hydrated granular materials
(b)	Weigh
(c)	Transport
(d)	Dump and compact
(e)	Return
WP3	**Placement and compaction of hot mix asphalt to form the binder layer**
(a)	Load hot asphalt mix
(b)	Weigh
(c)	Transport
(d)	Dump asphalt into paver, spread hot asphalt mix, and compact
(e)	Return
WP4	**Placement and compaction of hot mix asphalt to form the wearing course**
(a)	Load hot asphalt mix
(b)	Weigh
(c)	Transport
(d)	Dump asphalt into paver, spread hot asphalt mix, and compact
(e)	Return

Based on Polat et al. (2009)

subcontracted. Due to the repetitive nature of the activities in these 5 km sections, a repetitive scheduling approach will be adopted to ensure effective management of activities and resources. The resources required to complete one 5 km section of the road are shown in Table 13.4. Each 5 km section is further divided into six subsections, each 834 m long and 25 m wide. The length of these subsections is determined such that the duration of work for each WP for each subsection is approximately equal. This implies approximately equal productivity for the WPs, allowing roughly parallel display of WP progress lines. This has been indeed the case for the first three WPs. WP4, on the other hand, has higher productivity. Each section is completed in one period for first three WPs whereas it takes 2/3 periods for WP4 to complete one section.

To estimate the total time required to finish one 5 km section of the road, the PM asked his planning team to develop the corresponding repetitive scheduling model ignoring resource constraints. This schedule is illustrated in Fig. 13.8. All six sections are completed in 9 periods. The first three WPs can be scheduled in parallel

Table 13.4 Resources requirement across four WPs

Work packages	Resource	Quantity	Work packages	Resource	Quantity
WP1. Granular base course	Truck	10	WP2. Stabilized base layer	Truck	12
	Grader	1		Paver	1
	Water truck	1		Road roller	1
	Road roller	1		Rubber roller	1
	Flagman	1		Water truck	1
				Flagman	1
WP3. Binder layer	Truck	7	WP4. Wearing course	Truck	7
	Paver	1		Paver	1
	Steel wheel roller	1		Steel wheel roller	1
	Rubber roller	1		Rubber roller	1
	Worker	5		Worker	5
	Flagman	1		Flagman	1

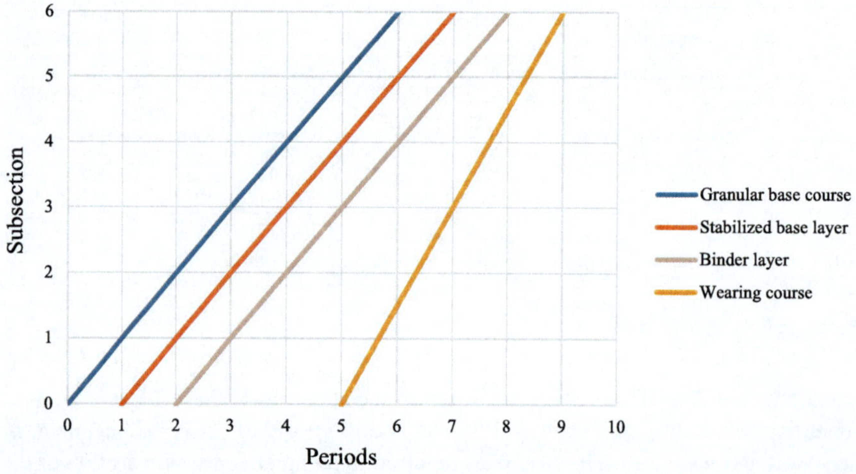

Fig. 13.8 WP progress lines diagram for the *unconstrained* schedule

without any break in between with a finish-to-finish time buffer of one period. As for WP4, in order to have a finish-to-finish time buffer of one period, the start-to-start time buffer between WP3 and WP4 is taken as 3 periods.

As per the schedule, one 5 km section can be completed in 9 days. However, this schedule assumes no resource constraints. In reality, resource availability is a critical factor that must be taken into consideration while scheduling the project through a repetitive project scheduling approach. Therefore, NOMAYG JV requested its subcontractor to share with them the list of their available resources, so that a

Table 13.5 Resources available with the subcontractor

#	Resource	Quantity
1	Flagman	3
2	Grader	1
3	Road roller	1
4	Water truck	1
5	Truck	17
6	Paver	1
7	Rubber roller	1
8	Steel wheel roller	1
9	Worker	5

more realistic completion time can be estimated for one 5 km section. The resource availability of the subcontractor is given in Table 13.5.

Examining the resource requirements of individual WPs, we observe that we can execute WP1 and WP3 simultaneously with the available resources. The same goes for WP1 and WP4. Furthermore, the WPs are not independent of each other. The following interdependencies were communicated by the Civil Engineering Department to the planning team working on the repetitive scheduling model;

(i) Only one WP can be carried out in one section at a time.
(ii) Different activities can be concurrently carried out in different sections, provided they do not require the same resources.
(iii) All WPs have a finish-to-start relationship and will be consecutively carried out in each section.
(iv) If a resource(s) is needed at the same time by two or more activities, the activity to be carried out in the preceding section has the priority.

After taking into consideration the constraints on resources and interdependencies among activities, a revised WP progress lines diagram, illustrated in Fig. 13.9, was presented to the PM. As per this model, one 5 km section of the road can be completed in 17 periods. Evaluating the WP progress lines diagram, the PM found it would be difficult to manage WP4 if the work is stopped with the subsection half completed and restarted after 2 periods. The PM would also like to avoid the stoppages of 1 and 2 periods for the first 3 WPs when moving from one subsection to the next. So, the PM asked the planning team to determine whether the schedule can be improved for easier implementation and whether the construction duration for 5 km sections could be further reduced if the subcontractor could provide additional resources.

The planning team came up with the following alternatives.

Alternative 1 The schedule for this alternative is displayed in Fig. 13.10. In this alternative, WP1, WP2, and WP3 are to be executed continuously with a finish-to-finish buffer of 1 period. WP4 is scheduled with a finish-to-start buffer of size 0 period with WP2. The construction for a 5 km section of the road will require

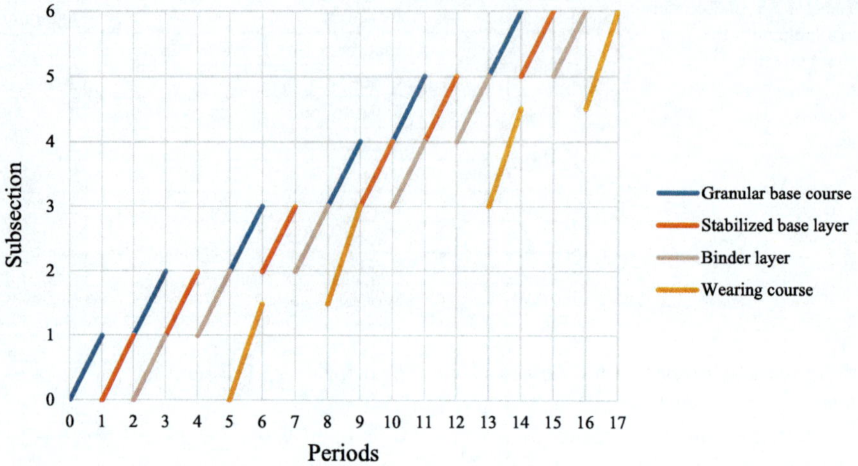

Fig. 13.9 WP progress lines diagram for the *revised* schedule

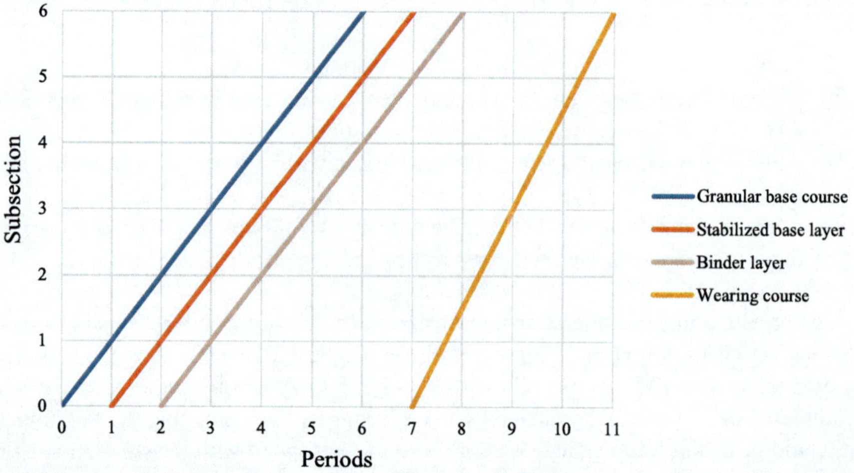

Fig. 13.10 Revised schedule based on *Alternative 1*

11 periods. The additional resources needed to implement this alternative are listed in Table 13.6.

Alternative 2 This alternative differs from Alternative 1 in that the starting time of WP4 is right-shifted by one period, eliminating the finish-to-start buffer with WP3. The resulting schedule is given in Fig. 13.11 with a construction duration for the 5 km segment of 12 periods. The subcontractor will need the additional resources listed in Table 13.7.

Table 13.6 Additional resources needed to implement Alternative 1

#	Resource	Quantity
1	Trucks	12
2	Paver	1
3	Rubber roller	1
4	Steel wheel roller	1
5	Laborer	5

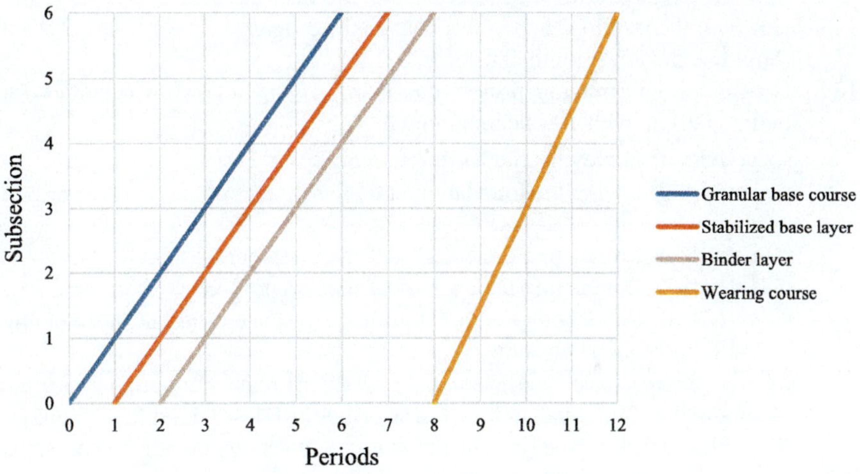

Fig. 13.11 Revised schedule based on *Alternative 2*

Table 13.7 Additional resources needed to implement Alternative 2

Resource	Quantity
Trucks	12
Paver	1
Rubber roller	1

Both alternatives permit a continuous flow of work for all WPs over all subsections and throughout the construction of the road section. The trade-off between the two alternatives is that in Alternative 1 we reduce the duration for construction by one period at the expense of 1 additional steel wheel roller and 5 additional workers compared to the requirements of Alternative 2. Although the cost of the additional resources needed in Alternative 1 is borne by the subcontractor, the subcontractor will reflect part of this cost in the total cost of the contract. This trade-off between the two alternatives is to be assessed by the PM and a final decision needs to be reached in consultation with the NOMAYG JV management.

Exercises

13.1 Prove or disprove the following statement: When for a given project duration *PD*, the constraint $ST_n + D_n \leq PD$ is satisfied as equality, then the minimum cost project duration is at least equal to *PD*.

13.2 State an algorithm for determining the minimum cost project duration.

13.3 Explain how the project duration and the risk of possible stoppage or slowdown of the preceding or succeeding WP are interrelated?

13.4 How will the learning curve effect change the progress of the activities? What should be the effect on the buffer B_i?

13.5 Consider a construction project consisting of the following activities for building 30 individual residential units.

- **Activity 1:** Surveying and earth removing.
- **Activity 2:** Laying the foundation and the structure of the first and second floors.
- **Activity 3:** Completing the exterior of the first floor.
- **Activity 4:** Completing the exterior of the second floor.
- **Activity 5:** Completing the roof, building side pavements and surrounding walls, and general cleaning.

Let the indirect cost coefficient C_{ind} be 50 TL/day. Furthermore, let the minimum time buffer values between activities be B1 = 8 days; B2 = 12 days; B3 = 7 days; B4 = 21 days. The direct cost and activity duration estimations are provided in the following table.

	Duration (days)			
Activity	Optimistic	Most likely	Pessimistic	Direct Cost (TL)
1	40	42	44	22
2	70	81	86	61
3	27	33	45	42
4	49	56	63	16
5	57	72	75	33

The engineering team provides 3 estimates for each activity duration as "optimistic", "most likely", and "pessimistic", which are then used to estimate the underlying Beta distribution of each activity duration, which in turn are used to obtain the expected project duration.

(a) Draw the WP progress lines diagram and determine the expected start and end times of each activity. Report the project duration and the total cost.

(b) Based on expert judgement, let the most likely durations in part (a) be the normal deterministic durations for all activities except activities 3 and 4. The durations the activities 3 and 4 can take on together with their associated probabilities are given in the table below.

Activity	Duration	Probability
3	33	0.2
	45	0.8
4	56	0.6
	63	0.4

Employing the data provided, use two different approaches to obtain the expected project duration and compare the results.

13.6 Consider the following production problem of producing N identical units, which can be modelled using the repetitive project scheduling approach. The data is provided in the following table.
$B_1 = 3$ days; $B_2 = 5$ days; $B_3 = 7$ days; Indirect cost per day is 45 TL.

	Duration (days)		
Activity	Crash	Normal	Cost slope (TL)
1	17	18	21
2	17	21	18
3	13	13	–
4	28	31	48

(a) Obtain the optimal project schedule.
(b) Suppose the unit crashing cost for activity 4 is reduced by 5 TL. How would the optimal schedule change?

13.7 Consider the following production problem of producing N identical units, which can be modeled using the repetitive project scheduling approach. The data is provided in the following table.

	Duration (days)		
Activity	Crash	Normal	Cost slope (TL)
1	20	28	19
2	20	36	22
3	30	40	26
4	32	37	11
5	31	35	23
6	29	33	21

The minimum time buffers between two consecutive activities are given as follows:
$B_1 = 10$ days; $B_2 = 10$ days; $B_3 = 10$ days; $B_4 = 8$ days and $B_5 = 8$ days. Indirect cost per day is 40 TL.

(a) Draw the WP progress lines diagram of activities with normal durations.
(b) Find the total duration and draw the WP progress lines diagram when you crash activities according to the net gain they provide.

13.8 Consider the following problem with 5 WPs whose data is provided below.

$B_1 = 8$	$D_1 = 42$	$d_1 = 36$	$C_{11} = 22$	$C_2 = 50$
$B_2 = 12$	$D_2 = 81$	$d_2 = 71$	$C_{12} = 61$	
$B_3 = 7$	$D_3 = 33$	$d_3 = 30$	$C_{13} = 42$	
$B_4 = 21$	$D_4 = 56$	$d_4 = 51$	$C_{14} = 16$	
	$D_5 = 72$	$d_5 = 66$	$C_{15} = 33$	

(a) Using the duration and cost data, find the optimum project duration.
(b) Determine the least cost shortest project duration.

13.9 Consider the following production problem of producing N identical units, which can be modelled using the repetitive project scheduling approach. The data is provided in the following table.

	Duration (days)		
Activity	Crash	Normal	Cost slope (TL)
1	36	36	–
2	69	76	26
3	66	66	–
4	53	53	–
5	66	72	44

The minimum time buffers between two consecutive activities are given as follows: $B_1 = 12$ days; $B_2 = 16$ days; $B_3 = 9$ days; $B_4 = 19$ days.
The indirect cost per day is 32 TL.

(a) Draw the WP progress lines diagram and find the optimal schedule with the data given in the table above considering the minimum total cost.
(b) Draw the WP progress lines diagram when the resulting schedule lasts for 142 days and report about the net gain for this schedule.
(c) Find the earliest due-date and report the net gain.

13.10 25 sleeping coaches are to be manufactured in batches of 5 units. The batches visit 4 sites in sequence and a WP is applied at each site, for a total of 4 WPs. For example, initially, before the first batch of the first WP is processed at the first site, site preparation is performed. Before the first batch of a WP starts at a site, a site preparation activity is executed which takes 3 days. The next batches of the same WP do not require any additional site preparation. The site preparation duration is the same at each site. The first site preparation starts at $t = 0$. The productivities of the WPs are given as follows:

- Productivity of WP A = 0.8 units/day
- Productivity of WP B = 1.4 units/day
- Productivity of WP C = 1.0 units/day
- Productivity of WP D = 2.2 units/day

(a) Determine starting and ending times of the WPs using only the minimum buffer times necessary. Draw the activity progression chart.
(b) Assume that the duration of a batch of 5 units of WP A is normally distributed with a mean equal to the deterministic duration given in part (a) and the variance is equal to (0.1) of its duration. Determine the buffer time between the WPs A and B such that it will not be exceeded with 0.96 probability.

13.11 30 units of a product are going to be produced. Each product is produced in 5 consecutive stages. In each stage, a WP of 30 units is completed. A worker can finish 50% of one unit in WP1 in a day. That is, the productivity of one worker in WP1 is 50%. The productivity of workers in the WPs is given as follows.

Work package	Productivity per worker(%)	Number of workers assigned
WP1	50	10
WP2	20	15
WP3	25	12
WP4	15	10
WP5	10	20

There is a 3 day minimum buffer time between any two WPs. To obtain the productivity of the worker group assume that the productivities of individual workers are linearly additive and are independent of the group size.

(a) Calculate the duration of each WP according to the given allocation and draw the WP progress lines diagram.
(b) The company can reallocate workers to different WPs thus expanding some WPs while crashing others. For this project find a reallocation of workers that will reduce the project duration using the repetitive project scheduling approach. Is that an optimal allocation?

13.12 A recurrent project consists of 5 WPs. According to the estimated durations, the following WP progress lines diagram is obtained for this project.

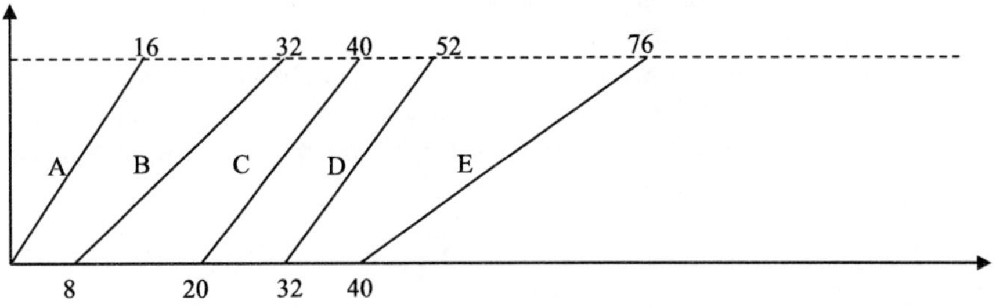

However, the activity duration of WP B cannot be estimated accurately. It is assumed to follow a normal distribution with a mean of 24 and a variance of 16.

(a) Calculate the probability that the WP progress lines diagram above will be infeasible.

(b) What change(s) in the buffer times can you suggest so that the WP progress lines diagram above will be feasible with a probability of at least 95%?

13.13 Consider a project dealing with the construction of 10 units. The work is organized into 5 WPs. On each working day of 8h shifts, 2 units are completed by each work package. Each WP employs two renewable resources to accomplish the work. The availabilities of resource 1 and resource 2 are 2 and 3 units, respectively. The resource requirements of the WPs are given in the following table.

Work package	Resource 1	Resource 2
1	1	1
2	1	2
3	2	1
4	1	1
5	2	1

Determine the project duration in working days.

13.14 (a) Write down the LP formulation for the repetitive project scheduling problem for the case where the objective function is the sum of the direct cost for activities and the indirect cost, where the indirect cost per unit time, C_1, is over all activities, i.e., for the whole project, and C_{2i} is linearization coefficient of the direct cost, in other words, the slope of the direct cost curve. Let the project duration be limited by a predefined value L. Use the following notation:

- ST_i = starting time of activity i.
- D_i = duration of activity i.
- B_i = minimum time buffer between activities i and $(i + 1)$.
- $D^C{}_i$ = crash duration of activity i.
- $D^N{}_i$ = normal duration of activity i.

(b) Consider the following production problem of producing 50 units, which can be modeled using the repetitive project scheduling problem approach. The data is provided in the following table concerning the normal durations.

Activity	Normal duration (days)
A	50
B	40
C	75
D	25
E	100

Determine the minimum time buffers between two consecutive activities and draw the WP progress lines diagram.

References

Miranda, E., & Bourque, P. (2010). Agile monitoring using the line of balance. *Journal of Systems and Software, 83*(7), 1205–1215.

Polat, G., Buyuksaracoglu, Y., Damci, A. (2009). *Scheduling asphalt highway construction operations using the combination of line-of-balance and discrete event simulation techniques*. Proceedings IEEM International Conference on Industrial Engineering and Engineering Management (IEEM09), Tsimshatsui, Hong Kong, 1126–1130.

Xu, X., Zhao, Y., & Chen, C. Y. (2016). Project-driven supply chains: Integrating safety-stock and crashing decisions for recurrent projects. *Annals of Operations Research, 241*(1–2), 225–247.

Yang, I. T., & Ioannou, P. G. (2004). Scheduling system with focus on practical concerns in repetitive projects. *Construction Management and Economics, 22*(6), 619–630.

14 Project and Portfolio Selection

Learning Outcomes
Upon successful completion of this Chapter, the reader will be able to:

1. Apply multi-criteria scoring models.
2. Formulate complex multi-attribute decision problems as AHP problems.
3. Formulate single objective mathematical programming models for project selection.
4. Formulate mathematical programming models for project and portfolio selection.
5. Apply multi-objective linear programming to generate a set of non-dominated solutions.
6. Apply goal programming to solve multi-objective project selection problems with preemptive priorities.
7. Solve multi-objective, multi-period project portfolio selection and scheduling problems using mathematical programming.

14.1 Introduction

An important topic in project management is project selection, which can require a rather complex decision-making process. Managers will generally try to choose projects that are expected to be more profitable and which will contribute to achieving the organization's goals. The project selection process uses careful screening and evaluation of the candidate projects to identify those best serving these purposes. A *screening process* seeks to evaluate the candidate projects through quick analysis to select those that meet the minimum requirements imposed by the organization and deserve further detailed study. However, a wide variety of external and internal factors including market conditions, availability of raw material and competent suppliers, probability of technical success, government regulations, conflicting preferences among the stakeholders greatly complicate the project selection. Exogenous factors such as government regulations and market conditions are not under the control of the organization, although they can sometimes influence

G. Ulusoy, Ö. Hazır, *An Introduction to Project Modeling and Planning*, Springer Texts in Business and Economics, https://doi.org/10.1007/978-3-030-61423-2_14

them to a certain extent. Others are endogenous factors that can reasonably be expected to be under the total control of the organization such as quality of the decision-making process and allocation of resources.

The selection of the right projects is crucial to the long-term survival of the organization and requires that the decision-making ability of top management be supported by the right selection methodology and high quality, informative data. Which units of the organization will participate in the project selection process must be assessed carefully for each project according to its needs. Different units are likely to have different criteria for identifying promising projects, and their reconciliation may not be an easy matter. Before making the final decision on which project to select, it is good management practice to consider all critical factors even if they are not included in the selection process regardless of whether these are exogenous or not.

There is a wide spectrum of different approaches to the project selection problem, so a complete treatment is not possible within the confines of a single chapter. To give the reader a flavor of the different types of approaches, we will present representative examples of two major families of such methods: *Scoring methods* and *optimization models*. For scoring methods, we will present a *multi-criteria scoring model*, and the *Analytic Hierarchy Process* (AHP). The optimization methods cover *mathematical programming* approaches for project selection such as integer programming, goal programming, and dynamic programming. Since the problem of project selection arises in a large number of different settings, a wide range of models have been developed to represent these different decision-making environments.

As has been stated earlier in Sect. 2.8, a *portfolio* is a collection of projects, programs, subsidiary portfolios, and operations managed as a group to achieve strategic objectives (PMI 2017). The *portfolio selection problem* deals with the selection of a subset of projects from a set of candidate projects so as to optimize the objective(s) of the organization. Mathematical programming models will be given for the portfolio selection problem. A mathematical programming model for the *portfolio selection and scheduling problem* will also be presented. A case study dealing with portfolio selection and scheduling will be provided.

14.2 A Multi-criteria Scoring Model

In the multi-criteria scoring model, the candidate projects are ranked relative to one another based on a set of criteria or attributes. Such models have been investigated rather widely for some time now.

The first task is to identify a set of criteria, which should encompass all important aspects of the selection process for the organization but should also be parsimonious to avoid unnecessary effort and detail. Since some criteria might be more important than others, a weight w_j in the range [0–1] is assigned to each criterion $j = 1,2,\ldots,N$, such that the sum of all weights equals 1. Once the set of criteria have been identified – by no means a trivial task – each candidate project $i = 1, 2, \ldots, n$ is given a score s_{ij} for each criterion j. These scores have to be dimensionless in order to be able to consider criteria expressed in incommensurable units of measure such as percentage, monetary unit, and kg. This is achieved by assigning each criterion a score in the range [1–10] or [1–100]. These individual criterion scores are aggregated with the corresponding weights to obtain the overall score S_i for the candidate project i as follows:

$$S_i = \sum_{j=1}^{N} s_{ij} w_j \tag{14.1}$$

where N is the number of criteria considered and the nonnegative weights w_j add up to 1.

The candidate project with the highest score is adopted for execution; ties can be broken by secondary objectives not yet considered. If the available budget allows more than one project proposal to be adopted, one can approach the problem as a knapsack problem (see, e.g., Winston 2003). The definition of the overall score S_i assumes the additivity of the individual criterion scores, implying independence of these criteria and lack of interaction among them.

It is important to acknowledge at the outset that the processes of determining the criteria, the corresponding weights, and the individual criterion scores are rather subjective processes which are usually performed by a group of experts. A widely recommended method to determine the weights of the scoring criteria is to employ the weight determination methodology of AHP. Since there is a strong subjective element in this whole decision-making process, the credibility of the process is critical to the acceptance of its recommendations by the stakeholders involved. Thus, pervasive transparency of the whole process as well as the results reported are an essential dimension of the process.

Let us consider the following example to demonstrate the application of the multiple criteria scoring model to a decision problem.

Example 14.1 A company wants to select a new Internet service provider (ISP) and its options are ISPA, ISPB, and ISPC. A panel of experts has determined that the major factors to be used to evaluate each option are *Connection Speed* (CS), *Service Quality* (SQ), and *Cost* (C). The cost criterion is composed of two components; *Fixed Setup Cost* (FSC) and *Monthly Installments* (MI). The weights associated with these criteria, as well as the scores over 100 points for each candidate project with respect to these criteria determined by the same expert panel are displayed in Table 14.1. The transparency mentioned above requires that the methods by which each value in Table 14.1 was obtained be openly and clearly defined and can be traced back if needed.

Table 14.1 Weights and scores for Example 14.1

	CS	SQ	FSC	MI
Weight	0.60	0.25	0.12	0.03
ISPA	60	16	40	10
ISPB	15	74	15	70
ISPC	25	10	45	20

We can now determine the scores for each project considered.

$$S_A = 0.60(60) + 0.25(16) + 0.12(40) + 0.03(10) = 45.1.$$

$$S_B = 0.60(15) + 0.25(74) + 0.12(15) + 0.03(70) = 31.4.$$

$$S_C = 0.60(25) + 0.25(10) + 0.12(45) + 0.03(20) = 23.5.$$

Based on the scores obtained, ISP_A is selected for implementation.

If two or more of the scores are close to each other, then it is appropriate to apply *sensitivity analysis* around the weights as well for some of the scores corresponding to this subset of projects. This might involve repeating the process for this subset of projects.

Let us continue with Example 14.1 and apply sensitivity analysis around the weights of CS and SQ to investigate whether the selection decision will change. The weights of these two criteria are reassessed and determined as 0.50 and 0.35, respectively. Then for this case, the scores for each project is determined as follows:

$$S_A = 0.50(60) + 0.35(16) + 0.12(40) + 0.03(10) = 40.7.$$

$$S_B = 0.50(15) + 0.35(74) + 0.12(15) + 0.03(70) = 37.3.$$

$$S_C = 0.50(25) + 0.35(10) + 0.12(45) + 0.03(20) = 22.0.$$

With these changes in the weights of CS and SQ, ISP_A is again selected for implementation but the difference in scores between the two alternatives ISP_A and ISP_B is reduced considerably.

14.3 The Analytic Hierarchy Process

The AHP was developed in the 1970s (Saaty 1980, 1990, 1994). It allows the decision-maker to structure a complex multi-attribute decision problem in the form of a hierarchy with projects or other alternatives such as products at the lowest level and the various objectives at higher levels.

After the hierarchy is established, the decision-makers compare the various elements at the same level of the hierarchy in pairs to obtain the priority vector $\underline{PV}$ of these elements. During these comparisons, data both about the elements and decision-makers' judgments about the element's relative importance is generated. To compare element i to element j, the following question is asked:

Table 14.2 The pairwise comparison scale

Relative importance	Value
Equal importance	1
Moderate importance of one over the other	3
Essential or strong importance	5
Very strong importance	7
Extreme importance	9
Intermediate values between the two adjacent judgments	2,4,6,8

"How important (preferable) is element i compared to element j?"

The answer to this question is given in terms of the pairwise comparison scale provided in Table 14.2. No fractional values are allowed. Let us denote the value of the answer to the above question by a_{ij}. This value actually represents the ratio of the weights attached to the respective elements i and j by the decision-maker: w_i/w_j. The pairwise comparison values a_{ij} constitute the entries of an $(n \times n)$ square matrix A, called the *judgment matrix*, where n is the number of elements under consideration. By the definition of a_{ij} that the judgement matrix A has the reciprocal property $a_{ij} = 1/a_{ji}$ and the diagonal entries are all equal to 1. Using the definition of a_{ij}, it can be shown that:

$$A\underline{w} = n\underline{w} \tag{14.2}$$

where $\underline{w}$ is the weight vector $(w_1, w_2, \ldots, w_n)^T$. If n is an eigenvalue of A, then $\underline{w}$ is the eigenvector associated with it. Note that A has rank 1 because every row is a constant multiple of the first row. Thus, all its eigenvalues except one are zero. The sum of the eigenvalues of a matrix is equal to its trace (the sum of its diagonal elements) and hence, for A it is equal to n. Therefore, n is the largest or principal eigenvalue of A.

In order to determine the weight vector $\underline{w}$, the sum of the entries for each column of A is calculated and each entry in a column is divided by the corresponding total column sum to normalize the values. Then the average of each row is computed to determine the weight vector components. The weights are in the range [0–1] and they add up to 1.

Another method to determine the weight vector $\underline{w}$ is as follows. For each row of the A matrix, the entries of the row are multiplied, and the nth root is taken. After this is done for each row of A, the resulting vector is normalized to obtain the weight vector $\underline{w}$.

Recall that A has the reciprocal property and is called a reciprocal matrix. Further, A is consistent, implying that the following also holds for A:

$$a_{jk} = a_{ik}/a_{ij} \quad i, j, k = 1, 2, \ldots, n. \tag{14.3}$$

Table 14.3 RI values for different A matrix sizes (Saaty and Kearns 1985)

Size of matrix	1	2	3	4	5	6	7	8	9	10
RI	0	0	0.58	0.90	1.12	1.24	1.32	1.41	1.45	1.49

The decision-maker might make some judgment errors when filling the judgement matrix A. It is known from eigenvalue theory that a small perturbation in n when A is consistent leads to an eigenvalue problem of the form:

$$A\underline{w} = \lambda_{max}\underline{w} \tag{14.4}$$

where $\lambda_{\max}$ is the principal eigenvalue of A, and A may no longer be consistent but remains reciprocal (Saaty and Kearns 1985).

Two results of importance are the following (Shiraishi et al. 1998):

Result 1: For a reciprocal matrix, $\lambda_{\max} \geq n$ holds always.
Result 2: A is consistent if and only if $\lambda_{\max} = n$.

Given that in general the judgement matrix A will not be consistent, we would like to assess the inconsistency level of A and to reject A if its inconsistency is beyond a certain level. For this purpose, *the consistency index* CI is defined as a measure of the deviation of the judgments from the consistent approximation, representing the negative average of the other roots of the characteristic polynomial of A:

$$\text{CI} = (\lambda_{\max} - n)/(n - 1). \tag{14.5}$$

CI is divided by the *random consistency number* RI to result in the *consistency ratio* CR. If CR is sufficiently small (say, 0.15 or less), we accept the estimate of $\underline{w}$. Otherwise, we try to improve the consistency of the judgement matrix A by going back to the decision-makers and asking them to reconsider and revise their judgments.

RI values are obtained as an average over a large number of reciprocal matrices of the same order whose entries are random. The RI values are given in Table 14.3.

Two methods are suggested to calculate $\lambda_{\max}$.

First approach: The set of equations resulting from $A\underline{w} = \lambda_{max}\underline{w}$ is employed to obtain n estimates for $\lambda_{\max}$ and then the average is taken as the representative $\lambda_{\max}$ value.

Second approach: The column sums of the A matrix are multiplied by the corresponding weight vector component and added to obtain the representative $\lambda_{\max}$ value.

In the final step of this method, called the *synthesis step*, numerical priorities are computed for each of the decision alternatives. Since these numbers represent the alternatives' relative ability to achieve the decision goal, they allow a straightforward consideration of the various courses of action. Let us define the criteria from which no further branching to other criteria occurs as terminal criteria. Note that the weights of the terminal criteria should add up to 1. Prior to the synthesis step, both the weights for all terminal criteria k and the weights for all alternatives j under all terminal criteria k are obtained. The jth component of the priority vector $\underline{\mathrm{PV}}$ associated with alternative j, (PV_j), is obtained as in (14.6):

$$PV_j = \sum_k w_k w_j^k \tag{14.6}$$

where w_k is the weight for terminal criterion k and w_k^j is the weight for alternative j under terminal criterion k.

Example 14.2 Let us return to Example 14.1 to demonstrate the application of AHP to a decision problem. The decision hierarchy is given as in Fig. 14.1. Assume here that if the consistency ratio $\mathrm{CR} \geq 0.150$, then the judgement matrix A is inconsistent.

The first level criteria; *Connection Speed*, *Service Quality*, and *Cost* are compared in pairs to obtain the entries of the judgement matrix A as in Table 14.4. In this example, the panel of experts decided that *Connection Speed* is somewhat more important than *Service Quality* (3) and extremely more important than *Cost* (9). *Service Quality* was much more important than *Cost* (7). Note that the diagonal elements of the comparison value matrix are equal to 1 by definition and for any comparison value a_{ij} the symmetric cell a_{ji} is filled with its reciprocal.

For determining the weights for the first level criteria we normalize the entries of the A matrix by dividing each entry by the corresponding column sum. The column

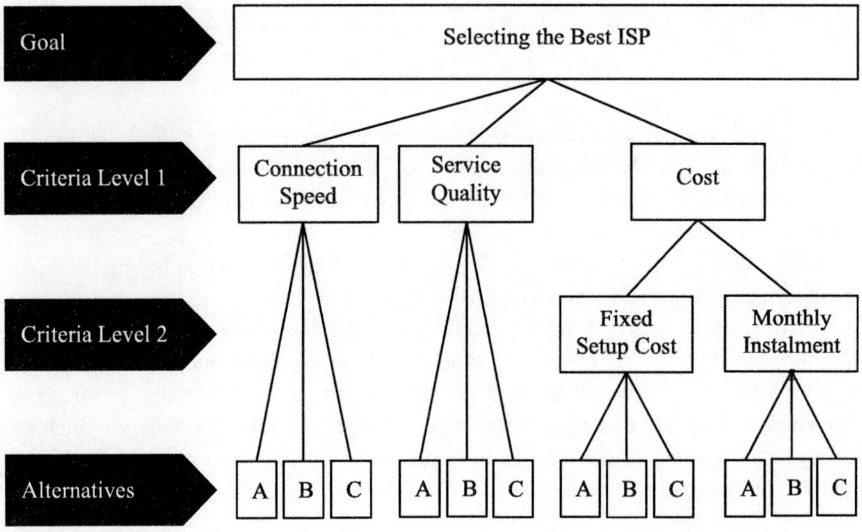

Fig. 14.1 AHP decision hierarchy

Table 14.4 The judgement matrix *A* for the first level criteria

Criteria	Connection speed	Service quality	Cost
Connection speed	1	3	9
Service quality	1/3	1	7
Cost	1/9	1/7	1
Column sum	1.444	4.143	17

Table 14.5 Obtaining the normalized *A* matrix and the weight vector $\underline{w}$

Criteria	Connection speed	Service quality	Cost	$\underline{w}$
Connection speed	0.692	0.724	0.529	0.648
Service quality	0.231	0.241	0.412	0.295
Cost	0.077	0.035	0.059	0.057

Table 14.6 The judgement matrix *A* for the second level criteria

Criteria	Fixed setup cost	Monthly installments
Fixed setup cost	1	5
Monthly installments	1/5	1
Column sum	1.200	6

sums are calculated as in Table 14.4. Each component of the weight vector is obtained by taking the average of the corresponding row entries of the normalized *A* matrix. The weight vector is given in Table 14.5.

For calculating λ_{max}, we will employ the second approach mentioned above.

$$\lambda_{max} = 1.444x0.648 + 4.143x0.295 + 17x0.057 = 3.127$$

CI = 0.064. Using RI = 0.58, we obtain the consistency ratio CR as:

$$CR = (0.064/0.58) = 0.110.$$

Recall that the pairwise comparisons for the judgment matrix *A* does not need to be repeated whenever its consistency ratio $CR \geq 0.15$. Since here $CR = 0.110 < 0.150$, we conclude that the panel is consistent in his/her judgments of relative importance among the first level criteria.

We obtain the weight vector for the second level criteria by employing the same approach we employed for the first level criteria.

The judgement matrix *A* for the second level criteria is obtained as in Table 14.6. The weights for the *Fixed Setup Cost* and the *Monthly Installments* in Table 14.7 represent the importance of these two criteria relative to each other. In order to determine their relative weights with respect to the criterion from which branching has occurred (parent criterion), we multiply these weights with the weight of the parent criterion:

Table 14.7 The weight vector $\underline{w}$ for the second level criteria

Criteria	Fixed setup cost	Monthly installments	$\underline{w}$
Fixed setup cost	0.833	0.833	0.833
Monthly installments	0.167	0.167	0.167

$$\text{Weight of the } \textit{Fixed Setup} \text{ Cost} = 0.057 \times 0.833 = 0.047.$$

$$\text{Weight of the } \textit{Monthly Installments} = 0.057 \times 0.167 = 0.010.$$

Note that the weights of the terminal criteria add up to 1.

The pairwise comparison of all criteria is completed, and the corresponding weight vectors are obtained. The pairs of alternative ISP options ISPA, ISPB and ISPC are compared in pairs in order to evaluate their priority vectors under the terminal criteria (Table 14.8).

The evaluations are performed by answering questions of the following type: "How much more preferable is alternative ISPA over ISPB on the basis of the *Connection Speed* ?" The decision maker is required to make the evaluation considering only the *Connection Speed*. The questioning continues until all alternative pairs are evaluated considering all terminal criteria. Recall that the terminal criteria are *Connection Speed*, *Service Quality*, *Fixed Setup Cost*, and *Monthly Installments*. The decision hierarchy with weights assigned is displayed in Fig. 14.2.

Since in the above calculations, all consistency ratios CR < 0.150, we assume all the associated judgement matrices A to be consistent.

In the synthesis step, these priority vectors calculated in Table 14.8 are added to the hierarchy and the priority values for these three ISP options are computed to find the most preferable one.

To demonstrate the calculation of the priority value of an alternative let us calculate the priority value for the alternative ISPA.

$$\begin{aligned}\text{Priority Value for ISPA} = {} & (\text{Weight of Connection Speed} \\ & \times \text{ISPA's Priority Vector Value under Connection Speed}) \\ & + (\text{Weight of Service Quality} \times \text{ISPA's Priority Vector Value under Service Quality}) \\ & + (\text{Weight of Fixed Setup Cost} \times \text{ISPA's Priority Vector Value under Fixed Setup Cost}) \\ & + (\text{Weight of Monthly Installments} \times \text{ISPA's Priority Vector Value under Monthly Installments})\end{aligned}$$

$$= 0.648 \times 0.633 + 0.295 \times 0.193 + 0.047 \times 0.429 + 0.010 \times 0.083$$

$$\text{Priority Value for ISPA} = 0.488.$$

The priority values for *ISPB* and *ISPC* are calculated similarly, and the results are displayed in Table 14.9. These values show that *ISPA* is far better than the other two ISPs and hence should be adopted as the project to be implemented.

Table 14.8 Determining the weight vectors for the alternatives under terminal criteria

Conn Speed	A	B	C
A	1	5	3
B	1/5	1	1/3
C	1/3	3	1
Column Sum	1.533	9.003	4.333

Conn Speed	A	B	C	$\underline{w}$
A	0.652	0.555	0.692	0.633
B	0.130	0.111	0.077	0.106
C	0.217	0.334	0.231	0.261

λ_{max}=3.056,

CI=0.028, CR=0.048

Service Quality	A	B	C
A	1	1/5	3
B	5	1	7
C	1/3	1/7	1
Column Sum	6.333	1.343	11

Service Quality	A	B	C	$\underline{w}$
A	0.158	0.149	0.273	0.193
B	0.789	0.745	0.636	0.723
C	0.053	0.106	0.091	0.083

λ_{max} = 3.106

CI = 0.053, CR = 0.091

Fixed Setup Cost	A	B	C
A	1	3	1
B	1/3	1	1/3
C	1	3	1
Column Sum	2.333	7	2.333

Fixed Setup Cost	A	B	C	$\underline{w}$
A	0.429	0.429	0.429	0.429
B	0.143	0.143	0.143	0.143
C	0.429	0.429	0.429	0.429

λ_{max}=3.003

CI = 0.002, CR = 0.003

Monthly Install	A	B	C
A	1	1/7	1/3
B	7	1	5
C	3	1/5	1
Column Sum	11.00	1.343	6.333

Monthly Install	A	B	C	$\underline{w}$
A	0.091	0.106	0.053	0.083
B	0.636	0.745	0.790	0.724
C	0.273	0.149	0.158	0.193

λ_{max}=3.108

CI = 0.054, CR = 0.093

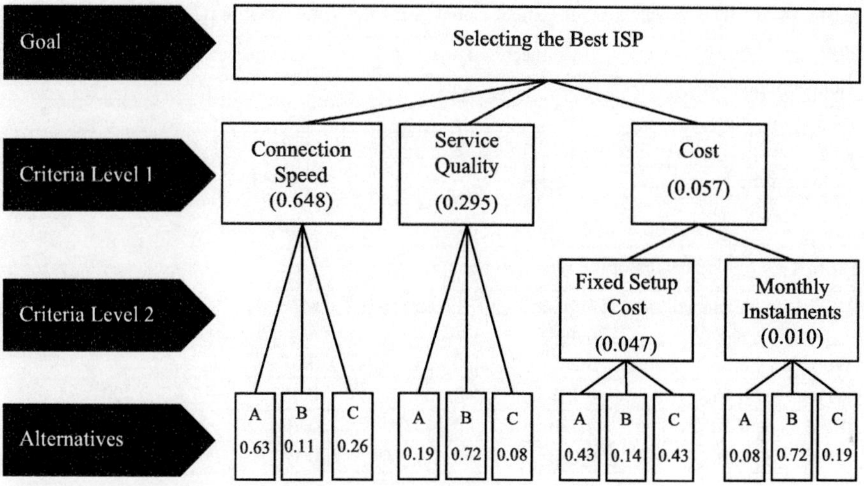

Fig. 14.2 AHP decision hierarchy with weights assigned

Table 14.9 Priority value table for all alternatives

	Connection speed	Service quality	Fixed Setup cost	Monthly installment	Priority Vector (PV)
	0.648	0.295	0.047	0.010	
ISP A	0.633	0.193	0.429	0.083	0.488
ISP B	0.106	0.723	0.143	0.724	0.296
ISP C	0.261	0.083	0.429	0.193	0.216

14.3.1 Sensitivity Analysis

As illustrated in Example 14.2, the priority scores of alternatives are heavily dependent on the criteria weights. These weights are subjective, and some level of uncertainty is always associated with them. The cause of this uncertainty in project portfolio domain could be due to inexperience with similar projects in the past, inability to accurately predict availability of required resources as well as other factors such as external competition, weather and labor disputes. As these uncertainties are resolved small changes in criteria weights are to be expected, but these can have a major impact on the final scores of available alternatives.

Sensitivity analysis can provide critical information on the stability of the priority vector under changes in the criteria weights. If priority vector is highly sensitive to criteria weights, the decision maker would be well advised to carefully review the process by which they are estimated. This will help participants to better understand the underlying model and furthermore, the decision maker will have more confidence in the decision-making process.

Table 14.10 The reevaluated judgement matrix A for the first level criteria

Criteria	Connection speed	Service quality	Cost
Connection speed	1	3	**5**
Service quality	1/3	1	**3**
Cost	**1/5**	**1/3**	1
Column sum	1.533	4.333	9

Table 14.11 The results of the sensitivity analysis

	Connection speed	Service quality	Fixed setup cost	Monthly installment	Priority Vector (PV)
Weights	0.633	0.260	0.088	0.018	
ISP A	0.633	0.193	0.429	0.083	0.490
ISP B	0.106	0.723	0.143	0.724	0.281
ISP C	0.261	0.083	0.429	0.193	0.228

Example 14.2 (Continued) Thinking that the cost criterion might be underrated, the management has decided to go through the process of obtaining the entries of the judgement matrix A again. The resulting judgement matrix A is given in Table 14.10.

Note that the pairwise comparisons of *Connection Speed with Cost* and *Service Quality* and *Cost* have been changed with more emphasis on *Cost* (Table 14.10).

Comparing the weights of the criteria with the previous ones (0.648, 0.295, 0.047, 0.010), we observe that the weights for the two Cost components have increased at the expense of *Connection Speed* and *Service Quality* (Table 4.11). The ranking of the alternatives, on the other hand, have not changed.

14.4 Mathematical Programming Models

One can formulate mathematical programming models of the project portfolio selection problem in a large number of different ways depending on the problem environment. We will distinguish between *single objective* and *multi-objective* mathematical programming models. Although multi-objective mathematical programming models are in general more involved, the use of such models is recommended due to the nature of the decision environment, which usually involves simultaneous consideration of multiple performance measures. A further classification of importance would be *single period* and *multi-period* models. Since the wide range of mathematical programming models for project selection precludes a comprehensive treatment, we will illustrate several typical approaches to demonstrate the basic approaches. Extensive reviews of this literature can be found in Archer and Ghasemzadeh (1996), Graves and Ringuest (2012).

We begin by presenting an example of a single objective mathematical programming model.

14.4.1 A Single Objective Mathematical Programming Model

In this Section, we present a *single objective* mathematical programming model for *portfolio selection* under resource limitations and for a *single period*. Without loss of generality, we present the model in a product development organization setting. In order to determine the next period's work plan, the product development organization accepts research proposals from their clients and selects a number of them for execution. There are n project proposals under consideration for selection, and the projects are assumed to be independent of each other. Each project can be realized in different versions with different resource requirements and leading to different benefits. The benefit of each project can be assessed in monetary terms or as a non-dimensional quantity. This non-dimensional benefit factor can also be interpreted as a score for the project involved. In the final project portfolio, at most one version of each project is allowed. The return from each version of each project is referred to as its benefit. For the product development environment the resources considered are limited to the most critical three resources: engineers, designers and available capital. The resource requirements for each version of each project are known. The objective is to select a subset of the n projects under consideration to maximize the total benefit subject to a set of resource constraints.

Let us define the notation:

i: project index $i = 1,\ldots,n$
N_i: number of versions of project i
b_{ij}: benefit expected from undertaking version j of project i
e_{ij}: number of engineers required for version j of project i
E: total number of engineers available
d_{ij}: number of designers required for version j of project i
D: total number of designers available
c_{ij}: amount of budget required for version j of project i
C: total budget available

The mathematical programming formulation is given as follows:

$$max\ Z = \sum_{i=1}^{n} \sum_{j=1}^{N_i} b_{ij} x_{ij} \tag{14.7}$$

subject to

$$\sum_{j=1}^{N_i} x_{ij} \leq 1 \quad i = 1, \ldots, n \tag{14.8}$$

$$\sum_{i=1}^{n} \sum_{j=1}^{N_i} e_{ij} x_{ij} \leq E \tag{14.9}$$

$$\sum_{i=1}^{n} \sum_{j=1}^{N_i} d_{ij} x_{ij} \leq D \tag{14.10}$$

$$\sum_{i=1}^{n}\sum_{j=1}^{N_i} c_{ij}x_{ij} \le C \tag{14.11}$$

$$x_{ij} = \begin{cases} 1 \text{ if project } i \text{ version } j \text{ is selected} \\ 0 \text{ otherwise} \end{cases} \quad i = 1, \ldots, n \quad j = 1, \ldots, N_i \tag{14.12}$$

The objective function represents the maximization of the total benefit accrued from the selection of a portfolio. That at most one version of a project can be selected is expressed in constraint set (14.8) for all projects. The constraints (14.9) and (14.10) impose the resource limits on the number of engineers and designers, respectively. Bound on the total budget available for the projects to be selected is given by the constraint (14.11). Finally, expression (14.12) defines the decision variables.

Example 14.3 Each year the sponsoring companies of the Manufacturing Systems Engineering Research Institute propose a number of projects in several different versions and the Institute decides which ones to accept to work on the following year. Three sponsoring companies each suggest a project with 2, 3, and 2 versions, respectively for next year's program. The Institute investigates each version of the suggested projects to estimate their staffing and budget requirements as well as the benefits expected. The estimations are summarized in Table 14.12. The available numbers of engineers and designers as well as the total budget the Institute has available to allocate to these projects next year are also included in the Table. e_{ij} and d_{ij} are represented as fractional values given by the ratio of the man-hours of work required for a project to the total number of man-hours available next year.

Determine the feasible portfolio of projects such that the benefit resulting is maximized.

Let us first write the model (14.7–14.12) using the data given in Table 14.12.

$$\begin{aligned} max\, Z = {} & (0.47x_{11} + 0.60x_{12}) + (0.44x_{21} + 0.58x_{22} + 0.69x_{23}) \\ & + (0.68x_{31} + 0.79x_{32}) \end{aligned}$$

subject to

$$x_{11} + x_{12} \le 1$$

$$x_{21} + x_{22} + x_{23} \le 1.$$

Table 14.12 Resource requirements and availabilities

	P11	P12	P21	P22	P23	P31	P32	Availability
Engineers	2.3	2.8	4.1	5.1	5.8	3.4	3.7	10
Designers	7.0	7.8	5.6	6.4	6.9	5.3	6.4	20
Budget req.	22.8	29.2	38.7	41.2	44.3	33.5	40.2	100
Benefit	0.47	0.60	0.44	0.58	0.69	0.68	0.79	–

$$x_{31} + x_{32} \leq 1$$

$$(2.3x_{11} + 2.8x_{12}) + (4.1x_{21} + 5.1x_{22} + 5.8x_{23}) + (3.4x_{31} + 3.7x_{32}) \leq 10$$

$$(7.0x_{11} + 7.8x_{12}) + (5.6x_{21} + 6.4x_{22} + 6.9x_{23}) + (5.3x_{31} + 6.4x_{32}) \leq 20$$

$$(22.8x_{11} + 29.2x_{12}) + (38.7x_{21} + 41.2x_{22} + 44.3x_{23}) + (33.5x_{31} + 40.2x_{32}) \leq 100$$

$$x_{ij} \in \{0,1\}\ i = 1,2,3; j = 1,2,3\ \textit{where}\ (i,j) \neq (1,3)\ \textit{and} \neq (3,3).$$

The optimal solution is found to be: $x_{11} = x_{22} = x_{31} = 1$, yielding a total benefit of 1.73. Thus the optimal solution recommends that the first version of project 1, the second version of project 2, and the first version of project 3 are accepted for implementation in next year's work plan.

In the *multi-period* version of the problem both the investment and the returns occur over several periods. Hence, the cash flows, technological coefficients and other parameters have to be estimated for each period. The budget not consumed in one period might or might not be transferable to the next period, and resource availabilities might change over time. The benefit accrued from a project might differ in different periods throughout the time horizon. Hence a mathematical programming formulation of the multi-period problem needs the inclusion of time indices in the definition of the decision variables.

14.4.2 Multi-objective Mathematical Programming Models

An important class of decision problems deals with multiple objectives. For a decision problem to be cast into a multi-objective format, all the objectives have to be *conflicting*, in the sense that improving one or more them is possible only at the cost of making others worse. If the objectives are not conflicting over the ranges of the decision variables considered, then the decision problem can be formulated as a single objective optimization problem. A decision environment with limited resources and multiple conflicting objectives forces the decision makers to make trade-offs between different objectives.

In the multi-objective mathematical programming problem, the different objective functions f_i are expressed in vector format: $\underline{\mathbf{f}} = (f_1, \ldots, f_m)$, where m is the number of objectives. A distinct feature of multi-objective mathematical programming is that in addition to the feasible domain defined by the decision variables we have an *m- dimensional objective space* defined by the objectives $f_1, \ldots, f_m$.

There are, in general, three categories of approaches to solving multi-objective mathematical programming problems. In the first of these, the decision maker assigns *a priori* a weight to each objective reflecting the trade-off structure among the objectives. Each weight is in the range [0,1] and their sum equals 1. By taking the weighted sum of the individual objective functions, transformed into non-dimensional format where necessary, reduces the multi-objective problem to a single objective one. The resulting objective function is often a linear combination of the non-dimensional objective functions of the original problem. By solving the single objective optimization problem, an optimal solution or alternative optima are obtained and presented to the decision-maker. The proper determination of weights is of crucial importance for the success of the solution process. It is good practice to determine intervals for the weights rather than point estimates and apply a sensitivity analysis to the results before reaching a final conclusion. The weight determination procedure of AHP can be employed to obtain a set of consistent weights.

The second category involves calculating a set of non-dominated (efficient) solutions to fully or approximately describe the *Pareto front* of non-dominated solutions represented in the objective space. Assuming *maximization* of all objectives, we define a non-dominated solution as follows:

Definition: A solution is said to be *non-dominated*, if an increase in one objective can only be obtained at the cost of a decrease of one or more other objectives. More formally, the solution vector $\underline{\mathbf{f}}^*$ is said to strictly dominate another solution vector $\underline{\mathbf{f}}$, if $f^*_i \geq f_i$ for all i and $f^*_i > f_i$ for some i. The set of solutions not strictly dominated by other points in the objective space (i.e., the set of non-dominated solutions) constitutes the *Pareto front*.

Examples of dominated and non-dominated solutions are demonstrated in Fig. 14.3a–b. Assume that we generate two solutions A and B and plot them as in Fig. 14.3a. Comparing solutions A and B following the *Definition* above, we see that A has a higher value in the objective f_2, whereas B has a higher value in f_1. Since none is dominating the other, both solutions A and B are said to be non-dominated. In Fig. 14.3b, we introduce new solutions C and D in addition to A and B. Considering 4 solutions together we observe that C dominates A and D dominates both A and B. Hence, only solutions C and D remain as the non-dominated solutions.

Once a representative set of non-dominated solutions has been generated, the decision maker makes a choice among these solutions. Under this approach we do not have a unique optimum solution; a set of non-dominated solutions is presented to the decision-maker from which they make their selection based on their preferences.

In the third category we have iterative interactive approaches in which the analyst provides the decision maker with a set of non-dominated solutions. Evaluating these results, the decision maker then guides the analyst to different domains in the solution space asking for further results. This process continues until the decision maker is satisfied with the solution(s) provided. The interactive approach can also be applied for the case where the decision maker provides the analyst with the weights. Once the analyst returns the single objective solution using the weights provided, the decision maker can revise the set of weights asking the analyst for the solution each time until a satisfactory solution is obtained.

In the following two sections, we will consider two approaches to multiple objective project selection models: (i) a *multi-objective linear programming* (MOLP) model that seeks a set of non-dominated solutions and (ii) a *goal programming* (GP) model. MOLP model belongs to the second category of approaches. The goal programming model, on the other hand, can be included in the first category of approaches.

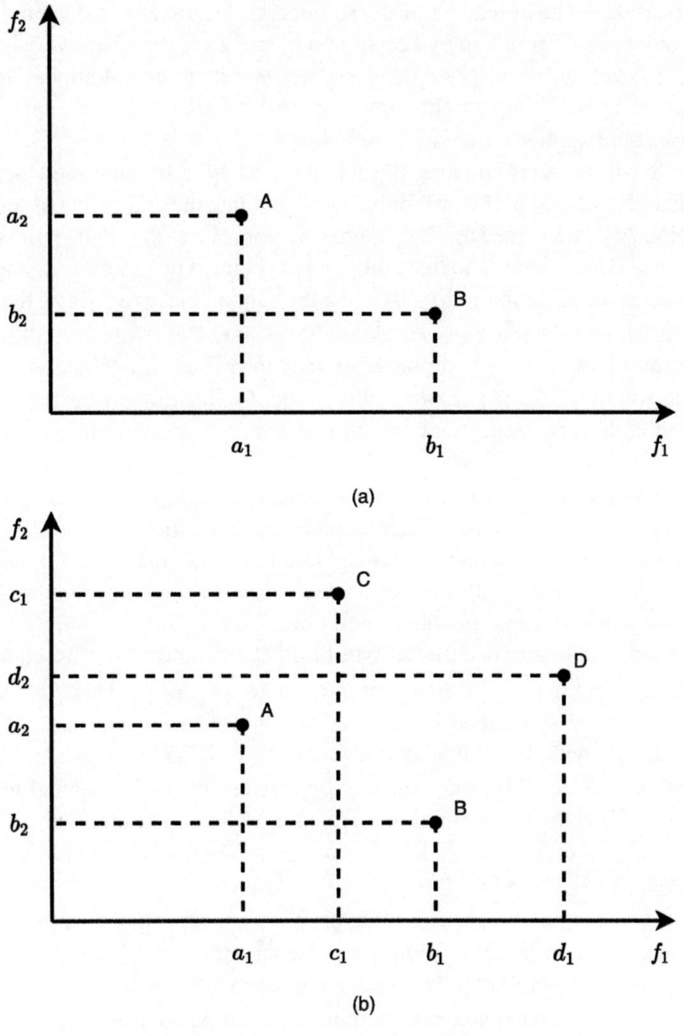

Fig. 14.3 (a–b) Examples of dominated and non-dominated solutions

14.4.3 Multi-objective Linear Programming Approach

As stated above the approach employed here seeks the generation of a set of non-dominated solutions. The success of this approach depends among others on how well this set approximates the actual Pareto front.

In the following, we will go through the steps of MOLP by considering two conflicting objective functions Z_1 and Z_2.

The objectives of MOLP as well as the constraints are of linear form. The solution procedure presented here makes use of this property and is based on the successive application of LP. The basic idea of the solution procedure is to reduce the problem to single objective LP problem by keeping only, say, Z_1 in the objective function and adding the other objective Z_2 to the constraint set as an additional inequality ($\geq$) constraint by assigning a right-hand side (*rhs*) value to it. The *rhs* value depends on the procedure applied as we will see below.

Since we will be approximating the actual Pareto front through sequential generation of non-dominated points, we must first determine the end points of the interval over which we will realize the approximation, i.e., the end points of the approximated Pareto front. The first end point is obtained by ignoring Z_2 and solving the problem as an LP problem to maximize the single objective Z_1. Let the first end point be designated as $\left(Z_1^1, Z_2^1\right)$. To obtain the second end point, we ignore Z_1 and solve the problem as an LP problem to maximize only Z_2, obtaining the non-dominated solution $\left(Z_1^2, Z_2^2\right)$. Since both Z_1 and Z_2 reach their maxima at the first and second end point, respectively. Both end points correspond to non-dominated solutions.

The coordinates of these two end points $\left(Z_1^1, Z_2^1\right)$ and $\left(Z_1^2, Z_2^2\right)$ also specify the end points of the intervals over which both objective functions Z_1 and Z_2 vary. One of these intervals is chosen and partitioned into (*m*-1) sub-intervals, not necessarily of equal length, so that (*m*-2) non-dominated solutions can be generated. Together with the end points we end up with *m* non-dominated solutions.

For demonstration purposes, let us partition $\left[Z_2^1, Z_2^2\right]$ into (*m*-1) sub-intervals and denote the end points of the first sub-interval as $\left[Z_2^1, Z_2^{11}\right]$. Then the first non-dominated solution is obtained by solving an LP problem to maximize Z_1 subject to the constraints with the additional constraint $Z_2 \geq Z_2^{11}$. This additional constraint will be satisfied as equality, since the two objectives are conflicting and in order to maximize Z_1, Z_2 is kept at its minimum value of Z_2^{11}. Let the optimum Z_1 value to this LP problem be denoted as Z_1^{11}. Then the non-dominated solution generated becomes $\left(Z_1^{11}, Z_2^{11}\right)$.

We now consider the second sub-interval $\left[Z_2^{11}, Z_2^{12}\right]$, repeating the non-dominated solution generation procedure adding the constraint $Z_2 \geq Z_2^{12}$ and deleting the previously added constraint on Z_2. Denote the optimal Z_1 value of this LP problem by Z_1^{12}. The newly generated non-dominated solution is then $\left(Z_1^{12}, Z_2^{12}\right)$. This procedure of generating non-dominated solutions is repeated until (*m*-2) non-dominated solutions are generated. Together with the two end points $\left(Z_1^1, Z_2^1\right)$ and $\left(Z_1^2, Z_2^2\right)$ we have *m* non-dominated solutions generated.

The solution procedure explained above will be presented below in algorithmic format.

An Algorithm for MOLP:

Step 1: Specify *m*, the number of non-dominated solutions to be generated; the conflicting objective functions Z_1, Z_2 and the constraints.

Step 2a: Maximize Z_1 subject to the constraints specified. Evaluate Z_2 at the optimal point obtained for maximizing Z_1. Specify $\left(Z_1^1, Z_2^1\right)$ as the first non-dominated solution.

Step 2b: Maximize Z_2 subject to the constraints specified. Evaluate Z_1 at the optimal point obtained for maximizing Z_1. Specify $\left(Z_1^2, Z_2^2\right)$ as the second non-dominated solution.

Step 3: Partition the interval $\left[Z_2^1, Z_2^2\right]$ into (m-1) sub-intervals not necessarily of equal length with the endpoints $Z_2^1, Z_2^{11}, Z_2^{12}, \ldots, Z_2^{m-2}, Z_2^2$.

Step 4: For each value of $rhs = Z_2^{11}, Z_2^{12}, \ldots, Z_2^{m-2}$, maximize Z_1 subject to the constraints specified with the additional constraint $Z_2 \geq rhs$. Specify each optimal solution obtained as a non-dominated solution.

Step 5: Having obtained all m non-dominated solutions terminate the algorithm.

Note that in case we decide to partition the interval $\left[Z_1^1, Z_1^2\right]$ rather than the interval $\left[Z_2^1, Z_2^2\right]$, then in Steps 3 and 4, the endpoints of the sub-intervals become $Z_2^1, Z_2^{11}, Z_2^{12}, \ldots, Z_2^{m-2}, Z_2^2$. The additional constraint becomes $Z_2 \geq rhs$, where $rhs = Z_1^{11}, Z_1^{12}, \ldots, Z_1^{m-2}$.

The set of non-dominated solutions generated is presented to the decision-maker, who either selects one of the points as the final solution or requests the generation of additional points leading to smaller sub-intervals, i.e., a search on the Pareto front with a finer mesh. Each solution contains implicitly a trade-off between the two objectives, captured by the gradient of the Pareto front approximation at the non-dominated solution.

In general, there is a trade-off between the computational effort spent on generating non-dominated solutions and the size and diversity of the portfolio of solutions presented to the decision-maker, which increases the likelihood of the decision-maker arriving at a satisfactory solution representing a better tradeoff between their preferences.

The following example illustrates this approach of generating non-dominated solutions.

Example 14.4 The PerfectWash Washing Machine Manufacturing Company has plans to introduce new models to improve its profit and market share. They have gone through the concept development phase of various model design projects and reduced the candidate projects to four models A, B, C, and D. Each model has several versions; i.e., their costs can be adjusted through the choice of the versions. The models will be subject to major changes or discontinuation after 4 years. PerfectWash formulates the problem as a 4-year time horizon problem. The estimated budget required for the full model of each of the product development projects is shown in Table 14.13.

All the projects are initiated prior to the first year. Their investment costs are assumed to be incurred beginning in the first year, while production and distribution

Table 14.13 Full investment amounts for the different models

Model	A	B	C	D
Investment	2.00	3.00	1.50	3.50

Table 14.14 Expected profit and market share for the different models

Model	A		B		C		D	
Year	Profit ($million)	Market share (%)	Profit ($million)	Market share (%)	Profit ($million)	Market share (%)	Profit ($million)	Market share (%)
1	2,80	8,00	3,20	4,27	2,20	9,80	7,60	2,09
2	3,30	9,25	3,80	5,07	2,40	10,60	8,20	2,60
3	3,60	10,00	4,10	5,47	2,50	11,00	8,80	3,11
4	3,10	8,75	4,00	5,33	2,30	10,20	8,60	2,94

Table 14.15 The minimum expected values for market share and profit for each year

Year	1	2	3	4
Profit ($million)	4.00	5.00	6.00	5.00
Market share (%)	18.00	22.00	22.00	20.00

of the models start with the first year. The budget available for investment is $6.5 million. The discount rate is taken as 8% per annum.

The estimated profit ($million) and market share (%) are given for the full investment version of the model in Table 14.14. It is assumed that investing less than the full amount for a particular model will reduce profit and market share in linear proportion to the reduction in the full investment.

The top management of PerfectWash wants to impose a constraint on the minimum expected market share for each year of the time horizon. Due to cash flow requirements and reinvestment plans, the projected profit for each year should not drop below the specified minimum given in Table 14.15.

The PerfectWash Company has two objectives of maximizing its profit and maximizing its market share subject to the constraints stated above. Determine and present the decision-maker with a set of 5 non-dominated solutions approximating the Pareto front.

Solution

We first specify the decision variables. The decision variables x_i, $i = A, B, C, D$ indicate for the ratio of model i's full investment to be realized. We then formulate the objective function Z_1 for the profit and Z_2 for the market share in terms of the decision variables.

$$
\begin{aligned}
Z_1 = &-(2x_A + 3x_B + 1.5x_C + 3.5x_D) \\
&+0.926(2.80x_A + 3.20x_B + 2.20x_C + 7.60x_D) \\
&+0.857(3.30x_A + 3.80x_B + 2.40x_C + 8.20x_D) \\
&+0.735(3.60x_A + 4.10x_B + 2.50x_C + 8.80x_D) \\
&+0.540(3.10x_A + 4.00x_B + 2.30x_C + 8.60x_D)
\end{aligned}
\tag{14.13}
$$

$$Z_2 = 0.25[(8.00x_A + 4.27x_B + 9.80x_C + 2.09x_D) + (9.25x_A + 5.07x_B + 10.60x_C + 2.60x_D) + (3.60x_A + 4.10x_B + 2.50x_C + 8.80x_D) + (3.10x_A + 4.00x_B + 2.30x_C + 8.60x_D)] \quad (14.14)$$

The MOLP problem is then stated as follows:

$$\begin{aligned} max\{Z_1, Z_2\} = max\{ & -(2x_A + 3x_B + 1.5x_C + 3.5x_D) \\ & + 0.926(2.80x_A + 3.20x_B + 2.20x_C + 7.60x_D) \\ & + 0.857(3.30x_A + 3.80x_B + 2.40x_C + 8.20x_D) \\ & + 0.735(3.60x_A + 4.10x_B + 2.50x_C + 8.80x_D) \\ & + 0.540(3.10x_A + 4.00x_B + 2.30x_C + 8.60x_D); \\ & 0.25\big[(8.00x_A + 4.27x_B + 9.80x_C + 2.09x_D) \\ & + (9.25x_A + 5.07x_B + 10.60x_C + 2.60x_D) \\ & + (3.60x_A + 4.10x_B + 2.50x_C + 8.80x_D) \\ & + (3.10x_A + 4.00x_B + 2.30x_C + 8.60x_D)\big]\} \end{aligned} \quad (14.15)$$

subject to

$$2.80x_A + 3.20x_B + 2.20x_C + 7.60x_D \geq 4 \quad (14.16)$$

$$3.30x_A + 3.80x_B + 2.40x_C + 8.20x_D \geq 5 \quad (14.17)$$

$$3.60x_A + 4.10x_B + 2.50x_C + 8.80x_D \geq 6 \quad (14.18)$$

$$3.10x_A + 4.00x_B + 2.30x_C + 8.60x_D \geq 5 \quad (14.19)$$

$$8.00x_A + 4.27x_B + 9.80x_C + 2.09x_D \geq 18 \quad (14.20)$$

$$9.25x_A + 5.07x_B + 10.60x_C + 2.60x_D \geq 22 \quad (14.21)$$

$$10.00x_A + 5.47x_B + 11.00x_C + 3.11x_D \geq 22 \quad (14.22)$$

$$8.75x_A + 5.33x_B + 10.20x_C + 2.94x_D \geq 20 \quad (14.23)$$

$$2.00x_A + 3.00x_B + 1.50x_C + 3.50x_D \geq 6.50 \quad (14.24)$$

$$0 \leq x_i \leq 1 \quad i = A, B, C, D \quad (14.25)$$

The multi-objective function of maximizing the profit and the market share is given in terms of the decision variables in (14.15). Constraints (14.16)–(14.19) enforce the lower bound on the yearly minimum profit and (14.20)–(14.23) the lower bound on the yearly minimum market share of the product portfolio. Constraint (14.24) puts a ceiling on the maximum amount of investment the PerfectWash

Company has allocated for the development of this product portfolio. Constraint (14.25) specifies the bounds [0,1] on the decision variables x_i, $i = A, B, C, D$.

We start the approximation of the Pareto front by determining the end points of the range over which we will compute the approximation. We accomplish this by solving two single objective optimization problems P_1 and P_2.

$$P_1 : \ max \ Z_1$$

subject to

(14.16)–(14.25)

The optimal solution is found as:

$$x_A^* = 1; x_B^* = 0; x_C^* = 1; x_D^* = 0.86 \text{ with } Z_1^* = 32.04$$

Evaluating Z_2 using the optimal x^*-values, we get: $Z_2 = 21.63$.

$$P_2 : \ max \ Z_2$$

subject to

(14.16)–(14.25)

The optimal solution is found as:

$$x_A^* = 0.09; x_B^* = 1; x_C^* = 1; x_D^* = 0 \text{ with } Z_2^* = 24.44.$$

Evaluating Z_1 using the optimal x^*-values, we get: $Z_1 = 21.81$.

So, the two end points, which are themselves non-dominated solutions, are determined as:

$$\left(Z_1^1, Z_2^1\right) = (32.04, 21.63) \text{ and } \left(Z_1^2, Z_2^2\right) = (21.81, 24.44).$$

To generate the remaining set of 3 non-dominated solutions we will partition Z_2 into 4 subintervals. For that purpose we set $Z_2 = 21.63, 22, 23, 24, 24.44$ as the end points of the sub-intervals. We will then solve P_1 with the additional constraint $Z_2 \geq rhs$ each time for each value of $rhs = 22, 23, 24$. With the three non-dominated solutions generated this way we have approximated the Pareto front with 5 non-dominated solutions as shown in Fig. 14.4.

$$\left(Z_1^{11}, Z_2^{11}\right) = (30.88, 22.00) \ x_A^* = 1; x_B^* = 0.11; x_C^* = 1; x_D^* = 0.76.$$

$$\left(Z_1^{12}, Z_2^{12}\right) = (27.16, 23.00) \ x_B^* = 1; x_B^* = 0.48; x_C^* = 1; x_D^* = 0.45.$$

$$\left(Z_1^{13}, Z_2^{13}\right) = (23.43, 24.00); x_A^* = 1; x_B^* = 0.84; x_C^* = 1; x_D^* = 0.17.$$

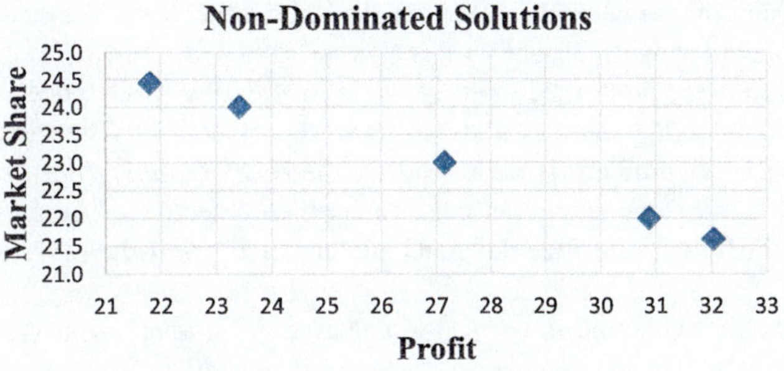

Fig. 14.4 Pareto front approximation with 5 non-dominated solutions

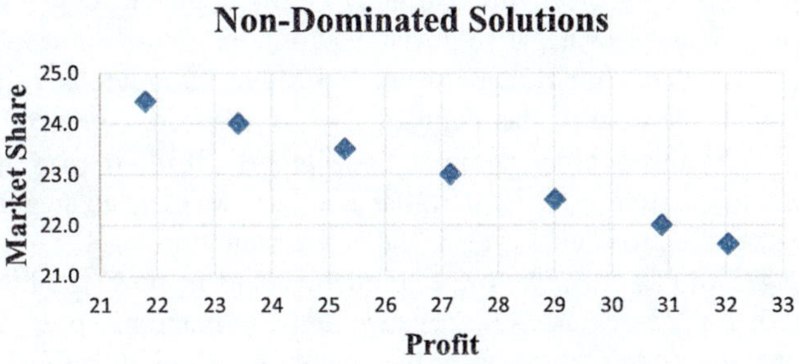

Fig. 14.5 Pareto front approximation with 7 non-dominated solutions

Clearly the Pareto front can better be approximated by generating further non-dominated solutions. Evaluating the distribution of the non-dominated solutions representing the Pareto front approximation in Fig. 14.4, we decide to solve for 2 additional non-dominated solutions by setting $rhs = 22.5, 23.5$. We solve P_1 twice, first adding the constraint $Z_2 \geq 22.5$ and then $Z_2 \geq 23.5$ and obtain the following non-dominated solutions:

$$\left(Z_1^{14}, Z_2^{14}\right) = (29.02, 22.50); x_A^* = 1; x_B^* = 0.29; x_C^* = 1; x_D^* = 0.61.$$

$$\left(Z_1^{15}, Z_2^{15}\right) = (30.88, 23.50); x_A^* = 1; x_B^* = 0.84; x_C^* = 1; x_D^* = 0.17.$$

The Pareto front approximation with 7 points is depicted in Fig. 14.5.

The decision-maker is presented with the set of 7 non-dominated solutions and will choose one as the final solution.

14.4.4 Goal Programming Model

There are decision environments in which one objective is preferred over the other objectives in an absolute way, meaning that this particular objective is infinitely more important for the decision maker than the other objectives. We can express this lexicographic ordering of these priorities G_i as $G_1 \gg G_2 \gg \ldots$, where $' \gg '$ indicates absolute preference and the ordering is in decreasing preference. Hence, the lexicographic ordering induces a hierarchy in the objectives. The resulting multi-objective decision problem is addressed by the *lexicographic goal programming* (GP). In this context we talk about goals rather than objectives. The decision maker needs to identify and prioritize the goals and attach an *aspiration level* -a value- to each goal.

The goals are expressed in terms non-negative *deviational variables* around the aspiration levels. There are *positive deviational variables* denoted by $p \geq 0$ and *negative deviational variables* denoted by $n \geq 0$. For example, a company might specify profit as its highest priority goal and attach an aspiration level of $20 million. If the company aims to achieve this goal of $20 million, then its objective would be minimizing any positive or negative deviation from this aspiration level, i.e., minimize $(p + n)$. Note that p and n cannot be both positive, although they can both take values of 0. If in the case of this company $p > 0$, then the company exceeds its aspiration level for this goal by an amount p. Similarly, if $n > 0$, then the company falls short of its aspiration level for this goal at a lower level by an amount n. If both p and n are zero, the profit goal is achieved at the aspiration level.

Continuing with this example, if the company wants its profit goal to be at least $20 million, then the associated objective would be to maximize p. If, on the other hand, the company prefers to minimize the amount by which its profit falls short of $20 million, n is to be minimized.

The objective functions for the deviational variables are expressed within the *achievement function a*. The achievement function a expresses the objective functions in terms of the deviational variables corresponding to the priority levels expressed in the lexicographic order. Consider, for example, a GP problem with 2 absolute priorities $G1$ and $G2$. The decision maker wants the deviations p_1 and n_1 from the aspiration level of $G1$ to be minimum and the positive deviation p_2 from the aspiration level of $G2$ to be maximum. The achievement function a becomes:

$$\underline{a} = \{(p_1 + n_1); p_2\}.$$

The optimization process takes place in *lexicographic order* starting with the first priority level $G1$ and continuing in the order of the remaining levels. In other words, the resources will be employed to optimize the first goal ignoring all others. Once this is accomplished, then the remaining resources will be used to optimize the second priority goal $G2$ without causing any deterioration in the optimum value obtained for the first goal. If any further resources remain and there is a third priority level $G3$, then the goal in the third priority level is optimized without causing any deterioration in the optimum value obtained for the first two priority level objectives. This process continues until either no feasible solution can be obtained at a priority level, or all priority levels are already considered or both.

Example 14.5 Consider the problem setting in Example 14.4 with the goals G_1 denoting profit and G_2 the average market share. The top management specifies the aspiration levels associated with these priorities as \$20 million in NPV for G_1 and for G_2, an average of 20% over the time horizon and considering all models. Management prefers increasing both goals above the aspiration level (by positive deviation). Formulate the problem as a GP problem and present the decision-maker with the GP solution.

Solution

Based on the problem statement above, the achievement function is the maximization of the positive deviations p_1 and p_2 from the aspiration levels specified for the goals G_1 and G_2, respectively.

$$\underline{a} = max\,\{p_1; p_2\}$$

subject to

$$\begin{aligned} &-(2x_A + 3x_B + 1.5x_C + 3.5x_D) \\ &\quad + 0.926(2.80x_A + 3.20x_B + 2.20x_C + 7.60x_D) \\ &\quad + 0.857(3.30x_A + 3.80x_B + 2.40x_C + 8.20x_D) \\ &\quad + 0.735(3.60x_A + 4.10x_B + 2.50x_C + 8.80x_D) \\ &\quad + 0.540(3.10x_A + 4.00x_B + 2.30x_C + 8.60x_D) - p_1 + n_1 \\ &\quad = 20 \end{aligned} \tag{14.26}$$

$$\begin{aligned} 0.25\big[&(8.00x_A + 4.27x_B + 9.80x_C + 2.09x_D) \\ &+ (9.25x_A + 5.07x_B + 10.60x_C + 2.60x_D) \\ &+ (3.60x_A + 4.10x_B + 2.50x_C + 8.80x_D) \\ &+ (3.10x_A + 4.00x_B + 2.30x_C + 8.60x_D)\big] - p_2 + n_2 = 20 \end{aligned} \tag{14.27}$$

(14.16)–(14.25)

$$p_1, p_2, n_1, n_2 \geq 0$$

We will solve this problem through two consecutive LP solutions each time taking into account the first and second components of the achievement function (Winston, 2003).

The first problem is stated as:

$$a^1 = max\,p_1$$

subject to

(14.26)
(14.16)–(14.25)

$$p_1, n_1 \geq 0$$

We obtain the following optimal solution:

$$a^{1*} = 12.04; p_1^* = 12.04, n_1^* = 0, x_A^* = 0.99; x_B^* = 0; x_C^* = 1; x_D^* = 0.86.$$

The sum of the aspiration level for profit and p_1^* becomes the profit value: 20 + 12.04 = 32.04.

The second problem is stated then as:

$$a^2 = max\, p_2$$

subject to

(14.26)–(14.27)
(14.16)–(14.25)

$$p_1 = 12.04$$

$$p_2, n_2 \geq 0$$

We obtain the following optimal solution:

$$a^{2*} = 1.63; p_1^* = 12.04, p_2^* = 1.63, n_2^* = 0, x_A^* = 0.99; x_B^* = 0; x_C^* = 1; x_D^* = 0.86.$$

The value of the achievement function together with the aspiration level indicates to the value of average market share: 20 + 1.63 = 21.63.

Combined, these two solutions correspond to $\left(Z_1^1, Z_2^1\right) = (32.04, 21.63)$; namely, to one end point of the interval enclosing the Pareto front approximation in the MOLP solution.

An Extension Let us now explore how the solution will change if top management changes its priority structure, setting G_1 to the average market share and G_2 to profit. When we apply the same solution procedure using the new priority structure and combine the resulting solutions we obtain the other end point $\left(Z_1^2, Z_2^2\right) = (21.81, 24.44)$ of the Pareto front. Under both priority structures the first component of the achievement function dominates, giving the maximum value for the corresponding goal and forcing the second goal to take on a value lower than its maximum value. Actually if both take on their maximum value at the optimum point, then the problem would not be a multi-objective problem since the objectives would not be conflicting.

14.4.5 Multi-objective, Multi-period Project Portfolio Selection and Scheduling Model

The mathematical models discussed in the previous sections assume that all projects start at time zero, which is usually not the case in practice. Organizations managing project portfolios would benefit considerably from a model representing the selection of a project portfolio from a set of candidate projects and the distribution of the selected projects over the time horizon so as to maximize a set of multiple objectives. The need for the distribution of the project portfolio over the time horizon stems from the need of more uniform distribution of resource requirements over time. In the following problem formulation, resource will imply finances with the only resource constraint being the bound on available finance in a period. The resource requirement for a given project is not constant over time and cumulative resource consumption over time is generally represented by an *S*-curve as depicted in Fig. 14.6 (Ghasemzadeh et al. 1999). Therefore, this assumption of the same starting time for all projects can lead to incorrect estimates of the resource requirement in each period for projects not included in the optimal portfolio due to perceived shortages, which may not occur if these projects are started in later periods. Therefore, it is important to find optimal starting points for all projects, which will result in a more desirable project portfolio selection. To address these issues, (Ghasemzadeh et al. 1999) proposed a zero-one integer linear programming model for project portfolio selection and optimal scheduling of the projects in the portfolio, which we will present below.

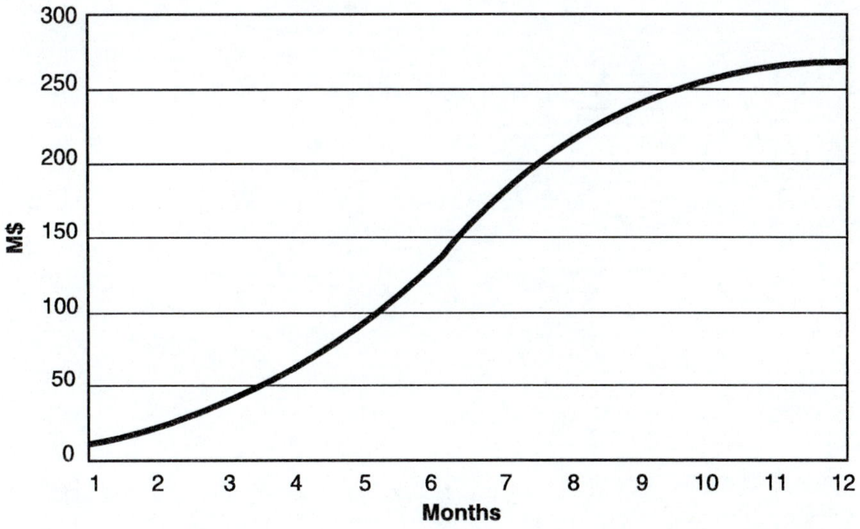

Fig. 14.6 Cumulative resource consumption for a project over time (Ghasemzadeh et al. 1999)

The relevant notation is defined as follows:

Z = The objective function
T = Total number of periods
N = Number of available project alternatives
V = Number of sets of different versions
M = Number of sets of mandatory projects
P = Number of sets of mutually exclusive projects
S_v = Set of different versions of a project
S_m = Set of mandatory projects
S_o = Set of ongoing projects that should be continued
S_p = Set of mutually exclusive projects
a_i = Normalized aggregate scores for each project alternative i
$C_{i,k+1-j}$ = Financing required for project i in period k
AF_k = Total financing available in period k
D_i = Duration of project i

The mathematical programming formulation is given below:

$$\max Z = \sum_{i=1}^{N} \sum_{j=1}^{T} a_i x_{ij} \tag{14.28}$$

Subject to

$$\sum\nolimits_{J=1}^{T} x_{ij} \leq 1 \quad i = 1, \ldots, N \tag{14.29}$$

$$\sum_{i=1}^{N} \sum_{j=1}^{k} C_{i,k+1-j} \quad x_{ij} \leq AF_k \quad k = 1, \ldots, T \tag{14.30}$$

$$\sum_{j=1}^{T} jx_{ij} + D_i \leq T + 1 \qquad i = 1, \ldots, N \tag{14.31}$$

$$\sum_{i \epsilon S_v} \sum_{j=1}^{T} x_{ij} \leq 1 \qquad v = 1, \ldots, V \tag{14.32}$$

$$\sum_{j=1}^{T} x_{ij} = 1 \quad i \in S_m, \quad m = 1, \ldots, M \tag{14.33}$$

$$x_{i1} = 1 \qquad i \in S_o \tag{14.34}$$

$$\sum_{j=1}^{T} x_{ij} \geq \sum_{J=1}^{T} x_{lj} \quad i \in P_l, \quad l = 1, \ldots, L \tag{14.35}$$

$$\sum_{j=1}^{T} jx_{lj} + (T+1)\left(1 - \sum_{j=1}^{T} x_{lj}\right) - \sum_{j=1}^{T} jx_{ij} \geq D_i \sum_{j=1}^{T} x_{ij} \quad i \in P_l, \tag{14.36}$$
$l = 1, \ldots, L$

$$\sum_{i \in S_p} \sum_{j=1}^{T} x_{ij} \leq 1 \quad p = 1, \ldots, P \tag{14.37}$$

$$x_{ij} = \begin{cases} 1 \; \textit{if project i is included in the portfolio and starts in period j} \\ 0 \quad \textit{otherwise} \quad i = 1, \ldots, N \; j = 1, \ldots, T \end{cases} \tag{14.38}$$

The objective function Z to be maximized represents the total score obtained by the set of projects selected from N alternative projects. The problem is a multi-criteria scoring problem. The score for each project i, a_i, can be determined using the procedure for determining the weights in AHP covered in Sect. 14.3. Hence, this problem formulation would fit the first category of multi-objective mathematical programming problems included in Sect. 14.4.2, where the multi-objective optimization problem is reduced to a single objective optimization problem using *a priori* assigned weights by the decision-maker to each objective.

The constraint set (14.29) ensures that each project starts only once in the given time horizon. The constraint set (14.30) ensures that the financing required over all projects in each time period does not exceed total financing available in that time period. It also ensures that a project that is initiated will continue to completion within the planning horizon. That all projects are finished within the planning horizon is ensured by the constraint set (14.31). A specific candidate project can be realized in different ways with different resource requirements resulting in different scores. The resulting projects might also differ slightly among themselves in project scope. These projects are categorized as different version of the same project and each becomes a candidate project on its own in the set of alternative projects. From within a set S_v of such different versions of a project at most one can be selected for implementation as expressed in the constraint set (14.32). There can be several such sets $v = 1, \ldots, V$. Projects may have various interdependence relations, which need to be incorporated in the formulation. Mutual exclusiveness is one such relation. Mutual exclusiveness is ensured by the constraint set (14.33). Mutually exclusive candidate projects are included in the set S_m and there are M such individual mutual exclusiveness relations. When the problem on hand at a point in time includes the selection and scheduling of ongoing projects in addition to new alternative projects, then a decision of interest would be which of the ongoing projects would be included in the newly generated project portfolio. The data for those included should be updated to reflect the current situation and the remaining part of that project. The decision of including an ongoing project in the project portfolio is fed into the formulation through the constraint set (14.34). This constraint set can be employed in making the decision whether to include or not an ongoing project through the use of scenarios designed for this purpose. A further interdependence relation can be in the form of precedence relations among the candidate projects. A candidate project l can have a set of predecessors P_l. Hence, candidate project l can be assigned only when all candidate projects in P_l are assigned and completed. These requirements are imposed by the constraint sets (14.35) and (14.36). L indicates the number of candidate projects with predecessor (s). The requirement that candidate project l can be assigned only when all candidate projects in P_l are assigned is met by (14.35) and that all preceding projects are finished before the start of the succeeding candidate project l by (14.36).

This formulation can be further extended depending on the specific project requirements. For example, some projects may have to meet strict deadlines, requiring additional constraints beyond the base model discussed above.

As mentioned above, the only resource constraint is the bound on available finance in a period. Bounds on other resources can also be added to the problem formulation at the expense of increased project complexity.

Another important aspect of project portfolio management is to balance portfolio selection and risk. The selected portfolio might consist of highly profitable projects with high-risk associated with them as well as low-risk projects with modest benefits. In general, high-risk projects demand relatively larger investment for a longer time period and depending on the strength of an organization and its strategic goals, an acceptable mix of projects balancing risk and profitability will be desired. Investment into long term projects might not be desirable, which might lead to short term cash flow problems. To achieve a balance between different kinds of projects, the mathematical formulation can be extended in such a way so as to limit the maximum allowed investment in high-risk and long-term projects.

In addition, post optimization sensitivity analysis through the generation of scenarios can also help decision makers choose the desired mix of projects. For that purpose, manual adjustments might be needed.

14.5 Conclusions, Recent Developments, and Some Future Research Directions

Today companies receive many project proposals from inside or outside of the organization to produce products or provide services in line with their needs. These proposals must pass through a scientific project selection process so as to use the company resources efficiently. In this Chapter, we present mathematical programming models and discuss multi-criteria decision making. As multi-criteria tools, we use multi-criteria scoring and also AHP. Several other tools can also be used for this purpose. In this regard, Data Envelopment Analysis (DEA), which is based on linear programming, is a good option to determine the relative efficiency of projects to achieve the organizational tools. It has been proven effective for ranking the projects and performance assessment. We refer to two applications, one for evaluating engineering design, the other R& D projects (Farris et al. 2006; Tohumcu and Karasakal 2010). It would be interesting to focus on different industries and present more case studies that incorporate using alternative multi-criteria decision-making methods, such as AHP and DEA. The comparison of these methods considering different industry settings, data requirements and availabilities would provide valuable insights to the practitioners.

Project selection research and practices have mostly focused on financial objectives up to now. Apart from the financial factors, compliance with the organizational strategy, sustainability, social impact, risks and robustness have become important criteria. However, quantifying these criteria is not an easy task. There is need for further studies that address measuring these criteria correctly and incorporating them to commonly used project selection tools.

Uncertainty in projects needs to be considered in project selection problem. In this regard, Hall (2016) suggests using robust optimization to model uncertainty in this problem. Uncertainty in resources, specifically variations in availabilities could be incorporated in these models (Hazır and Ulusoy 2020). Uncertainty in financial markets and their effect on the discount rates can be studied.

Hall (2016) also highlights the dynamic characteristics of real-life project selection problems.

In most of the existing studies, project selection is done only once at the beginning of the planning horizon (Herbots et al. 2007). However, project proposals continue to arrive and in case of acceptance, they need to share the resource pools with the ongoing projects. Queuing models, simulation, stochastic programming have been used to model dynamic project -acceptance and capacity-planning problems. However, dynamic models are much less compared to the static ones in the literature.

14.6 Case Study: Project Portfolio Selection for the Construction of a Series of Dams

The Government has established the Directorate of the Southeastern Anatolia Program (SEAP) under the Ministry of Infrastructure with the aim of significantly increasing the wealth of the region. Topological analysis and precipitation data in the region have resulted in eight possible locations for dams suitable for both irrigation and hydroelectric power generation. As an initial step in long range planning the Ministry has provided SEAP with the rough cut estimates of annual budget allocations for the next 20 years shown in Table 14.16 as a guideline. The Director of SEAP has been asked to generate a set of possible projects extending over several years in consultation with the stakeholders of the SEAP and then choose

Table 14.16 The annual budget allocation to SEAP ($million)

1980	1981	1982	1983	1984	1985	1986	1987	1988	1989
70	80	80	100	150	400	400	450	575	650
1990	1991	1992	1993	1994	1995	1996	1997	1998	1999
650	650	600	400	300	250	125	115	100	80

a subset from among these projects to implement over the next 20 years for presentation to the Ministry for further consideration.

The selected projects do not need to start simultaneously but must finish within the next 20 years. Each project i has a construction period D_i and will be evaluated based on the following five criteria:

- Total cost of the project
- Electricity generation capacity (MWs)
- Amount of land that can be irrigated
- Useful life
- Sustainability of current environmental conditions (SoCEC)

The SoCEC criterion is expressed in terms of the probability that the current environmental conditions will not change over the course of the next 40 years. Estimates of the data for the five criteria for each dam are presented in Table 14.17.

Karakaya Dams I and II are interrelated projects such that Karakaya Dam I, consisting of three 300 MW capacity turbines, can be built as an individual project. After its completion it can be extended to Karakaya Dam II by adding with three more 300 MW turbines, for a final generation capacity of 1800 MW. The increased electricity generation capability has no apparent impact on the useful life of Karakaya Dam I and hence, the useful life of Karakaya Dam II is the remaining useful life of Karakaya Dam I at the time of the expansion. Hence, Karakaya Dam I (project 4) is a predecessor of Karakaya Dam II (project 5). Kralkızı Dam (project 7) and Dicle Dam (project 8) are mutually exclusive as both these dams are located on Tigris Basin and the sites for these locations are only 32 km apart, too close for both

Table 14.17 The data for the five criteria for each dam

Name	Duration D_i (Years)	Electricity generation capacity (MWs)	Amount of land irrigated (Hectares)	Useful life (Years)	SoCEC (Probability)
Atatürk Dam	14	2400	727,700	40	0.85
Birecik Dam	06	672	134,000	55	0.75
Çamgazi Dam	04	150	25,350	40	0.65
Karakaya Dam I	08	900	250,000	55	0.95
Karakaya Dam II	05	900	250,000	–	0.95
Batman Dam	10	375	37,351	65	0.70
Kralkızı Dam	10	94	130,159	50	0.85
Dicle Dam	12	110	120,250	40	0.85

Table 14.18 Annual budget requirements for the construction of the dams ($million)

Name	1	2	3	4	5	6	7	8	9	10	11	12	13	14
Atatürk Dam	50	30	75	80	120	135	175	250	275	120	110	100	100	100
Birecik Dam	30	50	150	150	75	75								
Çamgazi Dam	50	150	170	100										
Karakaya Dam I	40	50	75	150	175	200	150	100						
Karakaya Dam II	30	75	100	50	50									
Batman Dam	30	40	50	50	50	155	250	100	95	95				
Kralkızı Dam	40	40	60	80	80	175	175	80	75	75				
Dicle Dam	50	30	80	100	125	225	225	150	125	80	80	70		

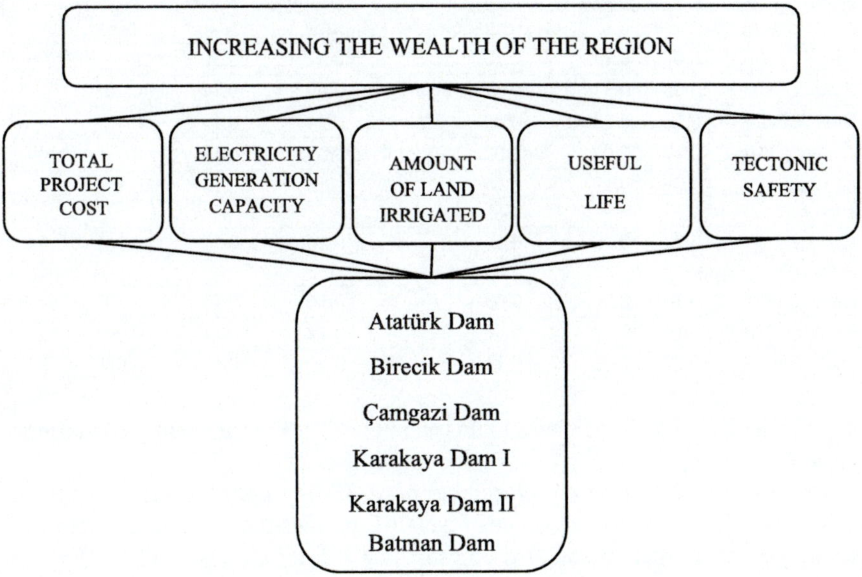

Fig. 14.7 AHP decision hierarchy for the dam project selection problem

to be viable. The annual budget required for the construction of all the eight projects mentioned above are tabulated in Table 14.18.

The annual discount rate is assumed to be uniform for the next 20 years at 2% and must be included in financial estimates. In other words, NPVs should be used while evaluating the cost of the dams.

Since the problem is a Multi-Criteria Decision Making (MCDM) problem, AHP is used to derive the aggregate scores v_i that represent the contribution of the corresponding project i to the overall objective of the Program; hence, the higher the value of v_i, the higher the contribution of project i. The hierarchy of the decision problem is illustrated in Fig. 14.7. Expert opinions will be utilized to perform

Table 14.19 NPVs (in $million) and the corresponding weights

Name	NPV	Weights
Atatürk Dam	1,494.99	0.05
Birecik Dam	501.76	0.16
Çamgazi Dam	454.69	0.17
Karakaya Dam I	865.53	0.09
Karakaya Dam II	292.95	0.27
Batman Dam	820.59	0.10
Kralkızı Dam	796.49	0.10
Dicle Dam	1,195.74	0.07

Table 14.20 Criterion weights for the alternative dam projects

Name	Cost	Capacity	Utility for irrigation	Useful life	SoCEC	Weights for dams
	0.06	0.14	0.20	0.31	0.30	v_i
Atatürk Dam	0.05	0.37	0.37	0.09	0.17	0.203
Birecik Dam	0.16	0.12	0.08	0.14	0.12	0.122
Çamgazi Dam	0.17	0.04	0.02	0.09	0.11	0.078
Karakaya Dam I	0.09	0.19	0.16	0.16	0.14	0.155
Karakaya Dam II	0.27	0.18	0.16	0.16	0.14	0.162
Batman Dam	0.10	0.07	0.03	0.16	0.10	0.101
Kralkızı Dam	0.10	0.02	0.06	0.12	0.11	0.092
Dicle Dam	0.07	0.02	0.11	0.09	0.10	0.088

pairwise comparisons at each level of this hierarchy. The Director wants to maximize the sum of the scores of all projects to be implemented.

The weight determination procedure of AHP is applied eliciting pairwise comparisons from experts at various levels of the hierarchy. However, note that the weight for a criterion can also be estimated if *hard data* is available for that criterion. For example, the total cost of the project is one of the critical factors for the Director of SEAP. Therefore, she utilizes NPV values of all projects to calculate required weights of each project with reference to the total cost of the project. Project cost has an inverse relation with the weight of that project therefore inverse of NPVs are used to estimate the required weight vector as illustrated in Table 14.19.

Priority/weight vectors calculated from pairwise comparison matrices and estimated from the give data and the corresponding aggregate scores representing the weights v_i for the projects are tabulated in Table 14.20.

Since the scores are a proxy for the wealth generation potential of each project, the objective is to maximize the sum of these scores while satisfying all the annual budget allocation and interdependence constraints. The mathematical programming formulation for the problem environment is stated below employing the notation given in Sect. 14.4.3.

$$\max Z = \sum_{i=1}^{8} \sum_{j=1}^{20} a_i x_{ij} \tag{14.28}$$

subject to

$$\sum\nolimits_{J=1}^{20} x_{ij} \leq 1 \quad i = 1, \ldots, 8 \tag{14.29}$$

$$\sum_{i=1}^{8} \sum_{j=1}^{k} C_{i,k+1-j} x_{ij} \leq AF_k \qquad k = 1, \ldots, 20 \tag{14.30}$$

$$\sum_{j=1}^{20} j x_{ij} + D_i \leq T + 1 \quad i = 1, \ldots, 8 \tag{14.31}$$

$$\sum\nolimits_{j=1}^{20} \left(x_{7j} + x_{8j} \right) \leq 1 \tag{14.32}$$

$$\sum_{j=1}^{20} x_{4j} \geq \sum_{J=1}^{5} x_{5j} \tag{14.35}$$

$$\sum_{j=1}^{20} 5x_{1j} + (T+1)\left(1 - \sum_{j=1}^{20} x_{5j}\right) - \sum_{j=1}^{20} 4x_{ij} \geq D_4 \sum_{j=1}^{20} x_{4j} \tag{14.36}$$

$$x_{ij} = \begin{cases} 1 \textit{ if project } i \textit{ is included in the portfolio and starts in period } j \\ 0 \quad \textit{otherwise} \quad i = 1, \ldots, N\, j = 1, \ldots, T \end{cases} \tag{14.38}$$

Atatürk Dam holds strategic importance amongst all the candidate dams; therefore, it was decided that Atatürk Dam must be included in the final portfolio. Furthermore, it has the longest duration and hence, any delay in this project will have severe impact. Thus, it will be started at time period 1. The formulation was accordingly updated by adding the constraint $x_{11} = 1$.

The optimal solution to the formulation yields an objective function value of 0.821 with the optimal project portfolio and schedule illustrated in Fig. 14.8. The budget utilization implied by this project schedule is graphically illustrated in Fig. 14.9. The overall average budget utilization is found to be 79%.

The SEAP Board convened to review the current solution and observed that due to the low utilization of the budget in some of the years, the Government might be reluctant to allocate funding as initially promised, asking the Director of SEAP to address this issue. Upon the instruction by the Director the SEAP Planning Office came up with a scenario for improving the budget utilization. The scenario was based on the redistribution of the budgetary allocation so that additional funding is granted in first 5 years. The Director approached the Government and presented the scenario. According to the scenario Karakaya Dam I would be started in the first year

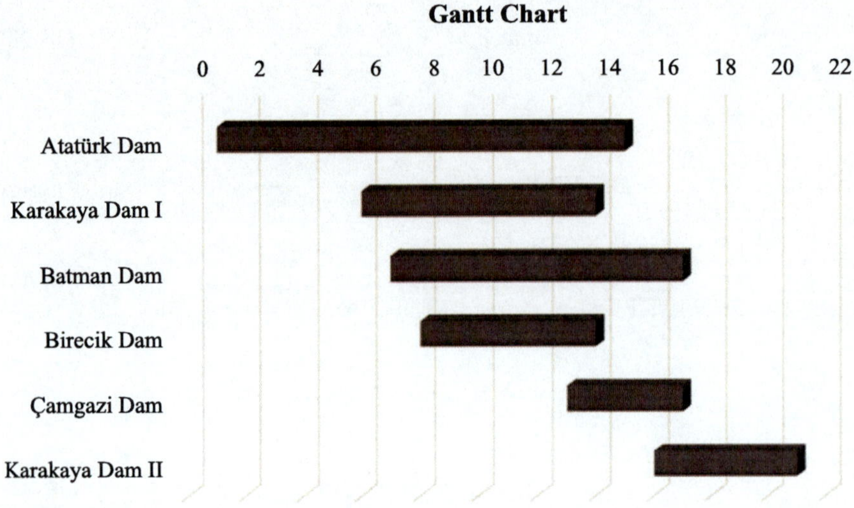

Fig. 14.8 Project portfolio and optimal schedule

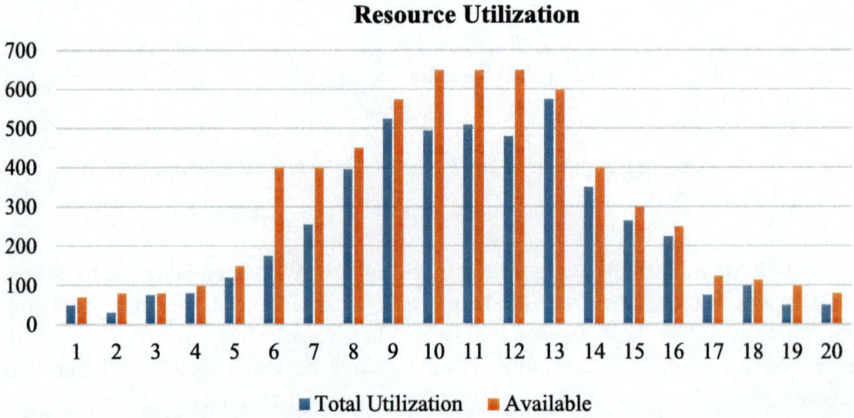

Fig. 14.9 Budget utilization with optimal project portfolio

and thus additional electricity generation capacity would be made available to the national grid earlier as well as start generating revenue stream earlier. The Director of SEAP argued that this proposal requires additional funding for the first 5 years, which can be obtained by transferring funds from later years but keeping the overall 20-year funding constant. After deliberations, the budgetary allocations were revised as tabulated in Table 14.21.

With these revised budgetary allocations and starting Karakaya Dam I in period 1 (i.e., besides $X_{41} = X_{11} = 1$), the new solution yielded a better objective function value of 0.834. The resulting project portfolio, schedule and budget utilization are

Table 14.21 Revised annual budget allocation to SEAP (in $million)

1980	1981	1982	1983	1984	1985	1986	1987	1988	1989
100	100	150	230	300	400	400	450	450	575
1990	1991	1992	1993	1994	1995	1996	1997	1998	1999
575	575	550	400	300	250	125	115	100	80

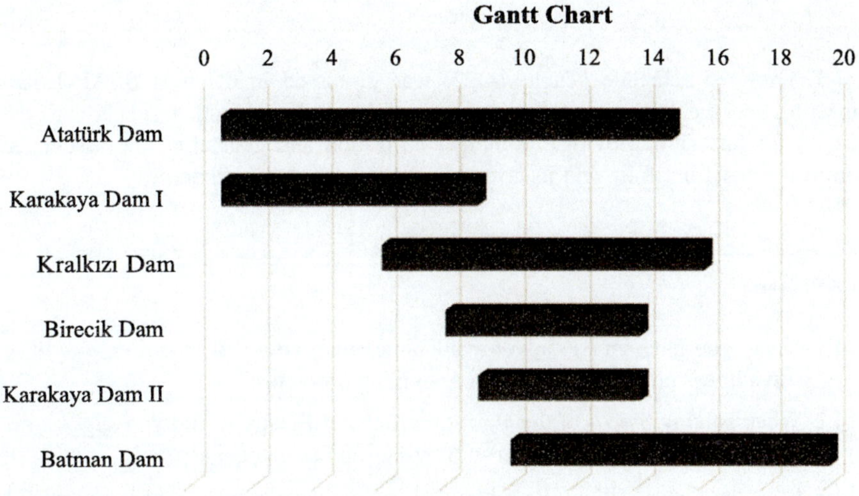

Fig. 14.10 Revised project portfolio and optimal schedule

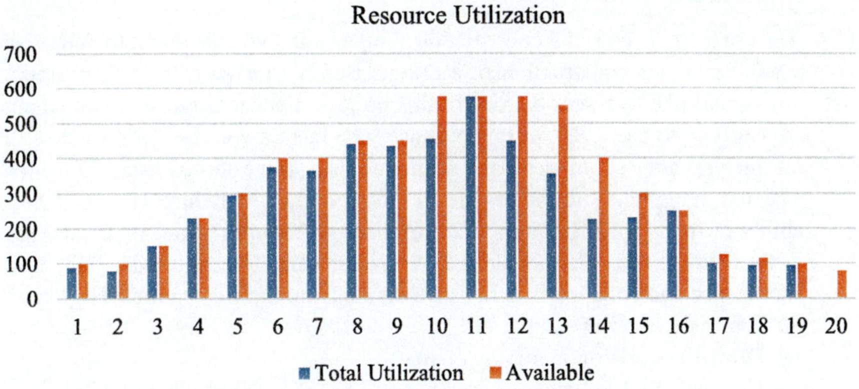

Fig. 14.11 Revised budget utilization with optimal project portfolio

provided in Figs. 14.10 and 14.11 and Table 14.22. The budget utilization has increased from 79% to 85% as a result of the revised annual budget allocation and the resulting schedule.

Table 14.22 Revised list of projects and their schedule

Name	Starting period
Atatürk Dam	1
Karakaya Dam I	1
Kralkızı Dam	6
Birecik Dam	8
Karakaya Dam II	9
Batman Dam	10

The revised schedule (Table 14.22) was discussed in the next SEAP Board meeting and the Director of SEAP was asked to submit another scenario to the Board for further improving the budget utilization and generating the associated optimal project portfolio and its corresponding construction schedule.

Exercises

14.1 Given you have an LP solver available explain how you would employ it to solve linear preemptive Goal Programming problems.

14.2 When solving Step 4 of the algorithm for MOLP, why would you suggest the additional constraint $Z_2 \geq rhs$ will be satisfied as an equality?

14.3 Consider the case of a GP problem where you have two absolute priorities G1 and G2. When solving this GP problem with G1 $\gg$ G2 first and then with G2 $\gg$ G1, you obtain the same solution. What would be your conclusion in such a case?

14.4 You are the manager of a construction company and you are about to authorize a project for the construction of a bridge. Before you give the authorization, you would like to assess the risk of this project. Your management consultant advises you to use AHP for this purpose. The factors you decide to include in the analysis are F1: Financial and economic risks; F2: Political risks; F3: Force Majeure items. The sub-factors are F11: Subcontractors failure; F12: Unavailability of funds; F13: Inflation; F21: Embargoes/Expropriation; F22: Changes in local laws; F31: Earthquakes; F32: Water damage and floods; F33: Soil subsistence and collapse. The risks are divided into three categories: High risk; medium risk; low risk.

(a) Draw the decision hierarchy chart.

(b) Calculate the relative importance of F1, F2, F3 with respect to the overall goal.

	F1	F2	F3
F1	1	3	6
F2	1/3	1	5
F3	1/6	1/5	1

Calculate λ_{max}; consistency index CI; consistency ratio CR. Take RI as 0.58.

(c) How would you phrase the questions for determining the judgment matrices at the lowest level of the hierarchy?

(d) Apply AHP to determine the level of risk for the project. In addition to the judgment matrix in part (a), the following judgment matrices are given, where L, M, H stand for low risk, medium risk and high risk, respectively.

Under F1

	F11	F12	F13
F11	1	3	1/3
F12	1/3	1	1/6
F13	3	6	1

Under F2

	F21	F22
F21	1	5
F22	1/5	1

Under F3

	F31	F32	F33
F31	1	1/7	1/9
F32	7	1	3
F33	9	1/3	1

Under F11

	L	M	C
L	1	3	1/3
M	1/3	1	1/6
C	3	6	1

Under F12

	L	M	H
L	1	5	6
M	1/5	1	3
H	1/6	1/3	1

Under F13

	L	M	H
L	1	1/6	1/4
M	6	1	5
H	4	1/5	1

Under F21

	L	M	H
L	1	6	1/3
M	1/6	1	1/6
H	3	6	1

Under F22

	L	M	H
L	1	4	6
M	1/4	1	3
H	1/6	1/3	1

Under F31

	L	M	H
L	1	6	1/4
M	1/6	1	5
H	4	1/5	1

Under F32

	L	M	H
L	1	4	7
M	1/4	1	3
H	1/7	1/3	1

Under F33

	L	M	H
L	1	1/3	1/2
M	3	1	2
H	2	1/2	1

14.5 (a) The Director of the Southeastern Anatolia Program (SAP) has been informed that the Program will be provided with relatively large annual budgets for the next T years. The budgets are also stated by the Government as part of a regional development plan. He is asked to generate a set of possible projects extending over several years in consultation with the stakeholders of the SAP and then choose among these projects a subset to implement over the time horizon T. The projects selected do not need to start simultaneously but they are required to finish within the next T years. Each project i has an execution period D_i, in general differing over the projects. Some of the projects have various versions out of which at most one is to be selected, if at all. For modeling purposes these versions are treated as individual projects. The annual expenditure for project i can vary depending on which year of its execution the project is in, i.e., if project i has been initiated in year j, then its annual expenditure in year $k \geq j$ is given as $C_{i,k+1-j}$. Let a_i be the score of project i determined to represent the contribution of project i to the Program; hence, the higher the better. The Director wants to maximize the sum of the scores of all projects implemented. Write down a mathematical programming model for the problem environment stated above.

(b) Let us assume that we have solved the above mathematical programming formulation and obtained an optimal solution; in other words, an optimal objective function value Z*, an optimal set of projects to be implemented together with their initiation period. Suggest a methodology for finding an alternative optimal solution, if there exists one. (*Hint*: Consult the paper by Ghasemzadeh F. Archer N. and Iyogun P., 1999)

14.6 A marketing agent wants to be promoted to the position of marketing manager of her agency. For this purpose, she wants to enroll into a graduate marketing certificate program. Once she enrolls into the program, she must complete the program within 5 semesters.

- There are 20 classes she can choose from.
- To complete the program, she must complete at least 12 courses.
- Each semester she can register for at most 4 classes.
- Courses 1,2,3 and 4 are mandatory.
- Course 5 is a prerequisite for course 6.
- If course 9 is taken, then course 7 and 8 must also be taken and they must be taken in 3 consecutive semesters in their given orders (i.e., in the order 7, 8, 9).
- Course 10 and Course 11 are not offered in the same semester.
- One can register for at most 2 of the courses in the set {12, 13, 14 and 15}
- There is an enrollment fee associated with each course (ci for course i).
- We assume that once the agent takes a course, she will pass the course. So, a course can be taken at most once during the 5 semestersThe costs for the courses are given as follows: (450, 450, 450, 450, 300, 300, 400, 400, 400, 400, 450, 450, 300, 300, 350, 300, 300, 350, 350, 350).

The agent wants to decide which courses to select in order to minimize the overall cost of the certificate program. Formulate this problem as an integer programming problem and solve using any IP solver.

14.7 A firm will decide on the percentage order-quantity allocation among three vendors, A, B or C. Each vendor is willing and has the capacity to satisfy any portion of the buyer's need at the quoted price. The criteria that are considered are quality, price, lead time, and service. Criteria are independent. The preference matrix is given as:

	Price	Quality	Lead time	Service
Price	1	1/7	3	5
Quality	7	1	8	9
Lead time	1/3	1/8	1	3
Service	1/5	1/9	1/3	1

From the pairwise comparisons the weights for each of the vendor with respect to criteria can be presented as:

	Vendor A	Vendor B	Vendor C
Price	0.195	0.717	0.088
Quality	0.731	0.081	0.188
Lead time	0.058	0.663	0.278
Service	0.731	0.188	0.081

(a) What is the desired solution for the firm? Check the preference matrix for consistency.
(b) When comparing the vendors under the price criterion, how do you phrase the question to the decision maker? Give an example.

14.8 Consider four different project proposals given to the State Waterworks Agency (SWA), which they will evaluate and decide on one for further analysis. The criteria the SWA employs in this initial assessment process together with their weights and the corresponding values of the individual projects for these criteria are given in the following Table.

	Proposal 1	Proposal 2	Proposal 3	Proposal 4	Weights
Number of employment created	128	206	214	167	0.10
Impact on environment	78	67	84	93	0.12
Amount of credit required	534	278	365	378	0.08

(continued)

Availability of credit	0.90	0.90	0.75	0.80	0.10
Amount of electricity generated	110	123	119	132	0.23
Amount of land irrigated	71500	78600	83500	69500	0.17
Cost of proposal	678	798	721	850	0.20

Impact on environment is expressed on scale of [1,100] with higher values indicating less damage on environment. Availability of credit is represented through subjective measures in the range [0,1], where higher values imply higher probability of securing the credit. Amount of electricity generated is expressed in MW and amount of land irrigated in ha. Cost of proposal is expressed in million $. Apply an initial screening to determine the proposal to be analyzed in detail.

14.9 A machine shop owner wants to buy a CNC turning machine and his options are models X, Y and Z. Through consultation with the experts the major factors to evaluate each option are determined as Cost (C), Quality of Surface Treatment (QST), and Setting Accuracy (SA). Cost criterion is considered in three parts; Fixed Cost (FC), Annual Maintenance Cost (AMC) and Operation Cost per minute (CC). The weights associated with these criteria are obtained using the associated procedure within AHP. The weights as well as the values for each candidate machine for these criteria are determined by this expert group as displayed in the table.
Determine which machine you would recommend to the shop owner using the data provided in the table below.

	QST	SA	FC(TL) $\times 10^3$	AMC(TL)	CC(TL)
Weight	0.31	0.22	0.18	0.09	0.20
Machine X	80×10^{-5}	$24\text{x}10^{-4}$	1222	38000	42
Machine Y	25×10^{-4}	$36\text{x}10^{-3}$	729	65000	38
Machine Z	55×10^{-4}	$40\text{x}10^{-3}$	662	77000	30

14.10 A firm will decide on the contractor for the execution of their factory building. Three contractors made it to the short list: firms A, B and C. The criteria that are considered are price quoted, quality of previous works, project management performance (PMP), and financial stability (FS) of the firm. Criteria are independent. The judgment matrix is given as:

	Price	Quality	PMP	FS
Price	1	1/7	3	5
Quality	7	1	8	9
PMP	1/3	1/8	1	3
FS	1/5	1/9	1/3	1

From the pairwise comparisons the weights for each of the contractors with respect to criteria can be presented as:

	ContractorA	ContractorB	ContractorC
Price	0.195	0.717	0.088
Quality	0.731	0.081	0.188
PMP	0.058	0.663	0.278
FS	0.731	0.188	0.081

(a) What is the desired solution for the firm? Check the preference matrix for consistency. Given we accept CR values 0.1 (10%) or less as an indication of consistency, what would be your conclusion? Would you ask for further analysis? Explain.

(b) When comparing the contractors under the quality criterion, how do you phrase the question to the decision maker? Give an example.

(c) Given the price quotations of the contractors A, B, and C are x, y, z ($), respectively, then how would the Price row under the contractors be specified?

Size of matrix	1	2	3	4	5	6	7	8	9	10
RI	0	0	0.58	0.90	1.12	1.24	1.32	1.41	1.45	1.49

14.11 Consider four different project proposals given to the State Waterworks Agency (SWA), which they will evaluate and decide which projects to accept. The criteria the SWA employs in this project selection process are the number of employments created, amount of electricity generated, and amount of land irrigated listed in the preemptive lexicographic order. The corresponding values of the individual projects for these criteria are given in the following Table together with the aspiration level for each criterion. The cost of each proposal is also included in the following table.

	Proposal 1	Proposal 2	Proposal 3	Proposal 4	Aspiration level
Number of employment created	128	206	214	167	320
Amount of electricity generated	110	123	119	132	225
Amount of land irrigated	71,500	78,600	83,500	69,500	150,000
Cost of proposal	678	798	721	850	

Amount of electricity generated is expressed in MW and amount of land irrigated in ha. Cost of each proposal is expressed in $ million. The total amount funds allocated by the Appropriation Committee in the Parliament is $1700 million. Assume all project proposals to have the same duration for realization and they all meet the environmental criteria. The decision-making body wants to maximize the amount by which the aspiration level of each criterion is exceeded. They also want not to exceed the total amount of funds allocated. Apply preemptive goal programming to determine the projects to be realized.

14.12 (a) A company is planning to start h new projects out of the n project proposals. The selected projects must be started and finished within the next T months. Each project i has an associated benefit denoted by b_i. The overall cost of completing the project i is given by c_i and the budget the company allocated for the selected projects is C. To start project i requires a set up time s_i given in months and the number of months required to complete project i is given by d_i. There are some restrictions in terms of which projects can be selected. Project 1 and 2 cannot be selected together. If project 3 is selected, then first project 4 and then project 5 must be completed before project 3 can be started. A project can be started at most once during the given T months and must be completed within the time horizon of T months. Formulate this problem so that the overall benefit for the company is maximized.

(b) The company is planning for the project portfolio to be realized within the next 24 months. There are 8 candidate projects and considering its management resources the company has decided to restrict its choice to 5 projects. A budget of $1,840,000 has been allocated for the projects selected. The set-up time s_i is given as 1 month for all candidate projects. The benefit ($'000), cost ($'000), and duration (months) for each candidate project are estimated as in the following table. The durations do not include the set-up time.

	Project #1	Project #2	Project #3	Project #4	Project #5	Project #6	Project #7	Project #8
Benefit	556	293	331	672	445	512	384	392
Cost	348	189	294	460	327	392	282	296
Duration	8	5	6	8	6	7	6	4

The company has decided to distribute the budget equally throughout the time horizon starting with the second month. Thus, an amount of $\bar{C}$ = $80,000 is allocated to each month j = 2,. . .,24. The cost incurred for each project is also equally distributed over the duration of the project. No cost is incurred during the set-up period.

Determine the project portfolio and the schedule of the selected projects for the problem stated above.

14.13 A clothing company is considering investing in opening new department stores in 8 different locations in Istanbul. The projected annual profit from a new department store in each location and the investment required for it is given in the table below. The overall budget available for investment is given by \$1,000,000. The company can invest in opening at most 5 department stores out of these 8 locations. They cannot open department stores in locations 3 and 5 both, and if they open a department store in location 4, then they must open department stores in locations 1 and 8 as well. Formulate this problem to maximize the total annual profit of the company and solve to optimality using any optimization solver.

Location	Projected annual profit	Investment amount required
1	\$70,000	\$250,000
2	\$30,000	\$180,000
3	\$45,000	\$220,000
4	\$62,000	\$280,000
5	\$140,000	\$430,000
6	\$38,000	\$160,000
7	\$60,000	\$240,000
8	\$180,000	\$520,000

14.14 You are the Director of the Project Office and you want to choose a project portfolio from among a set P of candidate projects to be implemented over a common time horizon of H years.

(a) Write down the mathematical programming formulation to determine the project portfolio from among a set P of candidate projects, which maximizes the NPV of the total return for the following H-year period. The discount rate is given by $\alpha > 0$ and you have a limited amount of budget B available initially. The projects are selected and implemented at time $t = 0$ and the first return is realized at the end of the first year. All cash flows except the initial investment occur at the end of each year. There is no possibility of borrowing money.

- r_i: annual net return on project i
- S_i: salvage value of project i after H years
- C_i: initial cost of project i

(b) Solve the following problem instance using a heuristic procedure. Explain your heuristic procedure explicitly.

$B = 5000$ (all units are in (\$'000)); $\alpha = 0.10$; $H = 10$ and the number of projects $|P| = 5$.

Project	1	2	3	4	5
r_i	300	140	300	500	170
S_i	360	230	170	760	320
C_i	1560	910	1480	2850	1000

(c) Solve the mathematical programming formulation given in part (a) using any IP solver and give an interpretation of the result. Compare with the solution in part (b).

14.15 Write down the mathematical programming formulation to determine the project portfolio from among a set P of candidate projects that maximizes the NPV of the total return in the following 10-year period. Assume a discount rate of α constant throughout the time horizon with continuous compounding being applied. You have a limited amount of budget B_0 available initially. At the end of each period your total return for that period is added to cash at hand. The initial cost is realized at the end of the time period in which the project is started. The first return is realized at the end of the next period the project has started. There is no possibility of borrowing money.

- r_{it}: annual net return on project i
- S_i: salvage value of project i at the end of 10 years
- C_i: initial cost of project i
- T_i: starting time of project i

(a) A project *i* can start in a given time period T_i; if at all, i.e., not all projects need to be realized.

(b) A project *i* can start in a given time period T_i or after the given time period T_i; if at all, i.e., not all projects need to be realized.

14.16 Excellence Electrical Motor Company needs a new assembly system. There are three different technologies to choose from. The properties of the technologies and their life cycle cost components are given below. There are three factors and several subfactors to rate a technology, which are given below.

- Operating system (OS)
  - Sensitivity and reliability of the system (SR)
  - Flexibility of the system (FS)
  - Quality of product (QP)
- Planning and control system (PC)
  - Responses to product design changes (RP)
  - Responses to demand changes (RD)
  - Responses to technological changes (RT)
- Conveyance system (CS)
  - Sensitivity and reliability of the system (SeR)
  - Flexibility of the system (FeS)
- The life cycle cost for each technology consists of the following cost items:
  - Research and development
  - Acquisition
  - Operation and support
  - Disposal and retirement

Note that all costs are incurred once except "operation and support" cost, which is incurred annually and for the whole life time of the technologies, which are 8 years for all alternatives. The company decides to apply AHP. The General Manager says, "let us apply the distributive mode". The following matrices have been obtained:

	OS	PC	CS
OS	1	2	6
PC	1/2	1	3
CS	1/6	1/3	1

	SR	FS	QP
SR	1	1/4	1/7
FS	4	1	1/3
QP	7	3	1

	RP	RD	RT
RP	1	1/2	4
RD	2	1	6
RT	1/4	1/6	1

	SeR	FeS
SeR	1	3
FeS	1/3	1

Technologies' scores under sub criteria of OS are:

	SR	FS	QP
T1	2	3	1/5
T2	4	2	4
T3	1/3	2	4

Technologies' scores under sub criteria of PC are:

	RP	RD	RT
T1	3	1	2
T2	7	3	4
T3	2	2	3

Technologies' scores under sub criteria of CS are:

	SeR	FeS
T1	3	2
T2	4	2
T3	3	1

	Tech #1	Tech #2	Tech #3
Research and development	100,000	180,000	30,000
Acquisition	1,800,000	2,500,000	800,000
Operation and support	350,000	750,000	180,000
Disposal and retirement	200,000	150,000	300,000

Through use of cost/benefit ratio, determine the best choice of technology among the three candidates. The discount rate is 10% annually.

14.17 A portfolio manager needs to decide which of three projects to postpone due to unplanned events that took place in his department. The worth of each project depends on the state of the company when the project is completed. He has come up with the following benefit matrix:

	State of the company		
	Above average	Average	Below average
Project A	85	53	24
Project B	90	50	32
Project C	75	70	65

The probability that the company will be in any of the above states is:

- P(above average) = 0.6
- P(average) = 0.3
- P(below average) = 0.1
  - (a) Determine which project should be postponed by the manager, and what is the expected benefit in such a case?
  - (b) Suppose that the manager can spend extra money and time and know more about the future state of the company. Calculate the value of such a perfect information.

14.18 In MSIE Company, they want to install an office automation system. They consider three alternatives. The Selection Committee has decided to apply Benefit/Cost analysis using AHP. In terms of benefit components, the Committee considered:

- Word processing capability
- Administrative capability
- Reliability

  In terms of cost, they considered:
- Initial cost
- Maintenance cost (including downtime)
- Learning

The initial costs are given as: CS1 = \$18,000; CS2 = \$26,000; CS3 = \$31,000. For benefit components, they ask to the decision maker questions like the following:

"How much more important is word processing than administrative support in an office automation system?"

For cost components, they ask to the decision maker questions like the following:

"How much more expensive is System 1 than System 2 under initial cost?"

Under WP Capability

	S1	S2	S3
S1	1	5	7
S2	1/5	1	3
S3	1/7	1/3	1

Under Adm. Capability

	S1	S2	S3
S1	1	1/3	6
S2	3	1	9
S3	1/6	1/9	1

Under Reliability

	S1	S2	S3
S1	1	1/2	1/5
S2	2	1	1/2
S3	5	2	1

The judgment matrix under the cost is to be constructed using the cost data provided above. The judgment matrix under the benefit is found to be:

	WP Cap	Adm. Cap	Rel.
WP Cap	1	1/3	1/7
Adm. Cap	3	1	1/3
Rel.	7	3	1

(a) Considering the pairwise matrices of systems under cost components given as in the following, make a choice based on benefit/cost ratio.

Under Maintenance Cost			
	S1	S2	S3
S1	1	2	4
S2	1/2	1	3
S3	1/4	1/3	1

Under Learning			
	S1	S2	S3
S1	1	7	3
S2	1/7	1	5
S3	1/3	1/5	1

(b) Considering the objective data for cost components given as below, make a choice based on benefit/cost ratio.
- Maintenance cost is usually 20% of the initial cost and learning cost is taken to be equal to the initial cost.
- For both (a) and (b) options of cost components' data, check consistency for each matrix. The random consistency number for 3×3 matrix is 0.58. Continue with the procedure even if some of the matrices turn out to be inconsistent.

References

Archer, N. P., & Ghasemzadeh, F. (1996). *Portfolio selection techniques: A review and a suggested integrated approach* (Innovation Research Center Working Paper 46). School of Business, McMaster University, Hamilton, ON.

Dean, B. V., & Nishry, M. (1965). Scoring and profitability models for evaluation and selecting engineering projects. *Operations Research, 13*(4), 550–569.

Farris, J. A., Groesbeck, R. L., Van Aken, E. M., & Letens, G. (2006). Evaluating the relative performance of engineering design projects: A case study using data envelopment analysis. *IEEE Transactions on Engineering Management, 53*(3), 471–482.

Ghasemzadeh, F., Archer, N., & Iyogun, P. (1999). A zero-one model for project portfolio selection and scheduling. *Journal of Operational Research, 50*, 745–755.

Graves, S. B., & Ringuest, J. L. (2012). Models and methods for project selection: Concepts from management science. In *Finance and information technology* (Vol. 58). New York: Springer.

Hall, N. G. (2016). Research and teaching opportunities in project management. In *Optimization challenges in complex, networked and risky systems* (pp. 329–388). Hanover: INFORMS.

Hazır, Ö., & Ulusoy, G. (2020). A classification and review of approaches and methods for modeling uncertainty in projects. *International Journal of Production Economics, 223, 107522.*

Herbots, J., Herroelen, W., & Leus, R. (2007). Dynamic order acceptance and capacity planning on a single bottleneck resource. *Naval Research Logistics (NRL), 54*(8), 874–889.

PMI. (2017). *The standard for portfolio management* (4th ed.). Newtown Square: Project Management Institute.

Saaty, T. L. (1980). *The analytic hierarchy process: Planning, priority setting, resource allocation*. New York: McGraw-Hill.

Saaty, T. L. (1990). How to make a decision: The analytic hierarchy process. *European Journal of Operational Research, 48*, 9–26.

Saaty, T. L. (1994). Highlights and critical points in the theory and application of the analytic hierarchy process. *European Journal of Operational Research, 74*, 426–447.

Saaty, T. L., & Kearns, K. P. (1985). *Analytical planning. The organization of systems*. Oxford: Pergamon Press.

Shiraishi, S., Obata, T., & Daigo, M. (1998). Properties of a positive reciprocal matrix and their application to AHP. *Journal of the Operations Research Society of Japan, 41*(3), 404–414.

Tohumcu, Z., & Karasakal, E. (2010). R&D project performance evaluation with multiple and interdependent criteria. *IEEE Transactions on Engineering Management, 57*(4), 620–633.

Winston, W. (2003). *Operations research: Applications and algorithms* (4th ed.). Boston: Duxbury Press.

15 Recent Developments and Some Promising Research Areas

Learning Outcomes
Upon successful completion of this Chapter, the reader will be able to:

1. Describe some of the recent developments and trends in project management area.
2. Discuss the relevance and importance of sustainability in project management practices.
3. Explain what big data refers to in project management
4. Discuss how developments in data analytics might change the project life cycle management.
5. Discuss how Industry 4.0 and the technological changes transform project management practices.

15.1 Sustainability and Project Management

The contemporary world is witnessing a change in production and consumption patterns as it is no longer possible to maintain existing economic growth paradigms without surpassing planetary boundaries. Protection of the environment and efficient use of natural resources are becoming more central elements of competitiveness for at least two main reasons. On the one hand, policy instruments such as extended producer responsibility laws, eco-labeling, industrial standards (energy efficiency, emission standards) and R&D grants all facilitate a transition towards a green economy. On the other hand, increasing public awareness of sustainable development and increasingly direct experience of the consequences of climate change are transforming the behavior and preferences of all stakeholders. Considering these facts, we address the sustainability aspects of project management in this Chapter. We first present the definition of sustainability and then discuss some sustainability issues in managing project life cycle management.

According to the Brundtland Commission Report (1987), sustainable development addresses meeting today's needs without sacrificing the needs of future

G. Ulusoy, Ö. Hazır, *An Introduction to Project Modeling and Planning*, Springer Texts in Business and Economics, https://doi.org/10.1007/978-3-030-61423-2_15

generations. In accordance with this definition of development, sustainability in business emphasizes the movement towards gaining economic wealth, preserving the environment and guaranteeing social justice (Silvius et al. 2012). In other words, it needs to be defined regarding three essential dimensions: economic, environmental and social. Broadly speaking, sustainability in business mainly seeks to achieve a balance and harmony in business results according to these criteria. More importantly, it has a great potential to improve the quality of life in our societies.

Considering its important role, businesses have lately begun to consider sustainability as an important component of strategic planning. There is a link between creating a social agenda, strategic positioning, and competitive advantage (Porter and Kramer 2006). Awareness of sustainability and its integration into decision making can enhance relations with stakeholders by emphasizing corporate social responsibility to differentiate the organization from its rivals. We will examine the impacts of this strategic role on the projects and their management.

The projects carried out in organizations are important means for achieving their strategic goals. In this regard, sustainability can be viewed as a component of project success. Despite this strategic importance, however, most current project management practices either neglect or barely consider sustainability issues. Especially in developing countries, they are overshadowed by economic considerations. Given the above mentioned strategic importance, sustainability needs to be included in project specifications and integrated as a project assessment criterion in addition to the more common time, cost and scope targets. Vis-à-vis the importance and realizations, currently, there is a gap both in literature and practice (Martens and Carvalho 2016), hence a need for further studies on this topic and efforts to be done by the practitioners.

Before discussing the integration of sustainability into project management practices, we first present some guidelines for sustainability assessment and reporting. The most prominent of these are the Sustainability Reporting Guidelines (Global Reporting Initiative [GRI] 2006), the United Nations Global Compact (UNGC 2008) reports, and the standards of International Organization for Standardization (ISO ISO/TS 21929-1 2006). Dumay et al. (2010) give a critical discussion of GRI standards and applications in the public sector.

These guidelines and development principles have also been studied in project management (Labuschagne et al. 2005; Morrissey et al. 2012; Gareis et al. 2013; Silvius and Schipper 2015) and indicators for assessing performance regarding sustainability have been developed (Fernández-Sánchez and Rodríguez-López 2010; Chawla et al. 2018). Adopting these indicators and performance measurement systems, redesigning the project life cycle and reexamining the managerial decision making is essential to develop sustainability awareness in projects (Labuschagne and Brent 2005). In this regard, Stanitsas et al. (2021) have recently conducted a systematic literature review and semi-structured interviews to identify the relevant indicators in construction projects and listed 27 economic, 18 environmental, and 37 social indicators. We consider that there is a need for further research to identify the indicators in other industries.

Another interesting research question addresses the factors that motivate project managers to focus or work on sustainability issues. Silvius and Schipper (2020) group the patterns of factors into three: “Pragmatic”, “Intrinsically motivated” and “Task driven”. Pragmatic managers are characterized to be stimulated by practical knowledge, tools, and results. Intrinsically motivated ones address sustainability because they feel and they care about it. Task driven ones focus on the characteristics, requirements, or objectives of the project and give importance to others’ opinions and rewards. We consider that this important research question could be further studied at different project, industry, organizational, and business culture contexts. We refer to recent review article of Chofreh et al. (2019) for a discussion on some other alternative research directions.

Silvius et al. (2015) interpret the relationship between sustainability and project management in terms of the sustainability of the project’s deliverables and processes (project delivery and management). Sustainability could be an important consideration while designing and planning the processes, including procurement and risk management (Peenstra and Silvius 2018, Silvius 2018). In this section, we emphasize the relationship with the processes and present a detailed examination of sustainability over the entire the project life cycle.

- It is essential to examine the environmental and social impacts of the projects in creating a project portfolio. In investment decisions, sustainability opportunities and risks need to be assessed.
- Sustainability goals need to be explicitly stated while defining the project,. In the selection of the PM and the team, sustainability awareness and experience could be integrated as selection criteria.
- During the planning phase, the sustainability impacts of the project activities and their outputs should be considered. Methods that eliminate or minimize the negative effects should be preferred in tasks such as activity mode selection, scheduling and resource assignments. Work conditions should be carefully examined; work health and safety should be emphasized. Estimates of activity time and resource requirements, should be reconsidered taking these conditions into account. Measures should be taken to avoid any kind of discrimination in recruiting the workforce at any level in the project.
- During the implementation phase, in addition to the time, cost and quality of the activities, their environmental and social impacts need to be evaluated. Benefits and damages created by project activities should be measured continuously. Suppliers, subcontractors, their services and processes need to be regularly monitored. In case of damages and problems, or low realization of the benefits, PMs should be able to take corrective actions such as re-planning the activities, requesting budget increases or in some severe cases, terminating the project, communicating the actions and results to the stakeholders.
- In the termination phase, project outputs are reported and lessons learned are recorded. At this point, projects need to be assessed regarding environmental and social impacts.

15.2 Project Management in the Era of Big Data

According to Choi et al. (2018), we are now living in the era of big data. What does it really mean? What does this era bring to us, human beings and our businesses? Researchers and practitioners of information systems have been stressing that "big data" does not simply refer to a large volume of data to be stored. Keeping this warning in mind, we will try to find answers to the following questions:

- What does big data refer to?
- What are the challenges and how could we overcome them? Which techniques could be used?
- What does it imply in projects?
- How can it change the project management practice?
- Which knowledge or decision-making areas of project management can benefit most from the emergence of big data?

15.2.1 Meaning and Techniques of Big Data Analytics

Big data has been commonly described by "the three Vs": volume, velocity and variety. They refer respectively to the huge volume, constant streaming and heterogeneous nature of data with different structures from different sources (Russom 2011; Ekambaram et al. 2018). Nowadays, two additional characteristics are also emphasized by integrating two more Vs: veracity and value. They respectively refer to the source and its authenticity, and contribution to intended analysis (Górecki 2018).

There has been rapid development in technology and the business environment in recent years. Olsson and Bull-Berg (2015) give some of the developments leading to the use of big data analytics in organizations as follows:

- Availability of large quantities of data, distributed via the Internet and obtained from sensor and tracking technologies.
- Increase in targeted advertising and pressures for making data available.
- Low cost access to storage and analysis systems
- Access to IT platforms to populate with data, such as digital maps for position data or Building Information Models (BIM) for construction data.

These developments have created a business environment with abundant digital data, streaming continuously from different sources (reliable or not), in huge amounts, mostly in an unstructured way. Managers have been reconsidering their decision making approaches since they can now obtain more information about business processes, and directly use that knowledge to improve their decision making in several areas. This big data hype brings challenges of managing the data, redefined across the 5 V dimensions. To overcome these challenges, Choi et al. (2018) list the following strategies:

- *Divide and Conquer*: Decompose the data or the problem into the smaller pieces so that we can manage to develop solutions or analyze the data.
- *Distributed and Parallel Processing*: Analyze the data simultaneously by using multiple distributed processors.
- *Incremental Learning Using New Cases*: This improves the training process and makes it more efficient, but requires large amounts of memory to store the knowledge acquired during the training process.
- *Statistical Inference*: It is based on sampling and making inferences about the population by using statistical theorems. It is commonly used in big data analytics applications.
- *Feature Selection:* It helps to reduce the dimensionality of the input variables by focusing on a subset of them expected to be the most relevant ones to predict the output variable. Chandrashekar and Sahin (2014) state that "feature selection methods provide a way of reducing the computation time, improving prediction performance and better understanding of the data in machine learning and pattern recognition applications."
- *Addressing Uncertainty with Learning*: It makes use of approaches like fuzzy learning to deal with problems occurring due to missing data that result in low veracity.
- *Scalability*: Developing flexible computing systems that can meet the computational requirements of big data analytics.
- *Heuristics*: Instead of searching for the optimal solutions, close to optimal solutions that can be generated in a reasonable time are sought.

They also mention several commonly used techniques and methods:

- *Statistics*: It encompasses scientific methods to make inferences using sampling and data analysis. Multivariate statistical analysis is an important tool for data analytics.
- *Machine Learning*: It makes use of algorithms to learn from the given data and make decisions using the acquired knowledge. Some methods of artificial intelligence such as neural networks are considered as a part of machine learning.
- *Data Mining*: It includes the techniques to acquire knowledge from a given data. These techniques usually make use of models and algorithms from statistics and machine learning.
- *Optimization*: It covers mathematical models and analytical approaches to find solutions to managerial decision-making problems. In many real life cases, rather than the optimal solutions, near-optimal ones are sought by developing effective heuristics.

To summarize, organizations today can make use of more automated data collection channels such as internet-based systems or sensor technologies. Through developments in storage and computing systems, such as cloud computing, they can access, store and process massive amount of data. Automation of data collection systems and new filtering and verification technologies may improve the reliability of data. As a result, access to high quantity and variety data offer managers opportunities to improve their decision making via data analysis and decision support tools. We now discuss how this progress can shape the future of project management practices.

15.2.2 Application Areas and Opportunities in Project Management

Applications of big data analytics in the domain of operations management have been rapidly increasing. Researchers working on forecasting, inventory management, revenue management, transportation and supply chain management problems have already adopted these techniques (Choi et al. 2018). However, applications of big data analytics in project management remain scarce. This can be partially explained by the fact that projects are unique, but many of the activities and processes involved are repetitive or already known. Nevertheless, organizations can benefit from big data analytics in their project management practices in several ways.

First, in the area of project risk management, big data analytics can help to identify relevant risks, and predict their probabilities and impacts more accurately. Owolabi et al. (2018) have recently studied predicting the completion time of construction projects using predictive data analytics techniques, such as machine learning and regression. Based on their data, random forest, a type of machine learning algorithm, was shown to be the most effective prediction technique. However, in practice, most companies and industries are not benefiting from these opportunities. To give an example, Górecki (2018) surveyed the Polish construction industry and observed a very low interest of professionals in implementing big data analytics.

Another risk based application area could be to eliminate or avoid some of the financial risks in investment decision-making. For example, the accuracy of the cost predictions is critically important in tender pricing in infrastructure projects such as highway/railway constructions. These projects require large amounts of investments in many cases. Lack of reliable data or appropriate prediction tools might result in low or high bidding, placing the budget goals of the project at risk. In some cases even the financial viability of the undertaking organizations might be jeopardized. However, use of big data analytics can help managers to perform a more effective pre-contract analysis and feasibility study, reducing the financial risks to be faced due to bidding high or low.

Obviously, risk management is critical not only in construction industry. In software development, for example, risk analysis needs to be used within the framework of agile management, which is based on continuous design and development. In this regard, Batarseh and Gonzalez (2018) addressed the agile development lifecycle. They investigated how failures can be predicted using data analytics in the software development phases.

In addition to managing risks, big data analytics can also support managers in resource planning and allocations. In this regard, N'Cho (2017) studied the recruitment of human resources and examined how to form the project teams in the aerospace industry. N'Cho suggests using data analytics tools to collect and analyze all information about the candidates. The analyses help to recruit qualified team members. Kusimo et al. (2019) focused on resource management in the construction industry. They interviewed experts working in the UK and identified the major problems that resulted mainly from poor data management. They noted that the creation and effective use of a resource database that collects data from all the projects undertaken by the organization would help to eliminate these problems.

Big data analytics can also be used to evaluate the projects delivered. In this regard, Olsson and Bull-Berg (2015) focused on building and transportation infrastructure projects and studied the performance indicators. They show how new data categories such as internet traffic or movement-related data can be exploited. Vanhoucke et al. (2016) focus on evaluation of time and cost performance of projects, particularly assessing performance periodically during the project life cycle. They underline that big data analytics enables PMs to better measure project progress better and predict the future more precisely. They can implement new data-intensive ideas on scheduling, risk analysis, and control. However, we strongly believe that applications in project management can be expanded to embrace several decision making areas, such as project selection and portfolio management, pre-project analysis and bidding, forecasting and performance assessment, resource analysis and planning, project quality planning and analysis.

15.3 Fourth Industrial Revolution and the New Age Project Management

Industry 4.0 has been commonly used to express the new automation technologies and development in manufacturing since the Hannover Fair in 2011 (Xu et al. 2018). These technologies include cyber-physical systems (CPS), internet of things (IOT), cloud computing, 3D printing, wireless sensor networks (WSN) and radio-frequency identification (RFID). The first three revolutions refer to the introduction of, respectively, steam power, electrical power, information and communication technologies; whereas the fourth enables implementing smart and connected manufacturing systems.

These new technologies change the business practices in manufacturing and service systems rapidly. As a consequence, both project environments and project management practices are changing.

15.3.1 Evolution of the Project Environments

The era of Industry 4.0 implies a rise in projects on technology development, new product/process designs, and creation of new start-up companies. According to Pajares et al. (2017), we will see an "avalanche of projects". Taking account of higher uncertainty in new product development and financial viability of high-tech start-ups, project environments can be expected to become more and more uncertain.

Projects are also becoming more complex. Complexity does not necessarily stem from size, often from strong interdependencies among project elements. These interdependencies make it difficult to break the overall task into smaller work-packages. Especially in new product development projects, complexity is coupled with lack of clarity in planning requirements and the scope, which makes planning very difficult (Pajares et al. 2017).

Projects have also become more and more technology oriented. As technology is changing rapidly, the way the tasks are performed are changing. New methods and techniques are replacing the traditional methods. For instance, using drones and building information models (BIM) for monitoring and controlling is becoming a common practice in construction projects. 3D printing technology has been changing the manufacturing industry and transforming the procurement strategies in projects.

The developments in information and communication technologies modify the importance of distance, communication patterns, the risk notion, etc., hence the institutional setting of projects. First, information sharing and storing technologies and interconnected devices now enable many of the project tasks to be performed or controlled anywhere and anytime. As a result, project teams have become more and more geographically distributed. Second, project teams have been composed of more and more new generations, who were grown in the information age. Their expectations, perceptions, attention spans, communication ways and habits, and technology aptness are different (Desmond 2018). Third, new categories of risks such as cyber-attacks and vulnerability of data have appeared (Tupa et al. 2017).

The revolution that we have defined as Industry 4.0 will also lead to changes in organizational structures and to new forms of governance. We will observe the shift from process-based structures to project-based management structures will accelerate in organizations. In addition, more and more startups, with project-oriented structures will be created. The skill sets of the project managers and team members that organizations, both established or startups, require will considerably change.

Lastly, the connotation of project success has been changing; speed and agility have become critical factors. Due to the fierce competitive business environments, product life cycles have been getting shorter and shorter. To give an example, the time difference between the launch of a model of smartphone and the replacement with the next model is now expressed in months, not years anymore.

15.3.2 Transformation of the Project Management Practices

Traditional project management emphasizes planning and control and adopts deliverables oriented practices. However, in today's highly uncertain and dynamic project environments emphasizing the preplanning and effective control might be often ineffective, mainly because of the need for frequent scope changes. Consequently, new project management practices are needed now. Agile and hybrid project management methodologies have been replacing the plan-focused traditional systems. The main difference is the emphasis on continuous designing and developing and adoption of iterative develop and deliver cycles. However, several critical elements of traditional approaches such as structural planning, optimization of the resources, monitoring and control of performance, adopting systematic approaches to problem solving, aspects of the traditional aspects will always be important since resources will stay scarce. Regarding uncertainty management, Isikli et al. (2018) and Sarvari et al. (2018) investigated the project portfolio management practices in the era of digital transformation. Considering the complex and highly uncertain new business environment, which requires managing many interrelated projects simultaneously, Isikli et al. (2018) propose to use simulation based optimization to model the environment and construct the resource efficient portfolio.

As a consequence of the increased complexity in projects, PMs require to solve several unstructured or semi-structured decision-making problems and consider many alternative solutions. The decision-making problems are suitable application areas for decision support systems (DSS). In addition to the inherent complexity of decision problems in a project, PMs will control more and more projects simultaneously, due to the increase in projects, especially on innovation and technology development projects. This increase in the number of projects assigned will lead the PMs to take decisions more quickly. As a result, they will need to find out and adopt reliable data analysis and decision support tools to improve their decision-making (see the review paper of Hazir 2015, on DSS).

Regarding managerial decision making, use of expert systems and artificial intelligence (AI) can enable agile management by facilitating a methodological change, and hence improve project execution and control in the cases where the requirements cannot be determined precisely in the planning phase. In this regard, Peña et al. (2019) make use of soft computing and machine learning techniques. Soft computing techniques such as the neural network theory and probabilistic reasoning make use of the accumulated knowledge and experience and try to solve complex problems by providing effective interpretation and estimation algorithms. Their method extracts the relevant data from the completed projects' database, implements machine learning techniques, and evaluates the alternative monitoring and control systems through calculation of statistical metrics. Kanakaris et al. (2019) propose to use machine learning to predict the unexpected events and provide early warnings to support project managers. They also add that the use of AI techniques in project management practice is currently very limited.

With the help of 3D design and printing technologies, we see that prototyping could be used more frequently in projects. New technologies make collecting customer feedback simple and enjoyable; as prototypes are more often developed and shared with the client, transform delivery of projects and support adoption of agile project management techniques, which is useful in design and development projects where requirements are not precisely defined. Prototyping will be more and more used with the increase in the adoption of agile project management techniques. On the other hand, implementation of new digital technologies such as BIM provides considerable benefits. Smith (2016) explores the new opportunities in project cost management and underlines the challenges to be faced because of the new data management requirements (see Sect. 15.2 for big data management).

With the emergence of new digital technologies, organizations have been increasingly transforming their work processes. In this regard, blockchain systems provide opportunities to enhance information management in projects through improving efficiency and traceability of transactions and providing secure data sharing systems. Hargaden et al. (2019) underline that use of these systems in construction projects is very limited but great potential exists for extensive use. As BIM systems are widely adopted in construction industry, more and more digital documents are produced and the security of sharing and processing these documents becomes an organizational issue to address. Blockchain systems could support this digital transformation and adoption of BIM systems (Hargaden et al. 2019).

In Chap. 9, we highlighted the importance of contracting in project management. In this regard, blockchain systems can support the creation of more efficient management control systems, where secure and more direct transactions can be carried out owing to the use of smart contracts. According to Hargaden et al. (2019), smart contracts could transform the construction industry by removing the use of intermediaries such as lawyers in the transactions.

In the era of Industy 4.0, in addition to the development of data management systems, communication systems and technologies advance rapidly as well. They support organizations to make collaborations and conduct projects simultaneously in many countries efficiently. As project teams become geographically distributed, managing the multi-company and multinational project team will bring additional difficulties. Intercultural communication skills and experiences of PMs and adoption of new communication technologies will become critically important for accomplishing project goals. We can also foresee that project teams will increasingly adopt more independent working methods and places in order to enhance creative and innovation. Home offices will be common and open innovation practices will be integrated. In such innovation-oriented environments, the project manager needs to enhance autonomy of the project team members and ensure that decisions are collectively taken. In addition to leading the team and establishing a reliable communication with the stakeholders, managing the project risks constitute an important role of the project managers. With the introduction of new digital technologies, new risks, such as cyberattacks concerning data security, have been occurring and risk management processes have been changing (Tupa et al. 2017). PMs need to reconsider their risk management practices from identification to the monitoring and control by integrating the new risk categories. In this regard, Rane et al. (2019) studied these risk categories in construction projects and developed a risk management framework based on Industry 4.0 technologies such as cyber-security and cloud computing or Internet of Things (IoT).

In the era of short product life cycles, only companies with flexibility and agility in design, production, and management processes can survive. Regarding projects, we need to reconsider the well-known, three dimensions of project objectives: time, cost, and scope. The importance of time dimension will increase in many projects

(Pajares et al. 2017). Robustness regarding time objectives will be emphasized (see Chap. 12 and Hazır et al. 2010 for definition and analysis of robustness).

The discussion above highlights the potential benefits of the use of Industry 4.0 technologies and transformation of project management practices in the digital age. However, project management literature is scarce in this regard. There is a need for further studies to investigate;

- Which knowledge or decision-making areas/problems of project management can benefit most from the developments of Industry 4.0 technologies?
- Which factors affect the adoption and implementation of these technologies?
- What are the sectoral differences in terms of adoption and implementation of these technologies in project management?

References

References for Section Sustainability and Project Management

Brundtland, G. (1987). *Our common future: Report of the 1987 World Commission on Environment and Development* (pp. 1–59). Oslo: United Nations.

Chawla, V., Chanda, A., Angra, S., & Chawla, G. (2018). The sustainable project management: A review and future possibilities. *Journal of Project Management, 3*(3), 157–170.

Chofreh, A. G., Goni, F. A., Malik, M. N., Khan, H. H. & Klemeš, J. J. (2019). The imperative and research directions of sustainable project management. *Journal of Cleaner Production, 238,* 117810.

Dumay, J., Guthrie, J., & Farneti, F. (2010). GRI sustainability reporting guidelines for public and third sector organizations: A critical review. *Public Management Review, 12*(4), 531–548.

Fernández-Sánchez, G., & Rodríguez-López, F. (2010). A methodology to identify sustainability indicators in construction project management—Application to infrastructure projects in Spain. *Ecological Indicators, 10*(6), 1193–1201.

Gareis, R., Huemann, M., & Martinuzzi, A. (2013). *Project management & sustainable development principles*. Newton Square, PA: Project Management Institute.

Global Reporting Initiative (GRI). (2006). *Sustainability reporting guidelines*. Amsterdam: Global Reporting Initiative.

ISO/TS 21929-1. (2006). *Sustainability in building construction – Sustainability indicators – Part 1: Framework for development of indicators for buildings*. Geneva: International Organization for Standardization (ISO).

Labuschagne, C., & Brent, A. C. (2005). Sustainable project life cycle management: The need to integrate life cycles in the manufacturing sector. *International Journal of Project Management, 23*(2), 159–168.

Labuschagne, C., Brent, A. C., & Claasen, S. J. (2005). Environmental and social impact considerations for sustainable project life cycle management in the process industry. *Corporate Social Responsibility and Environmental Management, 12*(1), 38–54.

Martens, M. L., & Carvalho, M. M. (2016). The challenge of introducing sustainability into project management function: Multiple-case studies. *Journal of Cleaner Production, 117*, 29–40.

Morrissey, J., Iyer-Raniga, U., McLaughlin, P., & Mills, A. (2012). A strategic project appraisal framework for ecologically sustainable urban infrastructure. *Environmental Impact Assessment Review, 33*(1), 55–65.

Peenstra, R. T., & Silvius, A. J. G. (2018). Considering sustainability in projects: Exploring the perspective of suppliers. *International Journal of Information Systems and Project Management, 6*(2), 5–22.

Porter, M. E., & Kramer, M. R. (2006). Strategy and society: The link between competitive advantage and corporate social responsibility. *Harvard Business Review, 85*(12), 78–92.

Silvius, A. J. G. (2018). Integrating sustainability into project risk management. In *Global business expansion: Concepts, methodologies, tools, and applications* (pp. 330–352). Hershey: IGI Global.

Silvius, A. J. G., & Schipper, R. (2015). Developing a maturity model for assessing sustainable project management. *Journal of Modern Project Management, 3*(1), 16–27.

Silvius, A. J. G. & Schipper, R. (2020). Exploring variety in factors that stimulate project managers to address sustainability issues. *International Journal of Project Management, 38*(6), 353–367.

Silvius, A. J. G., Van Den Brink, J., Schipper, R., Köhler, A., & Planko, J. (2012). *Sustainability in project management*. Farnham/Burlington: Gower.

Stanitsas, M., Kirytopoulos, K. & Leopoulos, V. (2020). Integrating sustainability indicators into project management: The case of construction industry. *Journal of Cleaner Production,* 123774

United Nations Global Compact. (2008). *Corporate citizenship in the world economy*. New York: United Nations.

References for Section Project Management in the Era of Big Data

Batarseh, F. A., & Gonzalez, A. J. (2018). Predicting failures in agile software development through data analytics. *Software Quality Journal, 26*(1), 49–66.

Chandrashekar, G., & Sahin, F. (2014). A survey on feature selection methods. *Computers and Electrical Engineering, 40*, 16–28.

Choi, T. M., Wallace, S. W., & Wang, Y. (2018). Big data analytics in operations management. *Production and Operations Management, 27*(10), 1868–1883.

Ekambaram, A., Sørensen, A. Ø., Bull-Berg, H., & Olsson, N. O. (2018). The role of big data and knowledge management in improving projects and project-based organizations. *Procedia Computer Science, 138*, 851–858.

Górecki, J. (2018). Big data as a project risk management tool. In *Risk management treatise for engineering practitioners*. London: IntechOpen. https://doi.org/10.5772/intechopen.79182.

Kusimo, H., Oyedele, L., Akinade, O., Oyedele, A., Abioye, S., Agboola, A., & Mohammed-Yakub, N. 2019. Optimisation of resource management in construction projects: a big data approach. *World Journal of Science, Technology and Sustainable Development, 16*(2), 82–93.

N'Cho, J. (2017). Contribution of talent analytics in change management within project management organizations: The case of the French aerospace sector. *Procedia Computer Science, 121,* 625–629.

Olsson, N. O., & Bull-Berg, H. (2015). Use of big data in project evaluations. *International Journal of Managing Projects in Business, 8*(3), 491–512.

Owolabi, H. A., Bilal, M., Oyedele, L. O., Alaka, H. A., Ajayi, S. O., & Akinade, O. O. (2018). Predicting completion risk in PPP projects using Big Data analytics. *IEEE Transactions on Engineering Management, 67*, 430. https://doi.org/10.1109/TEM.2018.2876321.

Russom, P. (2011). *Big Data Analytics* (4th Quarter). Renton, WA: TDWI Research.

Vanhoucke, M., Coelho, J., & Batselier, J. (2016). An overview of project data for integrated project management and control. *Journal of Modern Project Management, 3*(2), 6–21.

References for Section Fourth Industrial Revolution and the New Age Project Management

Desmond, C. (2018). What is the impact of today's technologies on project management? *IEEE Engineering Management Review, 46*(2), 24–25.

Hargaden, V., Papakostas, N., Newell, A., Khavia, A., & Scanlon, A. (2019). *The role of blockchain technologies in construction engineering project management.* In 2019 IEEE International Conference on Engineering, Technology and Innovation (ICE/ITMC): 1–6, IEEE.

Hazır, Ö. (2015). A review of analytical models, approaches and decision support tools in project monitoring and control. *International Journal of Project Management, 33*(4), 808–815.

Hazır, Ö., Haouari, M., & Erel, E. (2010). Robust scheduling and robustness measures for the discrete time/cost trade-off problem. *European Journal of Operational Research, 207*(2), 633–643.

Isikli, E., Yanik, S., Cevikcan, E., & Ustundag, A. (2018). Project portfolio selection for the digital transformation era. In *Industry 4.0: Managing the digital transformation* (pp. 105–121). Cham: Springer.

Kanakaris, N., Karacapilidis, N., Kournetas, G., & Lazanas, A. (2019, February). *Combining machine learning and operations research methods to advance the project management practice.* In International Conference on Operations Research and Enterprise Systems: 135–155, Springer, Cham.

Pajares, J., Poza, D., Villafañez, F., & López-Paredes, A. (2017). Project management methodologies in the fourth technological revolution. In *Advances in management engineering* (pp. 121–144). Cham: Springer.

Peña, A. B., Castro, G. F., Alvarez, D. M. L., Alcivar, I. A. M., Núñez, G. L., Cevallos, D. S., & Santa, J. L. Z. (2019). *Method for project execution control based on soft computing and machine learning.* In 2019 XLV Latin American Computing Conference (CLEI): 1–7, IEEE.

Rane, S. B., Potdar, P. R., & Rane, S. (2019). Development of project risk management framework based on Industry 4.0 technologies. *Benchmarking: An International Journal,* https://doi.org/10.1108/BIJ-03-2019-0123.

Sarvari, P. A., Ustundag, A., Cevikcan, E., Kaya, I., & Cebi, S. (2018). Technology roadmap for Industry 4.0. In *Industry 4.0: Managing the digital transformation* (pp. 95–103). Cham: Springer.

Smith, P. (2016). Project cost management with 5D BIM. *Procedia-Social and Behavioral Sciences, 226,* 193–200.

Tupa, J., Simota, J., & Steiner, F. (2017). Aspects of risk management implementation for Industry 4.0. *Procedia Manufacturing, 11,* 1223–1230.

Xu, L. D., Xu, E. L., & Li, L. (2018). Industry 4.0: State of the art and future trends. *International Journal of Production Research, 56*(8), 2941–2962.

Abbreviations

AOA	activity on arc
AON	activity on node
APM	association for project management
B&B	branch and bound
BOT	build-operate-transfer
cdf	cumulative distribution function
CAGR	cumulative average growth rate
CCPM	critical chain project management
CBS	cost breakdown structure
CI	criticality index
CP	critical path
CPL	critical path length
CPM	critical path method
CUB	current upper bound
CV	coefficient of variation
DTCTP	the discrete time/cost trade-off problem
DTCTP-B	the budget problem
DTCTP-D	the deadline problem
DTCTP-E	the efficiency problem
ECPL	expected critical path length
EMV	expected monetary value
ESS	early start schedule
EVM	earned value management
FS	free slack (float)
GERT	graphical evaluation review technique
GP	goal programming
GPR	generalized precedence relationships
GRD	greatest resource demand
GRPW	greatest ranked positional weight
IP	integer programming
IPMA	international project management association
IS	independent slack (float)

G. Ulusoy, Ö. Hazır, *An Introduction to Project Modeling and Planning*, Springer Texts in Business and Economics, https://doi.org/10.1007/978-3-030-61423-2

KPI	key performance indicator
LB	lower bound
LCL	lower control limit
LFT	late finish time
LSS	late start schedule
LST	late start time
LOB	line of balance
MIS	most immediate successors
MRCPSP	multi-mode resource-constrained project scheduling problem
MMRCPSP	multi-objective multi-mode resource-constrained project scheduling problem
MOLP	multi-objective linear programming
MSLK	minimum slack
MTS	most total successors
NPV	net present value
NSGA-II	non-dominated sorting genetic algorithm II
OBS	organizational breakdown structure
pdf	probability density function
PERT	program evaluation and review technique
PM	project manager
PMBOK	a guide to project management body of knowledge
PMI	project management institute
PMO	project management office
PMP	project management professional
PPM	project portfolio management
PRINCE2	projects in controlled environment
PSGS	parallel schedule generation scheme
R&D	research and development
RACP	resource availability cost problem
RBS	risk breakdown structure
RCPSP	resource-constrained project scheduling problem
RSEM	root-square error method
RSM	resource scheduling method
SDR	slack/duration ratio
SGS	schedule generation scheme
SPC	statistical process control
SPT	shortest processing time
SS	safety slack (float)
SSGS	serial schedule generation scheme
TOC	theory of constraints
TS	total slack (float)
UCL	upper control limit
WRUP	Weighted resource utilization ratio and precedence

Mathematical Notations

α	discount rate
δ	project deadline
δ_i	deadline for project i / activity i
$\lambda_{\max}$	principal eigenvalue of the judgement matrix A
σ	standard deviation
a	optimistic activity duration (PERT)
a_i	normalized aggregate scores for each project alternative i
a_r	number of resources acquired of resource type r
a_{ij}	direct cost of compressing per unit time for activity (i,j)
A	set of activities
A	(nxn) judgement matrix (AHP)
AF_k	total financing available in period k
b	pessimistic activity duration (PERT)
B	budget
B_i	minimum of the time buffer between work packages $\mathrm{WP_i}$ and $\mathrm{WP_{i+1}}$
c_i	unit direct cost of compressing $\mathrm{WP_i}$
c_{jm}	cost of activity j processed at mode m
$\overline{c_{jm}}$	upper bound on c_{jm}
$\underline{c_{jm}}$	lower bound on c_{jm}
c_{mode}	most likely cost estimate (PERT)
c_{opt}	optimistic cost estimate (PERT)
c_{pess}	pessimistic cost estimate (PERT)
$\mathrm{C_{max}}$	project duration (makespan)
$C(s,t)$	minimal cut set
C_j	total cost of activity j per period
C^{av}	budget amount available per period
C_{1i}	indirect cost per period for $\mathrm{WP_i}$
C_{2i}	direct cost per men allocated to $\mathrm{WP_i}$
C_t^{av}	budget amount available in period t
$C_{i,\, k+1-j}$	financing required for project i in period k

G. Ulusoy, Ö. Hazır, *An Introduction to Project Modeling and Planning*, Springer Texts in Business and Economics, https://doi.org/10.1007/978-3-030-61423-2

C_{ij}^C, C_i^C	crash direct cost of activity (*i,j*), crash direct cost of activity *i*
C_{ij}^N, C_i^N	normal direct cost of activity (*i,j*), normal direct cost of activity *i*
$C_k(u_{kt})$	performance measure for resource type *k* in period *t* at a resource level u_{kt}
CF_j	net cash flow amount for activity *j*
CI	consistency index (AHP)
CS_i	number of men assigned to WP_i
CR	consistency ratio (AHP)
d_{ij}, d_j	duration of activity (*i,j*), duration of activity *j*
D_{ij}^N, D_i^N	normal duration of activity (*i,j*), normal duration of activity *i*
D_{ij}^C, D_i^C	crash duration of activity (*i,j*), crash duration of activity *I*
D_i	duration of WP_i in repetitive scheduling
$e^{-\alpha t}$	continuous compounding factor
E	earliness
$\overline{E}$	average earliness
E_i	earliest occurrence time of event *i* / earliness of activity *i*
E_{max}	maximum earliness
EAS	eligible activity set
EF_{ij}	early finish time of activity (*i,j*)
EFT_j	early finish period of activity *j*.
ES	early start
EF	early finish
EV	earned value
f_t^α	discount factor for the cash flow in period *t*.
F_i	forbidden set at node *i*
FT_i	finish time of activity *i* / WP_i.
$FSij$	the free slack (or free float) for activity (*i,j*)
G (*N, A*)	AON project network, where *N* denotes the set of activities (nodes), and $A \subset N \times N$ the set of the arcs
G (*M, A*)	AOA project network, where *M* denotes the set of events (nodes), and $A \subset M \times M$ the set of the activities (arcs)
H	set of precedence relations
H	predetermined number of payments
IS_{ij}	independent slack for activity (*i,j*)
J	set of activities
$\lvert J \rvert$	number of activities including dummy activities, if any
k_{jr}	per period renewable resource requirement of resource type *r* of activity *j*
k_{jn}	non-renewable resource requirement of resource type *r* for activity *j*
K_r	limit on the usage of renewable resource *r* per period
K_n	amount of non-renewable resource type *n* available
L	lateness
$\overline{L}$	average lateness
L_i	latest occurrence time of event *i* / lateness of activity *i*

L_{max}	maximum lateness
LF	late finish
LF_{ij}	late finish time of activity (i,j)
LFT_j	late finish period of activity j
LS	late start
LS_{ij}	late start time of activity (i,j)
LST_j	late finish period of activity j
m	number of event nodes (AOA)
m	mode index
m	most likely activity duration (PERT)
M	set of events (AOA)
M	number of projects in a multiple project environment
M_j	set of modes of activity
n	number of activities on AOA project network
N	set of activities
N	set of non-renewable resource types
N	the batch size in repetitive projects
N	number of criteria of the overall score S_i
O	overhead cost per unit time / per period
$\tilde{p}_j$	random variable for the duration of activity j
p_j	production rate of $\mathrm{WP_j}$
p_{jm}	processing time of activity j at mode m
P_j	set of immediate predecessors to activity j
PD	project duration.
PL	set of distinct priority lists
PV	planned value
$\underline{\mathrm{PV}}$	priority vector (AHP)
res	remaining resources
R	set of renewable resource types
$\lvert R \rvert$	number of resource types
RF	resource factor
RI	random consistency number (AHP)
$\mathrm{RS_r}$	resource strength for renewable resource r
s	sample standard deviation
s_{ij}	score for candidate project i and criterion j
s_{ij}	slack variable for activity (i,j)
$S(i)$	set of successors of activity i
S_i	overall score for the candidate project i
S_{n+1}	project completion time
SS	set of schedulable activities
SS_{ij}	safety slack for activity (i,j).
ST_i	start time of activity i / $\mathrm{WP_i}$
T	time horizon in periods
T	tardiness

$\overline{T}$	average tardiness
T_i	occurrence time of event *i* / tardiness of activity *i*
T_{max}	maximum tardiness
T_w	weighted tardiness
TS_{ij}	the total slack for activity *(i,j)*
u_{kt}	level of resource usage of type *k* in period *t*
$\underline{w}$	weight vector (AHP)
w_j	weight for candidate project *i* in multi-criteria scoring
w_k^j	weight for alternative *j* under terminal criterion *k* (AHP)
WC_i	work content of WP_i

Index

A

Active schedule, *see* Schedule
Activity on arc type representation (AOA), 14, 63–69, 72, 74, 75, 95–98, 101, 103, 108, 110, 120, 121, 131, 135, 137–139, 147, 148, 154, 155, 160, 172, 194, 204, 214, 219, 221, 228, 230, 234, 235, 242, 244, 293, 305, 366, 367
Activity on node type representation (AON), 14, 63, 64, 68–69, 95–97, 101, 103, 108–110, 116, 134, 137, 139, 154, 155, 163, 207, 224, 225, 229, 257, 266, 361, 363
Activity-based costing (ABC), 84–86, 103, 375
Actual Cost of Work Performed (ACWP), 312, 320, 321, 326
Agile project management, 17–18, 23, 43–45, 325, 394, 465
Analog models, 61
Analytic Hierarchy Process (AHP), 407–418, 422, 435–437, 439, 440, 444, 445, 448, 453, 454

B

Backward passes, 104, 106–109, 116, 148–150, 172, 173, 208, 223, 227
Baseline schedules, 13, 17, 19, 357–359, 365
Beta distributions, 79, 82, 83, 168–170, 177, 178, 183, 192, 370, 400
Bid spread, 291
Bidding
  competitive, 290
  front-loaded, 293, 300
  negotiated, 290
Big data, 91, 326, 352, 457, 460–463, 466
Big data analytics, 90, 326, 352, 354, 460–463
Bill of quantities, 282, 284
Bonuses, 102, 120, 143, 205, 283, 285, 289, 300
Bottom-up costing, 59, 60
Branch and Bound (B&B) algorithm, 209, 245
Branching strategies, 209–212, 248, 269
Breadth first search, 247
Budget at completion (BAC), 323, 326, 329, 330
Budgeted Cost of Work Performed (BCWP), 312, 320, 321, 326
Budgeted Cost of Work Scheduled (BCWS), 312, 320
Budget problems, 156–158
Buffer management
  feeding buffer, 360
  project buffer, 360, 365, 366, 377
  resource buffer, 360
Building information models (BIM), 89, 460, 464, 465
Build-Operate-Transfer (BOT) Project, 362
Burgess and Killebrew algorithm, 254–257, 259, 260, 263, 278

C

Central Limit Theorem, 171, 177
Client-contractor bargaining problem, 297–298
Clients, 5, 6, 12, 18, 24, 27, 28, 32–34, 44, 51, 124, 132, 133, 141, 142, 281–291, 293, 297–300, 302–305, 349, 419, 465
Contract
  bills of quantities, 284
  cost plus, 284
  cost-plus fixed fee, 285
  fixed price, 285
  lump-sum payment, 133, 284
  price list, 284
  schedule of rates, 284, 286
  target contracts, 285

G. Ulusoy, Ön. Hazır, *An Introduction to Project Modeling and Planning*, Springer Texts in Business and Economics, https://doi.org/10.1007/978-3-030-61423-2

Contract (*cont.*)
  types, 142, 281–306
  variable price, 285
Contractors, 5, 13, 41, 69, 120, 125, 132, 133, 139, 142, 192, 275, 281–294, 296–304, 306, 320, 340, 349, 381, 394, 448, 449
Control, *see* Project control
Cost accounts, 59–61, 94
Cost aggregation, *see* Bottom-up costing
Cost breakdown Structure (CBS), 49, 52, 59–61, 81, 255, 256
Cost drivers, 85, 86, 99, 313
Cost Performance Indices
  cost performance index (CPI), 316
  cost variance (CV), 316
  deflated cost variance (DCV), 320, 321
  inflation variance (IV), 320, 321
Crash direct cost, 143
Crash durations, 143, 146, 154, 159, 160, 162, 163, 387, 404
Crashing
  crash buffers, 146, 147
  free slack buffer, 146, 147
Critical Chain Project Management (CCPM), 15, 359–360
Criticality index (CI), 171, 176, 180, 181, 188, 412, 414, 445
Critical path method (CPM), 14–16, 96, 103–111, 116, 117, 120, 132, 135, 139, 142, 148–152, 167, 170, 173, 174, 179, 180, 187, 203, 204, 208, 209, 223, 227, 293, 295, 318, 358, 365, 372
Cyclic project network
  checking for, 67–68, 75

D
Data analytics, 86, 90, 91, 457, 461, 462
Deadline problem, 156–158
Decision trees, 335, 347, 348, 351, 354
Depth-first search, 247
Deviational variable
  negative deviational variable, 430
  positive deviational variable, 430
Direct cost
  direct labor cost, 82
  direct material cost, 82
Discrete time/cost tradeoff problem (DTCTP), 17, 156, 361, 362
Disjunctive activities, 210
Dispatching rule, *see* Priority rule
Dominance rules, 210, 211, 247
Doubly constrained resources, 16, 17, 228

E
Earned schedule (ES), 108, 110, 135, 136, 148–152, 173, 236, 294, 295, 317, 318, 323, 324, 327–329
Earned Value Management (EVM)
  Actual Cost of Work Performed (ACWP), 312, 320, 321, 326
  Budgeted Cost of Work Performed (BCWP), 312, 320, 321, 326
  Budgeted Cost of Work Scheduled (BCWS), 312, 320
  Deflated Cost of Work Performed (DCWP), 320, 321
  Earned Value (EV), 44, 312–314, 316, 319, 323, 325, 329
  Planned Value (PV), 312
Equitable Payment Schedule Problem, 298
Estimate to completion, 323
Estimated at completion, 323
Estimating activity durations
  analogous estimating, 77, 78
  bottom-up estimating, 79
  documentation, 79
  expert judgment, 77
  labor productivity-based estimating, 78
  parametric estimating, 78
  three-point estimating, 79
  tools for estimating, 77
Estimating activity cost
  analogous estimating, 80, 81
  bottom-up estimating, 83
  documentation, 83
  expert judgment, 80
  labor productivity-based estimating, 82
  parametric estimating, 82
  three-point estimating, 82
  tools for estimating, 80
Estimating the percent complete, 313–314

F
Feeding buffers, 360
50% rule, 360
Financial objectives, 102, 124, 206, 374, 437
Fixed price contracts, 284, 286
Forward pass, 104, 106, 108, 109, 116, 148–150, 208, 223
Fourth Industrial Revolution, 463–467
Free slacks, 105–108, 110, 134, 138, 146, 147, 149–151, 159, 162, 172, 173, 271, 358, 376
Functional organizations, 24–25, 27
Fuzzy project scheduling, 359, 374

G

Gantt Charts, 14, 19, 55–57, 70, 117, 119, 121, 122, 302–304, 318, 333, 365
General overhead cost, *see* Overhead cost
Generalized precedence relationships (GPRs), 69–71, 97, 110–111, 153, 154
Geographically distributed project teams, 36, 37
Goal programming (GP), 407, 408, 422, 429–432, 444, 450
Graphical Evaluation Review Technique (GERT), 15

H

Hammock activities, 64
Heuristic solution procedures, 214–221, 223, 240, 251, 261

I

Iconic models, 61
Incentive plan
  bilateral, 283
  unilateral, 283
Industry 4.0, 463, 464
Inflation
  impact of, 320–321
Integer programming (IP), 102, 117, 121, 246, 307, 408, 447

J

Judgment matrices, 411, 414, 445, 448, 455

K

Knowledge management, 39, 49, 86–90

L

Left shift
  global, 214
  local, 214
  one period, 214
Level of aggregation, 49, 71–73
Lexicographic goal programming, 430
Linear programming (LP), 101, 102, 111–115, 132, 135, 136, 154–156, 159, 160, 163, 261, 385, 387–390, 394, 404, 424, 431, 433, 436, 444
Line of balance (LOB), 382
Lower bound (LB), 209, 211, 237, 246, 247, 269, 363, 369–371, 427

M

Makespan, 102, 111, 171, 204, 207, 209, 211, 224, 225, 229, 231, 233, 234, 236, 242, 244, 278
Make-to-order production, 70, 394
Mark-up, 133, 283, 287, 290, 291, 294, 303–306
Matrix organization
  Weak (lightweight), 26
  Balanced, 26
  Strong (heavyweight), 26
Maximum cash balance, 206
Meetings, *see* Project meetings
Milestones, 13, 14, 32, 49, 51, 72, 92, 120, 171, 290, 313, 382
Minimum LB search, 247
Mode
  resource-resource trade-off, 222, 223
  time-resource trade-off, 222
Monitoring, 6, 9–10, 13, 14, 27, 32, 45, 53, 58, 60, 62, 71, 86, 89, 92, 123, 285, 311, 312, 325, 326, 336, 337, 350, 352, 394, 464–466
Monte Carlo simulation, 167, 178–182, 347, 365, 368, 370–374
Multi-criteria scoring model, 408–410
Multi-mode Resource Constrained Project Scheduling Problem (MRCPSP), 16, 17, 222, 267, 268
Multi-objective linear programming (MOLP), 407, 422–429, 432, 444
Multi-objective mathematical programming models, 418, 421–422
Multi-objective, multi-period project portfolio selection and scheduling, 407, 433–436

N

Negotiations, 34, 80, 126, 184, 281, 283, 285, 286, 289, 290, 297, 298
Net present value (NPV), 101, 102, 115–121, 124, 132, 133, 138, 139, 171, 200, 203, 204, 206, 224–227, 229, 287, 293, 294, 296–298, 302–306, 328, 369, 374, 375, 431, 440, 451, 452
Network complexity (NC), 63, 268
Network construction rules, 64–66
Network modeling
  of multiple projects, 74
Non-delay schedule, *see* Schedule
Non-dominated solution, 422, 424, 425
Non-renewable resource, *see* Resource
Normal activity duration, 143

Normal direct cost, 143, 148
Normal distributions, 171, 172, 174, 189, 315, 370, 403

O
Objective function
financial, 124, 205–206
non-regular, 203–204, 225, 228
quality based, 206
regular, 203–205
resource based, 204, 207
time based, 13, 101, 102, 204–206
Organizational Breakdown Structure (OBS), 49, 52, 58–59, 256
Organizational structures
functional, 23
matrices, 23
project-based, 4, 23, 464
Overhead cost
general, 85, 86, 142
projects, 84–86, 99, 142, 153, 155, 160, 162, 226

P
Pareto front, 422–426, 428, 429, 432
Payment scheduling models
lump-sum payment, 286
payments at event occurrences, 286
payments at equal time intervals, 286, 287
progress payments, 286
Penalties, 35, 102, 120, 143, 154, 205, 206, 283, 285, 289, 290, 294, 299, 300, 349, 362
Percent complete
subjective assessment, 313
objective measured assessment, 313
rules of thumb, 314
indirect assessment, 314
PERT costing, 83, 177, 183
PERT-distribution, 178, 179
Planned Value (PV), 312, 314, 316, 319, 323, 324, 329
Portfolio management, 23, 45, 46, 91, 463
Portfolio selection and scheduling problems, 407, 408
Portfolio selection problem, 408, 418
Post-project analysis (PPA), 10, 11, 19, 33, 86, 88, 91
Precedence relations
start-to-start, 69
start-to-finish, 70, 71, 382
finish-to-start, 207, 242
finish-to-finish, 70, 71, 383, 396
PRINCE2, 37
Principle of optimality, 103, 104
Priority rules, 211, 217–218, 220, 227, 230–232, 234–236, 238, 239
Proactive-reactive scheduling, 358
Probability impact matrix, 353
Program Evaluation and Review Technique (PERT), 14–16, 79, 167–170, 174–180, 182, 183, 187, 188, 190, 262, 370
Program management, 40–41
Progress lines, 382, 387, 395–398, 400–405
Progress payments, 283, 285–287, 293, 297
Project buffer, 360, 365, 366, 377
Project charter, 33, 46, 49–51, 72
Project contract, *see* Contract
Project control
in contracts, 281, 287, 288
Project crashing, 145–154, 159, 161
Project life cycles, 1, 6, 7, 9, 10, 17–19, 88, 91, 339, 457–459, 463
Project management culture
competitive, 30
cooperative, 30
fragmented, 31
isolated, 30
non-cooperative, 30
Project Management Institute (PMI), 1, 3, 5, 10, 12, 18, 37, 38, 40–44, 49, 50, 52, 72, 77, 79, 80, 83, 87, 88, 285, 289, 335–337, 341, 349, 350, 408
Project management office (PMO), 38–40
Project management professional, 37
Project management team, 28
Project managers, 1, 3, 13, 15, 19, 23–25, 28, 29, 31, 33–35, 37, 42, 45, 46, 51, 63, 69, 91, 99, 101, 112, 125, 126, 128, 183, 184, 190, 193, 206, 226, 227, 229, 251, 253, 254, 256, 311, 317, 325, 328, 350, 351, 357, 360, 375, 377, 453, 464, 466
Project meetings, 38
Project networks, 14–16, 49, 62–64, 66–69, 71–75, 91, 95–98, 101–113, 117, 119–121, 123, 130, 131, 133–136, 138, 139, 147–155, 158–160, 164, 170, 172, 174–176, 179, 186, 187, 189–191, 193, 194, 204, 209–212, 214, 215, 219, 221, 225, 227, 229, 231, 233, 235–238, 240, 242, 254, 257, 266, 273, 276–278, 293–295, 304, 305, 328, 333, 363, 365, 366, 370, 371
Project organizations, 3, 7, 23, 25, 27, 28, 32, 33, 38, 82
Project overhead cost, *see* Overhead cost
Project portfolio management (PPM), 41–43, 45, 436, 465

Project scopes, 13, 18, 27, 44, 45, 49–51, 62, 123, 338, 350, 435
Project scheduling, *see* Scheduling
Project sponsor, 23, 28, 32–34, 36, 50, 51
Project team
  geographically distributed, 466
  team members, 8, 10, 19, 28, 29, 31, 36, 37, 89, 343, 466

Q
Q-GERT, 15
Qualitative risk analysis, 336
Quality robustness, 358, 359, 376
Quantitative risk analysis, 336, 348, 352, 369
Quantity surveyor, 283, 288, 332

R
Reactive scheduling, 357, 358
Regular performance measures, 203
Renewable resources, 16, 200–202, 207–209, 211, 219–221, 223, 225, 226, 228–230, 234, 237, 238, 244, 251, 252, 261, 267, 297, 298, 404
Repetitive projects
  production rates, 384
  repetitive scheduling, 382
  WP progress line, 382
Resource
  activity related, 200
  buffers, 15, 228, 360, 396
  doubly constrained, 16, 200
  non-renewable, 16, 199, 200, 207–209, 223, 228, 230, 237, 252, 268
  project related, 200
  partially non-renewable, 200
  renewable, 16, 17, 199, 200, 223, 230, 236, 237, 252, 268, 276
  resource related, 200
Resource availability cost problem (RACP), 17, 207, 251, 252, 265, 266, 269
Resource Constrained Project Scheduling Problem (RCPSP), 16, 17, 199, 200, 207–209, 211, 214, 216, 222, 225, 227, 229, 244, 247, 265, 267, 268, 297, 359, 360
Resource dedication policy (RDP), 252, 267, 268
Resource investment problem (RIP), 265
Resource leveling
  problems, 17, 251–278
Resource management policies, 74, 251, 252, 267–269
Resource portfolio management problem, 252, 268, 269
Resource sharing policy (RSP), 252, 267
Responsibility matrix, 58, 59
Retainage, 133, 283, 287, 304, 305
Risk analysis
  qualitative methods, 341
Risk assessment matrix, 341, 342
Risk breakdown structure (RBS), 337, 339, 340, 354
Risk identification, 336–339
Risk in contracts, 288, 289
Risk management, 13, 19, 88, 282, 335–350, 352, 354, 459, 462, 466
Risk monitoring and control, 336, 350
Risk registers, 88, 123, 337–339, 343
Risk response planning, 336, 348–350
Risk severity level
  probability of occurrence, 289, 342, 343, 345–347
  overall impact, 343, 346, 347
Risk severity matrix, 342, 345
Robust optimization, 183, 266, 361–365, 374, 376, 377, 437
Robust scheduling, 358–366
Root square error method (RSEM), 360

S
Scenario analyses, 182, 290, 326, 370, 374
Schedule
  active, 216, 219, 222, 266
  non-delay, 216, 222
  semi-active, 221
Schedule and cost performance index (SCI), 317, 322, 323, 326
Schedule generation scheme (SGS)
  parallel (PSGS), 220–222
  serial (SSGS), 221–222
Schedule performance indices
  schedule variance (SV), 314, 325, 328, 330, 333
  schedule performance index (SPI), 328, 330, 333
Schedule robustness
  quality robustness, 358, 359, 376
  solution robustness, 358, 376
Scheduling
  fuzzy, 357, 359, 374
  reactive, 269, 352, 358, 359, 374
  proactive-reactive, 358
  robust (proactive), 227, 262, 269, 352, 357–359, 374
  stochastic, 263, 292, 357, 359, 368, 374

Scoring methods
multi-criteria scoring model, 407–410
Semi-active schedule, *see* Schedule
Sensitivity analysis, 103, 111, 112, 290, 347, 357, 359, 369, 374, 375, 394, 410, 417–418, 422, 436
Serial production lines, 381, 384
Single objective mathematical programming models, 407, 418–421
Slack
events, 104, 106, 110, 131, 236
free, 105–108, 110, 134, 139, 146, 147, 149, 150, 151, 159, 162, 172, 173, 271, 358
independent, 105, 131
safety, 105, 130, 131
total, 108, 147, 173, 294
Solution robustness, 358, 376
Sources of risks, 337
Stability, 269, 358, 417, 448
Stakeholders
external, 28, 407
internal, 28, 407
management, 28, 336, 407, 458
Steering committee, 32, 33
Strength-Weakness-Opportunity-Threat (SWOT) analysis, 341
Stochastic project scheduling, 167–184, 187–194, 299, 359
Subcontractors, 41, 89, 184, 206, 282, 299, 300, 343, 349, 396–399, 444, 459
Summary activities, 72–73, 94
Sustainability, 25, 437, 438, 457–459, 466
Symbolic models, 61

T
Team, *see* Project management team
Tender, 282, 284–286, 290, 462
Theory of Constraints (TOC), 15, 103, 359
Time-based objectives, 13, 102, 204
Time-cost trade-off problem
continuous, 160
discrete, 141, 142, 156–158
Total adjustment cost problem (TACP), 17, 251, 263–265
Total bid spread, 291
Total slacks, 105–108, 110, 113, 131, 138, 146, 159, 173, 236, 358
Triangular distribution, 82, 179
Trust, 28–31, 34–36, 284, 289
Types of schedules, 214–217

V
Value of information, 292
Variable price contracts, 284, 285

W
Work breakdown structure (WBS), 19, 49, 51–62, 79, 83, 84, 86, 91–94, 98, 127, 130, 292, 306, 312, 313, 317, 318, 325, 337
Work package (WP), 52–54, 58–61, 82, 84, 86, 123, 125, 126, 159, 177, 181, 182, 194, 257, 270, 282, 284, 304, 307, 313, 353, 382–387, 392, 395–398, 400–405